# MANUAL FOR SPILLS OF HAZARDOUS MATERIALS

Technical Services Branch
Environmental Protection Service
Environment Canada
Ottawa, Ontario.

March 1984

©Minister of Supply and Services Canada 1984
Cat. No. Em40-320/1984E
ISBN 0-662-13186-X

**FOREWORD**

This manual provides qualitative and quantitative information for those responding to, or planning for, hazardous materials spills. It is unique in that it encompasses quantitative data on chemical and physical properties, fire properties, human health and toxicity, reactivity and environmental toxicity, as well as qualitative response information. According to criteria developed by the Environmental Emergencies Technology Division, 150 top priority chemical substances, as well as fuels, oils and other frequently spilled substances, are included in the manual. There are 220 listings, many of which encompass several different forms/isomers, solutions or formulations of the basic product. This first edition will be updated and expanded upon when sufficient new information is gathered.

## ACKNOWLEDGEMENTS

The publication of this manual results from the dedication and perseverance of M.F. Fingas and S. Lawrie, Environmental Protection Service (EPS), Environment Canada. J. de Gonzague (editor-in-chief), and M. Cloutier, EPS, as well as R. McKenney, J. Daka and P. Lahaye, under contract to EPS, assisted substantially in its preparation.

Extensive revisions to the initial version were effected by a Working Group consisting of some of the aforementioned and M. Purnarouskis, United States Coast Guard, J. Dick and R. Rye, Concordia University, and P. Mazerolle, EPS.

This latter version was reviewed by a wide range of specialists who provided additionial corrections and additions. Especially acknowledged are the Canadian Chemical Producers' Association and its members who provided comments and many new data.

A preliminary version of the manual (over 100 substances were later added) was prepared by ECO Research Ltd., under contract to EPS. It was reviewed by W. Carter, L. Solsberg and J. Bridgeland.

The many people who are not mentioned here, and who contributed to this manual, are gratefully acknowledged.

# TABLE OF CONTENTS

|  |  | Page |
|---|---|---|
| FOREWORD | | i |
| ACKNOWLEDGEMENTS | | ii |
| 1 | DESCRIPTION OF MANUAL'S CONTENTS | 1 |
| 1.1 | Main Listing | 1 |
| 1.1.1 | Name | 1 |
| 1.1.2 | Chemical Formula | 1 |
| 1.1.3 | UN Number | 1 |
| 1.2 | Identification | 1 |
| 1.2.1 | Common Synonyms | 1 |
| 1.2.2 | Observable Characteristics | 1 |
| 1.2.3 | Manufacturers/Suppliers | 1 |
| 1.2.4 | Transportation and Storage Information | 2 |
| 1.2.5 | Physical and Chemical Characteristics | 5 |
| 1.3 | Hazard Data | 9 |
| 1.3.1 | Human Health | 9 |
| 1.3.2 | Fire | 16 |
| 1.3.3 | Reactivity | 16 |
| 1.3.4 | Environment | 17 |
| 1.4 | Emergency Measures | 20 |
| 1.4.1 | Special Hazards | 20 |
| 1.4.2 | Immediate Responses | 20 |
| 1.4.3 | Protective Clothing and Equipment | 20 |
| 1.4.4 | Fire and Explosion | 20 |
| 1.4.5 | First Aid | 20 |
| 1.5 | Environmental Protection Measures | 21 |
| 1.5.1 | Response | 21 |
| 1.5.2 | Disposal | 21 |
| 2 | INDEX OF ENTRIES | 23 |
| 3 | AGRICULTURE CANADA PESTICIDE REGISTRATION NUMBERS | 55 |
| 4 | CONVERSION FACTORS | 65 |
| 4.1 | Abbreviations Used in Conversions | 68 |
| BIBLIOGRAPHY | | 69 |
| HAZARDOUS MATERIALS: TWO-PAGE ENTRIES | | 75 |

# 1 DESCRIPTION OF MANUAL'S CONTENTS

This section explains the various entries of each 2-page listing in the manual. The descriptions of the entries that follow are in the order that appear in a typical listing.

## 1.1 Main Listing

### 1.1.1 Name.

- In each case, the most commonly used "transport" name has been chosen as the name to list the substance.

### 1.1.2 Chemical Formula.

- A standardized semi-structural formula has been given; where the substance is not a pure chemical, a typical formula is given.

### 1.1.3 UN Number.

- This is the Inter-governmental Maritime Organization (IMO; formerly IMCO) number for the given substance.
- In some cases several numbers apply to the product and relate to the various formulations.

## 1.2 Identification

### 1.2.1 Common Synonyms.

- Alternative chemical names and commonly used names are given.
- Commercial or trade names are shown in a few cases.
- An index of all synonyms appears at the front of the manual.

### 1.2.2 Observable Characteristics.

- A description of the appearance is given, followed by a description of the substance's odour.
- It should be noted that the appearance and the odour may depend on the purity or form of the substance; thus, these characteristics should not be used exclusively when identifying the product.

### 1.2.3 Manufacturers/Suppliers.

- The manufacturers or suppliers of the given product are listed.
- In those cases where all Canadian firms cannot be listed, only selected manufacturers are given and, where possible, these were selected on the basis of production volume.

## 1.2.4 Transportation and Storage Information.

<u>Shipping State</u>
- The physical state(s) (liquid, solid or gas) of the substance when shipped.

<u>Classification</u>
- The classification of the substance under the Transport of Dangerous Goods Regulations.
- The following is a summary of these classifications:

| CLASS | DIVISION | |
|---|---|---|
| 1- Explosives | 1.1 | Capable of producing a mass explosion |
| | 1.2 | Projection hazard but not a mass explosion hazard |
| | 1.3 | Fire hazard and either a minor blast or minor projection hazard or both, but not a mass explosion hazard |
| | 1.4 | Minor hazard if ignited or initiated during transport, not a projection hazard |
| | 1.5 | Insensitive explosive substances that may represent a mass explosion risk |
| 2- Gases - compressed or deeply refrigerated | 2.1 | Flammable gases |
| | 2.2 | Nonpoisonous, nonflammable gases |
| | 2.3 | Poisonous gases |
| 3- Flammable and Combustible Liquids | 3.1 | Flash point <-18°C (c.c.) |
| | 3.2 | Flash point -18°C to <37.8°C (c.c.) |
| | 3.3 | Flash point 37.8°C to 93.3°C (c.c.) |
| 4- Flammable Solids | 4.1 | Readily combustible |
| | 4.2 | Combustible from spontaneous heating, or exposure to air |
| | 4.3 | Emit flammable gases, or become spontaneously combustible on contact with water or water vapour |
| 5- Oxidizing Substances; Organic Peroxides | 5.1 | Oxidizer |
| | 5.2 | Organic Peroxide |
| 6- Poisonous (toxic) and Infectious Substances | 6.1 | Substances that are poisonous by ingestion, inhalation or skin contact |
| | 6.2 | Infectious substances |
| 7- Radioactive Materials and Prescribed Substances | - | Within the meaning of the "Atomic Energy Control Act" and classified by the Atomic Energy Control Board |
| 8- Corrosives | - | Cause severe damage to living tissue by chemical action, may corrode or destroy freight or means of transport |
| 9- Miscellaneous Dangerous Substances | - | Substances not ascribed to any other class which from experience may present some danger warranting regulation in transport |
| 9.1 Miscellaneous Products, Substances or Organisms | | |
| 9.2 Environmentally Dangerous Substances | | |
| 9.3 Dangerous Wastes | | |
| Packing Group I | Very Dangerous | |
| Packing Group II | Dangerous | |
| Packing Group III | Moderately Dangerous | |

## Labels

- The labels/placards as required under the Transport of Dangerous Goods Regulations are described.
- These labels/placards are depicted below:

Class 1
Explosive

Class 2.1
Flammable Compressed Gas

Class 2.2
Nonflammable Gas

Class 2.3
Poisonous Gas

Class 3
Flammable or Combustible Liquid

Class 4.1
Flammable Solid

Class 4.2
Combustible Solid (on exposure to air)

Class 4.3
Combustible Solid (on expose to water or water vapour)

Class 5
Oxidizer (or Organic Peroxide)

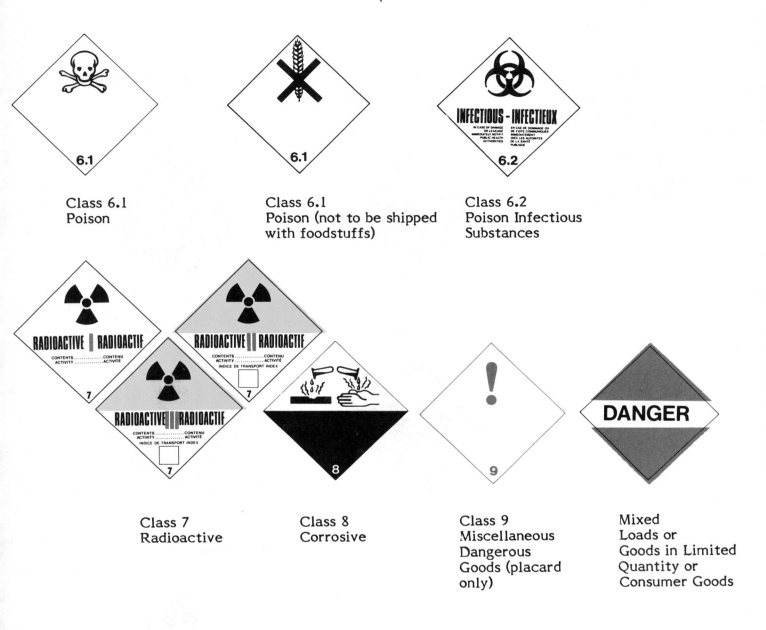

### Inert Atmosphere
- The description of the gas which should be used with materials which react with air, or as a general safety precaution.

### Venting
- The description of the venting requirements including open (no special requirements), closed (sealed container), safety-relief (a device which opens at a preset pressure) and pressure-vacuum (a device which opens at a preset pressure or vacuum).

Storage Temperature
- The temperature at which the substance is typically stored.

Pump Type; Hose Type
- Listing of the known types which are compatible with the material.

Grades or Purity
- A listing of the common commercial grades of the substance with the typical purity given.
- In some cases, the other ingredients of the commercial grade.
- Some solutions are given in °Bé (degrees Baumé), a specific gravity measurement.

Containers and Materials
- A listing of the common containers in which the product is stored or transported and the materials from which these containers are constructed.

### 1.2.5 Physical and Chemical Characteristics.

Physical State (at 20°C)
- The physical state (gas, liquid or solid) of the pure substance at 20°C and 1 atmosphere pressure.

Solubility (water)
- The solubility of the substance in grams per 100 millilitres of water (g/100 mL).
- g/100 mL is also equal to the percentage by weight.
- No standard scheme exists to describe solubility; the system used in this manual is as follows:

| Description | Solubility (g/100 mL water) |
| --- | --- |
| Soluble in all proportions | >>100 |
| Very Soluble | >50 |
| Soluble | 10 to 50 |
| Moderately Soluble | 1 to 10 |
| Slightly Soluble | 0.1 to 1 |
| Insoluble | <0.1 |

- Scales used in other literature include (direct comparison of solubility by scale alone is therefore difficult):

| Solubility Scale | Description | Solubility (g/100 mL water) |
| --- | --- | --- |
| A | Very Soluble<br>Soluble<br>Slightly Soluble | >50<br>5 to 50<br><5 |
| B | Soluble<br>Slight Solubility<br>Insoluble | >25<br>10 to 24<br><10 |
| C | Extremely Soluble<br>Highly Soluble<br>Moderately Soluble<br>Slightly Soluble<br>Practically Insoluble | >1<br>0.1 to 1<br>0.02 to 0.1<br>0.002 to 0.02<br><0.002 |

Molecular Weight
- The sum of the atomic weights of a molecule of the pure substance where the relative mass of each atom is based on a scale in which carbon-12 is assigned a mass value of 12.
- The molecular weight is useful in converting pressure, volume and temperature relationships for gases or vapours.

Vapour Pressure
- The pressure that a vapour exerts on its surroundings, given here in millimetres of mercury (mm Hg) at various temperatures.
- 760 mm Hg is 1 atmosphere (standard) pressure.
- A substance with a high vapour pressure "gives off" more vapours than a substance with a lower vapour pressure at the same temperature and thus would require consideration as a gas as well as a liquid or solid in a spill situation.

Boiling Point
- The temperature at which a substance boils, that is the temperature at which its vapour pressure equals the ambient air pressure.
- Given here in degrees Celsius (°C) and 1 atmosphere pressure (760 mm Hg) or in a few cases at lower pressures where these are the only data available.

Floatability
- The physical behaviour of the substance when spilled on water.

Odour
- A description of the odour and the odour threshold are given.
- Odour threshold as presented here is the entire range of measurements presented in literature from the 0% level - when the odour was first perceived by a person in a test panel - to the 100% recognition level - when all persons in a test panel recognized an odour.
- These odour threshold values cannot be relied upon to prevent over-exposure since human sensitivity to odours varies widely, odours could be masked by other odours, and some compounds rapidly deaden the sense of smell.

Flash Point
- The lowest temperature at which vapours above a volatile substance will ignite in air when exposed to a flame.
- Is given in degrees Celsius (°C) and followed by the method used -- closed cup (c.c.) or open cup (o.c.) -- the latter value generally is about 5 to 10°C higher than the closed cup value.
- Flash point values are often used to rate the flammability or combustibility of a substance; the criterion used in this manual is that any substance with a flash point (c.c.) under 60°C is flammable and those with a flash point at or greater than 60°C are combustible.
- The Canadian Transport Commission regulates as flammable liquids those with a flash point less than 23°C (73°F). Beginning January 1, 1984, those liquids having a flash point of less than 61°C (141°F) will also be regulated as flammable liquids. The United States Department of Transport regulates as flammable liquids those having flash points of less than 37.8°C (100°F) and as combustible liquids those having flash points of 37.8°C (100°F) or more but less than 93.3°C (200°F).
- For readers of this manual the following scale is proposed as a general guide to the flammability of materials:

| Rating | Flash point (c.c.) |
| --- | --- |
| Highly Flammable | <-10°C |
| Very Flammable | -10°C to 20°C |
| Flammable | 20°C to 60°C |

| | |
|---|---|
| Combustible | 60°C to 120°C |
| Low Combustibility | >120°C |

### Vapour Density
- The ratio of the weight of vapour to the weight of an equal volume of dry air at the same pressure and temperature.
- A vapour density of less than 1 implies that the vapour will be buoyant and rise in air, and vice versa; it should be noted that mixing of gas with air is very dependent on atmospheric conditions and should not be solely judged by its vapour density.
- Acetylene, ammonia, ethylene, hydrogen and methane are common gases that have a vapour density of less than 1.
- Substances with vapour densities greater than 1 have a tendency under most atmospheric conditions to reside in depressions for a time before mixing with the ambient air.

### Specific Gravity
- The ratio of the weight of a solid or liquid to the weight of an equal volume of water, at some specified temperature.
- If the specific gravity is less than 1.0 (or less than 1.03 in seawater), the chemical will float on water; if higher, it will sink. This, of course, presumes that the material does not entrain air or does not react.

### Colour
- Description of the colour(s) of the substance; it should be noted that the purity or form of the substance may alter its colour.

### Explosive Limits
- Same as "Flammability Limits" and is the concentration range of vapour in air between which propagation of a flame occurs on contact with a source of ignition.
- L.E.L. is the lower explosive limit and U.E.L. is the upper explosive limit.
- Generally, these values are given at 20°C and temperature increases extend the limits while temperature decreases narrow the limits.

### Melting Point
- The temperature at which a solid turns to a liquid.
- For many substances this is similar to freezing point, the temperature at which a liquid turns to a solid. Some substances, however, are not symmetric in this regard and a range has sometimes been presented.

- The purity of a substance affects the melting point; a range has sometimes been presented for this situation too.

## 1.3 Hazard Data

### 1.3.1 Human Health.

- This section provides data on the effects of the substance on humans and will distinguish by route of entry (inhalation, contact with skin and eyes, ingestion and absorption through the skin), as well as by the form of the substance (physical state, vapour, dust, fume or mist), where these are applicable.

<u>Symptoms</u>
- Summary of the symptoms shown in the literature for acute (periodic and short-term rather than continuous) exposure to the substance.
- Descriptions of the symptoms are given in common terms such as pain, irritation, dizziness, etc., where these terms can be used.
- The following terms are also employed:

| | |
|---|---|
| Abdominal | - referring to stomach and intestinal area |
| Alimentary Canal | - mouth, esophagus, stomach and intestines |
| Anaesthesia | - loss of sensation |
| Asphyxia (asphyxiation) | - breathing difficulty due to lack of oxygen |
| Bronchitis | - an inflammation (redness, swelling, etc.) of the bronchial tubes in the lungs |
| Central Nervous System Depression | - suppression of sensory and motor impulses to and from the brain and spinal cord as evidenced by: poor motor response, slower reflexes, slower breathing, etc. |
| Coma | - a state of deep unconsciousness |
| Convulsions | - abnormal, violent and involuntary series of contractions of the muscles |
| Cornea | - the transparent part of the covering of the eyeball |
| Corneal | - of or referring to the cornea |
| Cramps (stomach cramps) | - painful involuntary muscular contractions |
| Cyanosis | - bluish discolouration of the skin |
| Defatting | - removal of oils from skin tissue resulting in dry, parched skin |

| | |
|---|---|
| Delirium | - mental disturbance characterized by confusion, disordered speech and hallucinations |
| Dermatitis | - inflammation (redness, swelling, etc.) of the skin |
| Desiccant | - a moisture-absorbing (drying) agent |
| Distension (of stomach) | - an expansion or enlargement |
| Dullness | - slowness in responding to stimuli |
| Edema | - abnormal accumulation of fluid in body tissue |
| Euphoria | - abnormal feeling of well-being or elation |
| Extremities | - portions of the body away from the torso - e.g., hands and feet |
| Gastrointestinal | - relating both to the stomach and intestines |
| Hemorrhagic Gastritis | - inflammation and bleeding of the stomach |
| Impairment (mental) | - slowing or decreasing the mental ability |
| Inflammation | - swelling and redness of body tissue |
| Lesions | - injury, damage or abnormal change in a tissue or organ |
| Mucous Membranes | - membranes rich in mucous glands, such as those found in the nose, mouth and throat |
| Narcosis | - a state of stupor, unconsciousness or slowed activity produced by the influence of a chemical |
| Nausea | - general feeling of stomach sickness often resulting in vomiting |
| Opaqueness of Cornea | - whiteness or colouring of the normally transparent covering of the eye |
| Pallor | - paleness or loss of facial colour |
| Paralysis | - complete or partial loss of function in a portion of the body including loss of muscle control and/or loss of sensation |
| Perforation | - penetration or holing of a body tissue |
| Pulmonary | - relating to the lungs |
| Pulmonary Edema | - abnormal accumulation of fluid in lung tissue |
| Salivation | - excessive discharge of saliva |
| Shock | - deep depression of vital processes caused by a rapid fall in blood pressure |
| Spasm | - involuntary and sudden muscular contraction |
| Sputum | - spit and other expectorate matter |
| Ulceration (skin) | - break in skin with a loss or destruction of surface tissue |

### Toxicology

- Provides a relative rating (high, moderate or low) of the toxicity of the product by route (contact, inhalation or ingestion).

### TLV®

- Is the "Threshold Limit Value".
- A registered trademark of the American Conference of Governmental Industrial Hygienists; values presented here are their 1981 recommendations.
- Is a workplace number used as a guide to the maximum average exposure to a chemical for 8-hour days and 5 days per week.
- Is useful for spills as a guide for evacuation (see also Short-term Inhalation Limit, below).
- When "skin" is specified, this implies that the material is percutaneous - that it is absorbed through the skin at a rate similar to or greater than that absorbed by the body through inhalation.

### Short-term Inhalation Limit

- Taken here as the given STEL ("Short-term Exposure Limit") value unless otherwise noted.
- A workplace value similar in origin and meaning to the TLV® except that this is the maximum concentration to which workers can be continuously exposed for a period up to 15 minutes without suffering irritation or chronic and irreversible effects provided that no more than four of these exposures occur per day, and that there is at least 1 hour between exposure periods and that the TLV® is not exceeded (as an average).
- The value is also useful for spill situations representing a maximum exposure value without respiratory (or skin) protection.

### $LC_{50}$

- "Lethal Concentration Fifty".
- The calculated concentration of a substance in air, exposure to which caused the death of 50% of the test population in a specified time.

### $TC_{Lo}$

- "Toxic Concentration Low".
- The lowest concentration of a substance in air that has produced any toxic effect on the test population for any given period of time.

## $LC_{Lo}$

- "Lethal Concentration Low".
- The lowest concentration of a substance in air for a specified period of time that has produced death in the test population.

## $LD_{50}$

- "Lethal Dose Fifty".
- The calculated dose (in grams of substance per kilogram of body weight) that caused the death of 50% of the test population.
- The test population may have the substance administered by different routes:
  Oral - by feeding and then ingestion
  Skin - by contact with skin and subsequent absorption
  Intraperitoneal - by injection into the peritoneum (membrane lining the abdomen)
  Subcutaneous - by injection under the skin
  Intravenous - by injection into the bloodstream
  Cutaneous - by skin, exact means (skin, subcutaneous, etc.) not specified
- IMO has provided a toxicity scale which might be used as a relative guide:

| Scale | Description | $LD_{50}$ (g/kg) |
|---|---|---|
| 4 | highly hazardous | <0.005 |
| 3 | moderately hazardous | 0.005 to 0.05 |
| 2 | slightly hazardous | 0.05 to 0.5 |
| 1 | practically nonhazardous | 0.5 to 5.0 |
| 0 | not hazardous | >5.0 |

## $LD_{Lo}$

- "Lethal Dose Low".
- The lowest dose (g/kg) that caused death in a test population.
- The routes of administration are those as described in $LD_{50}$.

## $TD_{Lo}$

- "Toxic Dose Low".
- The lowest dose of a substance reported to have produced any toxic effect in a test population.
- The routes of administration are described under $LD_{50}$.

## Delayed Toxicity
- Provides information on delayed symptoms of exposure, possible long-term effects of exposure or information on suspected carcinogenicity.
- "Suspected carcinogen" means the substance was tested for carcinogenicity using animals and showed positive results, i.e., may have potential to cause cancer.

## Absorption by Skin
- Certain substances are readily absorbed through the skin necessitating the use of special protective clothing.
- This property is indicated by statements such as "readily absorbed through the skin", etc.; by a TLV® for the skin; or by an $LD_{50}$, $LD_{Lo}$ or $TD_{Lo}$ that is high via the skin or subcutaneous route.
- These substances are called percutaneous (through-the-skin) materials.
- Examples of chemicals which can enter the body at least as rapidly by skin absorption as by inhalation include the following (source of these is not dealt with in the manual):

| | | |
|---|---|---|
| Acetonitrile | Cyanides | Morpholine |
| Acrylonitrile | Epichlorohydrin | Pentachlorophenol |
| Allyl Alcohol | Ethyl Acrylate | Phenol |
| Aniline | Furfural | Propyl Alcohol |
| n-Butyl Alcohol | Furfuryl Alcohol | Tetraethyl lead |
| Carbon Disulfide | Hydrazine | Toluene |
| Carbon Tetrachloride | Malathion | Xylene |
| Chlordane | Methyl Acrylate | |
| Cresol | Methyl Alcohol | |

## IDLH
- "Immediately Dangerous to Life or Health".
- A value represents a maximum concentration from which one could escape within 30 minutes without irreversible health effects. It is used as the value at which air-supplied respiratory protection is required and at which filter or chemical cartridge protection would not be suitable.
- This value has not been included in the main listings as it is often misinterpreted as a "safe" value; however, a list of some IDLH values follows.
- A "practical" definition of IDLH is suggested as the concentration at which irreversible effects on health could be expected.

## A LIST OF SOME IDLH VALUES

| Substance | IDLH (ppm, unless otherwise specified) | Substance | IDLH (ppm, unless otherwise specified) |
|---|---|---|---|
| Acetaldehyde | 10 000 | Chlordane | 500 mg/m$^3$ |
| Acetic Acid | 1 000 | Chlorine | 25 |
| Acetic Anhydride | 1 000 | Chlorobenzene | 2 400 |
| Acetone | 20 000 | Chlorobromomethane | 5 000 |
| Acetonitrile | 4 000 | Chloroform | 1 000 |
| Acrolein | 5 | Chromium (as soluble Cr salts) | 250 mg/m$^3$ |
| Acrylonitrile | 4 | Cresol | 250 |
| Allyl Alcohol | 150 | Cumene | 8 000 |
| Allyl Chloride | 300 | Cyanide (KCN or NaCN) | 50 mg/m$^3$ |
| Ammonia | 500 | Cyclohexane | 10 000 |
| n-Amyl Acetate | 4 000 | Cyclohexanol | 3 500 |
| Aniline | 100 | Cyclohexanone | 5 000 |
| Arsine | 6 | Cyclohexene | 10 000 |
| Benzene | 2 000 | 2,4-D | 500 mg/m$^3$ |
| Benzyl Chloride | 10 | o-Dichlorobenzene | 1 700 |
| Boron Trifluoride | 100 | p-Dichlorobenzene | 1 000 |
| Bromine | 10 | Dichlorodifluoromethane | 50 000 |
| Butadiene | 20 000 | 1,1-Dichloroethane | 4 000 |
| 2-Butanone | 3 000 | 1,2-Dichloroethylene | 4 000 |
| t-Butyl Acetate | 8 000 | Dichloromonofluoromethane | 50 000 |
| Butyl Acetate | 10 000 | Dichlorotetrafluoroethane | 50 000 |
| sec-Butyl Alcohol | 10 000 | Diethylamine | 2 000 |
| t-Butyl Alcohol | 8 000 | Difluorodibromomethane | 2 500 |
| n-Butyl Alcohol | 8 000 | Diisobutyl Ketone | 2 000 |
| Butyl Amine | 2 000 | Dimethylamine | 2 000 |
| Butyl Mercaptan | 2 500 | Diisopropylamine | 1 000 |
| Calcium Oxide | 250 mg/m$^3$ | Diphenyl | 300 mg/m$^3$ |
| Carbon Dioxide | 50 000 | Epichlorohydrin | 100 |
| Carbon Disulfide | 500 | Ethanolamine | 1 000 |
| Carbon Monoxide | 1 500 | Ethyl Acetate | 10 000 |
| Carbon Tetrachloride | 300 | Ethylacrylate | 2 000 |

| Substance | IDLH (ppm, unless otherwise specified) | Substance | IDLH (ppm, unless otherwise specified) |
|---|---|---|---|
| Ethylbenzene | 2 000 | Methyl Isobutyl Carbinol | 2 000 |
| Ethyl Chloride | 20 000 | Methyl Isobutyl Ketone | 3 000 |
| Ethyl Ether | 19 000 | Methyl Mercaptan | 400 |
| Ethyl Mercaptan | 2 500 | Methyl Methacrylate | 4 000 |
| Ethylamine | 4 000 | Methylamine | 100 |
| Ethylene Dichloride | 1 000 | Methylene Chloride | 5 000 |
| Ethylene Oxide | 800 | Morpholine | 8 000 |
| Fluorine | 25 | Naphthalene | 500 |
| Formaldehyde | 100 | Naphtha | 10 000 |
| Formic Acid | 100 | Nitric Acid | 100 |
| Furfural | 250 | Nitric Oxide (NO) | 100 |
| Furfuryl Alcohol | 250 | Nitrobenzene | 200 |
| Heptane | 4 250 | Nitrogen Dioxide ($NO_2 + N_2O_4$) | 50 |
| Hexane | 5 000 | Octane | 3 750 |
| Hydrazine | 80 | Pentachlorophenol | 150 mg/m$^3$ |
| Hydrogen Bromide | 50 | Pentane | 5 000 |
| Hydrogen Chloride | 100 | Perchloroethane | 300 |
| Hydrogen Cyanide | 50 | Perchloroethylene | 500 |
| Hydrogen Fluoride | 20 | Phenol | 100 |
| Hydrogen Peroxide | 75 | Phosgene | 2 |
| Hydrogen Sulfide | 300 | Phosphine | 200 |
| Isoamyl Alcohol | 8 000 | Phthalic Anhydride | 10 000 |
| Isobutyl Alcohol | 8 000 | Propane | 20 000 |
| Isopropyl Alcohol | 20 000 | Propyl Alcohol | 4 000 |
| LPG (liquified petroleum gas) | 19 000 | Propylene Oxide | 2 000 |
| Malathion | 5 000 mg/m$^3$ | Pyridine | 3 600 |
| Mercury | 28 mg/m$^3$ | Sodium Hydroxide | 200 mg/m$^3$ |
| Methyl Acetate | 10 000 | Styrene | 5 000 |
| Methyl Acrylate | 1 000 | Sulfur Dioxide | 100 |
| Methyl Alcohol | 25 000 | Sulfuric Acid | 80 mg/m$^3$ |
| Methyl Butyl Ketone | 5 000 | Terphenyls | 3 500 mg/m$^3$ |
| Methyl Chloride | 10 000 | Tetraethyl Lead | 40 mg/m$^3$ |

| Substance | IDLH (ppm, unless otherwise specified) | Substance | IDLH (ppm, unless otherwise specified) |
|---|---|---|---|
| Tetrahydrofuran | 20 000 | 1,1,2-Trichloroethane | 500 |
| Toluene | 2 000 | Trichloroethylene | 1 000 |
| Toluene-2,4-diisocyanate | 10 | Turpentine | 1 900 |
| | | Xylene | 10 000 |
| 1,1,1-Trichloroethane | 1 000 | Zinc Chloride | 2 000 mg/m$^3$ |

### 1.3.2 Fire.

<u>Fire Extinguishing Agents</u>
- Lists those agents reported in the literature as suitable, or indicates which agents are not suitable.

<u>Behaviour in Fire</u>
- Description of unusual behaviour or properties.

<u>Ignition Temperature</u>
- Is also equivalent to the autoignition temperature or the minimum temperature at which the substance will ignite without a spark or flame being present.

<u>Burning Rate</u>
- The rate (in millimetres per minute) at which the depth of a pool of liquid decreases as the substance burns.

<u>Detonation Velocity</u>
- Given for explosives only. It is the velocity at which the explosion shock wave propagates in the material (given in metres per second).

### 1.3.3 Reactivity.

<u>With Water</u>
- Description of reaction is given.

<u>With Common Materials</u>
- Other chemicals or groups of chemicals with which the substance reacts are given.
- The nature of the reaction (e.g., produces an explosion, causes fire, etc.) is not given.
- "Reacts violently with" implies a serious situation; however, the severity is not specified.
- Tertiary (three-chemical) reactions are sometimes specified and are presented with brackets and a plus sign (e.g., Acetone + Acetic Acid).

Stability

- "Stable" means that the substance will not react or decompose in a hazardous way under the temperature, pressure, contact and mechanical conditions normally encountered in storage or transportation.
- This category is not applicable to fire or accident situations.

### 1.3.4 Environment.

Water

- Provides descriptive information, aquatic toxicity data and BOD data.

$LC_{50}$

- "Lethal Concentration Fifty" or abbreviation for "Median Lethal Concentration".
- Is the concentration (in milligrams of substance per litre of water, which is approximately equivalent to ppm (parts per million)) in water at which 50% of the test population died during a specified time period.
- tns, is time period not specified.

$LC_{100}$

- "Lethal Concentration One Hundred".
- Is equivalent to the above except that 100% of the test population died; it is used in this manual to describe fish kill observations.

$TL_m$

- "Median Tolerance Limit".
- Is the concentration (mg/L) in water at which 50% of the test population will show abnormal behaviour (including death).
- IMO has provided a rating scheme as follows, which might be used as a guide to aquatic toxicity:

| Rating | Description | $TL_m$ (mg/L or ppm) |
|---|---|---|
| 4 | highly toxic | <1 |
| 3 | moderately toxic | 1 to 10 |
| 2 | slightly toxic | 10 to 100 |
| 1 | practically non-toxic | 100 to 1000 |
| 0 | not hazardous | >1000 |

The following is a list of species which have been referred to in this manual:

| Common Name | Latin Name |
|---|---|
| American shad | *Alosa sapidissima* |
| Aquatic plant | *Elodea canadensis* |
| Bluegill | *Lepomis macrochirus* |
| Blue-green algae | *Anabaena sp.* |
| Brine shrimp | *Artemia salina* |
| Brook trout | *Salvelinus fontinalis* |
| Brown shrimp | *Crangon crangon* |
| Catfish (american) | *Ameiurus nebulosus* (Le Sueur) |
| Channel catfish | *Ictalurus punctatus* |
| Chub | *Squalius cephalus* (L.) |
| Cockle | *Cerastoderma edule* |
| Cod | *Gadus morrhua* |
| Creek chub | *Semotilus atromaculatus* |
| Fathead minnow | *Pimephales promelas* |
| Goldfish | *Carassius auratus* |
| Grass shrimp | *Hippolyte zostericola, Palaemonetes pugio* |
| Green algae | *Chlorella vulgaris, Scenedesmus quadricauda* |
| Guppy | *Lebistes reticulatus* |
| Marine diatom | *Nitzschia linearis* |
| Minnow | *Phoxinus phoxinus* |
| Mosquito fish | *Gambusia affinis* |
| Perch (American, yellow) | *Perca fluviatilis flavescens* (Mitchill) |
| Pinfish (threespine stickleback) | *Gasterosteus aculeatus* |
| Prawn | *Metapenaeus monoceros* |
| Rainbow trout | *Salmo gairdneri* |
| Salmon (Atlantic) | *Salmo salar* (L.) |
| Stickleback (12-spined) | *Pygosteus pungitius* (L.) |
| Stickleback (threespine) | *Gasterosteus aculeatus* |
| Striped bass | *Morone saxatilis* |
| Sunfish (common) | *Lepomis humilis* |
| Water flea | *Daphnia magna* |
| Water shrimp | *Gammarus pulex* |

| Latin Name | Common name |
|---|---|
| *Alosa sapidissima* | American shad |
| *Ameiurus nebulosus* (Le Sueur) | Catfish (American) |
| *Anabaena* sp. | Blue-green algae |
| *Artemia salina* | Brine shrimp |
| *Carassius auratus* | Goldfish |
| *Cerastoderma edule* | Cockle |
| *Chlorella vulgaris* | Green algae |
| *Crangon crangon* | Brown shrimp |
| *Daphnia magna* | Water flea |
| *Elodea canadensis* | Aquatic plant |
| *Gadus morrhua* | Cod |
| *Gambusia affinis* | Mosquito fish |
| *Gammarus pulex* | Water shrimp |
| *Gasterosteus aculeatus* | Pinfish, threespine stickleback |
| *Hippolyte zostericola* | Grass shrimp |
| *Ictalurus punctatus* | Channel catfish |
| *Lebistes reticulatus* | Guppy |
| *Lepomis humilis* | Common sunfish |
| *Lepomis macrochirus* | Bluegill |
| *Metapenaeus monoceros* | Prawn |
| *Morone saxatilis* | Striped bass |
| *Nitzschia linearis* | Marine diatom |
| *Palaemonetes pugio* | Grass shrimp |
| *Perca fluviatilis flavescens* (Mitchill) | American yellow perch |
| *Phonixus phoxinus* | Minnow |
| *Pimephales promelas* | Fathead minnow |
| *Pygosteus pungitius* (L.) | 12-spined stickleback |
| *Salmo gairdneri* | Rainbow trout |
| *Salmo salar* (L.) | Atlantic salmon |
| *Salvelinus fontinalis* | Brook trout |
| *Scenedesmus quadricauda* | Green algae |
| *Semotilus atromaculatus* | Creek chub |
| *Squalius cephalus* (L.) | Chub |

BOD
- "Biological Oxygen Demand" or sometimes "Biochemical Oxygen Demand".
- The quantity of oxygen dissolved in water which is consumed by biological oxidation of the chemical during a specified period of time.
- $BOD_5$ for example is the BOD in 5 days.
- It is given in this manual as percentage by weight (weight of oxygen versus the weight of the substance).
- ThOD or Theor. is the theoretical oxygen demand and is the percentage (by weight) of oxygen required to completely oxidize the spilled substance when viewed as a simple chemical reaction.

Land/Air
- Toxicity data relating to farm animals, fowl or other land species.

Food Chain Concentration Potential
- Description of bioaccumulation of the product.

### 1.4 Emergency Measures

### 1.4.1 Special Hazards.

- Summarizes immediate concerns.

### 1.4.2 Immediate Responses.

- Summarizes suggested "first" actions.

### 1.4.3 Protective Clothing and Equipment.

- Provides a summary of proper protective equipment, where this information is available.
- It should be noted that entry into a situation where the product and its concentration in air are both unknown must only be made with a totally encapsulated suit and a self-contained breathing apparatus (SCBA). Cartridge-type respirators are not to be worn in any situation where the concentrations are unknown or unpredictable.

### 1.4.4 Fire and Explosion.

- Summarizes fire-fighting information.

### 1.4.5 First Aid.

- Summarizes literature recommendations for first aid.

## 1.5 Environmental Protection Measures

### 1.5.1 Response.

- Summarizes actions which can be taken to minimize environmental damage.

### 1.5.2 Disposal.

- Brief recommendations for source of information, or procedures for disposal where this is possible without serious environmental consequences.

## 2 INDEX OF ENTRIES

Each main entry appears in this index in **CAPITALIZED, BOLD FACED TYPE.** Common synonyms and trade names are also included (in upper and lower case lettering) and are each followed by the name under which they appear in the text.

### A

Absolute Ethanol (Ethyl Alcohol)
Accothion (Fenitrothion)
**ACETALDEHYDE**
**ACETIC ACID**
Acetic Acid Amyl Ester (Amyl Acetate)
Acetic Acid, Anhydride (Acetic Anhydride)
Acetic Acid Ethenyl (Vinyl Acetate)
Acetic Acid Ethyl Ester (Ethyl Acetate)
Acetic Acid Pentyl Ester (Amyl Acetate)
Acetic Acid Vinyl Ester (Vinyl Acetate)
Acetic Aldehyde (Acetaldehyde)
**ACETIC ANHYDRIDE**
Acetic Ester (Ethyl Acetate)
Acetic Ether (Ethyl Acetate)
Acetic Oxide (Acetic Anhydride)
**ACETONE**
**ACETONITRILE**
**ACETYLENE**
Acetylenogen (Calcium Carbide)
Acetyl Oxide (Acetic Anhydride)
Acid Calcium Phosphate (Calcium Phosphate, Monobasic)
Acraldehyde (Acrolein)
**ACROLEIN**
Acrylaldehyde (Acrolein)
Acrylic Acid, Ethyl Ester (Ethyl Acrylate)
Acrylic Acid, Methyl Ester (Methyl Acrylate)
**ACRYLONITRILE**
**ADIPIC ACID**

Adipinic Acid (Adipic Acid)
Aerothene (1,1,1-Trichloroethane)
Agricultural Lime (Calcium Hydroxide)
Agricultural Limestone (Calcium Carbonate)
Agrothion (Fenitrothion)
Alcohol (Ethyl Alcohol)
Alcohol, Dehydrated (Ethyl Alcohol)
Aldehyde (Acetaldehyde)
Allyl Aldehyde (Acrolein)
Alum (Aluminum Sulfate)
Aluminatrihydrate (Aluminum Hydroxide)
**ALUMINUM ALKYL COMPOUNDS**
**ALUMINUM CHLORIDE**
Aluminum Chloride, Anhydrous (Aluminum Chloride)
Aluminum Hydrate (Aluminum Hydroxide)
**ALUMINUM HYDROXIDE**
Aluminum Hydroxide Gel (Aluminum Hydroxide)
Aluminum Sodium Fluoride (Cryolite)
**ALUMINUM SULFATE**
Aminobenzene (Aniline)
Aminocaproic Lactam (Caprolactam)
**AMINOCARB**
2-Aminoethanol (Ethanolamine)
β-Aminoethyl Alcohol (Ethanolamine)
Aminomethane (Methylamines)
Aminotriacetic Acid (Nitrilotriacetic Acid)
Amkil (MCPA)
**AMMONIA (Anhydrous)**
Ammonia Liquor (Ammonium Hydroxide)
Ammonia Solution (Ammonium Hydroxide)
**AMMONIUM SULFATE**
Ammonia Water (Ammonium Hydroxide)
Ammonium Acid Phosphate (Ammonium Phosphates, Monobasic)
Ammonium Biphosphate (Ammonium Phosphates, Monobasic)

**AMMONIUM CHLORIDE**

Ammonium Dihydrogen Phosphate (Ammonium Phosphates, Monobasic)

Ammonium Hydrate (Ammonium Hydroxide)

Ammonium Hydrogen Phosphate (Ammonium Phosphates, Monobasic)

**AMMONIUM HYDROXIDE**

Ammonium Muriate (Ammonium Chloride)

**AMMONIUM NITRATE**

**AMMONIUM PHOSPHATES**

Ammonium Phosphate, Dibasic (Ammonium Phosphates)

Ammonium Phosphate, Monobasic (Ammonium Phosphates)

Ammonium Phosphate, Primary (Ammonium Phosphates, Monobasic)

Ammonium Phosphate, Secondary (Ammonium Phosphates, Dibasic)

Amorphous Phosphorus (Phosphorus, Red)

**AMYL ACETATE**

Amylacetic Ester (Amyl Acetate)

**ANILINE**

Aniline Oil (Aniline)

Anilinobenzene (Diphenyl Amine)

Ant and Grub Killer (Chlordane)

Ant Oil (Furfural)

Antifreeze (Ethylene Glycol)

Antiknock (Methylcyclopentadienyl Manganese Tricarbonyl)

Aqua Ammonia (Ammonium Hydroxide)

Arconitrile Butadiene Rubber (Latex)

**ARSENIC**

Arsenic Hydride (Arsine)

Arsenic Sequioxide (Arsenic Trioxide)

Arsenic Trihydride (Arsine)

**ARSENIC TRIOXIDE**

Arsenious Oxide (Arsenic Trioxide)

Arseniuretted Hydrogen (Arsine)

Arsenous Anhydride (Arsenic Trioxide)

Arsenous Hydride (Arsine)

**ARSINE**

**ASPHALT**

Automotive Fuel (Gasoline)

Aviation Gasoline (Oils, Fuel - Aviation)

AV-gas (Gasoline)

Azacyclopropane (Ethyleneimine)

Aziridine (Ethyleneimine)

**B**

Banana Oil (Amyl Acetate)

Banex (Dicamba)

Banlen (Dicamba)

Barite (Barium Sulfate)

**BARIUM CARBONATE**

**BARIUM SULFATE**

Barytes (Barium Sulfate)

Basofor (Barium Sulfate)

Battery Acid (Sulfuric Acid)

Benzenamine (Aniline)

**BENZENE**

Benzene Chloride (Chlorobenzene)

Benzenecarboxylic Acid (Benzoic Acid)

1,4-Benzenedicarboxyl Acid (Terephthalic Acid)

p-Benzenedicarboxylic Acid (Terephthalic Acid)

1,4-Benzenedicarboxylic Acid, Dimethyl Ester (Dimethyl Terephthalate)

Benzene-Diphenyl (Terphenyls)

Benzin (Benzene)

Benzine (Naphtha Solvent)

Benzinoform (Carbon Tetrachloride)

**BENZOIC ACID**

Benzol (Benzene)

Bicalcium Phosphate (Calcium Phosphate, Dibasic)

Biethylene (1,3-Butadiene)

**BISPHENOL A**

Bitumen (Asphalt)

Bivinyl (1,3-Butadiene)

Blanc Fixe (Barium Sulfate)

Blasting Oil (Nitroglycerine)

Bleaching Powder (Calcium Hypochlorite, Hydrate)

Bluestone (Copper Sulfate)

Blue Vitriol (Copper Sulfate)

Bone Ash (Calcium Phosphate, Tribasic)

Boracic Acid (Boric Acid)

**BORAX**

**BORIC ACID**

Bran Oil (Furfural)

Brimstone (Sulfur)

Brush Killer (2,4-D)

Bunker A (Oils, Fuel - Residuals)

Bunker B (Oils, Fuel - Residuals)

Bunker C (Oils, Fuel - Residuals)

Bunker Fuel Oil (Oils, Fuel - Residuals)

Burnt Lime (Calcium Oxide)

**1,3-BUTADIENE**

n-Butanal (Butyraldehyde)

**BUTANE**

n-Butane (Butane)

1,4-Butanedicarboxylic Acid (Adipic Acid)

1-Butanol (Butyl Alcohol)

2-Butanol (Butyl Alcohol)

2-Butanone (Methyl Ethyl Ketone)

1-Butene (Butylene)

cis-2-Butene (Butylene)

trans-2-Butene (Butylene)

cis-Butenedioic Anhydride (Maleic Anhydride)

Butter of Zinc, Granular (Zinc Chloride)

**BUTYL ALCOHOL**

n-Butyl Alcohol (Butyl Alcohol)

s-Butyl Alcohol (Butyl Alcohol)

t-Butyl Alcohol (Butyl Alcohol)

2-sec-Butyl-4,6-Dinitrophenol (Dinoseb)

**BUTYLENE**

α-Butylene (Butylene)

Butyl Hydride (Butane)

**BUTYRALDEHYDE**

n-Butyraldehyde (Butyraldehyde)

Butyric Aldehyde (Butyraldehyde)

**C**

Cake Alum (Aluminum Sulfate)

Calcium Acetylide (Calcium Carbide)

Calcium Biphosphate (Calcium Phosphate, Monobasic)

**CALCIUM CARBIDE**

**CALCIUM CARBONATE**

**CALCIUM CHLORIDE**

Calcium Chloride, Anhydrous (Calcium Chloride)

Calcium Chloride Dihydrate (Calcium Chloride)

Calcium Chloride Hexahydrate (Calcium Chloride)

Calcium Chloride Monohydrate (Calcium Chloride)

**CALCIUM CYANIDE**

Calcium Hydrate (Calcium Hydroxide)

**CALCIUM HYDROXIDE**

**CALDIUM HYPOCHLORITE**

Calcium Orthophosphate (Calcium Phosphate, Tribasic)

**CALCIUM OXIDE**

Calcium Oxychloride (Calcium Hypochlorite, Anhydrous)

**CALCIUM PHOSPHATE**

Calcium Phosphate, primary (Calcium Phosphate, Monobasic)

Calmathion (Malathion)

Calx (Calcium Oxide)

**CAPROLACTAM**

Carbamide (Urea)

**CARBARYL**

Carbide (Calcium Carbide)

Carbinamine (Methylamines, Monomethylamine)

Carbinol (Methanol)

**CARBOFURAN**

Carbolic Acid (Phenol)

Carbon Bichloride (Perchloroethylene)

Carbon Bisulfide (Carbon Disulfide)

**CARBON DIOXIDE**

**CARBON DISULFIDE**

Carbonic Acid Gas (Carbon Dioxide)

Carbonic Anhydride (Carbon Dioxide)

Carbonic Dichloride (Phosgene)

**CARBON MONOXIDE**

Carbon Oxychloride (Phosgene)

Carbon Tet (Carbon Tetrachloride)

**CARBON TETRACHLORIDE**

Carbonyl Chloride (Phosgene)

Carbonyldiamide (Urea)

Carboxybenzene (Benzoic Acid)

N,N-bis (Carboxymethylglysine) (Nitrilotriacetic Acid)

Caustic (Sodium Hydroxide)

Caustic Lime (Calcium Hydroxide)

Caustic Soda (Sodium Hydroxide)

Celthion (Malathion)

Chalk (Calcium Carbonate)

Chile Saltpeter (Sodium Nitrate)

Chinese White (Zinc Oxide)

**CHLORDANE**

Chloride of Lime (Calcium Hypochlorite, Hydrate)

Chloride of Phosphorus (Phosphorus Trichloride)

**CHLORINE**

1-Chloro-2,3-Epoxypropane (Epichlorohydrin)

3-Chloro-1,2-Propylene Oxide (Epichlorohydrin)

**CHLOROBENZENE**

Chlorobenzol (Chlorobenzene)

Chlorodifluoromethane (Fluorochloromethanes, Freon 22)

Chloroethane (Ethyl Chloride)

Chloroethene (Vinyl Chloride)

Chloroethylene (Vinyl Chloride)

**CHLOROFORM**

Chloroformyl Chloride (Phosgene)

Chlorohydric Acid (Hydrochloric Acid)

Chloromethane (Methyl Chloride)

Chloromethyloxirane (Epichlorohydrin)

Chloropropylene Oxide (Epichlorohydrin)

**CHLOROSULFONIC ACID**

Chlorothene (1,1,1-Trichloroethane)

Chlorotrifluoromethane (Fluorochloromethanes, Freon 13)

Chlorox (Sodium Hypochlorite)

Chlorsulfonic Acid (Chlorosulfonic Acid)

Chlorsulfuric Acid (Chlorosulfonic Acid)

Chromic Acid (Chromic Anhydride)

Chromic Acid, Solid (Chromic Anhydride)

**CHROMIC ANHYDRIDE**

Chromium Oxide (Chromic Anhydride)

Chromium VI Oxide (Chromic Anhydride)

Chromium Trioxide (Chromic Anhydride)

Cinnamene (Styrene Monomer)

Coal Naphtha (Benzene)

Coal Oil (Kerosene)

**COBALT**

Cobalt Metal (Nonradioactive) (Cobalt)

Colamine (Ethanolamine)

**COPPER**

**COPPER CHLORIDE**

Copper II Chloride (Copper Chloride)

Copper Metal (Copper)

**COPPER NAPHTHENATE**

**COPPER SULFATE**

Copper Sulfate Pentahydrate (Copper Sulfate)

Copper 2 Sulfate Pentahydrate (Copper Sulfate)

Copper Uversol (Copper Naphthenate)

## CRESOL

2-Cresol (Cresol)

3-Cresol (Cresol)

4-Cresol (Cresol)

m-Cresol (Cresol)

o-Cresol (Cresol)

p-Cresol (Cresol)

Cresylic Acid (Cresol)

Crude Arsenic (Arsenic Trioxide)

## CRYOLITE

## CUMENE

## CUMENE HYDROPEROXIDE

Cumol (Cumene)

Cumyl Hydroperoxide (Cumene Hydroperoxide)

Cuprenol (Copper Naphthenate)

Cupric Chloride (Copper Chloride)

Cupric Cyanide (Cyanides)

Cupricin (Cyanides)

Cupric Sulfate (Copper Sulfate)

Cuprous Cyanide (Cyanides)

Curaterr (Carbofuran)

Cyanide of Calcium (Calcium Cyanide)

## CYANIDES (COPPER, POTASSIUM, ZINC)

Cyanoethylene (Acrylonitrile)

Cyanogas (Calcium Cyanide)

Cyanomethane (Acetonitrile)

## CYCLOHEXANE

Cythion (Malathion)

## D

## 2,4-D

Dandelion Killer (2,4-D)

Danex (Trichlorfon)

DCB (1,2-Dichlorobenzene)

DEAC (Aluminum Alkyl Compounds)

DEAI (Aluminum Alkyl Compounds)
Denatured Alcohol (Ethyl Alcohol)
DEZ (Aluminum Alkyl Compounds)
DFA (Diphenyl Amine)
1,2-Diaminoethane (Ethylenediamine)
Diammonium Hydrogen Phosphate (Ammonium Phosphates, Dibasic)
Diammonium Sulfate (Ammonium Sulfate)
DIBAC (Aluminum Alkyl Compounds)
DIBAH (Aluminum Alkyl Compounds)
Dibasic Calcium Phosphate (Calcium Phosphate)
1,2-Dibromoethane (Ethylene Dibromide)
Dicalcium Orthophosphate (Calcium Phosphate, Dibasic)
**DICAMBA**
**1,2-DICHLOROBENZENE**
o-Dichlorobenzene (1,2-Dichlorobenzene)
Dichlorodifluoromethane (Fluorochloromethanes, Freon 12)
1,2 Dichloroethane (Ethylene Dichloride)
Dichlorofluoromethane (Fluorochloromethanes, Freon 21)
Dichloromethane (Methylene Chloride)
2,4-Dichlorophenoxyacetic Acid (2,4-D)
Diesel Oil Light (Oils, Fuel - Distillates)
Diesel Oil Medium (Oils, Fuel - Distillates)
Diethylaluminum Chloride (Aluminum Alkyl Compounds)
Diethylaluminum Iodide (Aluminum Alkyl Compounds)
Diethylenimide Oxide (Morpholine)
Diethylmercapto Succinate S-Ester with o,o-Dimethyl Phosphorodithionate (Malathion)
Diethylzinc (Aluminum Alkyl Compounds)
Dihydroazirine (Ethyleneimine)
Diisobutylaluminum Chloride (Aluminum Alkyl Compounds)
Diisobutylaluminum Hydride (Aluminum Alkyl Compounds)
Dimethylamine (Methylamines)
4-Dimethylamino-3-Methylphenol Methylcarbamate (Aminocarb)
4-Dimethylamino-M-Tolyl Methylcarbamate (Aminocarb)
1,2-Dimethylbenzene (Xylenes)
1,3-Dimethylbenzene (Xylenes)

1,4-Dimethylbenzene (Xylenes)

Dimethyl 1,4-Benzenedicarboxylate (Dimethyl Terephthalate)

α,α-Dimethylbenzyl Hydroperoxide (Cumene Hydroperoxide)

Dimethylcarbinol (Isopropyl Alcohol)

Dimethylene Oxide (Ethylene Oxide)

Dimethylenimine (Ethyleneimine)

**DIMETHYL ETHER**

Dimethylketone (Acetone)

Dimethylphthalate (Dimethyl Terephthalate)

**DIMETHYL TEREPHTHALATE**

Dinocap (Dinoseb)

**DINOSEB**

**DIPHENYL AMINE**

Diphenylbenzene (Terphenyls)

**DIPHENYLMETHANE-4,4'-DIISOCYANATE**

Diphenylmethane Diisocyanate (Diphenylmethane-4,4'-Diisocyanate)

Dipping Acid (Sulfuric Acid)

Disodium Hydrogen Phosphate (Sodium Phosphate, Dibasic)

Disodium Monohydrogen Phosphate (Sodium Phosphate, Dibasic)

Disodium Phosphate (Sodium Phosphate, Dibasic)

Disodium Sulfate (Sodium Sulfate)

Disulphuric Acid (Oleum)

Dithionic Acid (Oleum)

Divinyl (1,3-Butadiene)

DMP (Dimethyl Terephthalate)

DMT (Dimethyl Terephthalate)

Dowicide 7 (Pentachlorophenol)

Dowicide-G (Sodium Pentachlorophenate)

Dowtherm-E (1,2-Dichlorobenzene)

DPA (Diphenyl Amine)

Dry Cleaners Naphtha (Naphtha Solvent)

Dry Cleaning Fluid (Perchloroethylene)

Dry Ice (Carbon Dioxide)

DSP (Sodium Phosphate, Dibasic)

Dutch Oil or Liquid (Ethylene Dichloride)

Dutox (Trichlorfon)

Dyanap (Dinoseb)

Dylox (Trichlorfon)

Dytop (Dinoseb)

# E

E14049 (Malathion)

EADC (Aluminum Alkyl Compounds)

EASC Aluminum Alkyl Compounds)

EB (Ethylbenzene)

ECH (Epichlorohydrin)

EDB (Ethylene Dibromide)

Elemental Arsenic (Arsenic)

Elgetol (Dinoseb)

Emmaton (Malathion)

EPI (Epichlorohydrin)

Epichlorhydrin (Epichlorohydrin)

**EPICHLOROHYDRIN**

1,2-Epoxyethane (Ethylene Oxide)

Epsilon-Caprolactam (Caprolactam)

Estakil (2,4-D)

Estemine (2,4-D)

Ethanal (Acetaldehyde)

1,2-Ethanediamine (Ethylenediamine)

1,2-Ethanediol (Ethylene Glycol)

Ethanenitrile (Acetonitrile)

Ethanoic Acid (Acetic Acid)

Ethanoic Anhydride (Acetic Anhydride)

Ethanol (Ethyl Alcohol)

**ETHANOLAMINE**

Ethene (Ethylene)

Ethenylbenzene (Styrene Monomer)

Ethine (Acetylene)

**ETHYL ACETATE**

**ETHYL ACRYLATE**

**ETHYL ALCOHOL**
Ethyl Aldehyde (Acetaldehyde)
Ethylaluminum Dichloride (Aluminum Alkyl Compounds)
Ethylaluminum Sesquichloride (Aluminum Alkyl Compounds)
**ETHYL BENZENE**
Ethyl Benzol (Ethylbenzene)
**ETHYL CHLORIDE**
**ETHYLENE**
Ethylene Bromide (Ethylene Dibromide)
Ethylene Chloride (Ethylene Dichloride)
**ETHYLENEDIAMINE**
**ETHYLENE DIBROMIDE**
**ETHYLENE DICHLORIDE**
Ethylene Dihydrate (Ethylene Glycol)
**ETHYLENE GLYCOL**
**ETHYLENEIMINE**
Ethyleneimine (Inhibited) (Ethyleneimine)
**ETHYLENE OXIDE**
Ethylene Tetrachloride (Perchloroethylene)
Ethyl Ethanoate (Ethyl Acetate)
Ethylethylene (Butylene)
**2-ETHYLHEXANOL**
2-Ethylhexyl Alcohol (2-Ethylhexanol)
2-Ethyl-1-Hexanol (2-Ethylhexanol)
Ethyl Hydrate (Ethyl Alcohol)
Ethyl Methyl Ketone (Methyl Ethyl Ketone)
Ethyl MMT (Methylcyclopentadienyl Manganese Tricarbonyl)
Ethyl Nitrile (Acetonitrile)
Ethylolamine (Ethanolamine)
Ethyl-2-Propenoate (Ethyl Acrylate)
Ethyne (Acetylene)

**F**

Far-Go (Triallate)
**FENITROTHION**

**FERRIC CHLORIDE**

Ferric Perchloride (Ferric Chloride)

Ferric Trichloride (Ferric Chloride)

Fertilizer Acid (Sulfuric Acid)

Filter Alum (Aluminum Sulfate)

Flea and Tick Spray (Malathion)

Flowers of Sulfur (Sulfur)

Fluohydric Acid (Hydrofluoric Acid)

**FLUORINE**

**FLUOROCHLOROMETHANES**

**FLUOSILICIC ACID**

Fluxing Lime (Calcium Oxide)

Fly Killer (Trichlorfon)

**FORMALDEHYDE**

Formaldehyde Solution (Formaldehyde)

Formalin (Formaldehyde)

Formalith (Formaldehyde)

**FORMIC ACID**

Formic Aldehyde (Formaldehyde)

Freon 11, 12, 13, 21, 22 (Fluorochloromethanes)

Fuel Oil No. 1 (Kerosene) (Oils, Fuel - Distillates)

Fuel Oil No. 2 (Oils, Fuel - Distillates)

Fuel Oil No. 2-D (Oils, Fuel - Distillates)

Fuming Sulfuric Acid (Oleum)

Furadan (Carbofuran)

Fural (Furfural)

2-Furaldehyde (Furfural)

2-Furan Carbonal (Furfural)

2,5-Furandione (Maleic Anhydride)

**FURFURAL**

Furfuraldehyde (Furfural)

**G**

Gas (Natural Gas)

**GASOLINE**

German Saltpeter (Ammonium Nitrate)

Glacial Acetic Acid (Acetic Acid)

Glauber's Salt (Sodium Sulfate)

**GLYCERINE**

Glycerol (Glycerine)

Glycerol, Nitric Acid Triester (Nitroglycerine)

Glyceryl Trinitrate (Nitroglycerine)

Glycinol (Ethanolamine)

Glycyl Alcohol (Glycerine)

Grain Alcohol (Ethyl Alcohol)

Granular Powder Zinc (Zinc)

Greenland Spar (Cryolite)

Grey Arsenic (Arsenic)

Grubex (Trichlorfon)

Gum Turpentine (Turpentine)

Gumthus (Turpentine)

H

1-Heptadecacarboxylic Acid (Stearic Acid)

Hexafluorosilicic Acid (Fluosilicic Acid)

Hexahydrobenzene (Cyclohexane)

Hexahydro-2H-Azepine-2-One (Caprolactam)

Hexamethylene (Cyclohexane)

Hexanaphthene (Cyclohexane)

**n-HEXANE**

Hexane (n-Hexane)

Hexanedioic Acid (Adipic Acid)

Home Heating Oil (Oils, Fuel - Distillates)

Hydrargyrum (Mercury)

Hydrated Alumina (Aluminum Hydroxide)

**HYDRAZINE**

**HYDROCHLORIC ACID**

Hydrochloric Acid, Anhydrous (Hydrogen Chloride)

Hydrochloric Ether (Ethyl Chloride)

**HYDROFLUORIC ACID**

Hydrofluoric Acid, Anhydrous (Hydrogen Fluoride)

Hydrofluosilicic Acid (Fluosilicic Acid)

Hydrofol Acid (Stearic Acid)

**HYDROGEN**

**HYDROGEN CHLORIDE**

Hydrogen Chloride, Aqueous (Hydrochloric Acid)

Hydrogen Dioxide (Hydrogen Peroxide)

**HYDROGEN FLUORIDE**

Hydrogen Fluoride, Aqueous (Hydrofluoric Acid)

**HYDROGEN PEROXIDE**

**HYDROGEN SULFIDE**

Hydrogen Sulphate (Sulfuric Acid)

Hydrogen Sulphide (Hydrogen Sulfide)

Hydroperoxide (Hydrogen Peroxide)

Hydrosilicofluoric Acid (Fluosilicic Acid)

Hydroxybenzene (Phenol)

1-Hydroxybutane (Butyl Alcohol)

Hydroxymethylbenzene (Cresol)

2,2-bis (Hydroxymethyl)-1,3-Propanediol (Pentaerythritol)

2,2-bis (4-Hydroxyphenyl) Propane (Bisphenol A)

Hydroxytoluene (Cresol)

**I**

Icetone (Cryolite)

Illuminating Oil (Kerosene)

Iron Chloride (Ferric Chloride)

Iron (III) Chloride (Ferric Chloride)

Iron Perchloride (Ferric Chloride)

Iron Trichloride (Ferric Chloride)

Isobutyl Methyl Ketone (Methyl Isobutyl Ketone)

Isopropanol (Isopropyl Alcohol)

**ISOPROPYL ALCOHOL**

Isopropylbenzene (Cumene)

Isopropylbenzene Hydroperoxide (Cumene Hydroperoxide)

4,4'-Isopropylidenediphenol (Bisphenol A)

p,p'-Isopropylidenediphenol (Bisphenol A)

## J

Jet A (Oils, Fuel - Aviation)

Jet A-1 (Oils, Fuel - Aviation)

Jet B (Oils, Fuel - Aviation)

Jet C (Oils, Fuel - Aviation)

Jet Fuel: JP-1 (Kerosene)

JP-4 (Oils, Fuel - Aviation)

JP-5 (Oils, Fuel - Aviation)

JP-6 (Oils, Fuel - Aviation)

## K

Kansel (Dicamba)

Karbophos (Malathion)

**KEROSENE**

Kerosene (Oils, Fuel - Distillates)

Kerosine (Kerosene)

2-Ketohexamethylenimine (Cryolite)

Killex (2,4-D)

Killex (Dicamba)

Kiloseb (Dinoseb)

Kil-Mor (2,4-D)

Kryolith (Cryolite)

## L

**LATEX**

Latex, Liquid Synthetic (Latex)

Lawn Weed Killer (Dicamba)

**LEAD ACETATE**

Lead Acetate Trihydrate (Lead Acetate)

**LEAD CHROMATE**

Lead Monoxide (Lead Oxides, Yellow)

**LEAD NITRATE**

**LEAD OXIDES**

Lead Oxide, Black (Lead Oxides)

Lead Oxide, Red (Lead Oxides)

Lead Oxide, Yellow (Lead Oxides)

Lead Tetroxide (Lead Oxides, Red)

Light Naphtha (Naphtha Solvent)

Lime (Calcium Hydroxide)

Lime Chloride (Calcium Hypochlorite, Hydrate)

Lime Hydrate (Calcium Hydroxide)

Liquid Alum (Aluminum Sulfate, Tetradecahydrate)

Liquid Bleach (Sodium Hypochlorite)

Liquid Oxygen (Oxygen)

Litharge, Leaded (Lead Oxides, Black)

LNG (Natural Gas)

LOX (Oxygen)

LPG (Butane)

LPG (Butylene)

Lye (Sodium Hydroxide)

## M

Magnesium Hydrate (Magnesium Hydroxide)

**MAGNESIUM HYDROXIDE**

Magnesium Magna (Magnesium Hydroxide)

Malaspray (Malathion)

**MALATHION**

Malathiozol (Malathion)

Maleic Acid, Anhydride (Maleic Anhydride)

**MALEIC ANHYDRIDE**

Marble (Calcium Carbonate)

Marsh Gas (Methane)

MASC (Aluminum Alkyl Compounds)

Matacil (Aminocarb)

MCB (Chlorobenzene)

**MCPA**

MCPA Amine (MCPA)

MDBA (Dicamba)

MDI (Diphenylmethane-4,4'-Diisocyanate)

MEA (Ethanolamine)

Mediben (Dicamba)

MEK (Methyl Ethyl Ketone)

Mercaptomethane (Methyl Mercaptan)

**MERCURY**

Meta-Cresol (Cresol)

Metallic Arsenic (Arsenic)

Metathion (Fenitrothion)

Methacrylate Monomer (Methyl Methacrylate)

Methacrylic Acid, Methyl Ester (Methyl Methacrylate)

Methanal (Formaldehyde)

Methanamine (Methylamines)

**METHANE**

Methane Carboxylic Acid (Acetic Acid)

Methane Dichloride (Methylene Chloride)

Methanethiol (Methyl Mercaptan)

**METHANOL**

Methenyl Trichloride (Chloroform)

**METHYL ACRYLATE**

Methyl Alcohol (Methanol)

Methylaluminum Sesquichloride (Aluminum Alkyl Compounds)

**METHYLAMINES**

Methylbenzene (Toluene)

Methylbenzol (Toluene)

**METHYL CHLORIDE**

Methylchloroform (1,1,1-Trichloroethane)

Methyl Cyanide (Acetonitrile)

**METHYLCYCLOPENTADIENYL MANGANESE TRICARBONYL**

Methylcyclopentadienylmanganesetricarbonyl
  (Methylcyclopentadienyl Manganese Tricarbonyl)

**METHYLENE CHLORIDE**

Methylene Oxide (Formaldehyde)

Methylenedichloride (Methylene Chloride)

Methylene-bis (Phenyl Isocyanate) (Diphenylmethane-4,4'-Diisocyanate)

Methylene-Diparaphenylene Isocyanate (Diphenylmethane-4,4'-Diisocyanate)
Methyl Ether (Dimethyl Ether)
Methylethylcarbinol (Butyl Alcohol)
**METHYL ETHYL KETONE**
Methyl Hydride (Methane)
Methyl Hydroxide (Methanol)
**METHYL ISOBUTYL KETONE**
Methyl Ketone (Acetone)
**METHYL MERCAPTAN**
**METHYL METHACRYLATE**
Methyl-α-Methacrylate (Methyl Methacrylate)
Methyl 2-Methyl-2-Propenoate (Methyl Methacrylate)
Methyl Oxide (Dimethyl Ether)
2-Methyl-4-Pentanone (Methyl Isobutyl Ketone)
4-Methyl-2-Pentanone (Methyl Isobutyl Ketone)
Methyl Phenol (Cresol)
2-Methyl-2-Propanol (Butyl Alcohol)
Methyl Propenate (Methyl Acrylate)
Methyl-2-Propenoate (Methyl Acrylate)
Methyltrinitrobenzene (Trinitrotoluene)
MIBK (Methyl Isobutyl Ketone)
Milk of Magnesia (Magnesium Hydroxide)
Minium (Lead Oxides, Red)
MME (Methyl Methacrylate)
MMT (Methylcyclopentadienyl Manganese Tricarbonyl)
Monocalcium Phosphate (Calcium Phosphate, Monobasic)
Monochlorobenzene (Chlorobenzene)
Monochloroethane (Ethyl Chloride)
Monochloromethane (Methyl Chloride)
Monoethanolamine (Ethanolamine)
Monoethylene Glycol (Ethylene Glycol)
Monomethylamine (Methylamines)
Mono PE, (Pentaerythritol)
Monosodium Dihydrogen Phosphate (Sodium Phosphate, Monobasic)
Monosodium Phosphate (Sodium Phosphate, Monobasic)

Monoxide (Carbon Monoxide)

**MORPHOLINE**

Moth Balls (Naphthalene)

Moth Flakes (Naphthalene)

Motor Fuel Antiknock Compound (Tetraethyl Lead)

MSP (Sodium Phosphate, Monobasic)

Muriatic Acid (Hydrochloric Acid)

**N**

**NAPHTHALENE**

Naphthalin (Naphthalene)

**NAPHTHA SOLVENT**

Naphthenic Acid, Copper Salt (Copper Naphthenate)

**NATURAL GAS**

Natural Gas (Methane)

Natural Rubber Latex (Latex)

Necatorina (Carbon Tetrachloride)

NG (Nitroglycerine)

Nitram (Ammonium Nitrate)

Nitratine (Sodium Nitrate)

**NITRIC ACID**

Nitric Acid, Lead II Salt (Lead Nitrate)

Nitric Acid Triester (Nitroglycerine)

**NITRILOTRIACETIC ACID**

**NITROGLYCERINE**

Nitroglycerol (Nitroglycerine)

Nitrol (Nitroglycerine)

Nitrophos (Fenitrothion)

**NONYL PHENOL**

Nonylphenol (Nonyl Phenol)

o-Nonyl Phenol (Nonyl Phenol)

p-Nonyl Phenol (Nonyl Phenol)

Norway Saltpeter (Ammonium Nitrate)

NTA (Nitrilotriacetic Acid)

## O

Octachlor (Chlordane)

Octadecanoic Acid (Stearic Acid)

n-Octadecylic Acid (Stearic Acid)

**OILS, CRUDE**

**OILS, FUEL (Aviation)**

**OILS, FUEL (Distillates)**

**OILS, FUEL (Residuals)**

Olefiant Gas (Ethylene)

**OLEUM**

Orthoboric Acid (Boric Acid)

Ortho-Cresol (Cresol)

Ortho-Dichlorobenzene (1,2-Dichlorobenzene)

Ortho-Klor (Chlordane)

Orthophosphoric Acid (Phosphoric Acid)

Oxane (Ethylene Oxide)

Oxirane (Ethylene Oxide)

Oxomethane (Formaldehyde)

**OXYGEN**

## P

Paper Maker's Alum (Aluminum Sulfate)

Para-Cresol (Cresol)

Patent Alum (Aluminum Sulfate)

PCP (Pentachlorophenol)

PE (Pentaerythritol)

Pear Oil (Amyl Acetate)

Pearl Alum (Aluminum Sulfate)

Pebble Lime (Calcium Oxide)

Penchlorol (Pentachlorophenol)

Penta (Pentachlorophenol)

**PENTACHLOROPHENOL**

Pentachlorophenol, Sodium Salt (Sodium Pentachlorophenate)

Pentaerythrite (Pentaerythritol)

**PENTAERYTHRITOL**

Pentek (Pentaerythritol)

Pentyl Acetate (Amyl Acetate)

"Per" (Perchloroethylene)

**PERCHLOROETHYLENE**

Perchloromethane (Carbon Tetrachloride)

Permatox (Pentachlorophenol)

Peroxide (Hydrogen Peroxide)

Petrol (Gasoline)

Petroleum Asphalt (Asphalt)

Petroleum Ether (Naphtha Solvent)

Petroleum Solvent (Naphtha Solvent)

Petroleum Spirits (Naphtha Solvent)

Phenic Acid (Phenol)

**PHENOL**

Phenyl Chloride (Chlorobenzene)

Phenyl Hydroxide (Phenol)

Phenylamine (Aniline)

N-Phenylaniline (Diphenyl Amine)

Phenylcarboxylic Acid (Benzoic Acid)

Phenylethane (Ethylbenzene)

Phenylethylene (Styrene Monomer)

Phenylformic Acid (Benzoic Acid)

Phenylhydride (Benzene)

Phenylic Acid (Phenol)

Phenylmethane (Toluene)

2-Phenylpropane (Cumene)

**PHOSGENE**

Phosphate of Soda (Sodium Phosphate, Dibasic)

**PHOSPHORIC ACID**

Phosphoric Sulphide (Phosphorus Pentasulfide)

Phosphorus Chloride (Phosphorus Trichloride)

**PHOSPHORUS PENTASULFIDE**

Phosphorus Persulfide (Phosphorus Pentasulfide)

**PHOSPHORUS, RED**

Phosphorus Sulphide (Phosphorus Pentasulfide)

**PHOSPHORUS TRICHLORIDE**

**PHOSPHORUS, WHITE**

p-Phthalic Acid (Terephthalic Acid)

**PHTHALIC ANHYDRIDE**

Pickle Alum (Aluminum Sulfate)

Plastic Latex (Latex)

Plumboplumic Oxide (Lead Oxides, Red)

Plumbous Acetate (Lead Acetate)

Plumbous Oxide (Lead Oxides, Yellow)

**POLYCHLORINATED BIPHENYLS (PCBs)**

**POTASH (POTASSIUM CHLORIDE)**

**POTASSIUM CARBONATE**

**POTASSIUM HYDROXIDE**

**POTASSIUM SULFATE**

Potato Top Killer (Dinoseb)

Precipitated Calcium Phosphate (Calcium Phosphate, Tribasic)

Premerge (Dinoseb)

**PROPANE**

1,2,3-Propanetriol (Glycerine)

1,2,3-Propanetriol Trinitrate (Nitroglycerine)

2-Propanol (Isopropyl Alcohol)

2-Propanone (Acetone)

2-Propenal (Acrolein)

2-Propenenitrile (Acrylonitrile)

2-Propenoic Ethyl Ester (Ethyl Acrylate)

Propenoic Acid, Methyl Ester (Methyl Acrylate)

sec-Propyl Alcohol (Isopropyl Alcohol)

**PROPYLENE**

**PROPYLENE GLYCOL**

**PROPYLENE OXIDE**

Pyromucic Aldehyde (Furfural)

**Q**

Quicklime (Calcium Oxide)

Quicksilver (Mercury)

**R**

Range Oil (Kerosene)

Red Fuming Nitric Acid (Nitric Acid)

Red Lead (Lead Oxides, Red)

Residual Fuel Oil No. 4 (Oils, Fuel - Residuals)

Residual Fuel Oil No. 5 (Oils, Fuel - Residuals)

Residual Fuel Oil No. 6 (Oils, Fuel - Residuals)

RFNA (Nitric Acid)

Rhomene (MCPA)

Roman Vitriol (Copper Sulfate)

Rubbing Alcohol (Isopropyl Alcohol)

**S**

Sal Ammoniac (Ammonium Chloride)

Salt Cake (Sodium Sulfate)

Sand Acid (Fluosilicic Acid)

Santophen 20 (Pentachlorophenol)

Secondary Calcium Phosphate (Calcium Phosphate, Dibasic)

Secondary Sodium Phosphate (Sodium Phosphate, Dibasic)

Sewage Gas (Methane)

Silicate of Soda (Sodium Silicate)

Silicofluoric Acid (Fluosilicic Acid)

Sinox (Dinoseb)

Slaked Lime (Calcium Hydroxide)

Soda Lye (Sodium Hydroxide)

Soda Niter (Sodium Nitrate)

**SODIUM**

Sodium Acid Chromate (Sodium Dichromate)

Sodium Acid Phosphate (Sodium Phosphate, Monobasic)

**SODIUM ALUMINATE**

**SODIUM ARSENITE**

Sodium Aluminum Fluoride (Cryolite)

Sodium Bichromate (Sodium Dichromate)

Sodium Bichromate Dihydrate (Sodium Dichromate)

Sodium Biphosphate (Sodium Phosphate, Monobasic)

Sodium Bisulfide (Sodium Hydrosulfide)

Sodium Bisulfite (Sodium Dithionite)

Sodium Borate, Decahydrate (Borax)

**SODIUM BOROHYDRIDE**

**SODIUM CARBONATE**

**SODIUM CHLORATE**

**SODIUM CHLORIDE**

**SODIUM CYANIDE**

**SODIUM DICHLOROISOCYANURATE**

**SODIUM DICHROMATE**

**SODIUM DITHIONITE**

Sodium Fluoaluminate (Cryolite)

Sodium Hydrate (Sodium Hydroxide)

Sodium Hydrogen Sulfide (Sodium Hydrosulfide)

**SODIUM HYDROSULFIDE**

Sodium Hydrosulfite (Sodium Dithionite)

**SODIUM HYDROXIDE**

**SODIUM HYPOCHLORITE**

Sodium Mercaptan (Sodium Hydrosulfide)

Sodium Metasilicate (Sodium Silicate)

**SODIUM NITRATE**

Sodium Orthophosphate (Sodium Phosphate, Tribasic)

Sodium Orthophosphate, Primary (Sodium Phosphate, Monobasic)

Sodium Orthophosphate, Secondary (Sodium Phosphate, Dibasic)

**SODIUM PENTACHLOROPHENATE**

Sodium Pentachlorophenol (Sodium Pentachlorophenate)

Sodium Pentachlorophenolate (Sodium Pentachlorophenate)

Sodium Pentachlorophenoxide (Sodium Pentachlorophenate)

**SODIUM PHOSPHATE** (Dibasic)
**SODIUM PHOSPHATE** (Monobasic)
**SODIUM PHOSPHATE** (Tribasic)
**SODIUM SILICATE**
**SODIUM SULFATE**
**SODIUM SULFITE**
Sodium Sulfoxylate (Sodium Dithionite)
Sodium Sulphydrate (Sodium Hydrosulfide)
Sodium Tetraborate, Decahydrate (Borax)
Spirits (Ethyl Alcohol)
Spirits of Turpentine (Turpentine)
Spurge (Dinoseb)
**STEARIC ACID**
Stearophanic Acid (Stearic Acid)
Stove (Range) Oil (Kerosene)
**STYRENE MONOMER**
Styrol (Styrene Monomer)
Sugar of Lead (Lead Acetate)
**SULFUR**
**SULFUR DIOXIDE**
Sulfur Hydride (Hydrogen Sulfide)
**SULFURIC ACID**
Sulfuric Chlorohydrin (Chlorosulfonic Acid)
Sulfurous Acid, Anhydride (Sulfur Dioxide)
Sulfurous Oxide (Sulfur Dioxide)
**SULFURYL CHLORIDE**
Sulphur (Sulfur)
Sulphuretted Hydrogen (Hydrogen Sulfide)
Sulphuric Acid, Fuming (Oleum)
Synklor (Chlordane)
Synthetic Rubber Latex (Latex)

**T**

**TALL OIL**
Tallol (Tall Oil)
Tar Camphor (Naphthalene)

TEA (Aluminum Alkyl Compounds)

TEL (Tetraethyl Lead)

**TEREPHTHALIC ACID**

Terephthalic Acid, Dimethyl Ester (Dimethyl Terephthalate)

**TERPHENYLS**

Tertiary Calcium Phosphate (Calcium Phosphate, Tribasic)

tert-Butanol (Butyl Alcohol)

Tetrachloroethylene (Perchloroethylene)

Tetrachloromethane (Carbon Tetrachloride)

**TETRAETHYL LEAD**

Tetraethyl Plumbane (Tetraethyl Lead)

Tetrahydro-1,4-Isoxazine (Morpholine)

Tetrahydro-p-Isoxazine (Morpholine)

Tetrahydro-1,4-Oxazine (Morpholine)

Tetrahydro-2H-1,4-Oxazine (Morpholine)

Tetrahydroxymethylmethane (Pentaerythritol)

Tetrakis (Hydroxymethyl) Methane (Pentaerythritol)

Tetramethyl Lead (Tetraethyl Lead)

Tetramethylolmethane (Pentaerythritol)

Tetraphosphorus Decasulphide (Phosphorus Pentasulfide)

Thiophosphoric Anhydride (Phosphorus Pentasulfide)

TIBA (Aluminum Alkyl Compounds)

Titanic Anhydride (Titanium Dioxide)

Titanic Oxide (Titanium Dioxide)

**TITANIUM DIOXIDE**

Titanium White (Titanium Dioxide)

TMA (Aluminum Alkyl Compounds)

TML (Tetraethyl Lead)

TNT (Trinitrotoluene)

**TOLUENE**

**TOLUENE DIISOCYANATE**

Toluol (Toluene)

2,4-Tolylene Diisocyanate (Toluene Diisocyanate)

Topichlor (Chlordane)

Toxilic Anhydride (Maleic Anhydride)

TPA (Terephthalic Acid)

Treflan (Trifluralin)

**TRIALLATE**

Tribasic Sodium Phosphate (Sodium Phosphate, Tribasic)

Tricalcium Orthophosphate (Calcium Phosphate, Tribasic)

Tricalcium Phosphate (Calcium Phosphate, Tribasic)

Tricarbonyl (Methylcyclopentadienyl Manganese Tricarbonyl)

Trichlorethane (1,1,1-Trichloroethane)

**TRICHLORFON**

Trichloroaluminum (Aluminum Chloride)

**1,1,1-TRICHLOROETHANE**

Trichlorofluoromethane (Fluorochloromethanes, Freon 11)

Trichloromethane (Chloroform)

Triethylaluminum (Aluminum Alkyl Comounds)

**TRIFLURALIN**

Triflurex (Trifluralin)

Triglycine (Nitrilotriacetic Acid)

Triglycollamic Acid (Nitrilotriacetic Acid)

1,2,3-Trihydroxypropane (Glycerine)

Triisobutylaluminum (Aluminum Alkyl Compounds)

Trimethylaluminum (Aluminum Alkyl Compounds)

Trimethylamine (Methylamines)

Trimethylcarbinol (Butyl Alcohol)

Trinitroglycerol (Nitroglycerine)

**TRINITROTOLUENE**

Trisodium Orthophosphate (Sodium Phosphate, Tribasic)

TSP (Sodium Phosphate, Tribasic)

**TURPENTINE**

Turps (Turpentine)

**U**

Unslaked Lime (Calcium Oxide)

Uranium Concentrate (Yellow Cake)

**UREA**

## V

Vanadic Acid Anhydride (Vanadium Pentoxide)

Vanadium Oxide (Vanadium Pentoxide)

VAC (Vinyl Acetate)

VAM (Vinyl Acetate)

**VANADIUM PENTOXIDE**

VCM (Vinyl Chloride)

Velsicol (Dicamba)

Vinegar Acid (Acetic Acid)

Vinegar Naphtha (Ethyl Acetate)

**VINYL ACETATE**

Vinyl Acetate Monomer (Vinyl Acetate)

**VINYL CHLORIDE**

Vinyl Chloride Monomer (Vinyl Chloride)

Vinyl Cyanide (Acrylonitrile)

Vinylbenzene (Styrene Monomer)

Vinylethylene (1,3-Butadiene)

Voranate T-80™ (Toluene Diisocyanate)

## W

Warble Killer (Trichlorofon)

Water Glass (Sodium Silicate)

Weed Killer (2,4-D)

Weed Preventer (Trifluralin)

Weed Stop (Trifluralin)

Weed-No-More (2,4-D)

Weedone (2,4-D)

White Arsenic (Arsenic Trioxide)

White Caustic (Sodium Hydroxide)

White Copperas (Zinc Sulfate, Heptahydrate)

White Dry Phosphorus (Phosphorus, White)

White Vitriol (Zinc Sulfate, Heptahydrate)

Wittox-C (Copper Naphthenate)

Wood Alcohol (Methanol)

Wood Ether (Dimethyl Ether)
Wood Naphtha (Methanol)
Wood Spirit (Methanol)
Wood Turpentine (Turpentine)

# X

**XYLENES**

Xylols (Xylenes)

# Y

Yaltox (Carbofuran)

**YELLOW CAKE**

Yellow Phosphorus (Phosphorus, White)

# Z

**ZINC**

**ZINC CHLORIDE**

Zinc Dichloride Solution (Zinc Chloride)

Zinc Dust (Zinc)

Zinc Muriate Solution (Zinc Chloride)

**ZINC OXIDE**

**ZINC SULFATE**

Zinc Sulfate Heptahydrate (Zinc Sulfate)

Zinc Sulfate Monohydrate (Zinc Sulfate)

Zinc Sulphate (Zinc Sulfate)

Zinc Vitriol (Zinc Sulfate, Heptahydrate)

Zinc White (Zinc Oxide)

Zincite (Zinc Oxide)

# 3 AGRICULTURE CANADA PESTICIDE REGISTRY NUMBERS
## (for those pesticides included in this manual)

1083 — carbaryl

2238 — 2,4-D (present as dimethylamine salt)
2283 — 2,4-D (present as mixed butyl esters)
2592 — chlordane
2687 — 2,4-D (present as diethanolamine salt)
2760 — chlordane
2833 — chlordane
2851 — 2,4-D (present as diethanolamine salt)

3132 — dinoseb (present as free form, alkanolamine salts or mixed amine salts)
3162 — chlordane
3186 — 2,4-D (present as dimethylamine salt)
3259 — chlordane
3277 — chlordane
3465 — chlordane
3479 — chlordane
3517 — 2,4-D (present as dimethylamine salt)
3518 — chlordane
3676 — 2,4-D (present as dimethylamine salt)
3749 — 2,4-D (present as isooctyl esters)
3927 — 2,4-D (present as dimethylamine salt)
3956 — 2,4-D (present as dimethylamine salt)
3959 — 2,4-D (present as isooctyl esters)

4058 — chlordane
4067 — MCPA (present as potassium or sodium salt)
4132 — chlordane
4155 — 2,4-D (present as diethanolamine salt)
4167 — dinoseb (present as free form, alkanolamine salts or mixed amine salts)
4253 — 2,4-D (present as dimethylamine salt)
4282 — malathion
4291 — 2,4-D (present as dimethylamine salt)
4343 — MCPA (present as potassium or sodium salt)
4383 — MCPA (present as esters)
4397 — chlordane
4478 — malathion
4486 — dinoseb (present in free form, alkanolamine salts or mixed amine salts)
4535 — dinoseb (present in free form, alkanolamine salts or mixed amine salts)
4588 — malathion
4590 — malathion

4638 — malathion
4657 — malathion
4709 — malathion
4728 — 2,4-D (present as mixed butyl esters)
4734 — 2,4-D (present as mixed butyl esters)
4741 — MCPA (present as diethanolamine salt, dimethylamine salt or mixed amine salts)
4748 — 2,4-D (present as mixed butyl esters)
4768 — 2,4-D (present as isooctyl esters)
4771 — 2,4-D (present as propylene glycol butyl ether ester)
4780 — 2,4-D (present as mixed butyl esters)
4860 — malathion
4916 — MCPA (present as diethanolamine salt, dimethylamine salt or mixed amine salts)
4928 — malathion
4980 — 2,4-D (present as isooctyl esters)
4982 — chlordane
4983 — 2,4-D (present as isooctyl esters)
4989 — 2,4-D (present as sodium salt)

5016 — chlordane
5125 — chlordane
5141 — malathion
5206 — chlordane
5212 — malathion
5276 — malathion
5316 — MCPA (present as potassium or sodium salt)
5362 — 2,4-D (present as dimethylamine salt)
5369 — chlordane
5429 — malathion
5441 — chlordane
5449 — malathion
5460 — MCPA (present as potassium or sodium salt)
5462 — MCPA (present as esters)
5504 — 2,4-D (present as mixed butyl esters)
5508 — MCPA (present as diethanolamine salt, dimethylamine salt or mixed amine salt)
5639 — 2,4-D (present as dimethylamine salt)
5780 — malathion
5821 — malathion
5931 — 2,4-D (present as dimethylamine salt)
5942 — MCPA (present as diethanolamine salt, dimethylamine salt or mixed amine salt)
5979 — MCPA (present as esters)
5981 — MCPA (present as diethanolamine salt, dimethylamine salt or mixed amine salt)

| | | | | |
|---|---|---|---|---|
| 6017 | — trichlorfon | | 7652 | — carbaryl |
| 6022 | — malathion | | 7667 | — trichlorfon |
| 6024 | — chlordane | | 7754 | — malathion |
| 6045 | — MCPA (present as esters) | | 7757 | — carbaryl |
| 6047 | — MCPA (present as diethanolamine salt, dimethylamine salt or mixed amine salts) | | 7811 | — MCPA (present as esters) |
| | | | 7812 | — MCPA (present as potassium or sodium salt) |
| 6140 | — 2,4-D (present as isooctyl esters) | | 7825 | — MCPA (present as esters) |
| 6190 | — 2,4-D (present as mixed butyl esters) | | 7835 | — 2,4-D (present as acid) |
| 6192 | — malathion | | 7839 | — malathion |
| 6274 | — MCPA (present as diethanolamine salt, dimethylamine salt or mixed amine salts) | | 7855 | — chlordane |
| | | | 7947 | — malathion |
| 6314 | — 2,4-D (present as mixed butyl esters) | | | |
| 6330 | — 2,4-D (present as propylene glycol butyl ether ester) | | 8021 | — malathion |
| | | | 8022 | — MCPA (present as potassium or sodium salt) |
| 6373 | — MCPA (present as diethanolamine salt, dimethylamine salt or mixed amine salts) | | 8026 | — carbaryl |
| | | | 8042 | — carbaryl |
| 6375 | — 2,4-D (present as dimethylamine salt) | | 8099 | — 2,4-D (present as isooctyl esters) |
| 6465 | — malathion | | 8125 | — MCPA (present as diethanolamine salt, dimethylamine salt or mixed amine salts) |
| 6526 | — 2,4-D (present as isooctyl esters) | | | |
| 6530 | — 2,4-D (present as dimethylamine salt) | | 8151 | — carbaryl |
| 6662 | — 2,4-D (present as mixed butyl esters) | | 8159 | — 2,4-D (present as diethanolamine salt) |
| 6680 | — 2,4-D (present as mixed butyl esters) | | 8167 | — triallate |
| 6698 | — 2,4-D (present as isooctyl esters) | | 8184 | — carbaryl |
| 6713 | — malathion | | 8211 | — MCPA (present as potassium or sodium salt) |
| 6718 | — 2,4-D (present as isooctyl esters) | | 8217 | — 2,4-D (present as mixed butyl esters) |
| 6745 | — chlordane | | 8253 | — MCPA (present as potassium or sodium salt) |
| 6771 | — MCPA (present as esters) | | 8354 | — malathion |
| 6799 | — chlordane | | 8372 | — malathion |
| 6839 | — carbaryl | | 8403 | — malathion |
| 6840 | — malathion | | 8425 | — 2,4-D (present as isooctyl esters) |
| 6860 | — malathion | | 8431 | — 2,4-D (present as isooctyl esters) |
| 6965 | — MCPA (present as esters) | | 8466 | — carbaryl |
| 6967 | — 2,4-D (present as dimethylamine salt) | | 8466 | — malathion |
| 6969 | — MCPA (present as diethanolamine salt, dimethylamine salt or mixed amine salts) | | 8469 | — 2,4-D (present as diethanolamine salt) |
| | | | 8469 | — 2,4-D (present as dimethylamine salt) |
| 6971 | — 2,4-D (present as mixed butyl esters) | | 8472 | — carbaryl |
| 6975 | — malathion | | 8480 | — malathion |
| | | | 8491 | — 2,4-D (present as dimethylamine salt) |
| 7012 | — 2,4-D (present as isooctyl esters) | | 8495 | — 2,4-D (present as isooctyl esters) |
| 7108 | — MCPA (present as potassium or sodium salt) | | 8503 | — MCPA (present as esters) |
| 7110 | — 2,4-D (present as isooctyl esters) | | 8524 | — 2,4-D (present as dimethylamine salt) |
| 7207 | — carbaryl | | 8543 | — 2,4-D (present as isooctyl esters) |
| 7264 | — malathion | | 8593 | — 2,4-D (present as diethanolamine salt) |
| 7401 | — 2,4-D (present as isooctyl esters) | | 8595 | — 2,4-D (present as diethanolamine salt) |
| 7446 | — carbaryl | | 8624 | — malathion |
| 7456 | — malathion | | 8631 | — dicamba (present as acid, diethanolamine salt, dimethylamine salt) |
| 7473 | — MCPA (present as esters) | | | |
| 7525 | — 2,4-D (present as isooctyl esters) | | 8651 | — 2,4-D (present as acid) |
| 7527 | — 2,4-D (present as isooctyl esters) | | 8657 | — 2,4-D (present as isooctyl esters) |
| 7571 | — 2,4-D (present as mixed butyl esters) | | 8662 | — chlordane |
| 7647 | — trichlorfon | | | |

8705 — dinoseb (present as free form, alkanolamine salts or mixed amine salts)
8725 — carbaryl
8765 — malathion
8810 — 2,4-D (present as dimethylamine salt)
8826 — malathion
8834 — carbaryl
8885 — dicamba (present as acid, diethanolamine salt or dimethylamine salt)
8885 — 2,4-D (present as dimethylamine salt)
8899 — 2,4-D (present as isooctyl esters)
8903 — 2,4-D (present as isooctyl esters)
8911 — MCPA (present as potassium or sodium salt)
8927 — 2,4-D (present as butoxyethyl ester)
8950 — trichlorfon
8959 — 2,4-D (present as isooctyl ester)

9001 — carbaryl
9007 — 2,4-D (present as other amine salts)
9017 — MCPA (present as potassium or sodium salt)
9023 — malathion
9031 — 2,4-D (present as mixed butyl esters)
9033 — 2,4-D (present as dimethylamine salt)
9042 — carbaryl
9061 — carbaryl
9081 — carbaryl
9083 — carbaryl
9088 — chlordane
9099 — carbaryl
9103 — 2,4-D (present as dimethylamine salt)
9150 — chlordane
9150 — malathion
9172 — carbaryl
9176 — carbaryl
9177 — 2,4-D (present as dimethylamine salt)
9178 — MCPA (present as diethanolamine salt, dimethylamine salt or mixed amine salts)
9232 — 2,4-D (present as dimethylamine salt)
9257 — trifluralin
9262 — 2,4-D (present as mixed butyl ester)
9265 — malathion
9268 — 2,4-D (present as isooctyl esters)
9272 — carbaryl
9284 — MCPA (present as diethanolamine salt, dimethylamine salt or mixed amine salts)
9337 — malathion
9342 — 2,4-D (present as acid)
9350 — 2,4-D (present as dimethylamine salt)
9350 — dicamba (present as acid, diethanolamine salt, dimethylamine salt)

9355 — 2,4-D (present as isooctyl esters)
9401 — malathion
9408 — trichlorfon
9419 — trichlorfon
9439 — 2,4-D (present as isooctyl esters)
9465 — 2,4-D (present as dimethylamine salt)
9465 — dicamba (present as acid, diethanolamine salt, dimethylamine salt)
9492 — carbaryl
9494 — chlordane
9506 — 2,4-D (present as dimethylamine salt)
9510 — MCPA (present as esters)
9513 — chlordane
9516 — MCPA (present as diethanolamine salt, dimethylamine salt, mixed amine salts)
9524 — 2,4-D (present as acid)
9528 — 2,4-D (present as dimethylamine salt)
9537 — carbaryl
9540 — 2,4-D (present as other amine salts)
9547 — 2,4-D (present as dimethylamine salt)
9548 — MCPA (present as esters)
9550 — 2,4-D (present as mixed butyl esters)
9560 — 2,4-D (present as isooctyl esters)
9561 — 2,4-D (present as isooctyl esters)
9581 — 2,4-D (present as isooctyl esters)
9587 — 2,4-D (present as isooctyl esters)
9606 — 2,4-D (present as dimethylamine salt)
9606 — dicamba (present as acid, diethanolamine salt, dimethylamine salt)
9625 — 2,4-D (present as isooctyl esters)
9661 — chlordane
9726 — carbaryl
9740 — dicamba (present as acid, diethanolamine salt, dimethylamine salt)
9740 — 2,4-D (present as dimethylamine salt)
9750 — chlordane
9777 — 2,4-D (present as dimethylamine salt)
9802 — malathion
9809 — chlordane
9811 — 2,4-D (present as dimethylamine salt)
9811 — dicamba (present as acid, diethanolamine salt, dimethylamine salt)
9824 — malathion
9827 — trichlorfon
9853 — MCPA (present as diethanolamine salt, dimethylamine salt, mixed amine salts)
9856 — MCPA (present as esters)
9858 — MCPA (present as potassium or sodium salts)
9876 — chlordane
9885 — 2,4-D (present as isooctyl esters)

9903 — 2,4-D (present as diethanolamine salt)
9903 — dicamba (present as acid, diethanolamine salt or dimethylamine salt)
9907 — 2,4-D (present as butoxyethyl ester)
9920 — malathion
9946 — malathion
9947 — malathion
9975 — malathion
9977 — 2,4-D (present as dimethylamine salt)
9986 — malathion
9986 — carbaryl

10020 — 2,4-D (present as diethanolamine salt)
10046 — malathion
10066 — MCPA (present as diethanolamine salt, dimethylamine salt or as mixed amine salts)
10067 — MCPA (present as esters)
10068 — 2,4-D (present as dimethylamine salt)
10069 — 2,4-D (present as isooctyl esters)
10070 — 2,4-D (present as mixed butyl esters)
10121 — 2,4-D (present as isooctyl esters)
10132 — malathion
10134 — malathion
10156 — carbaryl
10159 — carbaryl
10163 — 2,4-D (present as isooctyl esters)
10164 — malathion
10166 — 2,4-D (present as acid)
10174 — malathion
10184 — 2,4-D (present as dimethylamine salt)
10187 — MCPA (present as esters)
10195 — carbaryl
10196 — carbaryl
10215 — 2,4-D (present as isooctyl esters)
10308 — triallate
10313 — chlordane
10324 — malathion
10325 — dicamba (present as acid, diethanolamine salt or dimethylamine salt)
10325 — 2,4-D (present as dimethylamine salt)
10327 — trifluralin
10331 — 2,4-D (present as isooctyl ester)
10352 — carbaryl
10359 — carbofuran
10363 — carbofuran
10387 — carbaryl
10399 — 2,4-D (present as dimethylamine salt)
10401 — MCPA (present as diethanolamine salt, dimethylamine salt or as mixed amine salts)
10430 — 2,4-D (present as isooctyl esters)

10458 — dinoseb (present in free form, alkanolamine salts or mixed amine salts)
10460 — 2,4-D (present as dimethylamine salt)
10472 — 2,4-D (present as isooctyl esters)
10473 — 2,4-D (present as isooctyl esters)
10483 — MCPA (present as potassium or sodium salts)
10491 — 2,4-D (present as dimethylamine salt)
10509 — malathion
10559 — carbaryl
10559 — malathion
10565 — carbaryl
10565 — malathion
10567 — carbaryl
10567 — malathion
10568 — carbaryl
10568 — malathion
10579 — chlordane
10590 — 2,4-D (present as dimethylamine salt)
10590 — dicamba (present as acid, diethanolamine salt or dimethylamine salts)
10619 — carbaryl
10622 — 2,4-D (present as isooctyl esters)
10626 — chlordane
10629 — trifluralin
10637 — carbaryl
10638 — chlordane
10639 — malathion
10644 — carbaryl
10645 — carbaryl
10653 — chlordane
10658 — chlordane
10666 — carbofuran
10681 — chlordane
10684 — carbaryl
10685 — carbaryl
10687 — carbaryl
10709 — carbaryl
10711 — carbaryl
10715 — chlordane
10725 — malathion
10725 — malathion
10726 — carbaryl
10726 — malathion
10727 — carbaryl
10732 — dinoseb (present in free form, alkanolamine salts or mixed amine salts)
10734 — malathion
10742 — malathion
10757 — malathion
10758 — malathion

10776 — fenitrothion
10778 — 2,4-D (present as butoxyethyl ester)
10778 — dicamba (present as acid,
    diethanolamine salt or dimethylamine salt)
10793 — malathion
10794 — malathion
10802 — malathion
10817 — MCPA (present as diethanolamine salt,
    dimethylamine salt or as mixed amine salts)
10826 — carbofuran
10827 — carbofuran
10828 — carbofuran
10853 — 2,4-D (present as acid)
10856 — malathion
10862 — carbaryl
10880 — malathion
10912 — 2,4-D (present as diethanolamine salt)
10916 — 2,4-D (present as dimethylamine salt)
10919 — carbaryl
10920 — carbaryl
10930 — dicamba (present as acid,
    diethanolamine salt or dimethylamine salt)
10930 — 2,4-D (present as acid)
10946 — carbaryl
10949 — 2,4-D (present as dimethylamine salt)
10949 — dicamba (present as acid,
    diethanolamine salt or dimethylamine salt)
10951 — chlordane
10964 — carbaryl
10965 — carbaryl
10967 — carbaryl
10969 — MCPA (present as diethanolamine salt,
    dimethylamine salt or as mixed amine salts)
10970 — 2,4-D (present as dimethylamine salt)
10970 — 2,4-D (present as diethanolamine salt)
10971 — 2,4-D (present as dimethylamine salt)
10988 — 2,4-D (present as acid)

11003 — 2,4-D (present as other amine salts)
11013 — chlordane
11025 — dinoseb (present in free form,
    as alkanolamine salts or as mixed amine salts)
11032 — 2,4-D (present as isooctyl esters)
11037 — trichlorfon
11037 — malathion
11044 — carbaryl
11055 — 2,4-D (present as dimethylamine salt)
11055 — 2,4-D (present as diethanolamine salt)
11055 — 2,4-D (present as other amine salts)
11087 — MCPA (present as potassium or sodium salt)

11090 — carbaryl
11096 — carbaryl
11113 — 2,4-D (present as dimethylamine salt)
11115 — carbaryl
11121 — malathion
11130 — malathion
11132 — trichlorfon
11137 — fenitrothion
11138 — fenitrothion
11153 — 2,4-D (present as butoxyethyl ester)
11157 — dinoseb (present in free form,
    as alkanolamine salts or mixed amine salts)
11166 — carbaryl
11183 — 2,4-D (present as diethanolamine salt)
11213 — chlordane
11214 — trichlorfon
11215 — carbaryl
11218 — carbaryl
11223 — malathion
11224 — dicamba (present as acid,
    diethanolamine salt or dimethylamine salt)
11230 — chlordane
11231 — chlordane
11236 — 2,4-D (present as diethanolamine salt)
11242 — 2,4-D (present as dimethylamine salt)
11243 — MCPA (present as potassium or sodium salt)
11244 — MCPA (present as esters)
11245 — 2,4-D (present as mixed butyl esters)
11246 — 2,4-D (present as isooctyl esters)
11247 — MCPA (present as diethanolamine salt,
    dimethylamine salt or mixed amine salts)
11249 — malathion
11249 — carbaryl
11258 — 2,4-D (present as butoxyethyl ester)
11271 — 2,4-D (present as isooctyl esters)
11273 — 2,4-D (present as dimethylamine salt)
11278 — malathion
11281 — trichlorfon
11299 — dinoseb (present in free form,
    as alkanolamine salts or as mixed amine salts)
11314 — malathion
11319 — malathion
11321 — MCPA (present as potassium or sodium salt)
11333 — 2,4-D (present as isooctyl esters)
11335 — dinoseb (present in free form,
    as alkanolamine salts or as mixed amine salts)
11337 — 2,4-D (present as acid)
11358 — chlordane
11364 — chlordane
11373 — chlordane

11374 — chlordane
11380 — chlordane
11397 — trichlorfon
11398 — malathion
11415 — carbaryl
11441 — 2,4-D (present as dimethylamine salt)
11442 — dinoseb (present in free form,
    as alkanolamine salts or as mixed amine salts)
11448 — 2,4-D (present as dimethylamine salt)
11456 — carbaryl
11456 — malathion
11457 — carbaryl
11457 — malathion
11458 — 2,4-D (present as diethanolamine salt)
11461 — dinoseb (present in free form,
    as alkanolamine salts or as mixed amine salts)
11463 — MCPA (present as potassium or sodium salt)
11479 — chlordane
11483 — carbaryl
11483 — malathion
11485 — carbaryl
11493 — 2,4-D (present as isooctyl esters)
11495 — 2,4-D (present as dimethylamine salt)
11514 — carbaryl
11515 — carbaryl
11421 — 2,4-D (present as propylene glycol butyl ether ester)
11545 — dinoseb (present in free form,
    as alkanolamine salts or as mixed amine salts)
11547 — dicamba (present as acid,
    diethanolamine salt or dimethylamine salt)
11547 — 2,4-D (present as dimethylamine salt)
11551 — MCPA (present as diethanolamine salt,
    dimethylamine salt or as mixed amine salts)
11552 — MCPA (present as esters)
11562 — 2,4-D (present as dimethylamine salt)
11571 — 2,4-D (present as dimethylamine salt)
11574 — (present as dimethylamine salt)
11591 — malathion
11599 — carbaryl
11618 — MCPA (present as diethanolamine salt,
    dimethylamine salt or as mixed amine salts)
11621 — malathion
11641 — malathion
11651 — 2,4-D (present as mixed butyl esters)
11652 — trichlorfon
11681 — malathion
11708 — trichlorfon
11720 — 2,4-D (present as isooctyl ester)
11721 — 2,4-D (present as dimethylamine salt)
11726 — chlordane

11729 — malathion
11760 — dinoseb (present in free form,
    as alkanolamine salts or as mixed amine salts)
11762 — 2,4-D (present as isooctyl esters)
11779 — 2,4-D (present as dimethylamine salt)
11787 — 2,4-D (present as diethanolamine salt)
11803 — 2,4-D (present as mixed butyl esters)
11804 — 2,4-D (present as mixed butyl esters)
11810 — 2,4-D (present as dimethylamine salt)
11814 — MCPA (present as potassium and sodium salt)
11817 — dicamba (present as acid,
    diethanolamine salt or dimethylamine salt)
11817 — 2,4-D (present as dimethylamine salt)
11827 — 2,4-D (present as isooctyl esters)
11828 — MCPA (present as esters)
11829 — MCPA (present as diethanolamine salt,
    dimethylamine salt or as mixed amine salts)
11830 — MCPA (present as potassium or sodium salts)
11840 — malathion
11843 — chlordane
11851 — dicamba (present as acid,
    diethanolamine salt or dimethylamine salt)
11852 — 2,4-D (present as acid)

12073 — malathion
12087 — MCPA (esters)
12090 — chlordane
12133 — chlordane
12135 — carbaryl
12137 — chlordane
12146 — carbaryl
12176 — chlordane
12216 — malathion
12231 — carbaryl
12236 — carbaryl
12278 — carbaryl
12278 — malathion
12330 — malathion
12331 — malathion
12332 — chlordane
12357 — malathion
12358 — malathion
12381 — 2,4-D (present as isooctyl esters)
12438 — 2,4-D (present as dimethylamine salt)
12438 — 2,4-D (present as diethanolamine salt)
12445 — chlordane
12455 — carbaryl
12456 — malathion
12456 — chlordane
12525 — malathion

12525 — carbaryl
12527 — malathion
12527 — carbaryl
12536 — chlordane
12560 — carbaryl
12586 — 2,4-D (present as dimethylamine salt)
12586 — dicamba (present as acid,
   diethanolamine salt and dimethylamine salt)
12587 — dicamba (present as acid,
   diethanolamine salt and dimethylamine salt)
12589 — dicamba (present as acid,
   diethanolamine salt and dimethylamine salt)
12590 — malathion
12611 — trifluralin
12639 — chlordane
12645 — 2,4-D (present as sodium salt)
12646 — chlordane
12859 — 2,4-D (present as dimethylamine salt)
12865 — chlordane
12931 — trifluralin
12951 — 2,4-D (present as isooctyl esters)
12959 — malathion
12968 — carbaryl

13028 — malathion
13052 — carbaryl
13064 — carbaryl
13065 — carbaryl
13241 — 2,4-D (present as dimethylamine salt)
13248 — malathion
13332 — chlordane
13335 — 2,4-D (present as isooctyl esters)
13356 — 2,4-D (present as dimethylamine salt)
13451 — chlordane
13453 — carbaryl
13491 — carbaryl
13494 — chlordane
13509 — 2,4-D (present as dimethylamine salt)
13509 — dicamba (present as acid,
   diethanolamine salt or dimethylamine salt)
13510 — malathion
13548 — malathion
13570 — MCPA (present as diethanolamine salt,
   dimethylamine salt or other amine salts)
13605 — carbaryl
13606 — carbaryl
13607 — carbaryl
13616 — carbaryl
13620 — chlordane
13622 — chlordane

13641 — chlordane
13645 — 2,4-D (present as isooctyl esters)
13646 — 2,4-D (present as isooctyl esters)
13652 — carbaryl
13661 — carbaryl
13662 — carbaryl
13700 — 2,4-D (present as isooctyl esters)
13701 — carbaryl
13723 — carbaryl
13739 — 2,4-D (present as isooctyl esters)
13750 — MCPA (present as potassium or sodium salts)
13761 — MCPA (present as diethanolamine salts,
   dimethylamine salt, amine salts)
13761 — dicamba (present as acid,
   diethanolamine salt, dimethylamine salt)
13851 — trichlorfon
13883 — malathion
13884 — carbaryl
13890 — 2,4-D (present as isooctyl esters)
13900 — carbaryl
13901 — carbaryl
13929 — carbaryl
13967 — fenitrothion

14010 — carbaryl
14017 — malathion
14017 — carbaryl
14027 — carbaryl
14102 — chlordane
14123 — trifluralin
14124 — trifluralin
14127 — carbaryl
14128 — carbaryl
14130 — carbaryl
14144 — malathion
14150 — 2,4-D (present as dimethylamine salt)
14150 — dicamba (present as acid,
   diethanolamine salt or dimethylamine salt)
14151 — carbaryl
14160 — carbaryl
14162 — carbaryl
14167 — 2,4-D (present as other amine salts)
14170 — dicamba (present as acid,
   diethanolamine salt or dimethylamine salt)
14173 — carbaryl
14186 — aminocarb
14187 — MCPA (present as diethanolamine salt,
   dimethylamine salt or mixed amine salts)
14188 — MCPA (present as potassium or sodium salt)
14211 — malathion

14214 — malathion
14223 — 2,4-D (present as diethanolamine salt)
14258 — malathion
14265 — malathion
14268 — carbaryl
14269 — malathion
14284 — dinoseb
14299 — fenitrothion
14300 — 2,4-D (present as acid)
14302 — carbaryl
14307 — trichlorfon
14313 — MCPA (present as diethanolamine salt,
          dimethylamine salt or mixed amine salts)
14336 — 2,4-D (present as mixed butyl esters)
14342 — carbaryl
14343 — carbaryl
14360 — MCPA (present as diethanolamine salt,
          dimethylamine salt or mixed amine salts)
14362 — 2,4-D (present as dimethylamine salt)
14377 — carbaryl
14384 — 2,4-D (present as isooctyl esters)
14405 — 2,4-D (present as dimethylamine salt)
14407 — malathion
14411 — trichlorfon
14412 — trichlorfon
14414 — 2,4-D (present as other amine salts)
14424 — carbaryl
14483 — 2,4-D (present as isooctyl esters)
14489 — carbaryl
14490 — carbaryl
14525 — malathion
14527 — carbaryl
14529 — carbaryl
14537 — carbaryl
14537 — malathion
14545 — trifluralin
14566 — carbaryl
14573 — carbaryl
14574 — carbaryl
14584 — malathion
14593 — dicamba (present as acid,
          diethanolamine salt or dimethylamine salt)
14594 — 2,4-D (present as isooctyl esters)
14622 — 2,4-D (present as dimethylamine salt)
14623 — 2,4-D (present as isooctyl esters)
14626 — 2,4-D (present as isooctyl esters)
14631 — MCPA (present as potassium or sodium salt)
14637 — 2,4-D (present as butoxyethyl ester)
14650 — MCPA (present as diethanolamine salt,
          dimethylamine salt or mixed amine salts)
14656 — malathion
14662 — MCPA (present as ester)
14663 — 2,4-D (present as isooctyl esters)
14664 — 2,4-D (present as isooctyl esters)
14666 — 2,4-D (present as mixed butyl esters)
14672 — MCPA (present as diethanolamine salt,
          dimethylamine salt or mixed amine salts)
14675 — MCPA (present as diethanolamine salt,
          dimethylamine salt or mixed amine salts)
14706 — carbaryl
14710 — chlordane
14714 — 2,4-D (present as isooctyl esters)
14715 — 2,4-D (present as mixed butyl esters)
14718 — MCPA (present as diethanolamine salt,
          dimethylamine salt or mixed amine salts)
14722 — 2,4-D (present as dimethylamine salt)
14723 — 2,4-D (present as dimethylamine salt)
14725 — 2,4-D (present as dimethylamine salt)
14726 — 2,4-D (present as dimethylamine salt)
14729 — malathion
14730 — MCPA (present as diethanolamine salt,
          dimethylamine salt or mixed amine salts)
14732 — dinoseb
14733 — 2,4-D (present as dimethylamine salt)
14739 — 2,4-D (present as isooctyl esters)
14743 — 2,4-D (present as isooctyl esters)
14758 — 2,4-D (present as isooctyl esters)
14764 — MCPA (present as esters)
14769 — malathion
14778 — dinoseb
14785 — dinoseb
14796 — 2,4-D (present as butoxyethyl ester)
14798 — carbaryl
14800 — 2,4-D (present as butoxyethyl ester)
14803 — 2,4-D (present as isooctyl esters)
14850 — chlordane
14852 — carbaryl
14853 — malathion
14859 — carbaryl
14861 — carbaryl
14868 — malathion
14894 — malathion
14902 — dicamba (present as acid,
          diethanolamine salt or dimethylamine salt)
14902 — 2,4-D (present as dimethylamine salt)
14912 — carbaryl
14965 — carbaryl

15014 — trichlorfon
15015 — 2,4-D (present as dimethylamine salt)

15015 — 2,4-D (present as diethanolamine salt)
15016 — MCPA (present as diethanolamine salt, dimethylamine salt, mixed amine salt)
15030 — carbaryl
15075 — 2,4-D (dimethylamine salt)
15075 — dicamba (present as acid, diethanolamine salt, dimethylamine salt)
15086 — dinoseb (present in free form, alkanolamine salts, mixed amine salts)
15105 — 2,4-D (present as dimethylamine salt)
15110 — 2,4-D (present as isooctyl esters)
15112 — MCPA (present as esters)
15114 — 2,4-D (present as diethanolamine salt)
15118 — malathion
15127 — MCPA (present as esters)
15130 — malathion
15135 — carbaryl
15149 — 2,4-D (present as dimethylamine salt)
15156 — carbaryl
15158 — 2,4-D (present as dimethylamine salt)
15159 — 2,4-D (isooctyl esters)
15160 — 2,4-D (isooctyl esters)
15186 — malathion
15237 — 2,4-D (present as mixed butyl esters)
15240 — MCPA (present as diethanolamine salt, dimethylamine salt, or mixed amine salts)
15244 — 2,4-D (present as isooctyl esters)
15258 — trichlorfon
15271 — 2,4-D (present as dimethylamine salt)
15278 — 2,4-D (present as isooctyl esters)
15279 — 2,4-D (present as dimethylamine salt)
15308 — 2,4-D (present as butoxyethyl ester)

15319 — trichlorfon
15325 — 2,4-D (present as diethanolamine salt)
15335 — carbaryl
15365 — 2,4-D (present as dimethylamine salt)
15374 — 2,4-D (present as diethanolamine salt)
15377 — malathion
15380 — malathion
15386 — carbaryl
15389 — carbaryl
15391 — carbaryl
15393 — 2,4-D (present as acid)
15400 — 2,4-D (present as diethanolamine salt)
15401 — 2,4-D (present as isooctyl esters)
15404 — 2,4-D (present as diethanolamine salt)
15405 — 2,4-D (present as diethanolamine salt)
15441 — 2,4-D (present as isooctyl esters)
15504 — chlordane
15521 — 2,4-D (present as dimethylamine salt)
15668 — 2,4-D (present as dimethylamine salt)
15721 — carbaryl
15730 — 2,4-D (present as dimethylamine salt)
15851 — chlordane
15857 — carbaryl
15942 — 2,4-D (present as dimethylamine salt)
15950 — 2,4-D (present as dimethylamine salt)
15984 — 2,4-D (present as dimethylamine salt)
15985 — 2,4-D (present as dimethylamine salt)

16038 — 2,4-D (present as dimethylamine salt)
16149 — 2,4-D (present as dimethylamine salt)
16167 — 2,4-D (present as dimethylamine salt)

# 4  CONVERSION FACTORS

## AREA

| | | |
|---|---|---|
| acres | x 0.405 | = ha |
| | x 0.004 05 | = km$^2$ |
| | x 4 050 | = m$^2$ |
| | x 0.001 56 | = mi$^2$ |
| ft$^2$ | x 0.092 9 | = m$^2$ |
| | x 0.111 | = yd$^2$ |
| m$^2$ | x 10.8 | = ft$^2$ |
| | x 0.000 1 | = ha |
| | x 1.20 | = yd$^2$ |
| mi$^2$ | x 640 | = acres |
| | x 259 | = ha |
| | x 2.59 | = km$^2$ |

## CONCENTRATIONS (by weight)

| | | |
|---|---|---|
| mg/L (in water) | x 1 | = ppm |
| | x 0.000 1 | = % |
| mg/m$^3$ (in air) | x 24.1/(mol. wt.) | = ppm (at 20°C) |
| | x 20.8/(mol. wt.) | = ppm (at -20°C) |
| | x 0.002 41/(mol. wt.) | = % (at 20°C) |
| | x 0.002 08/(mol. wt.) | = % (at -20°C) |
| ppm (in water) | x 0.000 1 | = % |
| | x 1 | = mg/L |
| ppm (in air) | x 0.0415 (mol. wt.) | = mg/m$^3$ (at 20°C) |
| | x 0.0481 (mol. wt.) | = mg/m$^3$ (at -20°C) |
| | x 0.000 1 | = % |
| % (in water and air) | x 10 000 | = ppm |

## DENSITY

| | | |
|---|---|---|
| g/cm$^3$ | x 62.4 | = lb/ft$^3$ |
| | x 1 690 | = lb/yd$^3$ |
| kg/m$^3$ | x 0.001 | = g/cm$^3$ |
| | x 0.062 4 | = lb/ft$^3$ |
| | x 1.69 | = lb/yd$^3$ |
| lb/ft$^3$ | x 0.016 0 | = g/cm$^3$ |
| | x 16.0 | = kg/m$^3$ |
| | x 27 | = lb/yd$^3$ |

## FLOW

| | | |
|---|---|---|
| gal/s | x 0.160 | = ft$^3$/s |
| | x 60 | = gal/min |
| | x 72 | = gal (U.S.)/min |
| | x 4.55 | = L/s |
| | x 16.4 | = m$^3$/h |
| L/s | x 0.035 3 | = ft$^3$/s |
| | x 13.2 | = gal/min |
| | x 15.9 | = gal (U.S.)/min |

## LENGTH

| | | |
|---|---|---|
| cm | x 0.032 8 | = ft |
| | x 0.394 | = in |
| ft | x 30.5 | = cm |
| | x 0.305 | = m |
| in | x 2.54 | = cm |
| | x 25.4 | = mm |
| km | x 3 280 | = ft |
| | x 0.621 | = mi |
| | x 0.540 | = mi (naut) |
| | x 1 090 | = yd |
| m | x 3.28 | = ft |
| | x 1.09 | = yd |
| mi | x 5 280 | = ft |
| | x 1.61 | = km |
| | x 1 610 | = m |
| | x 0.869 | = mi (naut) |
| | x 1 760 | = yd |
| mi (naut) | x 6 080 | = ft |
| | x 1.85 | = km |
| | x 185 | = m |
| | x 1.15 | = mi |
| | x 2 030 | = yd |

## MASS/WEIGHT

| | | |
|---|---|---|
| g | x 0.002 20 | = lb |
| | x 0.035 2 | = oz |
| kg | x 2.20 | = lb |
| lb | x 454 | = g |
| | x 0.454 | = kg |
| | x 16 | = oz |
| oz | x 28.4 | = g |
| | x 0.028 4 | = kg |

## MASS/WEIGHT (cont'd)

| | | |
|---|---|---|
| ton (long) | x 1 020 | = kg |
| | x 2 240 | = lb |
| | x 1.12 | = ton (short) |
| | x 1.02 | = t |
| ton (short) | x 907 | = kg |
| | x 2 000 | = lb |
| | x 0.893 | = ton (long) |
| | x 0.907 | = t |
| t | x 1 000 | = kg |
| | x 2 200 | = lb |
| | x 1.10 | = ton (short) |
| | x 0.984 | = ton (long) |

## PRESSURE

| | | |
|---|---|---|
| atm | x 1.01 | = bars |
| | x 101 | = kPa |
| | x 760 | = mm Hg (0°C) |
| | x 14.7 | = psia |
| bars | x 0.987 | = atm |
| | x 100 | = kPa |
| | x 750 | = mm Hg (0°C) |
| | x 14.5 | = psia |
| kPa | x 0.009 87 | = atm |
| | x 0.01 | = bars |
| | x 7.50 | = mm Hg |
| | x 0.145 | = psia |
| mm Hg (0°C) | x 0.001 32 | = atm |
| | x 0.001 33 | = bars |
| | x 0.133 | = kPa |
| | x 0.019 3 | = psia |
| psia | x 0.068 0 | = atm |
| | x 0.068 9 | = bars |
| | x 6.89 | = kPa |
| | x 51.7 | = mm Hg (0°C) |
| | - 14.7 | = psig |

## SPILL VOLUMES and WEIGHTS

| | | |
|---|---|---|
| barrels | x 0.159 (s.g.) | = t |
| gal | x 0.004 55 (s.g.) | = t |
| gal (U.S.) | x 0.003 79 (s.g.) | = t |
| L | x 0.001 (s.g.) | = t |
| MMCF (gas) | x 36.6 (v.d.) | = t |
| $m^3$ (gas) | x 0.001 29 (v.d.) | = t |

## SOLUBILITY (in water)

| | | |
|---|---|---|
| g/100 mL | x 10 000 | = mg/L |
| | x 10 000 | = ppm |
| | x 1 | = % |
| mg/L | x 0.000 1 | = g/100 mL |
| | x 1 | = ppm |
| | x 0.000 1 | = % |
| ppm | x 0.000 1 | = g/100 mL (or %) |
| | x 1 | = mg/L |
| % | x 1 | = g/100 mL |
| | x 10 000 | = mg/L (or ppm) |

## TEMPERATURE

| | | |
|---|---|---|
| °C | x 1.8 + 32 | = °F |
| | + 273 | = K |
| °F | -32) x 0.556 | = °C |
| | -32) x 0.556 + 273 | = K |
| K | - 273 | = °C |
| | - 273) x 1.8) + 32 | = °F |

## VELOCITY

| | | |
|---|---|---|
| ft/s | x 1.10 | = km/h |
| | x 0.592 | = kn |
| | x 0.305 | = m/s |
| | x 0.682 | = mi/h |
| km/h | x 0.911 | = ft/s |
| | x 0.540 | = kn |
| | x 0.278 | = m/s |
| | x 0.621 | = mi/h |
| kn | x 1.69 | = ft/s |
| | x 1.85 | = km/h |
| | x 0.514 | = m/s |
| | x 1.15 | = mi/h |
| m/s | x 3.28 | = ft/s |
| | x 3.6 | = km/h |
| | x 1.94 | = kn |
| | x 2.24 | = mi/h |
| mi/h | x 1.47 | = ft/s |
| | x 1.61 | = km/h |
| | x 0.869 | = kn |
| | x 0.447 | = m/s |

## VOLUME

| | | |
|---|---|---|
| $cm^3$ | x 0.061 0 | = $in^3$ |
| $ft^3$ | x 1 730 | = $in^3$ |
| | x 0.028 3 | = $m^3$ |
| | x 0.037 0 | = $yd^3$ |
| $in^3$ | x 16.4 | = $cm^3$ |
| $m^3$ | x 35.3 | = $ft^3$ |
| | x 1.31 | = $yd^3$ |
| $yd^3$ | x 27 | = $ft^3$ |
| | x 0.765 | = $m^3$ |

## VOLUME (liquid measure)

| | | |
|---|---|---|
| barrel | x 35.0 | = gal |
| | x 42 | = gal (U.S.) |
| | x 159 | = L |
| | x 0.159 | = $m^3$ |
| gal | x 0.028 6 | = barrel |
| | x 1.20 | = gal (U.S.) |
| | x 4.55 | = L |
| gal (U.S.) | x 0.023 8 | = barrel |
| | x 0.832 | = gal |
| | x 3.79 | = L |
| L | x 0.006 29 | = barrel |
| | x 0.220 | = gal |
| | x 0.264 | = gal (U.S.) |

## 4.1 ABBREVIATIONS USED IN CONVERSIONS

| | | | | | |
|---|---|---|---|---|---|
| atm | - | atmosphere | m/s | - | metres per second |
| cm | - | centimetre | mi | - | mile (statute) |
| $cm^3$ | - | cubic centimetre | mi (naut) | - | nautical mile |
| $ft^3$ | - | cubic foot | mi/h | - | miles per hour |
| $in^3$ | - | cubic inch | MMCF | - | million cubic feet |
| $m^3$ | - | cubic metre | $mg/m^3$ | - | milligrams per cubic metre |
| $yd^3$ | - | cubic yard | mg/L | - | milligrams per litre |
| °C | - | degree Celsius | mm | - | millimetre |
| °F | - | degree Fahrenheit | mm Hg | - | millimetres of mercury |
| ft | - | foot | min | - | minute |
| ft/s | - | foot per second | oz | - | ounce (avdp.) |
| g | - | gram | ppb | - | parts per billion |
| gal (U.S.) | - | United States gallon | ppm | - | parts per million |
| g/100 mL | - | grams per 100 millilitres | ppt | - | parts per trillion |
| ha | - | hectare | % | - | percent |
| h | - | hour | lb | - | pound (avdp.) |
| gal | - | Canadian gallon | psia | - | pounds per square inch, absolute |
| in | - | inch | | | |
| K | - | kelvin (temperature) | psig | - | pounds per square inch, gauge |
| kg | - | kilogram | | | |
| km | - | kilometre | s | - | second |
| km/h | - | kilometres per hour | s.g. | - | specific gravity |
| kPa | - | kilopascal | t | - | tonne |
| kn | - | knot | T | - | temperature |
| L | - | litre | v.d. | - | vapour density |
| m | - | metre | yd | - | yard |
| MCF | - | thousand cubic feet | | | |

# BIBLIOGRAPHY

ACGIH, *Threshold Limit Values for Chemical Substances in the Work Environment for 1983-1984*, adopted by American Conference of Government and Industrial Hygienists (ACGIH), Cincinnati, Ohio (1983).

Aldrich, *The Source - The 1975-76 Aldrich Catalogue/Handbook of Organic and Biochemicals*, Aldrich Chemical Co. Inc., Milwakee, Wisconsin (1974).

Aldrich, *1981-1982 Aldrich Catalogue/Handbook of Fine Chemicals*, Catalog 20, Aldrich Chemical Co. Inc., Milwaukee, Wisconsin (1980).

Anonymous, *Directory of World Chemical Producers, 1980/81 Edition*, Chem. Info. Ser. Ltd. Pub., Oceanside, NY (1980).

Anonymous, *Hazardous Materials - Emergency Response Guidebook*, U.S. Dept. of Trans., Res. and Spec. Prog. Admin., Materials Trans. Bureau (1980).

Bennett, H. (ed.), *Concise Chemical and Technical Dictionary*, 3rd edition, Chem. Pub. Co., New York, NY (1974).

Bierkin, L.W., *Red Book on Transportation of Hazardous Materials*, CBI Pub. Co. Inc., Boston, MA (1978).

Braker, W. and A.L. Mossman, *Matheson Gas Data Book*, 5th edition, Matheson Gas Prod., Lyndhurst, New Jersey (1971).

Braker, W. and A.L. Mossman, *Matheson Gas Data Book*, 6th edition, Matheson Gas Prod., Lyndhurst, New Jersey (1980).

Braker, W., A.L. Mossman and D. Seigel, *Effects of Exposure to Toxic Gases - First Aid and Medical Treatment*, 2nd edition, Matheson Gas Prod., Lyndhurst, New Jersey (1979).

Bretherick, L., *Handbook of Reactive Chemical Hazards*, CRC Press Inc., Cleveland, Ohio (1975).

Bretherick, L., *Handbook of Reactive Chemical Hazards*, 2nd edition, Butterworth and Co. Ltd., London, England (1979).

Campbell, J.B., "Buyer's Guide", *Chemicalweek*, 779 pp. (October, 1980).

Cheremisinoff, P.N. and A.C. Morresi, *Benzene - Basic and Hazardous Properties*, Marcel Dekker, Inc., New York, NY (1979).

Clayton, G.D. and F.E. Clayton (eds.), *Patty's Industrial Hygiene and Toxicology*, third revised edition, Volume 2A, Toxicology, Wiley-Interscience Pub., John Wiley and Sons, NY (1981).

Considine, D.M. (editor-in-chief), *Chemical and Process Technology Encyclopedia*, McGraw-Hill Book Company California (1974).

Cotton F.A. and G. Wilkinson, <u>Advanced Inorganic Chemistry, A Comprehensive Text</u>, 3rd edition, Interscience Pub. (1972).

CPI, <u>CPI Product Profiles</u>, Corpus Info. Ser. Ltd., Don Mills, Ont. (1980).

DASE, <u>Handling Chemicals Safety</u>, 2nd edition, Dutch Assoc. of Safety Experts, Dutch Chem. Ind. Assoc. and the Dutch Safety Inst., Amsterdam (1980).

Dean, J.A., <u>Lange's Handbook of Chemistry</u>, 11th edition, McGraw-Hill Book Co. (1973).

Dept. of Transport, Coast Guard, <u>(CHRIS) Hazardous Chemical Data</u>, Vols. I and II, U.S. Coast Guard, Washington, D.C. (1978).

Driesbach, R.H., <u>Handbook of Poisoning: Prevention, Diagnosis and Treatment</u>, 10th edition, Lange Medical Publications, Los Altos, California (1980).

Driesbach, R.R., <u>Physical Properties of Chemical Compounds II</u>, No. 22, Advances in Chemistry Series, Am. Chem. Soc. (eds.), Am. Chem. Soc. Appl. Pub., Washington, D.C. (1959).

Driesbach, R.R., <u>Physical Properties of Chemical Compounds III</u>, No. 29, Advances in Chemistry Series, R.F. Gould (ed.), Am. Chem. Soc. (1961).

Eco Research Ltd., <u>Hazardous Materials Spill Manual</u>, Vols. I and II, M.A. Wilson (ed.), Eco Research, Pointe Claire, Que. (1977).

Environmental Health Directorate, <u>Aerial Application of Pesticides; Safety Manual</u>, 3rd edition, Minister of National Health and Welfare, Canada (1975).

Environmental Health Directorate, Health Protection Branch, <u>Amines in Steam</u>, National Health and Welfare, Pub. No. 79-EHD-39, Ottawa, Ontario (October, 1979).

Environmental Health Directorate, Health Protection Branch, <u>Methylcyclopentadienyl Manganese Tricarbonyl (MMT) - An Assessment of the Human Health Implications of its Use as a Gasoline Additive</u>, National Health and Welfare, Pub. No. 78-EHD-21 (1978).

EPA, <u>Hazardous Materials Spill Monitoring - Safety Handbook and Chemical Hazard Guide - Part B - Chemical Data</u>, Environmental Monitoring Series, Envir. Prot. Agency, Nat. Tech. Info. Ser. PB295854 (1979).

EPS, <u>Handbook on PCBs in Electrical Equipment</u>, de Gonzague, J. and S. Lawrie (eds.), Environmental Impact Control Directorate, Environmental Protection Service, Environment Canada (March, 1981).

Ethyl Corporation, <u>Handling Procedures for Aluminium Alkyl Compounds</u>, Ethyl Corporation of Canada Ltd., Toronto, Ontario (1971).

Fazzalari, F.A. (ed.), <u>Compilation of Odour and Taste Threshold Values Data</u>, American Society for Testing and Materials, DS48A, Philadelphia, PA (1978).

Gardner, W., E.I. Cooke and R.W.I. Cooke, <u>Handbook of Chemical Synonyms and Trade Names</u>, CRC Press, Cleveland, OH (1978).

General Electric, <u>Material Safety Data Sheets</u>, Material Information Services, Gen. El. Co., New York, Updated sheets from 1977 to 1982.

Gosselin, R.E., H.C. Hodge, R.P. Smith and M.N. Gleason, <u>Clinical Toxicology of Commercial Products - Acute Poisoning</u>, 4th edition, Williams and Wilkins Co., Baltimore (1979).

Grant, J., <u>Hach's Chemical Dictionary</u>, 4th edition, McGraw-Hill Book Co., New York (1972).

Guthrie, V.B., <u>Petroleum Products Handbook</u>, 1st edition, McGraw-Hill Book Co., New York (1960).

Hawley, G., <u>The Condensed Chemical Dictionary</u>, 9th edition, Van Nostrand-Reinhold Co. (1977).

Hawley, G., <u>The Condensed Chemical Dictionary</u>, 10th edition, Van Nostrand-Reinhold Co. (1981).

Hooker Chemical, <u>Hooker Caustic Soda (NaOH) - Product Information Manual</u>, Hooker Chem. Div., Canadian Occidental Petroleum Ltd., Vancouver, B.C. (1980).

IMCO, <u>International Maritime Dangerous Goods Code</u>, Inter-Governmental Maritime Consultative Organization, Vols. I to IV (1977).

ITC, <u>Canadian Chemical Register</u>, Chemicals Branch, Dept. of Industry, Trade and Commerce, Ottawa, Ontario (1979).

ITII, <u>Toxic and Hazardous Industrial Chemicals Safety Manual - for Handling and Disposal with Toxicity and Hazard Data</u>, The International Tech. Info. Inst. (ITII), Japan (1981).

Johnson, W.W. and M.T. Finley, <u>Handbook of Acute Toxicity of Chemicals to Fish and Invertebrates</u>, U.S. Dept. of the Interior, Pub. No. 137, Washington, D.C. (1980).

Katz, D.L., D. Cornell, R. Kobayashi, F.H. Poettmann, J.A. Vary, J.R. Elenbars and C.F. Weinaug, <u>Handbook of Natural Gas Engineering</u>, McGraw-Hill Book Co., New York, NY (1959).

Kirk-Othmer, <u>Encyclopedia of Chemical Technology</u>, 3rd edition, Vols. 1 to 16, Wiley Interscience Pub., John Wiley & Sons, New York, NY (1979).

Lefèvre, M.J. and E.O. Becker, <u>First Aid Manual for Chemical Accidents, for Use with Nonpharmaceutical Chemicals</u>, Dowden, Hutchinson and Ross Inc., Stroudsburg, Pennsylvania (1980).

Linde, <u>Linde, Specialty Gases - Safety Precautions and Emergency Procedures</u>, Union Carbide Corp. Pub. (1976).

Linke, W.F., <u>Solubilities of Inorganic and Metal-Organic Compounds</u>, Vol. I, 4th edition, Am. Chem. Soc., Washington, D.C. (1958).

Linke, W.F., Solubilities of Inorganic and Metal-Organic Compounds, Vol. II, 4th edition, Am. Chem. Soc., Washington, D.C. (1965).

Mackison, F.W., R.S. Stricoff, L.J. Partridge Jr. and A.D. Little Inc., NIOSH/OSHA Pocket Guide to Chemical Hazards, U.S. Dept. of Health, Education and Welfare, Nat. Inst. for Occup. Safety and Health (NIOSH) and U.S. Dept. of Labour, Occup. Safety and Health Admin. (OSHA) (September, 1978).

Marler, E.E.J., Pharmacological and Chemical Synonyms, 6th edition, Excerpta Medica (1978).

MCA, Laboratory Waste Disposal Manual (revised Sept. 1973), Manu. Chem. Assoc., Washington, D.C. (1973).

MCA, Chemical Safety Data Sheets, Manu. Chem. Assoc. (MCA), Washington, D.C.

MCA, Guide for Safety in the Chemical Laboratory, 2nd edition, Mana. Chem. Assoc., Van Nostrand-Reinhold Co. (1972).

Meister, R.T. (ed. director), 1981 Farm Chemicals Handbook, Farm Chemicals Magazine, Meister Pub. Co., Willoughby, Ohio (1981).

Mellan, I., Industrial Solvents Handbook, 2nd edition, Noyes Data Corp., New Jersey (1977).

Meyer, R., Explosives, Verlag Chemie, Weinheim, New York (1977).

Muir, G.D. (ed.), Hazards in the Chemical Laboratory, 2nd edition, The Chem. Soc. London, England (1977).

National Research Council of Canada, NRCC Associate Commitee on Scientific Criteria for Environmental Quality, The Effects of Alkali Halides in the Canadian Environment, Subcommittee on Heavy Metals and Certain Other Compounds, Environmental Secretariat, Pub. No. 15019, Ottawa, Ontario (1977).

National Research Council (U.S.), The Alkyl Benzenes, Committee on Alkyl Benzene Derivatives, Board on Toxicology and Environmental Health Hazards, Assembly of Life Sciences, National Academy Press, Washington, D.C. (1981).

NFPA, Fire Protection Guide on Hazardous Materials, 7th edition, Nat. Fire Prot. Assoc., Boston, MA (1978).

NHTSA, Emergency Action Guide for Selected Hazardous Materials, U.S. Dept. of Trans., Res. and Spec. Prog. Admin. and Nat. Highway Traffic Safety Admin. (NHTSA), Washington, D.C. (1978).

NIOSH, Current Intelligence Bulletins, U.S. Dept. of Health and Human Services, Nat. Inst. for Occup. Safety and Health (NIOSH) and U.S. Dept. of Labour, Occup. Safety and Health Admin. (OSHA), various bulletins (1975-1980).

NIOSH, Registry of Toxic Effects of Chemical Substances, 1978 edition, edited by Lewis, R.J. Sr. and R.L. Tatken, Nat. Inst. for Occup. Safety and Health (NIOSH) (January, 1979).

NIOSH, <u>1979 Registry of Toxic Effects of Chemical Substances</u>, Vols. I and II, edited by Lewis, R.J. Sr. and R.L. Tatken, U.S. Dept. of Health and Human Services, Cincinnati, Ohio (September, 1980).

NIOSH/OSHA, <u>Pocket Guide to Chemical Hazards</u>, Nat. Inst. for Occup. Safety and Health (NIOSH)/Occupational Safety and Health Administration (OSHA), Pub. No. 78-210 (January, 1980).

NRC (United States), <u>Prudent Practices for Handling Hazardous Chemicals in Laboratories</u>, National Academy Press, Washington, D.C. (1981).

Plunkett, E.R., <u>Handbook of Industrial Toxicology</u>, Chem. Pub. Co. Inc., New York, NY (1976).

Que Hee, S.S. and Sutherland, R.G., <u>The Phenoxyalkanoic Herbicides, Vol. 1 - Chemistry, Analysis and Environmental Pollution</u>, CRC Press, Inc., Boca Raton, Florida (1981).

Rabben, E.P. (ed.), <u>Hazardous Materials Regulations Excerpted for Railroad Employees</u>, Bureau of Explosives, Washington, D.C., Be Pamphlet 20 (1981).

Rao, K.R. (ed.), <u>Pentachlorophenol - Chemistry, Pharmacology, and Environmental Toxicology</u>, Plenum Press, New York, NY (1978).

Reid, R.C., J.M. Prausnitz and T.K. Sherwood, <u>The Properties of Gases and Liquids</u>, 3rd edition, McGraw-Hill Book Co., New York, NY (1977).

Robinson J.S., <u>Hazardous Chemical Spill Cleanup</u>, Noyes Data Corp., New Jersey (1979).

Sax, N.I., <u>Dangerous Properties of Industrial Materials</u>, 5th edition, Van Nostrand Reinhold Co., New York, NY (1979).

Sittig, M. (ed.), <u>Pesticide Manufacturing and Toxic Materials Control Encyclopedia</u>, Noyes Data Corp. Pub., New Jersey (1980).

Sittig, M. (ed.), <u>Priority Toxic Pollutants - Health Impacts and Allowable Limits</u>, Noyes Data Corp. Pub., New Jersey (1980).

Steere, N.V. (ed.), <u>CRC Handbook of Laboratory Safety</u>, 2nd edition, CRC Press Inc., Boca Raton, Florida (1980).

Student, P.J. (ed.), <u>Emergency Handling of Hazardous Materials in Surface Transportation</u>, Bureau of Explosives, Assoc. of Am. Railroads, Washington, D.C. (1981).

Transport Canada, <u>Emergency Response Guide for Dangerous Goods</u>, Can. Gov. Pub. Centre, Ottawa, Ont. (1979).

Transport Canada, <u>Transportation of Dangerous Goods Code</u>, Transport Canada, Can. Gov. Pub. Centre, Ottawa, Ont., TP 1050 (1980-81).

U.S. Dept. of Transport, <u>Emergency Action Guide for Selected Hazardous Materials</u>, U.S. DOT, Washington, D.C. (1978).

U.S. Dept. of Transport, <u>Hazardous Materials: Emergency Response Guidebook</u>, U.S. DOT, Washington, D.C., DOT P5800-2 (1980).

U.S. Environmental Protection Agency, <u>Oil and Hazardous Materials - Technical Assistance Data System</u> - use of "OHM-TADS" data system, U.S. Environmental Protection Agency, Oil and Special Materials Control Division, Office of Water Program Operations, Washington, D.C. (1981 version).

Verschueren, K., <u>Handbook of Environmental Data on Organic Chemicals</u>, Van Nostrand Reinhold Co., New York, NY (1977).

Weast, R.C. and M.J. Astle (eds.), <u>CRC Handbook of Chemistry and Physics</u>, 60th edition, Chemical Rubber Co. (CRC) Pub. (1979).

Weiss, G. (ed.), <u>Hazardous Chemicals Data Book</u>, Noyes Data Corp., New Jersey (1980).

Wilhoit, R.C. and B.J. Zwolinski, <u>Handbook of Vapour Pressures and Heats of Vaporization of Hydrocarbons and Related Compounds</u>, Thermodynamics Research Center, Dept. of Chem., Texas A & M University, Texas (1971).

Windholz, M., S. Budavari, L. Stroumtsos and M. Noether Fertig (eds.), <u>The Merck Index</u>, 9th edition, Merck and Co., Inc., New Jersey (1976).

Worthing, C.R. (ed.), <u>The Pesticide Manual - A World Compendium</u>, 6th edition, British Crop Protection Council (1979).

WSSA Herbicide Handbook Committee, <u>Herbicide Handbook of the Weed Science Society of America</u>, 4th edition, WSSA, Champaign, Illinois (1979).

Yaws, C.L., <u>Physical Properties - A Guide to the Physical, Thermodynamic and Transport Property Data of Industrially Important Chemical Compounds</u>, Chemical Engineering, McGraw-Hill Book Co., New York, NY (1977).

# HAZARDOUS MATERIALS: TWO-PAGE ENTRIES

# ACETALDEHYDE   CH₃CHO

## IDENTIFICATION

### Common Synonyms
ETHANAL
ACETIC ALDEHYDE
ETHYL ALDEHYDE
ALDEHYDE

### Observable Characteristics
Clear, colourless, liquid. Sharp fruity odour.

### UN No. 1089

### Manufacturers
Celanese Canada Limited, Edmonton, Alberta.

### Transportation and Storage Information
**Shipping State:** Liquid (boiling).
**Classification:** Flammable liquid.
**Inert Atmosphere:** Inerted.
**Venting:** Safety relief.
**Pump Type:** Centrifugal; stainless steel.

**Label(s):** Red label - FLAMMABLE LIQUID; Class 3.1, Group I.
**Storage Temperature:** Ambient.
**Hose Type:** Neoprene, polyethylene, polypropylene, PVC. Rubber not suitable.

**Grades or Purity:** Technical (>99%).
**Containers and Materials:** Drums, tank cars, tank trucks; lined or treated carbon steel, aluminum, stainless steel.

### Physical and Chemical Characteristics
**Physical State** (20°C, 1 atm): Liquid.
**Solubility** (Water): Miscible in all proportions.
**Molecular Weight:** 44.1
**Vapour Pressure:** 740 mm Hg (20°C).
**Boiling Point:** 20.4°C.

**Floatability** (Water): Floats and mixes.
**Odour:** Sharp fruity odour, penetrating, pungent (0.01 to 0.21 ppm, odour threshold).
**Flash Point:** -50°C (o.c.); -38°C (c.c.).
**Vapour Density:** 1.5
**Specific Gravity:** 0.78 (20°C).

**Colour:** Colourless.
**Explosive Limits:** 4 to 60%.
**Melting Point:** -123.5°C.

## HAZARD DATA

### Human Health
**Symptoms: Contact:** exposure to vapours causes severe irritation of mucous membranes, blurred vision, reddening of skin, coughing, pulmonary edema and narcosis. **Ingestion:** nausea, vomiting, diarrhea, narcosis and respiratory failure.
**Toxicology:** Moderately toxic by ingestion, inhalation and contact.
TLV®- 100 ppm; 180 mg/m³.   $LC_{50}$ - 20,000 ppm/30 min.   $LD_{50}$ - Oral: rat = 1.93 g/kg
Short-term Inhalation Limits - 150 ppm; 270 mg/m³   $LC_{Lo}$ - Inhalation: rat = 4 000 ppm/4 h (15 min).
Delayed Toxicity - No information.

### Fire
**Fire Extinguishing Agents:** Use dry chemical, carbon dioxide, alcohol foam. Water may be ineffective, but may be used to cool fire-exposed containers. Water spray may be used to control vapours.
**Behaviour in Fire:** Flashback may occur along vapour trail.   **Burning Rate:** 3.3 mm/min.
**Ignition Temperature:** 185°C.

### Reactivity
**With Water:** No reaction; soluble.
**With Common Materials:** Can react vigorously with acid anhydrides, phenols, anhydrous ammonia, halogens, phosphorus, acetic acid and strong alkalis. May form explosive peroxide mixtures.
**Stability:** Stable.

### Environment
**Water:** Prevent entry into water intakes and waterways. Harmful to aquatic life in low concentrations. **Fish toxicity:** 53 ppm/96 h/sunfish/TLm/freshwater; 237 to 249 mg/L/96 h/Nitzschia linearis (alga) /$LC_{50}$/freshwater; 70 ppm/24 h/pin perch/TLm/saltwater; BOD: 93 to 127%, 5 days.
**Land-Air:** No information.
**Food Chain Concentration Potential:** None.

## EMERGENCY MEASURES

### Special Hazards
FLAMMABLE. Low boiling point.

### Immediate Responses
Keep non-involved people away from spill site. Issue warning: "FLAMMABLE". Call fire department. Eliminate all sources of ignition. Avoid contact and inhalation. Stay upwind and use water spray to control vapour. Stop or reduce discharge, if this can be done without risk. Contact supplier for guidance. Dike to prevent runoff. Notify environmental authorities.

### Protective Clothing and Equipment
Respiratory protection - self-contained breathing apparatus. Acid suit - (jacket and pants) or coveralls. Boots - high, rubber (pants worn outside boots). Gloves - rubber or plastic.

### Fire and Explosion
Maintain safe distance when fighting fire. Use dry chemical, carbon dioxide or alcohol foam. Water may be ineffective, but may be used to cool fire-exposed containers. Water spray may be used to control vapours. Flashback may occur along vapour trail.

### First Aid
Move victim out of spill site to fresh air. Call for medical assistance, but start first aid at once. Inhalation: give artificial respiration if breathing has stopped. Give oxygen if breathing is laboured. Contact: remove contaminated clothing. Wash eyes and skin with plenty of warm water for at least 15 minutes. Ingestion: give water to conscious victim to drink. If medical assistance is not immediately available, transport victim to hospital, clinic or doctor.

## ENVIRONMENTAL PROTECTION MEASURES

### Response

**Water**
1. Stop or reduce discharge if safe to do so.
2. Contact manufacturer or supplier for advice.
3. If possible, contain discharge by damming or water diversion.
4. Dredge or vacuum pump to remove contaminants, liquids and contaminated bottom sediments.
5. Notify environmental authorities to discuss disposal and cleanup of contaminated materials.

**Land-Air**
1. Stop or reduce discharge if safe to do so.
2. Contact manufacturer or supplier for advice.
3. Dike to prevent runoff from rainwater or water application.
4. Remove material with pumps or vacuum equipment and place in appropriate containers.
5. Recover undamaged containers.
6. Remove contaminated soil for disposal.
7. Notify environmental authorities to discuss cleanup and disposal of contaminated materials.

### Disposal
1. Contact manufacturer or supplier for advice on disposal.
2. Contact environmental authorities for advice on disposal.
3. Incinerate (approval of environmental authorities required).

ACETALDEHYDE   $CH_3CHO$

# ACETIC ACID   CH₃COOH

## IDENTIFICATION

**Common Synonyms**
GLACIAL ACETIC ACID
ETHANOIC ACID
METHANE CARBOXYLIC ACID
VINEGAR ACID

**Observable Characteristics**
Clear, colourless, liquid. Strong pungent vinegar odour.

**UN No.** 2789 (glacial >80%)

**Manufacturers**
Celanese Canada Limited, Edmonton, Alberta.
Caledon Lab. Ltd. (glacial), Georgetown, Ontario.

### Transportation and Storage Information

**Shipping State:** Liquid or solid.
**Classification:** Corrosive liquid.
**Inert Atmosphere:** No requirement.
**Venting:** Open.
**Pump Type:** Centrifugal; stainless steel or plastic.

**Label(s):** Black and white label – CORROSIVE; Class 8, Group II.
**Storage Temperature:** Ambient.
**Hose Type:** Polyethylene, polypropylene, PVC.

**Grades or Purity:** Technical grades, 28%; 56%; 80%; 92%; glacial C.P.; U.S.P. Glacial, 99.4%
**Containers and Materials:** Plastic containers, polylined steel drums; tank trucks, tank cars; aluminum; stainless steel; various plastics. For strengths greater than 98%, aluminum.

### Physical and Chemical Characteristics

**Physical State** (20°C, 1 atm): Liquid.
**Solubility** (Water): Miscible in all proportions.
**Molecular Weight:** 60.1
**Vapour Pressure:** 11.4 mm Hg (20°C); 20 mm Hg (30°C).
**Boiling Point:** 118°C (glacial); 103°C (85% solution).

**Floatability** (Water): Sinks and mixes.
**Odour:** Strong, sharp, vinegar odour (0.21 to 1.0 ppm, odour threshold).
**Flash Point:** 43°C (o.c.); 40°C (c.c.).
**Vapour Density:** 2.1 (glacial); 2.0 (85%).
**Specific Gravity:** 1.05 (20°C) (glacial); 1.1 (85%).

**Colour:** Colourless.
**Explosive Limits:** 5.4 to 16%.
**Melting Point:** 16.6°C (glacial); -15°C (85%).

## HAZARD DATA

### Human Health

**Symptoms:** <u>Contact:</u> solutions are very corrosive and can cause severe burns to tissue. <u>Inhalation:</u> concentrated vapour can cause coughing, chest pain, irritation of nose and throat, nausea. <u>Ingestion:</u> burning sensation, nausea and vomiting, convulsions.
**Toxicology:** Corrosive upon inhalation, ingestion and contact.
TLV® - 10 ppm; 25 mg/m³.   LC₅₀ - Inhalation: mouse = 5 620 ppm/1 h   LD₅₀ - Oral: rat = 3.3 g/kg
Short-term Inhalation Limits - 15 ppm; 37 mg/m³   Delayed Toxicity - No information.
(15 min).

### Fire

**Fire Extinguishing Agents:** Use water spray, alcohol foam, dry chemical or carbon dioxide. Water spray may be used to cool fire-exposed containers, to control vapours, and to protect men effecting shut off.
**Behaviour in Fire:** At high temperature and in absence of O₂, decomposes releasing toxic fumes.
**Ignition Temperature:** 427°C (glacial).   **Burning Rate:** 1.6 mm/min.

### Reactivity

**With Water:** No reaction; soluble.
**With Common Materials:** Can react vigorously with oxidizing materials including: acetaldehyde, chlorosulfonic acid, chromates, chromic acid, hydrogen peroxide, nitric acid, oleum, permanganates, sodium hydroxide and sodium peroxide.
**Stability:** Stable.

### Environment

**Water:** Prevent entry into water intakes and waterways. Harmful to aquatic life in low concentrations. Fish toxicity: 75 ppm/96 h/bluegill/TLm/freshwater; 47 mg/L/124 h/Daphnia magna (water flea)/LC₅₀/freshwater; 251 mg/L/96 h/mosquito fish/TLm/freshwater; BOD: 52 to 62%, 5 days.
**Land-Air:** No information.
**Food Chain Concentration Potential:** None.

## EMERGENCY MEASURES

**Special Hazards**

CORROSIVE.

**Immediate Responses**

Keep non-involved people away from spill site. Issue warning; "CORROSIVE". Call Fire Department. Avoid contact and inhalation. Eliminate all ignition sources. Stay upwind and use water spray to control vapours. Stop or reduce discharge, if this can be done without risk. Contact supplier for guidance. Dike or dam to prevent runoff. Notify environmental authorities.

**Protective Clothing and Equipment**

Respiratory protection: self-contained breathing apparatus. Acid suit - (jacket and pants), rubber or plastic. Boots - high, rubber (pants worn outside boots). Gloves - rubber or plastic.

**Fire and Explosion**

Use water spray, dry chemical, alcohol foam or carbon dioxide. Cool fire-exposed containers with water.

**First Aid**

Move victim out of spill area to fresh air. Call for medical assistance, but start first aid at once. Inhalation: give artificial respiration if breathing has stopped (not mouth-to-mouth method). Give oxygen if breathing is laboured. Contact: remove contaminated clothing. Wash eyes and skin with plenty of warm water for at least 15 minutes. Ingestion: give milk or water to conscious victim. Do not induce vomiting. If vomiting occurs, give more water to further dilute the chemical. Keep patient warm and quiet. If medical assistance is not immediately available, transport victim to hospital, clinic or doctor.

## ENVIRONMENTAL PROTECTION MEASURES

**Response**

**Water**
1. Stop or reduce discharge if safe to do so.
2. Contact manufacturer or supplier for advice.
3. If possible, contain discharge by damming or water diversion.
4. Dredge or vacuum pump to remove contaminants, liquids and contaminated bottom sediments.
5. Notify environmental authorities to discuss disposal and cleanup of contaminated materials.

**Land-Air**
1. Stop or reduce discharge if safe to do so.
2. Contact manufacturer or supplier for advice.
3. Contain spill by diking with earth or other barrier.
4. Remove material with pumps or vacuum equipment and place in appropriate containers.
5. Recover undamaged containers.
6. Neutralize contaminated area with lime.
7. Notify environmental authorities to discuss disposal and cleanup of contaminated materials.

**Disposal**
1. Contact manufacturer or supplier for advice on disposal.
2. Contact environmental authorities for advice on disposal.
3. Incinerate (approval of environmental authorities required).

ACETIC ACID   $CH_3COOH$

# ACETIC ANHYDRIDE   $CH_3CO \cdot O \cdot COCH_3$

## IDENTIFICATION

UN No. 1715

### Common Synonyms
ETHANOIC ANHYDRIDE
ACETIC OXIDE
ACETYL OXIDE
ACETIC ACID, ANHYDRIDE

### Observable Characteristics
Clear, colourless, liquid. Strong pungent vinegar-like odour.

### Manufacturers
Celanese Canada Limited, Edmonton, Alberta.

### Transportation and Storage Information
**Shipping State:** Liquid.
**Classification:** Corrosive liquid.
**Inert Atmosphere:** No requirement.
**Venting:** Pressure-vacuum.
**Pump Type:** Centrifugal; stainless steel.

**Label(s):** Black and white label - CORROSIVE; Class 8, Group II.
**Storage Temperature:** Ambient.
**Hose Type:** Polyethylene, polypropylene, PVC.

**Grades or Purity:** Pure, 99% min. Technical, 75 to 98.5%.
**Containers and Materials:** Drums, tank cars, tank trucks; aluminum, stainless steel.

### Physical and Chemical Characteristics
**Physical State (20°C, 1 atm):** Liquid.
**Solubility (Water):** Reacts to form acetic acid.
**Molecular Weight:** 102.1
**Vapour Pressure:** 3.5 mm Hg (20°C); 7 mm Hg (30°C); 10 mm Hg (36°C).
**Boiling Point:** 140°C.

**Floatability (Water):** Sinks and reacts, producing acetic acid.
**Odour:** Strong, pungent, vinegar-like odour, (0.14 to 0.36 ppm, odour threshold).
**Flash Point:** 58°C (o.c.), 49°C (c.c.).
**Vapour Density:** 3.5
**Specific Gravity:** 1.08 (20°C).

**Colour:** Colourless.
**Explosive Limits:** 2.7 to 10.3%.
**Melting Point:** -73.1°C.

## HAZARD DATA

### Human Health
**Symptoms:** Contact: acetic anhydride is very corrosive and can cause severe burns to tissue. Inhalation: concentrated vapour can cause coughing, chest pain, irritation of nose and throat, nausea and vomiting. Ingestion: burning sensation, nausea, vomiting and convulsions.
**Toxicology:** Corrosive upon inhalation, ingestion and contact.
TLV®- 5 ppm; 20 mg/m³.        $LC_{50}$ - No information.        $LD_{50}$ - Oral: rat = 1.78 g/kg
Short-term Inhalation Limits - No information.        $LC_{Lo}$ - Inhalation: rat = 1 000 ppm/4 h
                                                      Delayed Toxicity - No information.

### Fire
**Fire Extinguishing Agents:** Use water spray, dry chemical and alcohol foam or carbon dioxide. Water spray may be used to cool fire-exposed containers, to control vapours, and to protect men effecting shut off.
**Behaviour in Fire:** No information.
**Ignition Temperature:** 390°C.        **Burning Rate:** 3.3 mm/min.

### Reactivity
**With Water:** Reacts; forming acetic acid. Reaction is exothermic and may progress rapidly, causing a violent reaction after 15-30 minutes. Reacts violently if strong acid present.
**With Common Materials:** Reacts violently with alkalis, chlorosulfonic acid, chromic anhydride, hydrochloric acid, nitric acid, oleum, hydrogen peroxide, hydrofluoric acid, permanganates, sodium hydroxide, glycerol and sulfuric acid.
**Stability:** Stable.

### Environment
**Water:** Prevent entry into water intakes and waterways. Harmful to aquatic life in low concentrations. Fish toxicity: 75 ppm/96 h/bluegill/TLm/freshwater; 100 to 300 ppm/4 h/shrimp/$LC_{50}$/saltwater; BOD: 53%, 1 to 5 days.
**Land-Air:** No information.
**Food Chain Concentration Potential:** None.

## EMERGENCY MEASURES

**Special Hazards**

CORROSIVE.

**Immediate Responses**

Keep non-involved people away from spill site. Issue warning: "CORROSIVE". Call Fire Department. Eliminate all ignition sources. Avoid contact and inhalation. Stay upwind and use water spray to control vapours. Stop or reduce discharge, if this can be done without risk. Contact supplier for guidance. Dike or dam to prevent runoff. Notify environmental authorities.

**Protective Clothing and Equipment**

Respiratory protection - self contained breathing apparatus. Acid suit - (jacket and pants) rubber or plastic. Boots - high, rubber (pants worn outside boots). Gloves - rubber or plastic.

**Fire and Explosion**

Fight fire from safe distance. Use water spray, dry chemical, alcohol foam or carbon dioxide. Cool fire-exposed containers with water.

**First Aid**

Move victim out of spill site to fresh air. Call for medical assistance, but start first aid at once. Inhalation: give artificial respiration if breathing has stopped (not mouth-to-mouth method); give oxygen if breathing is laboured. Contact: remove contaminated clothing, wash eyes and skin with plenty of warm water for at least 15 minutes. Ingestion: give milk or water to conscious victim to drink. Do not induce vomiting. If vomiting occurs, give more water to further dilute the chemical. If medical assistance is not immediately available, transport victim to hospital, clinic or doctor. Keep patient warm and quiet.

## ENVIRONMENTAL PROTECTION MEASURES

**Response**

**Water**

1. Stop or reduce discharge if safe to do so.
2. Contact manufacturer or supplier for advice.
3. If possible, contain discharge by damming or water diversion.
4. Dredge or vacuum pump to remove contaminants, liquids and contaminated bottom sediments.
5. Notify environmental authorities to discuss disposal and cleanup of contaminated materials.

**Land-Air**

1. Stop or reduce discharge if safe to do so.
2. Contact manufacturer or supplier for advice.
3. Contain spill by diking with earth or other barrier.
4. Remove material with pumps or vacuum equipment and place in appropriate containers.
5. Recover undamaged containers.
6. Neutralize contaminated area with lime.
7. Notify environmental authorities to discuss disposal and cleanup of contaminated materials.

**Disposal**

1. Contact manufacturer or supplier for advice on disposal.
2. Contact environmental authorities for advice on disposal.
3. Incinerate (approval of environmental authorities required).

ACETIC ANHYDRIDE  $CH_3CO \cdot O \cdot COCH_3$

# ACETONE   $CH_3COCH_3$

## IDENTIFICATION

UN No. 1090

### Common Synonyms
DIMETHYLKETONE
2-PROPANONE
METHYL KETONE

### Observable Characteristics
Clear, colourless, liquid. Sweet, fragrant odour.

### Manufacturers
Gulf Canada Limited, Montreal East, Quebec.
Shell Canada Limited, Montreal, Quebec.

### Transportation and Storage Information
**Shipping State:** Liquid.
**Classification:** Flammable liquid.
**Inert Atmosphere:** No requirement.
**Venting:** Through flame arrester or pressure-vacuum
**Pump Type:** Centrifugal, gear.

**Label(s):** Red label - FLAMMABLE LIQUID; Class 3.1, Group II.
**Storage Temperature:** Ambient.
**Hose Type:** Polyethylene, Hypalon, natural rubber, butyl.

**Grade or Purity:** Technical and reagent - 99.5% + 0.5% water.
**Containers and Materials:** Drums, tank cars, tank trucks. Steel, stainless steel and aluminum.

### Physical and Chemical Characteristics
**Physical State (20°C, 1 atm):** Liquid.
**Solubility (Water):** Soluble in all proportions.
**Molecular Weight:** 58.1
**Vapour Pressure:** 89 mm Hg (5°C); 182 mm Hg (20°C); 270 mm Hg (30°C).
**Boiling Point:** 56.2°C.

**Floatability (Water):** Floats and mixes.
**Odour:** Sweet, fragrant (0.46 to 140 ppm, odour threshold).
**Flash Point:** -15°C (o.c.); -18°C (c.c.).
**Vapour Density:** 2.0
**Specific Gravity:** 0.79 (20°C).

**Colour:** Colourless.
**Explosive Limits:** 2.2% to 13%.
**Melting Point:** -94.3°C.

## HAZARD DATA

### Human Health
**Symptoms: Contact:** skin - slight reddening and dryness; eyes - redness and irritation. **Inhalation:** coughing, headache, nausea, dizziness and narcosis. **Ingestion:** sore throat, headache, dizziness, narcosis and coma.
**Toxicology:** Only slightly toxic upon contact, inhalation and ingestion.
TLV® - 750 ppm; 1 780 mg/m³.       $LC_{50}$ - No information.       $LD_{50}$ - Oral : rat = 9.75 g/kg
Short-term Inhalation Limits - 1 000 ppm;       $LC_{Lo}$ - Inhalation: rat = 64 000 ppm/4 h
2 375 mg/m³ (15 min).       Delayed Toxicity - No information.

### Fire
**Fire Extinguishing Agents:** Use dry chemical, carbon dioxide or alcohol foam. Water in straight hose stream should not be used as it will scatter and spread fire. Water may be used to cool fire-exposed containers, control vapours and protect men effecting shut off.
**Behaviour in Fire:** Extremely flammable. Flash back may occur along vapour trail.       **Burning Rate:** 3.9 mm/min.
**Ignition Temperature:** 465°C.

### Reactivity
**With Water:** No reaction; soluble.
**With Common Materials:** Reacts violently with chromic oxide, chloroform, hydrogen peroxide (nitric and sulfuric acids), (nitric and acetic acids).
**Stability:** Stable.

### Environment
**Water:** Prevent entry into water intakes and waterways. Harmful to aquatic life. Fish toxicity: 14 250 ppm/24 h/sunfish/killed/tap water; 11 493 to 11 727 ppm/120 h/Nitzchia linearis (alga)/$LC_{50}$/freshwater; 10 mg/L/36 h/TLm/Daphnia magna/freshwater; BOD: 38 to 81%, 5 days.
**Land-Air:** No information.
**Food Chain Concentration Potential:** None.

## EMERGENCY MEASURES

### Special Hazards
FLAMMABLE.

### Immediate Responses
Keep non-involved people away from spill site. Issue warning: "FLAMMABLE". CALL FIRE DEPARTMENT. Eliminate all ignition sources. Notify manufacturer. Avoid contact or inhalation. Stop or reduce discharge, if safe to do so. Contain spill by diking. In fire, stay upwind and use water spray to control vapours. Notify environmental authorities.

### Protective Clothing and Equipment
In fires and confined spaces: Respiratory protection - self-contained breathing apparatus. Goggles - (mono), tight fitting (or face shield). Gloves - rubber. Boots - high, rubber (pants worn outside boots). Coveralls - of impervious material.

### Fire and Explosion
Use dry chemical, carbon dioxide or alcohol foam. Do not use water in straight hose stream, as it may scatter and spread the fire. Water may be used to cool fire-exposed containers. Flash back may occur along vapour trail.

### First Aid
Move victim out of spill site to fresh air. Call for medical assistance, but start first aid at once. Inhalation: give artificial respiration if breathing has stopped; give oxygen if breathing is laboured. Contact: skin - remove contaminated clothing and wash affected areas; eyes - flush and irrigate thoroughly water. Ingestion: give plenty of warm water to conscious victim to drink and induce vomiting. If medical assistance is not immediately available, transport to hospital, doctor or clinic.

## ENVIRONMENTAL PROTECTION MEASURES

### Response

**Water**
1. Stop or reduce discharge if safe to do so.
2. Contact manufacturer or supplier for advice.
3. If possible, contain discharge by damming or water diversion.
4. Dredge or vacuum pump to remove contaminants, liquids and contaminated bottom sediments.
5. Notify environmental authorities to discuss disposal and cleanup of contaminated materials.

**Land-Air**
1. Stop or reduce discharge if safe to do so.
2. Contact manufacturer or supplier for advice.
3. Contain spill by diking with earth or other barrier.
4. Remove material with pumps or vacuum equipment and place in appropriate containers.
5. Recover undamaged containers.
6. Absorb residual liquid on natural or synthetic sorbents.
7. Notify environmental authorities to discuss disposal and cleanup of contaminated materials.

### Disposal
1. Contact manufacturer or supplier for advice on disposal.
2. Contact environmental authorities for advice on disposal.
3. Incinerate (approval of environmental authorities required).

ACETONE    $CH_3COCH_3$

# ACETONITRILE  $CH_3CN$

## IDENTIFICATION

UN No. 1638

### Common Synonyms
METHYL CYANIDE
CYANOMETHANE
ETHANENITRILE
ETHYL NITRILE

### Observable Characteristics
Clear, colourless, liquid. Sweet, ethereal odour.

### Manufacturers
Caledon Laboratories,
Georgetown, Ont.
U.S. Manufacturer:
Eastman Chemical Products, Inc.,
Kingston, TN.

### Transportation and Storage Information
**Shipping State:** Liquid.
**Classification:** Flammable liquid; poison.
**Inert Atmosphere:** No requirement.
**Venting:** Pressure-vacuum.
**Pump Type:** No information.

**Label(s):** Red label - FLAMMABLE LIQUID; Class 3.2, Group II.
White label - POISON; Class 6.1, Group II.
**Storage Temperature:** Ambient.
**Hose Type:** No information.

**Grades or Purity:** Technical.
**Containers and Materials:** Drums, tank cars, tank trucks.

### Physical and Chemical Characteristics
**Physical State** (20°C, 1 atm): Liquid.
**Solubility (Water):** Miscible in all proportions.
**Molecular Weight:** 41.1
**Vapour Pressure:** 74 mm Hg (20°C); 115 mm Hg (30°C).
**Boiling Point:** 81.6°C.

**Floatability (Water):** Floats and mixes.
**Odour:** Sweet, etheral (39.8 ppm, odour threshold).
**Flash Point:** 5.6°C (o.c.); 12.8°C (c.c.).
**Vapour Density:** 1.4
**Specific Gravity:** 0.79 (20°C).

**Colour:** Colourless.
**Explosive Limits:** 4.4% to 16%.
**Melting Point:** -44 to -47°C.

## HAZARD DATA

### Human Health
**Symptoms: Contact:** skin - may be absorbed yielding symptoms similar to inhalation; eyes - redness. **Inhalation:** nose and throat irritation, headache, dizziness, laboured breathing. **Ingestion:** headache, dizziness, delirium, convulsions, paralysis and death due to central nervous system depression.
**Toxicology:** Toxic by contact, ingestion and inhalation.
$TLV^®$ - (skin) 40 ppm; 70 $mg/m^3$.   $LC_{50}$ - No information.   $LD_{50}$ - Oral: rat = 3.8 g/kg
Short-term Inhalation Limits - (skin) 60 ppm;   $LC_{Lo}$ - Inhalation : rat = 8 000 ppm/4 h
105 $mg/m^3$ (15 min).   Delayed Toxicity - No information.

### Fire
**Fire Extinguishing Agents:** Use dry chemical, alcohol foam or carbon dioxide. Can react with hot water or steam to produce toxic cyanide gas, therefore, water should not be employed. Water spray may be used, however, to cool fire-exposed containers and knock vapour down.
**Behaviour in Fire:** Flash back may occur along vapour trail.
**Ignition Temperature:** 524°C.   **Burning Rate:** 2.7 mm/min.

### Reactivity
**With Water:** No reaction with cold water; soluble. Can react with hot water or steam to produce toxic cyanide gas.
**With Common Materials:** Reacts with acids to produce cyanide gas. Can react violently with sulfuric acid, oleum, chlorosulfonic acid, and perchlorates.
**Stability:** Stable.

### Environment
**Water:** Prevent entry into water intakes and waterways. Harmful to aquatic life. Fish toxicity: 1 150 ppm/24 h/fathead minnow/TLm/hard water; 1 850 mg/L/96 h/bluegill/TLm/freshwater; BOD: 17%, 5 days.
**Land-Air:** No information.
**Food Chain Concentration Potential:** No information.

## EMERGENCY MEASURES

### Special Hazards
FLAMMABLE; POISON. Poisonous gases produced.

### Immediate Responses
Keep non-involved people away from spill site. Issue warning: "FLAMMABLE; POISON". CALL FIRE DEPARTMENT. Eliminate all ignition sources. Call supplier for guidance. Avoid contact and inhalation. Stop or reduce discharge, if this can be done without risk. Dike to prevent runoff. Notify environmental authorities.

### Protective Clothing and Equipment
Respiratory protection - self-contained breathing apparatus. Protective clothing - suit, coveralls; rubber. Gloves - rubber. Boots - rubber.

### Fire and Explosion
Use dry chemical, alcohol foam or carbon dioxide. Can react with hot water or steam to produce toxic cyanide gases, therefore, water should not be used. Flash back may occur along vapour trail.

### First Aid
Move victim out of spill area to fresh air. Call for medical assistance, but start first aid at once. Inhalation: if breathing has stopped give artificial respiration. If breathing is laboured give oxygen. Contact: remove contaminated clothing. Wash eyes and affected skin with plenty of water for at least 15 minutes. Ingestion: give water or milk to conscious victim to drink. Keep warm and quiet. If medical assistance is not immediately available, transport victim to hospital, doctor or clinic.

## ENVIRONMENTAL PROTECTION MEASURES

### Response

**Water**
1. Stop or reduce discharge if safe to do so.
2. Contact manufacturer or supplier for advice.
3. If possible, contain discharge by damming or water diversion.
4. Notify environmental authorities to discuss disposal and cleanup of contaminated materials.

**Land-Air**
1. Stop or reduce discharge if safe to do so.
2. Contact manufacturer or supplier for advice.
3. Contain spill by diking with earth or other barrier.
4. Remove material with pumps or vacuum equipment and place in appropriate containers.
5. Recover undamaged containers.
6. Absorb residual liquid on natural or synthetic sorbents.
7. Remove contaminated soil for disposal.
8. Notify environmental authorities to discuss disposal and cleanup of contaminated materials.

### Disposal
1. Contact manufacturer or supplier for advice on disposal.
2. Contact environmental authorities for advice on disposal.

ACETONITRILE  $CH_3CN$

# ACETYLENE $C_2H_2$

## IDENTIFICATION

UN No. 1001

### Common Synonyms
ETHYNE
ETHINE

### Observable Characteristics
Pure acetylene is a colourless, odourless gas. Typical, commercial acetylene has a garlic-like odour, due to impurities.

### Manufacturers
Canadian Liquidair Ltd., Montreal, Que., London, Ont., Winnipeg, Man., Regina, Sask., Union Carbide Canada Ltd., Oakville, Ontario; Montreal, Que. Gulf Oil Canada Ltd., Shawinigan, Que., Varennes, Que.

### Transportation and Storage Information

**Shipping State:** Gas or acetylene dissolved in acetone.
**Classification:** Flammable gas.
**Inert Atmosphere:** Not required.
**Venting:** Closed.
**Pump Type:** Not pertinent.

**Label(s):** Red label - FLAMMABLE GAS; Class 2.1.
**Storage Temperature:** Ambient.
**Hose Type:** Special - high pressure.

**Grades or Purity:** Commercial (dissolved in acetone), Technical 98% acetylene <0.05% by volume of phosphine or hydrogen sulfide.
**Containers and Materials:** Cylinders; steel.

### Physical and Chemical Characteristics

**Physical State** (20°C, 1 atm): Gas.
**Solubility** (Water): 0.11g/100mL (0°C).
**Molecular Weight:** 26.0
**Vapour Pressure:** 33 600 mm Hg (21°C).
**Boiling Point:** -84°C.

**Floatability** (Water): Floats (liquefied).
**Odour:** Garlic-like (commercial product); odourless (pure).
**Flash Point:** -17.8°C (c.c.)
**Vapour Density:** 0.91
**Specific Gravity:** 0.62 (-80°C, liquid).

**Colour:** Colourless.
**Explosive Limits:** 2.5 to 82%.
**Melting Point:** -81.8°C.

## HAZARD DATA

### Human Health
**Symptoms:** Inhalation: asphyxiant, 10% in air - no symptoms; 25%, reversible narcosis; 40%, collapse.
**Toxicology:** Relatively nontoxic but can cause asphyxiation by exclusion of oxygen.
TLV® - No information.  LC50 - No information.  LD50 - No information.
Short-term Inhalation Limits - No information.  Delayed Toxicity - None.
OSHA Standard - CL 2 500 ppm (based on pure acetylene).

### Fire
**Fire Extinguishing Agents:** Allow escaping acetylene to burn if flow cannot be shut off safely. Carbon dioxide, water spray or dry chemical may be used. Cool cylinders exposed to heat and flame with water spray.
**Behaviour in Fire:** The heat from a small fire can melt the cylinder fusible plug, resulting in increased flow and much larger fire. Exposed containers may rupture.
**Ignition Temperature:** 305°C.   **Burning Rate:** No information.

### Reactivity
**With Water:** No reaction.
**With Common Materials:** Under certain conditions, acetylene forms explosive compounds with copper, silver, mercury and their compounds. May react violently with fluorine, chlorine, bromine, iodine, and nitric acid. Can react vigorously with oxidizing materials.
**Stability:** Stable.

### Environment
**Water:** Prevent entry into water intakes and waterways. Harmful to aquatic life. Fish toxicity: 1 000 cc/L/1 h/sunfish/not killed/freshwater; 200 mg/L/33 h/rainbow trout/TLm/freshwater.
**Land-Air:** No information.
**Food Chain Concentration Potential:** No information.

## EMERGENCY MEASURES

**Special Hazards**

FLAMMABLE. Reactive with many compounds.

**Immediate Responses**

Keep non-involved people away and upwind from spill site. Issue warning: "FLAMMABLE". CALL FIRE DEPARTMENT. Notify manufacturer or supplier. Extinguish all sources of ignition. Allow cylinder to burn unless shutoff can be effected. Call environmental authorities.

**Protective Clothing and Equipment**

In fires or enclosed spaces: Respiratory protection - use self-contained breathing apparatus.

**Fire and Explosion**

Allow to burn unless shutoff can be effected. Fight fires from safe distance. Acetylene forms explosive mixture in confined space. Danger of container rupture in fire. Use water to cool fire-exposed, adjacent containers. Use carbon dioxide, dry chemical to extinguish.

**First Aid**

Move victim out of spill site to fresh air. Call for medical assistance, but start first aid at once. Inhalation: if breathing has stopped, give artificial respiration; if laboured, give oxygen. If medical assistance is not immediately available, transport victim to hospital, doctor or clinic.

## ENVIRONMENTAL PROTECTION MEASURES

**Response**

**Water**
1. Stop or reduce discharge if safe to do so.
2. Contact manufacturer or supplier for advice.
3. Notify environmental authorities to discuss disposal and cleanup of contaminated materials.

**Land-Air**
1. Stop or reduce discharge if safe to do so.
2. Contact manufacturer or supplier for advice.
3. Notify environmental authorities to discuss disposal and cleanup of contaminated materials.

**Disposal**
1. Contact manufacturer or supplier for advice on disposal.
2. Contact environmental authorities for advice on disposal.

ACETYLENE   $C_2H_2$

# ACROLEIN   CH₂CHCHO

## IDENTIFICATION

UN No. 1092 inhibited
2607 dimer

### Common Synonyms
ACRYLALDEHYDE
2-PROPENAL
ALLYL ALDEHYDE
ACRALDEHYDE

### Observable Characteristics
Colourless to yellow liquid. Pungent, acrid odour.

### Manufacturers
No Canadian manufacturer.
Selected U.S. manufacturers:
Shell Chemical Company, Houston, Texas.
Union Carbide Corp., New York, NY.

### Transportation and Storage Information

**Shipping State:** Liquid.
**Classification:** Flammable liquid.
**Inert Atmosphere:** No requirement.
**Venting:** Pressure-vacuum.
**Pump Type:** (material for seals) Garlock 233, 929; Duraplastic 22, Durametallic 10, Raybestos (Raybestos-Manhattan Inc.)

**Label(s):** Red label - FLAMMABLE LIQUID; dimer, Class 3.3, Group II; inhibited, Class 3.1, Group I.
**Storage Temperature:** Ambient.
**Hose Type:** Natural rubber, Thiokol 262T, Silastic 180, Garlock 7021, butyl. DO NOT USE NEOPRENE.

**Grades or Purity:** Industrial (92+%) inhibited with hydroquinone.
**Containers and Materials:** Drums, tank cars.

### Physical and Chemical Characteristics

**Physical State (20°C, 1 atm):** Liquid.
**Solubility (Water):** 20.6 g/100 mL (20°C).
**Molecular Weight:** 56.1
**Vapour Pressure:** 220 mm Hg (20°C); 330 mm Hg (30°C).
**Boiling Point:** 52.5°C.

**Floatability (Water):** Floats.
**Odour:** Pungent, acrid (0.05 to 1.5 ppm, odour threshold).
**Flash Point:** -17°C (o.c.); -26°C (c.c.).
**Vapour Density:** 1.94
**Specific Gravity:** 0.84 (20°C).

**Colour:** Colourless to yellow.
**Explosive Limits:** 2.8 to 31%.
**Melting Point:** -87°C.

## HAZARD DATA

### Human Health

**Symptoms: Contact:** skin-smarting, burning sensation, inflammation and blistering; eyes - burning sensation, watering, inflammation, loss of sight. **Inhalation:** irritation of mucous membranes, difficulty breathing, headache, nausea, muscular weakness and chemical bronchitis. **Ingestion:** sore throat, intense thirst, abdominal cramps, nausea and vomiting, difficulty breathing, convulsions.
**Toxicology:** Toxic by contact, ingestion and inhalation.
TLV® 0.1 ppm; 0.25 mg/m³.                                    LD₅₀ - Oral: rat = 0.046 g/kg
Short-term Inhalation Limits - 0.3 ppm; 0.8 mg/m³ (15 min).
LC₅₀ - No information.
LC_Lo - Inhalation: human = 153 ppm (10 min).
Delayed Toxicity - No information.

### Fire

**Fire Extinguishing Agents:** Use dry chemical, alcohol foam or carbon dioxide. Water may be ineffective but may be used to keep fire-exposed containers cool and knock down vapours.
**Behaviour in Fire:** In fire, polymerization may occur which may cause violent container rupture. Flash back may occur along vapour trail.
**Ignition Temperature:** 234°C.                                  **Burning Rate:** 3.8 mm/min.

### Reactivity

**With Water:** No reaction.
**With Common Materials:** Extremely violent polymerization reaction results when in contact with alkaline materials, such as caustics, ammonia or amines, or with strong acids, such as sulfuric or nitric acids. Reacts with oxidizing materials.
**Stability:** Unstable if not inhibited (0.1 to 0.25% hydroquinone).

### Environment

**Water:** Prevent entry into water intakes and waterways. Harmful to aquatic life. Fish toxicity: 0.5 ppm/24 h/Elodea sp. (plant)/kills all cells/freshwater; 0.08 pm/24 h/salmon/TLm/freshwater; 0.065 ppm/24 h/rainbow trout/TLm/freshwater; BOD: Not available.
**Land-Air:** No information.
**Food Chain Concentration Potential:** No information.

## EMERGENCY MEASURES

### Special Hazards
FLAMMABLE. POISONOUS. CAN POLYMERIZE VIOLENTLY.

### Immediate Responses
Keep non-involved people away from spill site. Issue warning: "FLAMMABLE; POISON". CALL FIRE DEPARTMENT. Eliminate all ignition sources. Call manufacturer or supplier for advice. Dike to prevent runoff. Avoid inhalation or contact. Shutoff discharge if this can be done without risk. Notify environmental authorities.

### Protective Clothing and Equipment
Respiratory protection - self-contained breathing apparatus. Gloves - rubber. Outer protective clothing as required - acid suit, coveralls. Boots - high, rubber (pants worn outside boots).

### Fire and Explosion
Dry chemical, alcohol foam or carbon dioxide may be used. Water may be ineffective but may be used to keep fire-exposed containers cool and knock down vapours. Flash back may occur along vapour trail.

### First Aid
Move victim out of spill area to fresh air. Call for medical assistance, but start first aid at once. Inhalation: if not breathing, apply artificial respiration (not mouth-to-mouth method). If breathing is laboured, give oxygen. Contact: wash eyes and affected skin with plenty of water for at least 15 minutes. Remove contaminated clothing at same time. Ingestion: if conscious, victim should drink large amounts of water. If medical assistance is not immediately available, transport victim to hospital, doctor or clinic.

## ENVIRONMENTAL PROTECTION MEASURES

### Response

**Water**
1. Stop or reduce discharge if safe to do so.
2. Contact manufacturer or supplier for advice.
3. If possible, contain discharge by damming or water diversion.
4. Dredge or vacuum pump to remove contaminants, liquids and contaminated bottom sediments.
5. Notify environmental authorities to discuss disposal and cleanup of contaminated materials.

**Land-Air**
1. Stop or reduce discharge if safe to do so.
2. Contact manufacturer or supplier for advice.
3. Contain spill by diking with earth or other barrier.
4. Remove material with pumps or vacuum equipment and place in appropriate containers.
5. Recover undamaged containers.
6. Absorb residual liquid on natural or synthetic sorbents.
7. Remove contaminated soil for disposal.
8. Notify environmental authorities to discuss disposal and cleanup of contaminated materials.

### Disposal
1. Contact manufacturer or supplier for advice on disposal.
2. Contact environmental authorities for advice on disposal.
3. Incinerate (approval of environmental authorities required).

ACROLEIN   $CH_2CHCHO$

# ACRYLONITRILE   CH$_2$CHCN

## IDENTIFICATION                        UN No. 1093

### Common Synonyms
VINYL CYANIDE
CYANOETHYLENE
2-PROPENENITRILE

### Observable Characteristics
Colourless to light yellow liquid.
Irritating peach-like odour.

### Manufacturers
No Canadian manufacturers.
**Canadian Supplier:**
Du Pont Canada, Maitland, Ont.

Monsanto Canada, Sarnia, Ont.
Borg-Warner, Cobourg, Ont.
Polysar, Sarnia, Ont.

Originating from:
E.I. Du Pont de Nemours & Co. Inc.
Wilmington, DE
Monsanto Co., St. Louis, MO
Vistron Corp.,
Cleveland, OH

### Transportation and Storage Information

**Shipping State:** Liquid.
**Classification:** Flammable; poison.
**Inert Atmosphere:** No requirement.
**Venting:** Pressure-vacuum.
**Pump Type:** No information.

**Label(s):** Red label - FLAMMABLE LIQUID; Class 3.2, Group I. White label - POISON; Class 6.1, Group I.
**Storage Temperature:** Ambient.
**Hose Type:** No information.

**Grades or Purity:** Technical, 98 to 100%.
**Containers and Materials:** Barrels, tank cars, tank trucks; steel, stainless steel (no copper or copper alloys).

### Physical and Chemical Characteristics

**Physical State** (20°C, 1 atm): Liquid.
**Solubility** (Water): 7 g/100 mL (20°C).
**Molecular Weight:** 53.1
**Vapour Pressure:** 100 mm Hg (22.8°C); 137 mm Hg (30°C).
**Boiling Point:** 77.4°C.

**Floatability** (Water): Floats.
**Odour:** Irritating, pungent, peach-like, (1.7 to 23 ppm, odour threshold).
**Flash Point:** 0°C (o.c.); -4°C (c.c.).
**Vapour Density:** 1.8
**Specific Gravity:** 0.80 (20°C).

**Colour:** Colourless to light yellow.
**Explosive Limits:** 3 to 17%.
**Melting Point:** -83.6°C.

## HAZARD DATA

### Human Health

**Symptoms: Ingestion:** sore throat, difficulty breathing, nausea, vomiting, abdominal pain. **Inhalation or skin contact:** nausea, vomiting, weakness, shortness of breath, collapse and loss of consciousness.
**Toxicology:** Extremely toxic by inhalation, ingestion or skin contact.
TLV® (inhalation) 2 ppm; 4.5 mg/m$^3$.   LC$_{50}$ - Inhalation: guinea pig = 576 ppm/4 h    LD$_{50}$ - Oral: rat = 0.082 g/kg
Short-term Inhalation Limits - No information.    Delayed Toxicity - Potential carcinogen.

### Fire

**Fire Extinguishing Agents:** Use alcohol foam, carbon dioxide or dry chemical. Water spray may be ineffective, but may be used to keep fire-exposed containers cool and knock down vapours.
**Behaviour in Fire:** Flashback may occur along vapour trail. Thermal decomposition may produce oxides of nitrogen, carbon monoxide and carbon dioxide. Exposed containers may rupture.
**Ignition Temperature:** 481°C.    **Burning Rate:** No information.

### Reactivity

**With Water:** No reaction.
**With Common Materials:** Can react violently with strong acids, alkalis, and bromine. Can react vigorously with oxidizing agents.
**Stability:** Stable.

### Environment

**Water:** Prevent entry into water intakes and waterways. Harmful to aquatic life. Fish toxicity: 11.8 mg/L/96 h/bluegill/TLm/soft water; 25 mg/L/24 h/sunfish/TLm/soft water; 14.3 mg/L/96 h/fathead minnnow/TLm/freshwater; BOD: 70%, 5 days.
**Land-Air:** No information.
**Food Chain Concentration Potential:** No information.

## EMERGENCY MEASURES

### Special Hazards
FLAMMABLE. POISON.

### Immediate Responses
Keep non-involved people away from spill site. Issue warning: "FLAMMABLE; POISON". CALL FIRE DEPARTMENT. Eliminate all ignition sources. Call manufacturer for advice. Avoid contact and inhalation. Stay upwind and use water spray to control vapour. Dike spill area. Stop or reduce discharge, if this can be done without risk. Notify environmental authorities.

### Protective Clothing and Equipment
Respiratory protection - self-contained breathing apparatus and full protective clothing.

### Fire and Explosion
Use dry chemical, alcohol foam or carbon dioxide to extinguish. Water may be ineffective, but may be used to cool fire-exposed containers. Burning acrylonitrile may release cyanide gases. Flash back may occur along vapour trail.

### First Aid
Move victim out of spill area to fresh air. Call for medical assistance, but start first aid at once. Contact: flush eyes and skin with plenty of water and remove contaminated clothing. Inhalation: if breathing has stopped, give artificial respiration (not mouth-to-mouth method), if laboured give oxygen. Ingestion: give plenty of water to conscious victim to drink, and induce vomiting. If medical assistance is not immediately available, transport victim to hospital, doctor or clinic.

## ENVIRONMENTAL PROTECTION MEASURES

### Response

**Water**
1. Stop or reduce discharge if safe to do so.
2. Contact manufacturer or supplier for advice.
3. If possible, contain discharge by damming or water diversion.
4. Notify environmental authorities to discuss disposal and cleanup of contaminated materials.

**Land-Air**
1. Stop or reduce discharge if safe to do so.
2. Contact manufacturer or supplier for advice.
3. Contain spill by diking with earth or other barrier.
4. Remove material with pumps or vacuum equipment and place in appropriate containers.
5. Recover undamaged containers.
6. Absorb residual liquid on material or synthetic sorbents.
7. Remove contaminated soil for disposal.
8. Notify environmental authorities to discuss disposal and cleanup of contaminated materials.

### Disposal
1. Contact manufacturer or supplier for advice on disposal.
2. Contact environmental authorities for advice on disposal.

ACRYLONITRILE   $CH_2CHCN$

# ADIPIC ACID   COOH $(CH_2)_4$ COOH

## IDENTIFICATION

| Common Synonyms | Observable Characteristics | UN No. Na 9077 |
|---|---|---|
| ADIPIC ACID<br>1,4-BUTANEDICARBOXYLIC ACID<br>HEXANEDIOIC ACID | White crystals or powder. Odourless. | **Manufacturers**<br>Du Pont Canada,<br>Maitland, Ontario. |

### Transportation and Storage Information

**Shipping State:** Solid.
**Classification:** None.
**Inert Atmosphere:** No requirement.
**Venting:** Open.

**Label(s):** Not regulated. Class 9.2, Group III.
**Storage Temperature:** Ambient.

**Grades or Purity:** Commercial, 99.8%.
**Containers and Materials:** Bottles, tins, multi-wall paper bags and drums.

### Physical and Chemical Characteristics

**Physical State** Solid.
**Solubility** (Water): 1.5 g/100 mL (15°C); 1.6 g/100 mL (20°C).
**Molecular Weight:** 146.1
**Vapour Pressure:** 0.28 mm Hg (47°C).
**Boiling Point:** 337°C (decomposes).

**Floatability** (Water): Sinks.
**Odour:** Odourless.
**Flash Point:** 196°C (c.c.).
**Vapour Density:** 5.0
**Specific Gravity:** 1.36 (20°C).

**Colour:** White.
**Explosive Limits:** 10 to 15 mg/L (dust).
**Melting Point:** 151 to 153°C.

## HAZARD DATA

### Human Health

**Symptoms:** Contact (liquid): eyes - irritation; skin - pronounced drying effect. Inhalation (dust, vapour): irritation of mucous membranes causing sneezing and coughing. Ingestion: sore throat and abdominal pain.
**Toxicology:** Relatively non-toxic.
**TLV®** No information.
**Short-term Inhalation Limits** - No information.

$LC_{50}$ - No information.
**Delayed Toxicity** - No information.

$LD_{50}$ - Oral: mouse = 1.9 g/kg

### Fire

**Fire Extinguishing Agents:** Use foam, water fog, dry chemical or carbon dioxide.
**Behaviour in Fire:** Melts and may decompose to produce volatile acidic vapours of valeric acid and other substances. Dust may form explosive mixture with air.
**Ignition Temperature:** 420°C.

**Burning Rate:** No information.

### Reactivity

**With Water:** No reaction.
**With Common Materials:** Can react with oxidizing materials.
**Stability:** Stable.

### Environment

**Water:** Prevent entry into water intakes and waterways. Harmful to aquatic life. Fish toxicity: <330 ppm/24 h/bluegill/TLm/freshwater; BOD: 36 to 60%, 5 days.
**Land-Air:** No information.
**Food Chain Concentration Potential:** No information.

## EMERGENCY MEASURES

**Special Hazards**

**Immediate Responses**

Keep non-involved people away from spill site. Call Fire Department. Avoid contact or inhalation of dust or fumes. Dike to prevent runoff. Call manufacturer for advice. Contact environmental authorities.

**Protective Clothing and Equipment**

In fires or confined spaces, Respiratory protection - self-contained breathing apparatus. Otherwise, protective clothing as required.

**Fire and Explosion**

Use foam, water fog, dry chemical or carbon dioxide to extinguish. Dust may form explosive mixture with air.

**First Aid**

Move victim out of spill area to fresh air. Call for medical assistance, but start first aid at once. Contact: eyes and skin - flush with large amounts of water and remove contaminated clothing. Ingestion: give water to conscious victim to drink. If medical assistance is not immediately available, transport victim to hospital, doctor or clinic.

## ENVIRONMENTAL PROTECTION MEASURES

**Response**

**Water**
1. Stop or reduce discharge if safe to do so.
2. Contact manufacturer or supplier for advice.
3. If possible, contain discharge by damming or water diversion.
4. Dredge or vacuum pump to remove contaminants, liquids and contaminated bottom sediments.
5. Notify environmental authorities to discuss disposal and cleanup of contaminated materials.

**Land-Air**
1. Stop or reduce discharge if safe to do so.
2. Contact manufacturer or supplier for advice.
3. Dike to prevent runoff from rainwater or water application.
4. Remove material by manual or mechanical means.
5. Recover undamaged containers.
6. Notify environmental authorities to discuss disposal and cleanup of contaminated materials.

**Disposal**
1. Contact manufacturer or supplier for advice on disposal.
2. Contact environmental authorities for advice on disposal.
3. Incinerate (approval of environmental authorities required).

ADIPIC ACID   COOH $(CH_2)_4$ COOH

# ALUMINUM ALKYL COMPOUNDS

## IDENTIFICATION

UN No. 1101 diethylaluminum chloride
1925 EASC
2220 solution       ) aluminum alkyl halides
2221 pure           )

### Common Synonyms

TMA $(CH_3)_3Al$ Trimethylaluminum;
MASC $(CH_3)_3Al_2Cl_3$ Methylaluminum Sesquichloride;
TEA $(C_2H_5)_3Al$ Triethylaluminum*;
DEAC $(C_2H_5)_2AlCl$ Diethylaluminum Chloride;
EASC $(C_2H_5)_3Al_2Cl_3$ Ethylaluminum Sesquichloride;
EADC $C_2H_5AlCl_2$ Ethylaluminum Dichloride;
DEAI $(C_2H_5)_2AlI$ Diethylaluminum Iodide;
TIBA $(iC_4H_9)_3Al$ Triisobutylaluminum*;
DIBAH $(iC_4H_9)_2AlH$ Diisobutylaluminum Hydride;
DIBAC $(iC_4H_9)_2AlCl$ Diisobutylaluminum Chloride;
DEZ $(C_2H_5)_2Zn$ Diethylzinc.
(*most common).

### Observable Characteristics

Typically clear liquids, which fume or burn on contact with air.

### Manufacturers

Ethyl Corporation of Canada Ltd.
Corunna, Ontario
Toronto, Ontario
U.S. manufacturer:
Ethyl Corporation,
Houston, Texas.

### Transportation and Storage Information

**Shipping State:** Liquid.
**Classification:** Flammable liquid.
**Inert Atmosphere:** Nitrogen.
**Venting:** No venting, special container.
**Pump Type:** Pumped by nitrogen gas pressure.

**Label(s):** Red, black and white label - FLAMMABLE LIQUID; Class 4.2.
**Storage Temperature:** Ambient.
**Hose Type:** Stainless steel, flexible stainless steel, KEL-F, graphite - asbestos.

**Grades or Purity:** 88 to 94%.
**Containers and Materials:** Special containers. Alkyltainer (1 gallon); Dual valve (3, 5, 26 gallon); portable tank (250, 430, 1 980 gallon); tank trailer (6 200 to 7 200 gallon) and tank car (11 100 gallon).

### Physical and Chemical Characteristics

**Physical State** (20°C, 1 atm): Liquid.
**Solubility:** (Water): Reacts violently.
**Molecular Weight:** TMA, 72.1; TEA, 114.2; TIBA, 198.3
**Vapour Pressure:** 0.76 mm Hg (60°C) (TMA).
**Boiling Point:** TMA, 126°C; TEA, 187°C; TIBA, 114°C (decomposes).

**Floatability** (Water): Reacts violently and ignites.
**Odour:** No information.
**Flash Point:** TEA, -53°C; TIBA, 0°C; TMA <-18°C.
**Vapour Density:** TMA, 2.0
**Specific Gravity:** TMA, 0.6; TEA, 0.83; TIBA, 0.78 (25°C).

**Colour:** Colourless.
**Explosive Limits:** No information.
**Melting Point:** TMA, 15.4°C; MASC, 22.8°C; TEA, -46°C; DEAC, -74°C; DASC, -21°C; EADC, 32°C; DEAI, -43°C; TIBA, 1°C; DIBA, -80°C; DIBAC, -39°C; DEZ, -28°C.

## HAZARD DATA

### Human Health

**Symptoms:** Contact: skin - produce severe chemical and thermal burns. Inhalation: sore throat, laboured breathing, may cause metal fume fever.
**Toxicology:** Relatively toxic by contact and inhalation.
TLV® 2 mg/m³.          LC$_{50}$ - No information.          LD$_{50}$ - No information.
Short-term Inhalation Limits - No information.          Delayed Toxicity - No information.

### Fire

**Fire Extinguishing Agents:** Stop or reduce discharge. Use dry chemical extinguishers. DO NOT USE WATER.
**Behaviour in Fire:** PYROPHORIC - IGNITES ON CONTACT WITH AIR. REACTS EXPLOSIVELY WITH WATER. Residue may reignite when disturbed.
**Ignition Temperature:** TMA, 190°C; TIBA, <4°C.     **Burning Rate:** No information.

### Reactivity

**With Water:** Reacts explosively with water.
**With Common Materials:** Reacts with air, acids, alcohols, oxidizers, carbon tetrachloride and other halogenated hydrocarbons.
**Stability:** Stable, when not exposed to water or air.

### Environment

**Water:** Prevent entry into water intakes and waterways. Aquatic toxicity rating: 100 to 1 000 mg/L/TLm/freshwater (for TEA).
**Land-Air:** No information.
**Food Chain Concentration Potential:** No information.

## EMERGENCY MEASURES

**Special Hazards**

PYROPHORIC - IGNITES ON CONTACT WITH AIR, REACTS EXPLOSIVELY WITH WATER.

**Immediate Responses**

Keep non-involved people away from spill site. Issue warning: "FLAMMABLE". CALL FIRE DEPARTMENT (but warn them about the pyrophoric properties - NO WATER). Contact manufacturer for assistance. Do not attempt to control discharge without advice or assistance from the manufacturer, or other knowledgeable person. Notify environmental authorities.

**Protective Clothing and Equipment**

Respiratory protection - self-contained breathing apparatus. Gloves - loose fitting, rubber, asbestos or leather. Other protective equipment - full aluminized proximity type suit when transferring material from one container to another on large scale.

**Fire and Explosion**

PYROPHORIC - ignites on contact with air. Reacts explosively with water. Extinguishing agent - dry powder extinguisher. DO NOT USE WATER. Avoid - water, air, acids, alcohols, oxidizers, carbon tetrachloride and other halogenated compounds. Also avoid high temperature and shock.

**First Aid**

Move victim out of spill area to fresh air. Call for medical assistance, but start first aid at once. Contact: eyes - irrigate gently with water for at least 15 minutes; skin - wash with plenty of water for at least 15 minutes and remove contaminated clothing. Inhalation: remove from exposure to smoke and fumes of combustion. Ingestion: wash out mouth promptly. Get medical attention at once. If medical attention is not immediately available, transport victim to hospital, doctor or clinic.

## ENVIRONMENTAL PROTECTION MEASURES

**Response**

Water
1. Stop or reduce discharge if safe to do so.
2. Contact manufacturer or supplier for advice.
3. Notify environmental authorities to discuss disposal and cleanup of contaminated materials.

Land-Air
1. Stop or reduce discharge if safe to do so.
2. Contact manufacturer or supplier for advice.
3. Notify environmental authorities to discuss disposal and cleanup of contaminated materials.

**Disposal**

1. Contact manufacturer or supplier for advice on disposal.
2. Contact environmental authorities for advice on disposal.

ALUMINUM ALKYL COMPOUNDS

# ALUMINUM CHLORIDE  AlCl₃

## IDENTIFICATION

UN No. 1726 anhydrous
2581 solution

### Common Synonyms
ALUMINUM CHLORIDE, ANHYDROUS TRICHLOROALUMINUM

### Observable Characteristics
Orange to yellow, through grey to white crystalline solid or powder. Hydrogen chloride odour.

### Manufacturers
Welland Chemical of Canada Ltd., Sarnia, Ontario.

### Transportation and Storage Information
**Shipping State:** Solid.
**Classification:** Corrosive.
**Inert Atmosphere:** No information.
**Venting:** No information.

**Label(s):** Black and white label - CORROSIVE; Class 8, Group III.
**Storage Temperature:** Ambient.

**Grades or Purity:** Technical, 98.5% AlCl₃ min.
**Containers and Materials:** Drums, flow bins, trucks.

### Physical and Chemical Characteristics
**Physical State** (20°C, 1 atm): Solid.
**Solubility** (Water): 69.9 g/100 mL (15°C) (reacts).
**Molecular Weight:** 133.5
**Vapour Pressure:** No information.
**Boiling Point:** 183°C.

**Floatability** (Water): Sinks and mixes with violent exothermic reaction, evolving hydrogen chloride gas.
**Odour:** Hydrogen chloride odour. (1 to 10 ppm, HCl odour threshold).
**Flash Point:** Not flammable.
**Vapour Density:** No information.
**Specific Gravity:** 2.4 (25°C).

**Colour:** Orange to yellow, through grey to white.
**Explosive Limits:** Not flammable.
**Melting Point:** 194°C.

## HAZARD DATA

### Human Health
**Symptoms:** Contact: eyes - redness, pain, blurred vision; skin - redness and pain. Inhalation: sore throat, coughing, laboured breathing. Ingestion: burning sensation, stomach cramps, dizziness, diarrhea, shock, convulsions, coma.
**Toxicology:** Moderately toxic, corrosive by ingestion and contact.
TLV® 5 ppm; 7 mg/m³ (as HCl).    LC₅₀ - No information.    LD₅₀ - Oral: rat = 3.7 g/kg
Short-term Inhalation Limits - No information.    Delayed Toxicity - No information.

### Fire
**Fire Extinguishing Agents:** Not flammable or combustible. Most fire extinguishing agents may be used, but a violent reaction will result (hydrogen chloride evolution) if streams of water are directed on aluminum chloride.
**Behaviour in Fire:** May react in fires to produce chloride gases.
**Ignition Temperature:** Not flammable.    **Burning Rate:** Not flammable.

### Reactivity
**With Water:** Reacts exothermically evolving hydrogen chloride gas.
**With Common Materials:** Old containers can explode upon opening. Can react violently with allyl chloride, ethylene, ethylene oxide, sodium and potassium.
**Stability:** Stable.

### Environment
**Water:** Prevent entry into water intakes and waterways.    **BOD:** No information.
**Land-Air:** No information.
**Food Chain Concentration Potential:** No information.

## EMERGENCY MEASURES

**Special Hazards**

CORROSIVE. REACTS WITH WATER.

**Immediate Responses**

Keep non-involved people away from spill site. Issue warning: "CORROSIVE". If Fire Department is called, warn them against use of water. Avoid contact or inhalation. If dry, shovel or scoop up as much as possible. If wet or on fire, contain spill by diking with dry earth or other available material. Stop or reduce discharge, if this can be done without risk. Notify manufacturer. Notify environmental authorities.

**Protective Clothing and Equipment**

Respiratory protection - use self-contained breathing apparatus. Gloves - rubber. Boots - rubber, safety (pants worn outside boots). Outerwear - as required; with aquatic spills, acid suits may be needed.

**Fire and Explosion**

Most fire extinguishing agents (except water) may be used on fires involving aluminum chloride.
Aluminum chloride is not explosive or flammable, but a violent exothermic reaction results (hydrogen chloride evolution) on contact with water.

**First Aid**

Move victim out of spill area to fresh air. Call for medical assistance, but start first aid at once. Contact: eyes - irrigate immediately with plenty of water for at least 15 minutes; skin: flush with plenty of water for at least 15 minutes. Contaminated clothing should be removed at once. (When no moisture is present, minor skin irritation only may occur.) Ingestion: is most unlikely, but if it occurred, the immediate local reaction would cause severe burns, not dependent on any toxic action. If medical assistance is not immediately available, transport victim to hospital, doctor or clinic.

## ENVIRONMENTAL PROTECTION MEASURES

**Response**

**Water**
1. Stop or reduce discharge if safe to do so.
2. Contact manufacturer or supplier for advice.
3. Notify environmental authorities to discuss disposal and cleanup of contaminated materials.

**Land-Air**
1. Stop or reduce discharge if safe to do so.
2. Contact manufacturer or supplier for advice.
3. Contain spill by diking with earth or other barrier.
4. Remove material by manual or mechanical means.
5. Recover undamaged containers.
6. Notify environmental authorities to discuss disposal and cleanup of contaminated materials.

**Disposal**
1. Contact manufacturer or supplier for advice on disposal.
2. Contact environmental authorities for advice on disposal.

ALUMINUM CHLORIDE  $AlCl_3$

# ALUMINUM HYDROXIDE    $Al_2O_3 \cdot 3H_2O$ or $Al(OH)_3$

## IDENTIFICATION

### Common Synonyms
ALUMINUM HYDROXIDE GEL
ALUMINUM HYDRATE
HYDRATED ALUMINA
ALUMINATRIHYDRATE

### Observable Characteristics
White, crystals, powder or granules.
Odourless.

### Manufacturers
Aluminum Co. of Canada Ltd., Montreal, Quebec.
Exolon Co. of Canada Ltd., Thorold, Ontario.
General Abrasive (Can.) Ltd., Niagara Falls, Ont.
Norton Co., Niagara Falls, Ontario.
Welland Chemical Ltd., Mississauga, Ontario.

### Transportation and Storage Information
**Shipping State:** Solid.
**Classification:** None.
**Inert Atmosphere:** No requirement.
**Venting:** Open.

**Label(s):** Not regulated.
**Storage Temperature:** Ambient.

**Grades or Purity:** Technical.
**Containers and Materials:** Bags, drums, bulk lots.

### Physical and Chemical Characteristics
**Physical State** (20°C, 1 atm): Solid.
**Solubility** (Water): 0.01 g/100 mL (18°C).
**Molecular Weight:** 78
**Vapour Pressure:** No information.
**Boiling Point:** Loses $H_2O$ at 300°C.

**Floatability** (Water): Sinks.
**Odour:** Odourless.
**Flash Point:** Not flammable.
**Vapour Density:** No information.
**Specific Gravity:** 2.4 (20°C).

**Colour:** White.
**Explosive Limits:** Not flammable.
**Melting Point:** Loses $H_2O$ at 300°C.

## HAZARD DATA

### Human Health
**Symptoms:** Inhalation (dust): aluminum containing dust or fumes can cause pulmonary reactions and long-term effects.
**Toxicology:** Relatively nontoxic material.
$TLV^®$ - 10 $mg/m^3$ (dust).
Short-term Inhalation Limits - (dust) 20 $mg/m^3$ (15 min).

$LC_{50}$ - No information.
Delayed Toxicity - Chronic exposure to dust can cause respiratory problems.

$LD_{50}$ - No information.
$LD_{Lo}$ - Intraperitoneal: rat = 0.15 g/kg

### Fire
**Fire Extinguishing Agents:** Most fire extinguishing agents may be used on fires involving aluminum hydroxide.
**Behaviour in Fire:** Not flammable or combustible.
**Ignition Temperature:** Not combustible.
**Burning Rate:** Not combustible.

### Reactivity
**With Water:** No reaction.
**With Common Materials:** No known reactions.
**Stability:** Stable.

### Environment
**Water:** Prevent entry into water intakes and waterways. Toxic to aquatic life; but of low solubility.
**Land-Air:** No information.
**Food Chain Concentration Potential:** No information.

## EMERGENCY MEASURES

### Special Hazards

### Immediate Responses
Keep non-involved people away from spill site. Notify manufacturer. Notify environmental authorities.

### Protective Clothing and Equipment
Dust respirator may be required in confined areas. Protective clothing as required.

### Fire and Explosion
Not flammable or combustible. Most fire extinguishing agents may be used on fires involving aluminum hydroxide.

### First Aid
Move victim out of spill area. Contact: skin and eyes - flush with water and remove contaminated clothing. <u>Ingestion</u>: induce vomiting by giving salt water to drink. Transport to medical aid if unsure of victim's condition upon inhalation.

## ENVIRONMENTAL PROTECTION MEASURES

### Response

**Water**
1. Stop or reduce discharge if safe to do so.
2. Contact manufacturer or supplier for advice.
3. If possible, contain discharge by damming or water diversion.
4. Dredge or vacuum pump to remove contaminants, liquids, and contaminated bottom sediments.
5. Notify environmental authorities to discuss disposal and cleanup of contaminated materials.

**Land-Air**
1. Stop or reduce discharge if safe to do so.
2. Contact manufacturer or supplier for advice.
3. Contain spill by diking with earth or other barrier.
4. Remove material by manual or mechanical means.
5. Recover undamaged containers.
6. Notify environmental authorities to discuss disposal and cleanup of contaminated materials.

### Disposal
1. Contact manufacturer or supplier for advice on disposal.
2. Contact environmental authorities for advice on disposal.

**ALUMINUM HYDROXIDE** $Al_2O_3 \cdot 3H_2O$ or $Al(OH)_3$

# ALUMINUM SULFATE   $Al_2(SO_4)_3$

## IDENTIFICATION

UN No. 1750 anhydrous

### Common Synonyms

ALUM
CAKE ALUM

FILTER ALUM
PAPER MAKER'S ALUM
PATENT ALUM
PICKLE ALUM
PEARL ALUM
(tetradecahydrate also known as Liquid Alum), $Al_2(SO_4)_3 \cdot 14H_2O$).

### Observable Characteristics

Granular white powder. Odourless.

### Manufacturers

Allied Chemical,
Valleyfield, Que., Dalhousie, N.B., Thorhold, Ont., Barnett, B.C.
Inland Chemicals Canada Ltd.,
Fort Saskatchewan, Alta.
Alcan Smelters and Chemicals Ltd.,
Jonquière, Que.

### Transportation and Storage Information

**Shipping State:** Solid and liquid (aqueous solution).
**Classification:** Corrosive.
**Inert Atmosphere:** No requirement.
**Venting:** Open.

**Label(s):** Black and white label - CORROSIVE; Class 8, Group I.
**Storage Temperature:** Ambient.

**Grades or Purity:** Technical, 45% solution.
**Containers and Materials:** Bags, fibre drums, multi-wall paper sacks, bulk by truck or train. Solution (45% $Al(SO_4)_3 \cdot 14\ H_2O$ in water); drums, tank trucks.

### Physical and Chemical Characteristics

**Physical State** (20°C, 1 atm): Solid.
**Solubility** (Water): 27.8 g/100 mL (25°C); anhydrous; 87 g/100 mL (0°C) hydrate.
**Molecular Weight:** 342 anhydrous; 594 hydrate.
**Vapour Pressure:** No information.
**Boiling Point:** Anhydrous 770°C (decomposes); hydrate solution 101°C.

**Floatability** (Water): Sinks and mixes.
**Odour:** Odourless.
**Flash Point:** Not combustible.
**Vapour Density:** No information.
**Specific Gravity:** 2.7 anhydrous (20°C); 1.7 hydrate (17°C); solution 1.34 (20°C)

**Colour:** White.
**Explosive Limits:** Not combustible.
**Melting Point:** Anhydrous 770°C (decomposes); hydrate solution -16°C.

## HAZARD DATA

### Human Health

**Symptoms:** Contact: skin and eyes - irritated by dust. Ingestion: nausea, vomiting. Inhalation: sore throat, coughing, irritation of nose and throat.
**Toxicology:** Relatively non-toxic.
**TLV®** - No information.
**Short-term Inhalation Limits** - No information.

$LC_{50}$ - No information.
**Delayed Toxicity** - No information.

$LD_{50}$ - Intraperitoneal: mouse = 0.27 g/kg
$LD_{50}$ - Oral: mouse = 6.1 g/kg

### Fire

**Fire Extinguishing Agents:** Most fire extinguishing agents may be used on fires involving aluminum sulfate.
**Behaviour in Fire:** Not flammable or combustible. May evolve toxic $SO_3$ gas at high temperatures (above 760°C).
**Ignition Temperature:** Not combustible.   **Burning Rate:** Not combustible.

### Reactivity

**With Water:** Reacts with water to produce sulfuric acid.
**With Common Materials:** May corrode metals in presence of moisture.
**Stability:** Stable.

### Environment

**Water:** Prevent entry into water intakes and waterways. Fish toxicity: 240 ppm/48 h/mosquito fish/TLm/freshwater; BOD: Not available.
**Land-Air:** No information.
**Food Chain Concentration Potential:** No information.

## EMERGENCY MEASURES

**Special Hazards**
CORROSIVE.

**Immediate Responses**
Keep non-involved people away from spill site. Issue warning: "CORROSIVE". Avoid contact. Stop or reduce discharge, if this can be done without risk. Notify supplier. Notify environmental authorities.

**Protective Clothing and Equipment**
Goggles - (mono) type, tight fitting. Coveralls, gloves and boots - all acid resistant.

**Fire and Explosion**
Not flammable or combustible. Most fire extinguishing agents may be used on fires involving aluminum sulfate.

**First Aid**
Move victim away from spill site to fresh air. Call for medical assistance, but start first aid at once. Contact - skin with water and remove contaminated clothing; eyes - flush with plenty of warm water for at least 15 minutes. Ingestion - give milk or water to conscious victim. Keep victim warm and quiet. If medical assistance is not immediately available, transport victim to hospital, clinic or doctor.

## ENVIRONMENTAL PROTECTION MEASURES

**Response**

**Water**
1. Stop or reduce discharge if safe to do so.
2. Contact manufacturer or supplier for advice.
3. If possible, contain discharge by damming or water diversion.
4. Dredge or vacuum pumps to remove contaminants, liquids, and contaminated bottom sediments.
5. Notify environmental authorities to discuss disposal and cleanup of contaminated materials.

**Land-Air**
1. Stop or reduce discharge if safe to do so.
2. Contact manufacturer or supplier for advice.
3. Contain spill by diking with earth or other barrier.
4. Remove material by manual or mechanical means.
5. Recover undamaged containers.
6. Notify environmental authorities to discuss disposal and cleanup of contaminated materials.

**Disposal**
1. Contact manufacturer or supplier for advice on disposal.
2. Contact environmental authorities for advice on disposal.

ALUMINUM SULFATE  $Al_2(SO_4)_3$

# AMINOCARB $C_{11}H_{16}N_2O_2$

## IDENTIFICATION

**Common Synonyms**
4-DIMETHYLAMINO-M-TOLYL METHYLCARBAMATE
4-DIMETHYLAMINO-3-METHYLPHENOL METHYLCARBAMATE
**Common Trade Names**
MATACIL
(An insecticide often used for spruce bud worm control.)

**Observable Characteristics**
Light brown solid or liquid.

**UN No. 2757**
**Danger Group According to Active Substance**
Group II >60 to 100%
Group III    Solid 6 to 60%
           Liquid 1 to 60%

**Manufacturers**
Chemagro Corp., Kansas City, Mo.

### Transportation and Storage Information

**Shipping State:** Solid or liquid (formulations).
**Classification:** None.
**Inert Atmosphere:** No requirement.
**Venting:** Open.
**Pump Type:** No information.

**Label(s):** Not regulated.
**Storage Temperature:** Ambient.
**Hose Type:** No information.

**Grades or Purity:** Solution of typically 18%.
**Containers and Materials:** Drums; steel.

### Physical and Chemical Characteristics

**Physical State** (20°C, 1 atm): Solid (technical).
**Solubility (Water):** Slightly soluble (technical).
**Molecular Weight:** 208.3
**Vapour Pressure:** No information.
**Boiling Point:** No information.

**Floatability (Water):** Sinks.
**Odour:** No information.
**Flash Point:** Not flammable.
**Vapour Density:** No information.
**Specific Gravity:** >1.0 (technical).

**Colour:** Light brown.
**Explosive Limits:** Not flammable.
**Melting Point:** 93 to 94°C.

## HAZARD DATA

### Human Health

**Symptoms:** Inhalation, Ingestion or Contact (absorbed by skin): dizziness nausea, salivation, tearing, abdominal cramps, vomiting, sweating, slow pulse, impairment of visual acuity.
**Toxicology:** Highly toxic by all routes.
**TLV®:** No information.
**Short-term Inhalation Limits** - No information.

**LC$_{50}$** - No information.
**Delayed Toxicity** - No information.

**LD$_{50}$** - Oral: rat = 0.030 g/kg
**LD$_{50}$** - Intraperitoneal: mouse = 0.007 g/kg

### Fire

**Fire Extinguishing Agents:** Use carbon dioxide, foam or dry chemical.
**Behaviour in Fire:** Releases toxic fumes in fires.
**Ignition Temperature:** No information.

**Burning Rate:** No information.

### Reactivity

**With Water:** No reaction.
**With Common Materials:** No information.
**Stability:** Stable.

### Environment

**Water:** Prevent entry into water intakes and waterways. Fish toxicity: 0.13 - 0.16 ppm/rainbow trout/LC$_{50}$/96h/freshwater.
**Land-Air: LD$_{50}$:** Oral: Wild bird = 0.05 g/kg
**Food Chain Concentration Potential:** Unknown.

## EMERGENCY MEASURES

### Special Hazards
POISON.

### Immediate Responses
Keep non-involved people away from spill site. Stop or reduce discharge if safe to do so. Notify manufacturer or supplier. Dike to contain material or water runoff. Notify environmental authorities.

### Protective Clothing and Equipment
In fires or confined spaces - Respiratory Protection - self-contained breathing apparatus and totally encapsulated suit. Otherwise, approved pesticide respirator and impervious outer clothing.

### Fire and Explosion
Use carbon dioxide, foam or dry chemical to extinguish. Releases toxic fumes in fires.

### First Aid
Move victim out of spill site to fresh air. Call for medical assistance, but start first aid at once. Inhalation: if breathing has stopped, give artificial respiration (not mouth-to-mouth method); if laboured, give oxygen. Contact: skin - remove contaminated clothing and flush affected areas with plenty of water; eyes - irrigate with plenty of water. Ingestion: give water to conscious victim to drink and induce vomiting; in the case of petroleum distillates, do not induce vomiting for fear of aspiration and chemical pneumonia. If medical assistance is not immediately available, transport victim to hospital, doctor, or clinic.

## ENVIRONMENTAL PROTECTION MEASURES

### Response

**Water**
1. Stop or reduce discharge if safe to do so.
2. Contact manufacturer or supplier for advice.

Floats    Sinks or mixes

3. If possible contain discharge by damming or water diversions.
4. If floating, skim and remove.

3. If possible contain discharge by damming or water diversions.
4. Dredge or vacuum pump to remove contaminants, liquids and contaminated bottom sediments.

5. Notify environmental authorities to discuss disposal and cleanup of contaminated materials.

**Land-Air**
1. Stop or reduce discharge if safe to do so.
2. Contact manufacturer or supplier for advice.
3. Contain spill by diking with earth or other barrier.
4. If liquid, remove material with pumps or vacuum equipment and place in appropriate containers.
5. If solid, remove material by manual or mechanical means.
6. Recover undamaged containers.
7. Absorb residual liquid on natural or synthetic sorbents.
8. Remove contaminated soil for disposal.
9. Notify environmental authorities to discuss cleanup and disposal of contaminated materials.

### Disposal
1. Contact manufacturer or supplier for advice on disposal.
2. Contact environmental authorities for advice on disposal.

### Available Formulations
**Technical Grade:** Purity: 99%
Properties: combustible, sinks in water.

**SN Solution**

Purity:
- typically 18%, remainder is organic, polar solvent

Properties:
- combustible, possibly miscible in water

AMINOCARB    $C_{11}H_{16}N_2O_2$

# AMMONIA (anhydrous) NH$_3$

## IDENTIFICATION

UN No. 1005

### Common Synonyms

### Observable Characteristics
Colourless gas. Sharp, irritating, ammonia odour.

### Manufacturers
Canadian Industries Limited, Courtright, Ontario. Cominco Limited, Carseland, Alta. Cyanamid of Canada Limited, Niagara Falls, Ont. Canadian Fertilizers Ltd, Medicine Hat, Alta.

### Transportation and Storage Information
**Shipping State:** Liquid (compressed gas).
**Classification:** Poisonous gas.
**Inert Atmosphere:** No requirement.
**Venting:** Relief valve (250 psi) on all pressure containers except cylinders.
**Pump Type:** No information.

**Label(s):** White label - POISONOUS GAS; Class 2.3.
**Storage Temperature:** Ambient.
**Hose Type:** No information.

**Grades or Purity:** Commercial 99.5%; refrigeration 99.95%.
**Containers and Materials:** Cylinders, tank cars, tank trucks; most metals except tin, copper, aluminum or lead.

### Physical and Chemical Characteristics
**Physical State** (20°C, 1 atm): Gas.
**Solubility** (Water): 89.9 g/100 mL (0°C); 36.4 g/100 mL (20°C); 7.4 g/100 mL (100°C).
**Molecular Weight:** 17.0
**Vapour Pressure:** 6657 mm Hg (21.1°C); 4 800 mm Hg (15.5°C).
**Boiling Point:** -33.4°C.

**Floatability** (Water): Floats, boils and mixes.
**Odour:** Sharp, irritating, pungent. (47 ppm, odour threshold).
**Flash Point:** Not flammable.
**Vapour Density:** 0.60 (0°C).
**Specific Gravity:** 0.68 (-33.4°C) (liquid); 0.6 (25°C).

**Colour:** Colourless.
**Explosive Limits:** 16 to 25% by volume.
**Freezing Point:** -77.7°C.

## HAZARD DATA

### Human Health
**Symptoms:** Contact: eyes - extremely irritating, can cause loss of vision; skin - burning and blistering, liquid ammonia causes frostbite. Inhalation: suffocation, difficulty breathing, coughing.
**Toxicology:** Toxic by inhalation.
TLV®- 25 ppm; 18 mg/m$^3$.
Short-term Inhalation Limits - 35 ppm, 27 mg/m$^3$ (15 min).

LC$_{50}$ - Inhalation: rat = 4 837 ppm/1 h
Delayed Toxicity - No information.

LD$_{50}$ - Oral: rat = 0.35 g/kg

### Fire
**Fire Extinguishing Agents:** Stop discharge before attempting to extinguish fire. Water may be used to cool fire-exposed containers and knock down vapours.
**Behaviour in Fire:** Low fire hazard. Slight danger of explosion (gas) in enclosed space at 16 to 25% by volume with air, under certain conditions.
**Ignition Temperature:** 650°C.
**Burning Rate:** 1 mm/min.

### Reactivity
**With Water:** No reaction; soluble.
**With Common Materials:** Forms explosive compounds in contact with silver, acetaldehyde, acrolein, halogens, chlorates, chlorites, chromates, ethylene oxide, nitric acid, nitrogen tetroxide and silver chloride. Corrodes copper, tin, lead, brass, bronze and galvanized steel.
**Stability:** Stable.

### Environment
**Water:** Prevent entry into water intakes and waterways. Harmful to aquatic life in very low concentrations. Aquatic toxicity rating = 1 to 10 ppm/96 h/TLm/freshwater; 0.66 ppm/48 h/Daphnia magna/LC$_{50}$/freshwater; 8.2 ppm/96 h/fathead minnow/TLm/freshwater; 0.5 ppm/24 h/rainbow trout/TLm/freshwater.
**Land-Air:** Widely used for direct application to soil as source of nitrogen.
**Food Chain Concentration Potential:** None.

## EMERGENCY MEASURES

**Special Hazards**

POISONOUS GAS.

**Immediate Responses**

Keep non-involved people away from spill site. Issue warning: "POISONOUS GAS". Call Fire Department. Evacuate area in case of large leaks or tank rupture. Stay upwind. Stop or reduce discharge if this can be done without risk. Contact manufacturer for guidance. Use water spray to knock vapours down. Notify environmental authorities.

**Protective Clothing and Equipment**

Respiratory protection - self-contained breathing apparatus and gastight suit.

**Fire and Explosion**

Stop discharge before attempting to extinguish fire. Water may be used to cool fire-exposed containers and knock down vapours.

**First Aid**

Move victim out of spill area to fresh air. Call for medical assistance, but start first aid at once. Inhalation: if breathing has stopped, start artificial respiration immediately; if breathing is laboured, oxygen may be given. Contact: eyes - hold lids open and irrigate constantly for at least 30 min with warm water; skin: remove clothing immediately and flood affected area for at least 15 min with warm water. If medical aid is not immediately available, transport victim to hospital doctor, or clinic.

## ENVIRONMENTAL PROTECTION MEASURES

**Response**

**Water**
1. Stop or reduce discharge if safe to do so.
2. Contact manufacturer or supplier for advice.
3. If possible, contain discharge by damming or water diversion.
4. Notify environmental authorities to discuss disposal and cleanup of contaminated materials.

**Land-Air**
1. Stop or reduce discharge if safe to do so.
2. Contact manufacturer or supplier for advice.
3. Dike to prevent runoff from rainwater or water application.
4. Notify environmental authorities to discuss disposal and cleanup of contaminated materials.

**Disposal**
1. Contact manufacturer or supplier for advice on disposal.
2. Contact environmental authorities for advice on disposal.

AMMONIA (anhydrous) $NH_3$

# AMMONIUM CHLORIDE  $NH_4Cl$

## IDENTIFICATION

UN No. Na 9085

### Common Synonyms
SAL AMMONIAC
AMMONIUM MURIATE

### Observable Characteristics
White crystals. Odourless.

### Manufacturers
No Canadian manufacturers.
Canadian supplier:
Allied Chemical
Mississauga, Ontario.
Chemical Industries Ltd.,
Montreal, Quebec.

Originating from:
Allied Chemical, USA

### Transportation and Storage Information
**Shipping State:** Solid.
**Classification:** Not regulated.
**Inert Atmosphere:** No requirement.
**Venting:** Open.
**Label(s):** None. Class 9.2, Group II
**Storage Temperature:** Ambient.
**Grades or Purity:** Technical 99+%.
**Containers and Materials:** Multiwall paper bags and barrels.

### Physical and Chemical Characteristics
**Physical State** (20°C, 1 atm): Solid.
**Solubility** (Water): 30 g/100 mL (0°C); 37.2 g/100 mL (20°C); 77 g/100 mL (100°C).
**Molecular Weight:** 53.5
**Vapour Pressure:** 1 mm Hg (160.4°C).
**Boiling Point:** 520°C.
**Floatability** (Water): Sinks and mixes.
**Odour:** Odourless.
**Flash Point:** Not flammable.
**Vapour Density:** 1.9
**Specific Gravity:** 1.53 (20°C).
**Colour:** White.
**Explosive Limits:** Not flammable.
**Melting Point:** Sublimes at 340°C.

## HAZARD DATA

### Human Health
**Symptoms: Contact:** eyes and skin – irritating. **Inhalation:** irritates nose, throat and respiratory passages. **Ingestion:** nausea and vomiting.
**Toxicology:** Moderately toxic upon ingestion and inhalation.
TLV® – 10 mg/m³ (fume).   $LC_{50}$ – No information.   $LD_{50}$ – Oral: rat = 1.65 g/kg
Short-term Inhalation Limits – 20 mg/m³ (fume)   Delayed Toxicity – No information.
(15 min).

### Fire
**Fire Extinguishing Agents:** Most fire extinguishing agents may be used on fires involving ammonium chloride. Water spray may be used to knock down fumes.
**Behaviour in Fire:** Not flammable or combustible; at very high temperatures evolves ammonia and hydrogen chloride.
**Ignition Temperature:** Not combustible.   **Burning Rate:** Not combustible.

### Reactivity
**With Water:** No reaction; soluble.
**With Common Materials:** Can react violently with ammonium nitrate and potassium chlorate.
**Stability:** Stable.

### Environment
**Water:** Prevent entry into water intakes and waterways. **Fish toxicity:** 6 ppm/96 h/sunfish/TLm/freshwater.
**Land-Air:** No information.
**Food Chain Concentration Potential:** No information.

## EMERGENCY MEASURES

**Special Hazards**

**Immediate Responses**

Keep non-involved people away from spill site. Dike to prevent runoff from water application. Call manufacturer or supplier. Call environmental authorities.

**Protective Clothing and Equipment**

In fires; Respiratory protection - self-contained breathing apparatus. Goggles. Gloves - rubber. Boots, coveralls - rubber.

**Fire and Explosion**

Not flammable or combustible. Most fire extinguishing agents may be used on fires involving ammonium chloride. Water spray may be used to knock down fumes.

**First Aid**

Move victim from spill site to fresh air. Call for medical assistance, but start first aid at once. Contact: skin - remove contaminated clothing and wash skin thoroughly with warm water; eyes - irrigate with plenty of warm water. Ingestion: give milk or water to conscious victim. Keep victim warm and quiet. If medical aid is not immediately available, transport victim to doctor, hospital or clinic.

## ENVIRONMENTAL PROTECTION MEASURES

**Response**

**Water**
1. Stop or reduce discharge if safe to do so.
2. Contact manufacturer or supplier for advice.
3. If possible, contain discharge by damming or water diversion.
4. Dredge or vacuum pump to remove contaminants, liquids and contaminated bottom sediments.
5. Notify environmental authorities to discuss disposal and cleanup of contaminated materials.

**Land-Air**
1. Stop or reduce discharge if safe to do so.
2. Contact manufacturer or supplier for advice.
3. Contain spill by diking with earth or other barrier.
4. Remove material by manual or mechanical means.
5. Recover undamaged containers.
6. Notify environmental authorities to discuss disposal and cleanup of contaminated materials.

**Disposal**
1. Contact manufacturer or supplier for advice on disposal.
2. Contact environmental authorities for advice on disposal.

AMMONIUM CHLORIDE $NH_4Cl$

# AMMONIUM HYDROXIDE  NH₄OH

## IDENTIFICATION

UN No. 2672

### Common Synonyms

AQUA AMMONIA
AMMONIUM HYDRATE
AMMONIA SOLUTION
AMMONIA WATER
AMMONIA LIQUOR

### Observable Characteristics

Colourless liquid. Sharp, irritating, ammonia odour.

### Manufacturers

Canadian Industries Ltd., Courtright, Ont.
Canadian Fertilizers Ltd., Medicine Hat, Alta.
Cominco Ltd., Carseland, Alta.
Cyanamid Canada Ltd., Niagara Falls, Ont.
Simplot Chemical Ltd., Brandon, Man.
Western Co-op Fertilizers Ltd., Calgary, Alta.

### Transportation and Storage Information

**Shipping State:** Liquid.
**Classification:** Corrosive liquid.
**Inert Atmosphere:** No requirement.
**Venting:** Pressure - vacuum.
**Pump Type:** Centrifugal; all iron or stainless steel. No copper alloys, brass or bronze.

**Label(s):** Black and white label - CORROSIVE LIQUID; Class 8, Group II.
**Storage Temperature:** Ambient.
**Hose Type:** PVC, rubber, polyethylene (steel or SS fittings only).

**Grades or Purity:** Grade A: 29.4% $NH_3$; B: 25%; C: 15%; USP: 27 to 29%; CP: 28%.
**Containers and Materials:** Plastic bottles, drums, tank trucks, tank cars (steel, steel polylined).

### Physical and Chemical Characteristics

**Physical State** (20°C, 1 atm): Liquid.
**Solubility** (Water): Soluble in all proportions.
**Molecular Weight:** 35.1 (solute).
**Vapour Pressure:** 10% $NH_3$, 31 mm Hg (0°C); 159 mm Hg (27°C). 20% $NH_3$, 88 mm Hg (0°C); 310 mm Hg (27°C); 30% $NH_3$, 238 mm Hg (0°C); 786 mm Hg (27°C).
**Boiling Point:** 25% $NH_3$, 36°C.

**Floatability** (Water): Floats and mixes.
**Odour:** Sharp, characteristic. (50 ppm, odour threshold).
**Flash Point:** Flammable as $NH_3$.
**Vapour Density:** 0.6
**Specific Gravity:** 0.90 (15.5°C).

**Colour:** Colourless.
**Explosive Limits:** Mixtures of air and $NH_3$ (16% to 25% $NH_3$) by volume may ignite or explode in an enclosed space if sparked or exposed to temperatures exceeding 649°C (1 200°F).
**Melting Point:** 25% $NH_3$, 77°C.

## HAZARD DATA

### Human Health

**Symptoms:** Contact: skin - redness, pain, burns; eyes - pain, watering, burns and damage. Inhalation: (of $NH_3$ vapours) irritation of nose, eyes and throat, difficulty breathing, coughing, chemical bronchitis. Ingestion: burning sensation in mouth, throat and stomach, pain in swallowing, nausea and vomiting of blood, stomach cramps, rapid breathing, diarrhea.
**Toxicology:** Relatively toxic by ingestion and inhalation.
TLV® 25 ppm; 18 mg/m³ (as $NH_3$).      $LC_{50}$ - No information.      $LD_{50}$ - Oral: rat = 0.35 g/kg
Short-term Inhalation Limits - 35 ppm, 27 mg/m³      Delayed Toxicity - No information.
(as $NH_3$) (15 min).

### Fire

**Fire Extinguishing Agents:** Not flammable. Most fire extinguishing agents may be used on fires involving ammonium hydroxide.
**Behaviour in Fire:** Not flammable. Mixtures of ammonia and air in an enclosed space with an ignition source could be explosive. When heated, releases ammonia gas.
**Ignition Temperature:** Not flammable as solution;      **Burning Rate:** Not flammable.
flammable as $NH_3$, 649°C.

### Reactivity

**With Water:** No reaction; soluble.
**With Common Materials:** Copper, tin, zinc and alloys are readily corroded. Reacts with acrolein, acrylic acid, chlorosulfonic acid, dimethyl sulfate, halogens, hydrochloric acid, hydrofluoric acid, nitric acid, sulfuric acid, oleum, propylene oxide and silver nitrate.
**Stability:** Stable.

### Environment

**Water:** Prevent entry into water intakes and waterways. Fish toxicity: 6.25 ppm/24 h/rainbow trout/lethal/freshwater; 20 ppm/96 h/Daphnia magna/static bioassay/TLm/acute/freshwater; Aquatic toxicity rating = 10 to 100 ppm/96 h/TLm/freshwater; BOD: Not available.
**Land-Air:** Ammonium hydroxide is widely used as a source of nitrogen for application to farm land.

## EMERGENCY MEASURES

**Special Hazards**

CORROSIVE.

**Immediate Responses**

Keep non-involved people away from spill site. Issue warning: "CORROSIVE". Call Fire Department (keep upwind). Call manufacturer for guidance. Avoid contact and inhalation of vapours. Stop or reduce discharge, if this can be done without risk. Dike spill to prevent runoff. Use water spray to knock down vapours, and to cool containers exposed to fire. Notify environmental authorities.

**Protective Clothing and Equipment**

Respiratory protection - self-contained breathing apparatus. Acid suit. Gloves - gauntlet type - rubber or plastic coated. Boots - high, rubber (pants worn outside boots).

**Fire and Explosion**

Not flammable, but when heated releases ammonia gas. Most fire extinguishing agents may be used on fires involving ammonium hydroxide. Use water spray to knock down vapours and cool containers.

**First Aid**

Move victim out of spill area to fresh air. Call for medical assistance, but start first aid at once. Inhalation: if breathing has stopped, start artificial respiration at once (not mouth-to-mouth method). If breathing is laboured, give oxygen. Contact: eyes - hold lids open and irrigate constantly for at least 30 minutes; skin - remove clothing immediately and flood affected area with water for at least 15 minutes. Ingestion: if victim is conscious, give large quantity of water to dilute the ammonia. Do not induce vomiting. If medical assistance is not quickly available, transport victim to hospital, doctor or clinic.

## ENVIRONMENTAL PROTECTION MEASURES

**Response**

Water
1. Stop or reduce discharge if safe to do so.
2. Contact manufacturer or supplier for advice.
3. If possible, contain discharge, by damming or water diversion.
4. Dredge or vacuum pump to remove contaminants, liquids and contaminated bottom sediments.
5. Notify environmental authorities to discuss disposal and cleanup of contaminated materials.

Land-Air
1. Stop or reduce discharge if safe to do so.
2. Contact manufacturer or supplier for advice.
3. Dike to prevent runoff from rainwater or water application.
4. Remove material with pumps or vacuum equipment and place in appropriate containers.
5. Recover undamaged containers.
6. Absorb residual liquid on natural or synthetic sorbents.
7. Notify environmental authorities to discuss disposal and cleanup of contaminated materials.

**Disposal**

1. Contact manufacturer or supplier for advice on disposal.
2. Contact environmental authorities for advice on disposal.

AMMONIUM HYDROXIDE $NH_4OH$

# AMMONIUM NITRATE   NH$_4$NO$_3$

## IDENTIFICATION

| Common Synonyms | Observable Characteristics | | |
|---|---|---|---|
| GERMAN SALTPETER<br>NORWAY SALTPETER<br>NITRAM | White to grey or brown.<br>Odourless. | UN No. 1942 <0.2% combustible substances<br>0222 ammonium nitrate<br>0223, 2068, 2069, 2070, 2071 fertilizers | |

**Manufacturers**

Canadian Industries Ltd., Courtright, Ontario; Nobel, Ont., Carseland, Alta., McMasterville, Que. Cominco Limited, Calgary, Alta. Cyanamid of Canada Ltd., Niagara Falls, Ontario.

Esso Chemical Canada, Redwater, Alta. Du Pont Canada, North Bay, Ont.

### Transportation and Storage Information

**Shipping State:** Solid.
**Classification:** Oxidizing material.
**Inert Atmosphere:** No requirement.
**Venting:** Open.

**Label(s):** Yellow label - OXIDIZER; Class 5.1, Group III.
**Storage Temperature:** Ambient.

**Grades or Purity:** Reagent grade. Fertilizer and explosive grades.
**Containers and Materials:** Bags (poly); bulk, trucks, rail cars.

### Physical and Chemical Characteristics

**Physical State** (20°C, 1 atm): Solid.
**Solubility** (Water): 118 g/100 mL (0°C); 192 g/100 mL (20°C).
**Molecular Weight:** 80.1
**Vapour Pressure:** No information.
**Boiling Point:** Decomposes >210°C.

**Floatability** (Water): Sinks and mixes.
**Odour:** Odourless.
**Flash Point:** Detonates under certain circumstances.
**Vapour Density:** No information.
**Specific Gravity:** 1.7 (20°C).

**Colour:** White to grey or brown.
**Explosive Limits:** Detonates under certain circumstances.
**Melting Point:** 170°C.

## HAZARD DATA

### Human Health

**Symptoms:** Contact: skin and eyes - irritation of eyes and mucous membranes. Inhalation: sore throat, coughing, shortness of breath. Ingestion: large amounts cause dizziness, cramps and vomiting.
**Toxicology:** Moderately toxic by contact and ingestion.
**TLV®** - No information.   **LC$_{50}$** - No information.   **LD$_{50}$** - No information.
**Short-term Inhalation Limits** - No information.   **Delayed Toxicity** - No information.

### Fire

**Fire Extinguishing Agents:** Use flooding amounts of water in early stages of fire. Exercise caution in application of water on molten material to stop spread of fire.
**Behaviour in Fire:** In decomposition or burning, generates poisonous NO$_x$ fumes. May explode if heated in container.
**Ignition Temperature:** Detonates under certain   **Burning Rate:** No information.
circumstances.

### Reactivity

**With Water:** No reaction; soluble.
**With Common Materials:** Ammonium nitrate is an oxidizing material and can cause any organic materials to burn. Can react with powdered metals, chlorides, phosphorus, sodium and sulfur.
**Stability:** Stable, within the limits of the foregoing.

### Environment

**Water:** Prevent entry into water intakes and waterways. Harmful to aquatic life. Aquatic toxicity rating = 10 to 100 ppm/96 h/TLm/freshwater; Fish toxicity: 800 µg/L/3.9 h/bluegill/killed/tapwater; 4 545 µg/L/90 h/goldfish/killed/distilled water.
**Land-Air:** Ammonium nitrate is widely used as a fertilizer. Livestock toxicity: 400 ppm (water).
**Food Chain Concentration Potential:** None.

## EMERGENCY MEASURES

### Special Hazards
OXIDIZER. Can detonate under certain circumstances.

### Immediate Responses
Keep non-involved people away from spill site. Issue warning: "OXIDIZER". Call Fire Department. Evacuate hazard area. Fight fires by water flooding. Notify manufacturer. Notify environmental authorities.

### Protective Clothing and Equipment
Use self-contained breathing apparatus and gastight suit, if involved in a fire. Chemical goggles - (tight fitting). Gloves - rubber or plastic. Acid suit - (jacket and pants) or coveralls (if gastight suit not available). Boots - high, rubber (pants worn over boots). When not involved with fire - gloves, boots and coveralls.

### Fire and Explosion
Apply water immediately in as large a volume as possible. Cool any fire-exposed containers with water and continue after fire is out.

### First Aid
Move victim out of spill area to fresh air. Call for medical assistance, but start first aid immediately. Inhalation: if breathing has stopped, give artificial respiration. If breathing is laboured, give oxygen. Contact: eyes - rinse eyes thoroughly with plenty of water; skin - remove contaminated clothing and wash affected areas thoroughly with water. Ingestion: give milk or water to conscious victim. If medical assistance is not immediately available, transport victim to hospital, doctor or clinic.

## ENVIRONMENTAL PROTECTION MEASURES

### Response

**Water**
1. Stop or reduce discharge if safe to do so.
2. Contact manufacturer or supplier for advice.
3. If possible, contain discharge by damming or water diversion.
4. Dredge or vacuum pump to remove contaminants, liquids and contaminated bottom sediments.
5. Notify environmental authorities to discuss disposal and cleanup of contaminated materials.

**Land-Air**
1. Stop or reduce discharge if safe to do so.
2. Contact manufacturer or supplier for advice.
3. Dike to prevent runoff from rainwater or water application.
4. Remove material by manual or mechanical means.
5. Recover undamaged containers.
6. Notify environmental authorities to discuss disposal and cleanup of contaminated materials.

### Disposal
1. Contact manufacturer or supplier for advice on disposal.
2. Contact environmental authorities for advice on disposal.

AMMONIUM NITRATE $NH_4NO_3$

# AMMONIUM PHOSPHATES

## IDENTIFICATION

### Common Synonyms
AMMONIUM PHOSPHATE, monobasic $NH_4H_2PO_4$
AMMONIUM ACID PHOSPHATE
AMMONIUM BIPHOSPHATE
AMMONIUM DIHYDROGEN PHOSPHATE
AMMONIUM PHOSPHATE, primary

AMMONIUM PHOSPHATE, dibasic $(NH_4)_2HPO_4$
AMMONIUM PHOSPHATE, secondary
DIAMMONIUM HYDROGEN PHOSPHATE

### Observable Characteristics
White crystals or powder. Odourless.

### Manufacturers
Esso Chemical Canada Ltd., Redwater, Alta.
Simplot Chemical Ltd., Brandon, Man.
Belledune Fertilizer Ltd., Belledune, NB
Cominco Ltd., Kimberley; Trail, B.C.

### Transportation and Storage Information
**Shipping State:** Solid.
**Classification:** None.
**Inert Atmosphere:** No requirement.
**Venting:** Open.

**Label(s):** None.
**Storage Temperature:** Ambient.

**Grades or Purity:** Technical; fertilizer (dibasic).
**Containers and Materials:** Multiwall paper bags, barrels, bulk lots.

### Physical and Chemical Characteristics
**Physical State** (20°C, 1 atm): Solid.
**Solubility** (Water): monobasic; 22.7 g/100 mL (0°C); 173.2 g/100 mL (100°C). dibasic, 57 g/100 mL (10°C); 106 g/100 mL (70°C).
**Molecular Weight:** monobasic, 115; dibasic, 132.1
**Vapour Pressure:** No information.
**Boiling Point:** monobasic decomposes at >150°C; dibasic decomposes at 155°C.

**Floatability** (Water): Sinks and mixes.
**Odour:** Odourless.
**Flash Point:** Not flammable.
**Vapour Density:** No information.
**Specific Gravity:** dibasic: 1.6 at 20°C; monobasic 1.8 at 20°C.

**Colour:** White.
**Explosive Limits:** Not flammable.
**Melting Point:** Monobasic 190°C; dibasic decomposes at 155°C.

## HAZARD DATA

### Human Health
**Symptoms:** Inhalation: monobasic form causes irritation of mucous membranes; dibasic form; ammonia vapours in closed area can cause lung irritation and asphyxia. Contact: causes skin and eye irritation.
**Toxicology:** Low order of toxicity by inhalation, contact or ingestion.
**TLV®** - No information.  **$LC_{50}$** - No information.  **$LD_{50}$** - No information.
**Short-term Inhalation Limits** - No information.  **Delayed Toxicity** - No information.

### Fire
**Fire Extinguishing Agents:** Not combustible. Most fire extinguishing agents may be used on fires involving ammonium phosphates.
**Behaviour in Fire:** Not combustible. Toxic and irritating fumes of ammonia and nitrogen oxides may form in fires.
**Ignition Temperature:** Not combustible.  **Burning Rate:** Not combustible.

### Reactivity
**With Water:** No reaction; soluble.
**With Common Materials:** Monobasic reacts violently with sodium hypochlorite.
**Stability:** Stable.

### Environment
**Water:** Prevent entry into water intakes and waterways. Harmful to aquatic life. Fish toxicity: 155 ppm/96 h/fathead minnow/$LC_{50}$/freshwater; Aquatic toxicity rating = 100 to 1 000 ppm/96 h/TLm/freshwater.
**Land-Air:** No information.
**Food Chain Concentration Potential:** None.

## EMERGENCY MEASURES

**Special Hazards**

**Immediate Responses**

Keep non-involved people away from spill site. Dike to prevent water runoff. Notify manufacturer. Notify environmental authorities.

**Protective Clothing and Equipment**

In fires or confined areas Respiratory protection - self-contained breathing apparatus. Otherwise, protective clothing as required.

**Fire and Explosion**

Not combustible. Most fire extinguishing agents may be used on fires involving ammonium phosphates.

**First Aid**

Remove victim from spill site to fresh air. Call for medical assistance, but start first aid at once. Inhalation: if breathing has stopped give artificial respiration; if laboured, give oxygen. Contact: skin and eyes - wash with large amounts of water. Ingestion: give water or milk to conscious victim. If medical assistance is not immediately available, transport victim to doctor, clinic or hospital.

## ENVIRONMENTAL PROTECTION MEASURES

**Response**

**Water**
1. Stop or reduce discharge if safe to do so.
2. Contact manufacturer or supplier for advice.
3. If possible, contain discharge by damming or water diversion.
4. Dredge or vacuum pump to remove contaminants, liquids and contaminated bottom sediments.
5. Notify environmental authorities to discuss disposal and cleanup of contaminated materials.

**Land-Air**
1. Stop or reduce discharge if safe to do so.
2. Contact manufacturer or supplier for advice.
3. Contain spill by diking with earth or other barrier.
4. Dike to prevent runoff from rainwater or water application.
5. Recover undamaged containers.
6. Absorb residual liquid on natural or synthetic sorbents.
7. Notify environmental authorities to discuss disposal and cleanup of contaminated materials.

**Disposal**
1. Contact manufacturer or supplier for advice on disposal.
2. Contact environmental authorities for advice on disposal.

AMMONIUM PHOSPHATES

# AMMONIUM SULFATE  $(NH_4)_2SO_4$

## IDENTIFICATION

| Common Synonyms | Observable Characteristics | Manufacturers |
|---|---|---|
| DIAMMONIUM SULFATE | White to brownish-grey crystals. Odourless. | Cominco Ltd., Trail, B.C. Sherritt Gordon Mines, Fort Saskatchewan, Alta. |

### Transportation and Storage Information

**Shipping State:** Solid.
**Classification:** None.
**Inert Atmosphere:** No requirement.
**Venting:** Open.

**Grades or Purity:** Commercial, technical.
**Containers and Materials:** Multiwall paper bags, drums, railroad cars, trucks.

**Label(s):** None.
**Storage Temperature:** Ambient.

### Physical and Chemical Characteristics

**Physical State** (20°C, 1 atm): Solid.
**Solubility** (Water): 41 g/100 mL (0°C); 43.4 g/100 mL (25°C); 50.4 g/100 mL (100°C).
**Molecular Weight:** 132.1
**Vapour Pressure:** No information.
**Boiling Point:** Decomposes at 230°C.

**Floatability** (Water): Sinks and mixes.
**Odour:** Odourless.
**Flash Point:** Not flammable.
**Vapour Density:** No information.
**Specific Gravity:** 1.77 (20°C).

**Colour:** White to brownish-grey.
**Explosive Limits:** Not flammable.
**Melting Point:** Decomposes at 230°C.

## HAZARD DATA

### Human Health

**Symptoms:** Inhalation: irritation from dust. Ingestion: abdominal pain and nausea.
**Toxicology:** Relatively nontoxic.
**TLV®:** No information.
**Short-term Inhalation Limits** - No information.

$LD_{50}$ - Oral: rat = 3 g/kg

$LC_{50}$ - No information.
**Delayed Toxicity** - No information.

### Fire

**Fire Extinguishing Agents:** Not combustible. Most fire fighting agents may be used on fires involving ammonium sulfate. Water spray or flooding may be effective.
**Behaviour in Fire:** When heated above 235°C releases toxic gases such as ammonia, $SO_2$ and $SO_3$.
**Ignition Temperature:** Not combustible.
**Burning Rate:** Not combustible.

### Reactivity

**With Water:** No reaction; soluble.
**With Common Materials:** Reacts violently with potassium chlorate, potassium nitrate and (potassium and ammonium nitrate).
**Stability:** Stable.

### Environment

**Water:** Prevent entry into water intakes and waterways. Harmful to aquatic life. Fish toxicity: 1 290 ppm/96 h/mosquito fish/TLm/freshwater; 292 ppm/96 h/Daphnia magna/TLm/freshwater; BOD: None.
**Land-Air:** No information.
**Food Chain Concentration Potential:** None.

## EMERGENCY MEASURES

### Special Hazards

### Immediate Responses
Keep non-involved people away from spill site. Dike to prevent water runoff. Notify manufacturer. Notify environmental authorities.

### Protective Clothing and Equipment
In fires or confined areas Respiratory protection - self-contained breathing apparatus. Otherwise, protective clothing as required.

### Fire and Explosion
Decomposition above 235°C releases toxic gases; ammonia, $SO_2$ and $SO_3$.

### First Aid
Move victim out of spill site to fresh air. Call for medical assistance, but start first aid at once. Inhalation: if breathing has stopped, give artificial respiration; if laboured, give oxygen. Contact: skin and eyes - wash with large amounts of water. Ingestion: give water or milk to conscious victim. If medical aid is not immediately available, transport victim to doctor, hospital or clinic.

## ENVIRONMENTAL PROTECTION MEASURES

### Response

**Water**
1. Stop or reduce discharge if safe to do so.
2. Contact manufacturer or supplier for advice.
3. If possible, contain discharge by damming or water diversion.
4. Dredge or vacuum pump to remove contaminants, liquids and contaminated bottom sediments.
5. Notify environmental authorities to discuss disposal and cleanup of contaminated materials.

**Land-Air**
1. Stop or reduce discharge if safe to do so.
2. Contact manufacturer or supplier for advice.
3. Contain spill by diking with earth or other barrier.
4. Dike to prevent runoff from rainwater or water application.
5. Recover undamaged containers.
6. Absorb residual liquid on natural or synthetic sorbents.
7. Notify environmental authorities to discuss disposal and cleanup of contaminated materials.

### Disposal
1. Contact manufacturer or supplier for advice on disposal.
2. Contact environmental authorities for advice on disposal.

AMMONIUM SULFATE   $(NH_4)_2SO_4$

# AMYL ACETATE  $CH_3COOC_5H_{11}$

## IDENTIFICATION

**UN No. 1104**

### Common Synonyms

ACETIC ACID AMYL ESTER
AMYLACETIC ESTER
BANANA OIL
PEAR OIL
PENTYL ACETATE
ACETIC ACID PENTYL ESTER

### Observable Characteristics

Clear, colourless to yellow liquid. Fruity, banana-like or pear-like odour.

### Manufacturers

Cosmos Chemlac Limited, Port Hope, Ontario.

### Transportation and Storage Information

**Shipping State:** Liquid.
**Classification:** Flammable liquid.
**Inert Atmosphere:** No requirement.
**Venting:** Open.
**Pump Type:** Gear or centrifugal (grounded).
**Label(s):** Red label - FLAMMABLE LIQUID; Class 3.2, Group II.
**Storage Temperature:** Ambient.
**Hose Type:** Polyethylene, polypropylene, etc.
**Grades or Purity:** Commercial 85 to 88%; technical 90 to 95%; pure 95 to 99%.
**Containers and Materials:** Drums, tank cars, tank trucks.

### Physical and Chemical Characteristics

**Physical State** (20°C, 1 atm): Liquid.
**Solubility** (Water): 0.18 g/100 mL (20°C).
**Molecular Weight:** 130.2
**Vapour Pressure:** 9 mm Hg (25°C); 10 mm Hg (35.2°C).
**Boiling Point:** 146 to 149°C.
**Floatability** (Water): Floats.
**Odour:** Banana-like or pear-like (0.002 to 0.86 ppm, odour threshold).
**Flash Point:** 27°C (o.c.); 25°C (c.c.).
**Vapour Density:** 4.5
**Specific Gravity:** 0.88 (20°C).
**Colour:** Colourless to yellow.
**Explosive Limits:** 1.0-7.5%
**Melting Point:** -100°C (pure); (-71°C lesser purities).

## HAZARD DATA

### Human Health

**Symptoms:** Inhalation: high concentrations cause irritation of eyes, nose and throat; anaesthetic effect; dizziness, nausea and vomiting, coughing, chest pain and shortness of breath. Contact: causes skin and eye irritation. Ingestion: can cause nausea, vomiting, narcosis, drowsiness, loss of consciousness.
**Toxicology:** Moderately toxic by inhalation and ingestion.
TLV® 100 ppm; 530 mg/m³.
Short-term Inhalation Limits - 150 ppm, 800 mg/m³ (15 min).
$LC_{50}$ - No information.
$LC_{Lo}$ - Inhalation: rat = 5 200 ppm/8 h
Delayed Toxicity - No information.
$LD_{50}$ - Oral: rat = 6.5 g/kg

### Fire

**Fire Extinguishing Agents:** Use alcohol foam, dry chemical or carbon dioxide. Water may be ineffective, but may be used to keep fire-exposed containers cool.
**Behaviour in Fire:** Containers may rupture violently when exposed to heat or flame. Flash back may occur along vapour trail.
**Ignition Temperature:** 360°C.
**Burning Rate:** 4.1 mm/min.
**Electrical Hazard:** Class 1, Group D.

### Reactivity

**With Water:** No reaction.
**With Common Materials:** Can react with oxidizing materials.
**Stability:** Stable.

### Environment

**Water:** Prevent entry into water intakes and waterways. Fish toxicities: 120 ppm/48 h/Daphnia magna /TLm/turbid water; 65 mg/L/48 h/mosquito fish/TLm/freshwater; 180 ppm/96 h/Scenedesmus (algae)/TLm/freshwater; Aquatic toxicity rating = 10 to 100 ppm/96 h/TLm/freshwater; BOD: 0.3 to 0.8 lb/lb, 5 days.
**Land-Air:** No information.
**Food Chain Concentration Potential:** None.

## EMERGENCY MEASURES

**Special Hazards**

FLAMMABLE.

**Immediate Responses**

Keep non-involved people away from spill site. Issue warning: "FLAMMABLE". CALL FIRE DEPARTMENT. Eliminate all ignition sources. Notify manufacturer or supplier. Avoid contact and inhalation. Stop or reduce discharge, if safe to do so. Contain spill by diking. Notify environmental authorities.

**Protective Clothing and Equipment**

Respiratory protection - In fires or confined spaces, use self-contained breathing apparatus. Otherwise, Goggles - (mono), tight fitting. Gloves - rubber. Boots - high, rubber (pants worn outside boots). Outer protective clothing as required.

**Fire and Explosion**

Use dry chemical, alcohol foam or carbon dioxide. Water may be ineffective, but may be used to cool fire-exposed containers. Flash back may occur along vapour trail.

**First Aid**

Move victim out of spill area, to fresh air. Call for medical assistance, but start first aid at once. Inhalation: give artificial respiration if breathing has stopped. Give oxygen if breathing is laboured. Contact: skin and eyes - irrigate eyes with water for at least 15 minutes. Flush skin with plenty of water, while removing contaminated clothes. Ingestion: give milk or water to conscious victim. If medical assistance is not quickly available, transport victim to hospital, doctor or clinic.

## ENVIRONMENTAL PROTECTION MEASURES

**Response**

Water
1. Stop or reduce discharge if safe to do so.
2. Contact manufacturer or supplier for advice.
3. If possible, contain discharge by booming.
4. If floating, skim and remove.
5. Notify environmental authorities to discuss disposal and cleanup of contaminated materials.

Land-Air
1. Stop or reduce discharge if safe to do so.
2. Contact manufacturer or supplier for advice.
3. Contain spill by diking with earth or other barrier.
4. Remove material with pumps or vacuum equipment and place in appropriate containers.
5. Recover undamaged containers.
6. Absorb residual liquid on natural or synthetic sorbents.
7. Notify environmental authorities to discuss disposal and cleanup of contaminated materials.

**Disposal**

1. Contact manufacturer or supplier for advice on disposal.
2. Contact environmental authorities for advice on disposal.
3. Incinerate (approval of environmental authorities required).

AMYL ACETATE   $CH_3COOC_5H_{11}$

# ANILINE  $C_6H_5NH_2$

## IDENTIFICATION

UN No. 1547

### Common Synonyms
ANILINE OIL
PHENYLAMINE
AMINOBENZENE
BENZENAMINE

### Observable Characteristics
Colourless to yellow-brown, oily liquid.
Amine-line odour.

### Manufacturers
No Canadian manufacturer.
Canadian supplier:
Uniroyal Chemical, Elmira, Ont.

Originating from:
Rubicon, USA

### Transportation and Storage Information
**Shipping State:** Liquid.
**Classification:** Poisonous liquid.
**Inert Atmosphere:** No requirement.
**Venting:** Pressure-vacuum.
**Pump Type:** Positive displacement, centrifugal; carbon steel, stainless steel.

**Label(s):** White label - POISONOUS LIQUID; Class 6.1, Group II.
**Storage Temperature:** Ambient.
**Hose Type:** Viton, butyl, polyethylene.

**Grades or Purity:** Commercial, 99.5% min.
**Containers and Materials:** Drums, tank cars, tank trucks; steel, stainless steel.

### Physical and Chemical Characteristics
**Physical State** (20°C, 1 atm): Liquid.
**Solubility** (Water): 3.7 g/100 mL (30°C); 6.4 g/100 mL (100°C).
**Molecular Weight:** 93.1
**Vapour Pressure:** 0.3 mm Hg (20°C); 1 mm Hg (35°C); 2.4 mm Hg (50°C).
**Boiling Point:** 184.2°C.

**Floatability** (Water): Sinks. (May not sink in very saline water.)
**Odour:** Amine-like (0.095 to 1.0 ppm, odour threshold).
**Flash Point:** 75.6°C (o.c.); 70.0°C (c.c.).
**Vapour Density:** 3.2
**Specific Gravity:** 1.02 (20°C).

**Colour:** Colourless to yellow-brown.
**Explosive Limits:** 1.3% (LEL).
**Melting Point:** -6.1°C.

## HAZARD DATA

### Human Health
**Symptoms:** Toxic by inhalation, ingestion and skin absorption: headache, nausea, dizziness, cyanosis, abdominal pain, convulsions.
**Toxicology:** High order of toxicity through skin absorption, ingestion or inhalation.
TLV® (inhalation and skin) 2 ppm; 10 mg/m³     $LC_{50}$ - Inhalation: mouse = 175 ppm/7 h     $LD_{50}$ - Oral: rat = 0.44 g/kg
Short-term Inhalation Limits - 5 ppm; 20 mg/m³     Delayed Toxicity - Damage to red blood cells, liver and kidney. May cause embryo or fetal damages.
(15 min).

### Fire
**Fire Extinguishing Agents:** Use dry chemical, foam or carbon dioxide. Use water to cool fire-exposed containers and knock down vapours.
**Behaviour in Fire:** Flammable toxic $NO_x$ and CO may be given off in a fire.
**Ignition Temperature:** 650°C.     **Burning Rate:** 3.0 mm/min.

### Reactivity
**With Water:** No reaction.
**With Common Materials:** Reacts vigorously with oxidizing materials. Reacts moderately to violently with acetic anhydride, chlorosulfonic acid, nitric acid, sulfuric acid, oleum, perchromates, perchromates, performic acid and silver perchlorate.
**Stability:** Stable.

### Environment
**Water:** Prevent entry into water intakes and waterways. Fish toxicity: 1 020 ppm/1 h/sunfish/killed/freshwater; Aquatic toxicity rating = 10 to 100 ppm/96 h/ TLm/freshwater; BOD: 150%, 5 days.
**Land-Air:** No information.
**Food Chain Concentration Potential:** No information.

## EMERGENCY MEASURES

### Special Hazards
POISON.

### Immediate Responses
Keep non-involved people away from spill site. Issue warning; "POISON". Call Fire Department. Avoid contact and inhalation. Contact manufacturer or supplier for advice. Stay upwind and use water spray to control vapour. Dike runoff. Stop or reduce discharge, if safe to do so. Notify environmental authorities.

### Protective Clothing and Equipment
Respiratory protection – use only self-contained breathing apparatus. Totally encapsulated protective clothing. Boots – high, rubber (pants worn outside boots). Gloves – rubber.

### Fire and Explosion
Use dry chemical, alcohol foam, carbon dioxide. Water may be ineffective, but may be used to cool fire-exposed containers.

### First Aid
Move victim out of spill area to fresh air. Call for medical assistance, but start first aid at once. **Inhalation:** if not breathing give artificial respiration (not mouth-to-mouth method), if breathing is laboured give oxygen. **Contact:** remove contaminated clothing; wash affected areas with warm water and soap if available for at least 15 minutes; eyes – irrigate, holding lids apart for at least 15 minutes. **Ingestion:** induce vomiting, but only in conscious victim. If medical assistance is not immediately available, transport victim to hospital, doctor or clinic.

## ENVIRONMENTAL PROTECTION MEASURES

### Response

**Water**
1. Stop or reduce discharge if safe to do so.
2. Contact manufacturer or supplier for advice.
3. If possible, contain discharge by damming or water diversion.
4. Dredge or vacuum pump to remove contaminants, liquids and contaminated bottom sediments.
5. Notify environmental authorities to discuss disposal and cleanup of contaminated materials.

**Land-Air**
1. Stop or reduce discharge if safe to do so.
2. Contact manufacturer or supplier for advice.
3. Contain spill by diking with earth or other barrier.
4. Remove material with pumps or vacuum equipment and place in appropriate containers.
5. Remove material by manual or mechanical means.
6. Absorb residual liquid on natural or synthetic sorbents.
7. Remove contaminated soil for disposal.
8. Notify environmental authorities to discuss disposal and cleanup of contaminated materials.

### Disposal
1. Contact manufacturer or supplier for advice on disposal.
2. Contact environmental authorities for advice on disposal.

ANILINE   $C_6H_5NH_2$

# ARSENIC   As

## IDENTIFICATION   UN No. 1557

| Common Synonyms | Observable Characteristics | Manufacturers |
|---|---|---|
| GREY ARSENIC<br>METALLIC ARSENIC<br>ELEMENTAL ARSENIC | Silver-grey lustrous, crystalline solid or black amorphous powder. Odourless. | No Canadian manufacturers.<br>U.S. manufacturer and supplier:<br>ASARCO Incorporated,<br>120 Broadway,<br>New York, NY<br><br>Major Canadian users:<br>Degussa,<br>Burlington, Ontario.<br>International Chemical,<br>Brampton, Ontario.<br>Kingsley and Keith Canada Ltd.,<br>Montreal, Quebec. |

### Transportation and Storage Information

**Shipping State:** Solid.
**Classification:** Poisonous solid.
**Inert Atmosphere:** No requirement.
**Venting:** No requirement.

**Label(s):** White label - POISONOUS SOLID; Class 6.1, Group II.
**Storage Temperature:** Ambient.

**Grades or Purity:** Technical, refined, crude 90 to 95%.
**Containers and Materials:** Drums; steel.

### Physical and Chemical Characteristics

**Physical State** (20°C, 1 atm): Solid.
**Solubility** (Water): <0.003 g/100 mL (20°C).
**Atomic Weight:** 74.9
**Vapour Pressure:** 1 mm Hg (372°C).
**Boiling Point:** Sublimes at 613°C.

**Floatability** (Water): Sinks.
**Odour:** Odourless.
**Flash Point:** Not flammable.
**Vapour Density:** No information.
**Specific Gravity:** 5.7 (20°C) (metallic); 4.7 (20°C) (black amorphous).

**Colour:** Silver-grey to black.
**Explosive Limits:** Not flammable (dust concentrations may explode).
**Melting Point:** 814°C (28 atm); sublimes 613°C.

## HAZARD DATA

### Human Health

**Symptoms: Contact:** skin - burning and stinging sensation, tightness in chest, nausea, bronzing of skin. **Inhalation:** (dust) restlessness, difficulty breathing, cyanosis, cough and foamy sputum. **Ingestion:** nausea, vomiting, abdominal pain, diarrhea, convulsions and coma.
**Toxicology:** Extremely toxic upon inhalation of dust, fumes; ingestion and skin contact.
TLV® - 0.2 mg/m$^3$.   $LC_{50}$ - Information not available.   $LD_{50}$ - No information.
Short-term Inhalation Limits - No Information.   Delayed Toxicity - Liver and kidney damage.   $LD_{Lo}$ - Intramuscular: rat = 0.02 g/kg

### Fire

**Fire Extinguishing Agents:** Not combustible. Most fire extinguishing agents may be used on fires involving arsenic.
**Behaviour in Fire:** Not combustible; but highly toxic As, and As$_x$ fumes released.
**Ignition Temperature:** Not combustible.   **Burning Rate:** Not combustible.

### Reactivity

**With Water:** No reaction.
**With Common Materials:** Emits highly toxic gas (arsine) on contact with hydrogen gas and also with acids plus reducing metals (e.g. Zn, Fe); burns spontaneously with gaseous chlorine. Reacts with bromates, chlorates, iodate, peroxides, potassium permanganate, silver nitrate.
**Stability:** Stable.

### Environment

**Water:** Prevent entry into water intakes and waterways. Harmful to aquatic life. EPA criterion = 57 mg/L/24 h/average; maximum of 130 mg/L at any time (freshwater); 29 mg/L/24 h/average should not exceed 67 mg/L at anytime (saltwater).
**Land-Air:** No information.
**Food Chain Concentration Potential:** Arsenic may be accumulated in food chain.

## EMERGENCY MEASURES

**Special Hazards**

POISON.

**Immediate Responses**

Keep non-involved people away from spill site. Issue warning: "POISONOUS". Avoid contact and inhalation. Prevent dust dispersion. Use full protective clothing and self-contained breathing apparatus. If possible, stop discharge, if safe to do so. Notify manufacturer. Notify environmental authorities.

**Protective Clothing and Equipment**

Respiratory protection - self-contained breathing apparatus, and totally encapsulated protective clothing.

**Fire and Explosion**

Not combustible, but most fire extinguishing agents may be used on fires involving arsenic.

**First Aid**

Move victim out of spill area to fresh air. Call for medical assistance, but start first aid at once. Contact: skin - remove contaminated clothing and wash eyes and affected skin thoroughly with plenty of water. Ingestion: keep victim warm and quiet. Treat as a severe emergency. Transport victim to hospital, doctor or clinic immediately.

## ENVIRONMENTAL PROTECTION MEASURES

**Response**

**Water**

1. Stop or reduce discharge if safe to do so.
2. Contact manufacturer or supplier for advice.
3. If possible, contain discharge by damming or water diversion.
4. Dredge or vacuum pump to remove contaminants, liquids and contaminated bottom sediments.
5. Notify environmental authorities to discuss disposal and cleanup of contaminated materials.

**Land-Air**

1. Stop or reduce discharge if safe to do so.
2. Contact manufacturer or supplier for advice.
3. Contain spill by diking with earth or other barrier.
4. Remove materials by manual or mechanical means.
5. Recover undamaged containers.
6. Remove contaminated soil for disposal.
7. Notify environmental authorities to discuss disposal and cleanup of contaminated materials.

**Disposal**

1. Contact manufacturer or supplier for advice on disposal.
2. Contact environmental authorities for advice on disposal.

ARSENIC   As

# ARSENIC TRIOXIDE $As_2O_3$

## IDENTIFICATION

### Common Synonyms
WHITE ARSENIC
CRUDE ARSENIC
ARSENIOUS OXIDE
ARSENOUS ANHYDRIDE
ARSENIC SESQUIOXIDE

### Observable Characteristics
White powder. Odourless.

### Manufacturers
No known Canadian producer or supplier.
U.S. producer:
ASARCO Incorporated,
120 Broadway,
New York, NY.
Major Canadian Users:
Degussa, Burlington, Ontario.
International Chemical, Brampton, Ontario.
Kingsley and Keith Canada Ltd.,
Montreal, Quebec.

**UN No. 1561**

### Transportation and Storage Information
**Shipping State:** Solid.
**Classification:** Poisonous solid.
**Inert Atmosphere:** No requirement.
**Venting:** No requirement.

**Label(s):** White label - POISONOUS SOLID; Class 6.1, Group II.
**Storage Temperature:** Ambient.

**Grades or Purity:** Crude, 95%; refined, 99%.
**Containers and Materials:** Drums, barrels, truck lots.

### Physical and Chemical Characteristics
**Physical State** (20°C, 1 atm): Solid.
**Solubility** (Water): 1.2 g/100 mL (0°C); 2.1 g/100 mL (25°C).
**Molecular Weight:** 197.8
**Vapour Pressure:** No information.
**Boiling Point:** 457°C.

**Floatability** (Water): Sinks (finely divided powder may temporarily float).
**Odour:** Odourless.
**Flash Point:** Not flammable.
**Vapour Density:** No information.
**Specific Gravity:** 3.87 (25°C).

**Colour:** White.
**Explosive Limits:** Not flammable.
**Melting Point:** 312 to 315°C; sublimes at 193°C.

## HAZARD DATA

### Human Health
**Symptoms: Contact:** skin - may be absorbed, redness, pain, serious skin burns, other symptoms as for inhalation may appear. **Inhalation:** sore throat, coughing, vomiting, weakness, thirst, shallow breathing, convulsions. **Ingestion:** nausea, vomiting, severe abdominal pain, diarrhea.
**Toxicology:** Extremely toxic upon inhalation of dust, fumes; ingestion and skin contact.
TLV® 0.2 mg/m3 (as arsenic).   LC$_{50}$ - No information.   LD$_{50}$ - Oral: rat = 0.02 g/kg
Short-term Inhalation Limits - No information.   Delayed Toxicity - Carcinogen.   LD$_{50}$ - Oral: mouse = 0.045 g/kg
   LD$_{50}$ - Oral: human = 0.0014 g/kg

### Fire
**Fire Extinguishing Agents:** Not combustible, but most fire extinguishing agents may be used on fires involving arsenic trioxide.
**Behaviour In Fire:** At high temperatures, $As_2O_3$ will volatilize, giving off dangerous fumes, such as arsenic trioxide, arsine.
**Ignition Temperature:** Not combustible.   **Burning Rate:** Not combustible.

### Reactivity
**With Water:** No reaction.
**With Common Materials:** Metals and other substances containing $As_2O_3$ in contact with acids will liberate arsine ($AsH_3$), a colourless, highly toxic gas. Reacts vigorously with fluorine and chlorate.
**Stability:** Stable.

### Environment
**Water:** Prevent entry into water intakes and waterways. Harmful to aquatic life in very low concentrations. Aquatic toxicity rating = 1 to 10ppm/96 h/ TLm/freshwater; BOD: No information.
**Land-Air:** No information.
**Food Chain Concentration Potential:** Accumulation in shellfish is a known example.

## EMERGENCY MEASURES

### Special Hazards
POISON.

### Immediate Responses
Keep non-involved people away from spill site. Issue warning: "POISON". Avoid contact. Stop or reduce discharge, if safe to do so. If water is being used to fight fire, dike area to contain toxic runoff. Notify manufacturer. Notify environmental authorities.

### Protective Clothing and Equipment
Respiratory protection - self-contained breathing apparatus and fully encapsulated protective clothing.

### Fire and Explosion
Not combustible; most fire extinguishing agents may be used on fires involving arsenic trioxide.

### First Aid
Move victim out of spill area to fresh air. Call for medical assistance, but start first aid at once. <u>Contact</u>: skin and eyes - remove contaminated clothing and wash eyes and affected skin thoroughly with plenty of water. <u>Ingestion</u>: keep victim warm and quiet. Treat as a severe emergency. Transport victim to hospital, doctor or clinic immediately.

## ENVIRONMENTAL PROTECTION MEASURES

### Response

**Water**
1. Stop or reduce discharge if safe to do so.
2. Contact manufacturer or supplier for advice.
3. If possible, contain discharge by damming or water diversion.
4. Dredge or vacuum pump to remove contaminants, liquids and contaminated bottom sediments.
5. Notify environmental authorities to discuss disposal and cleanup of contaminated materials.

**Land-Air**
1. Stop or reduce discharge if safe to do so.
2. Contact manufacturer or supplier for advice.
3. Dike to prevent runoff from rainwater or water application.
4. Remove materials by manual or mechanical means.
5. Recover undamaged containers.
6. Remove contaminated soil for disposal.
7. Notify environmental authorities to discuss disposal and cleanup of contaminated materials.

### Disposal
1. Contact manufacturer or supplier for advice on disposal.
2. Contact environmental authorities for advice on disposal.

ARSENIC TRIOXIDE   $As_2O_3$

# ARSINE  AsH$_3$

## IDENTIFICATION

UN No. 2188

### Common Synonyms
ARSENIC HYDRIDE
ARSENIURETTED HYDROGEN
ARSENIC TRIHYDRIDE
ARSENOUS HYDRIDE

### Observable Characteristics
Colourless gas. Garlic-like odour.

### Manufacturers

### Transportation and Storage Information
**Shipping State:** Liquid (compressed gas).
**Classification:** Poison, flammable.
**Inert Atmosphere:** No requirement.
**Venting:** Safety, relief.
**Pump Type:** No information.

**Label(s):** White label - POISON; Red label - FLAMMABLE.
**Storage Temperature:** Ambient.
**Hose Type:** No information.

**Grades or Purity:** Technical.
**Containers and Materials:** Cylinders; steel.

### Physical and Chemical Characteristics
**Physical State (20°C), 1 atm):** Gas.
**Solubility (Water):** 0.07 g/100 mL (20°C).
**Molecular Weight:** 78.0
**Vapour Pressure:** 765 mm Hg (-62°C); 11 360 mm Hg (21°C).
**Boiling Point:** -62.5°C.

**Floatability (Water):** Sinks and boils.
**Odour:** Garlic-like.
**Flash Point:** No information (flammable).
**Vapour Density:** 2.7
**Specific Gravity:** 1.65 (liquid) (-73.2°C).

**Colour:** Colourless.
**Explosive Limits:** No information (flammable).
**Melting Point:** -117°C.

## HAZARD DATA

### Human Health
**Symptoms:** Inhalation: headache, dizziness, nausea, difficulty breathing, cyanosis, dry cough, abdominal pain, convulsions, bronzing of skin; Contact: skin - frostbite, painful irritation, inflammation, blisters - symptoms similar to inhalation; eyes - irritation, watering, inflammation and burning.
**Toxicology:** Highly toxic by inhalation.
TLV®- 0.05 ppm; 0.2 mg/m$^3$.
**Short-term Inhalation Limits** - No information.

LD$_{50}$ - No information.
LCL$_o$ - Inhalation: human = 25 ppm/30 min
LCL$_o$ - Inhalation: rat = 300 mg/m$^3$/15 min
Delayed Toxicity - Symptoms may be delayed up to two days.

LD$_{50}$ - No information.

### Fire
**Fire Extinguishing Agents:** Shut off leak before attempting to extinguish fire. Most fire extinguishing agents may be used in fires involving arsine.
**Behaviour in Fire:** Releases toxic fumes in fire. Flash back may occur along vapour trail.
**Ignition Temperature:** No information.  **Burning Rate:** No information.

### Reactivity
**With Water:** No reaction.
**With Common Materials:** Can react vigorously with oxidizing materials. Reacts violently with chlorine, nitric acid and (potassium and ammonia).
**Stability:** Stable.

### Environment
**Water:** Prevent entry into water intakes and waterways. 57 µg/L freshwater EPA criterion 24 h average (as Arsenic); 29 µg/L saltwater EPA criterion 24 h average (as arsenic).
**Land-Air:** No information.
**Food Chain Concentration Potential:** No information.

## EMERGENCY MEASURES

**Special Hazards**
POISON. FLAMMABLE.

**Immediate Responses**
Keep non-involved people away from spill site. Issue warning "POISON; FLAMMABLE". CALL FIRE DEPARTMENT. Avoid contact and inhalation. Evacuate from downwind. Notify manufacturer or supplier. Notify environmental authorities.

**Protective Clothing and Equipment**
Respiratory protection - self-contained breathing apparatus and totally-encapsulated suit.

**Fire and Explosion**
Shut off leak before attempting to extinguish fire. Most fire extinguishing agents may be used on fires involving arsine. Releases toxic fumes in fire.

**First Aid**
Move victim out of spill area to fresh air. Call for medical assistance, but start first aid at once. Inhalation: if breathing has stopped, give artificial respiration (not mouth-to-mouth method); if laboured, give oxygen. Contact: skin - remove contaminated clothing and flush affected areas with plenty of water; treat as for frostbite; eyes - irrigate with plenty of water. If medical assistance is not immediately available, transport victim to hospital, doctor or clinic.

## ENVIRONMENTAL PROTECTION MEASURES

**Response**

**Water**
1. Stop or reduce discharge if safe to do so.
2. Contact manufacturer or supplier for advice.
3. Notify environmental authorities to discuss disposal and cleanup of contaminated materials.

**Land-Air**
1. Stop or reduce discharge if safe to do so.
2. Contact manufacturer or supplier for advice.
3. Recover undamaged containers.
4. Notify environmental authorities to discuss disposal and cleanup of contaminated materials.

**Disposal**
1. Contact manufacturer or supplier for advice on disposal.
2. Contact environmental authorities for advice on disposal.

ARSINE   $AsH_3$

# ASPHALT

## IDENTIFICATION

**Common Synonyms**
PETROLEUM ASPHALT
BITUMEN

**Observable Characteristics**
Dark brown to black solid (unheated) or thick liquid (heated). Strong tarry odour.

**UN No. 1993**

**Manufacturers**
Most oil refineries.

### Transportation and Storage Information

**Shipping State:** Solid; liquid (heated).
**Classification:** None.
**Inert Atmosphere:** No requirement.
**Venting:** Open.
**Label(s):** Not regulated.
**Storage Temperature:** Ambient or elevated.
**Grades or Purity:** Various grades. Industrial, special, paving, rapid curing, medium curing, slow curing and emulsified.
**Containers and Materials:** Drums, bulk lots, tank cars, tank trucks, steel.

### Physical and Chemical Characteristics

**Physical State (20°C, 1 atm):** Solid.
**Solubility (Water):** Insoluble.
**Molecular Weight:** Variable.
**Vapour Pressure:** Variable.
**Boiling Point:** 190 to 400°C.
**Floatability (Water):** May float or sink.
**Odour:** Tarry.
**Flash Point:** 10 to 200°C.
**Vapour Density:** No information.
**Specific Gravity:** Variable; 0.9 to 1.2 (20°C).
**Colour:** Brown to black.
**Explosive Limits:** No information.
**Melting Point:** Variable; >40°C.

## HAZARD DATA

### Human Health

**Symptoms:** Inhalation: irritates respiratory tract. Contact: skin and eyes - fumes irritate.
**Toxicology:** Low order of toxicity.
TLV - (petroleum fumes) 5 mg/m$^3$.
Short-term Inhalation Limits - 10 mg/m$^3$ (15 min) (petroleum fumes).
$LC_{50}$ - No information.
Delayed Toxicity - No information.
$LD_{50}$ - No information.
$TD_{Lo}$ - Intramuscular: rat = 5.4 g/kg

### Fire

**Fire Extinguishing Agents:** Dry chemical or carbon dioxide. Water or foam may cause frothing.
**Behaviour in Fire:** No information.
**Ignition Temperature:** 300 to 485°C.
**Burning Rate:** No information.

### Reactivity

**With Water:** No reaction.
**With Common Materials:** Can react with oxidizing materials. Reacts with fluorine.
**Stability:** Stable.

### Environment

**Water:** Prevent entry into water intakes or waterways. Fouling to shoreline is major hazard.
**Land-Air:** Fouling to landscape.
**Food Chain Concentration Potential:** No information.

## EMERGENCY MEASURES

**Special Hazards**
COMBUSTIBLE.

**Immediate Responses**
Keep non-involved people away from spill site. Call Fire Department. Avoid contact. Stop or reduce discharge, if this can be done without risk. Dike to contain spill or runoff. Notify supplier and environmental authorities.

**Protective Clothing and Equipment**
Goggles or face shield. Coveralls. Gloves - rubber. Boots - rubber.

**Fire and Explosion**
Use dry chemical or carbon dioxide to extinguish. Water or foam may cause frothing.

**First Aid**
Move victim out of spill area to fresh air. Call for medical assistance, but start first aid at once. Contact: skin - cover burns with sterile dressing if possible. Victim should be taken to hospital, doctor or clinic, if medical attention is required.

## ENVIRONMENTAL PROTECTION MEASURES

**Response**

**Water**
1. Stop or reduce discharge if safe to do so.
2. Contact manufacturer or supplier for advice.
3. If possible, contain discharge by booming.
4. If floating, skim and remove.
5. If possible contain discharge by damming or water diversion.
6. Dredge or vacuum pump to remove contaminants, liquids and contaminated bottom sediments.
7. Notify environmental authorities to discuss disposal and cleanup of contaminated materials.

**Land-Air**
1. Stop or reduce discharge if safe to do so.
2. Contact manufacturer or supplier for advice.
3. Contain spill by diking with earth or other barrier.
4. Remove material by manual or mechanical means.
5. Recover undamaged containers.
6. Notify environmental authorities to discuss disposal and cleanup of contaminated materials.

**Disposal**
1. Contact manufacturer or supplier for advice on disposal.
2. Contact environmental authorities for advice on disposal.
3. Incinerate (approval of environmental authorities required).

ASPHALT

# BARIUM CARBONATE   $BaCO_3$

## IDENTIFICATION

UN No. 1564 Barium Compounds

| Common Synonyms | Observable Characteristics | Manufacturers |
|---|---|---|
| | White crystals or powder. Odourless. | No Canadian manufacturer. Canadian suppliers: FMC of Canada Ltd., Vancouver, B.C. Philipp Brothers (Canada) Ltd., Montreal, Quebec. | Originating from: FMC Corp. USA |

### Transportation and Storage Information

**Shipping State:** Solid.
**Classification:** Poison.
**Inert Atmosphere:** No requirement.
**Venting:** Open.

**Label(s):** White label – POISON; Class 6.1, Group I.
**Storage Temperature:** Ambient.

**Grades or Purity:** Technical.
**Containers and Materials:** Bags, bulk by rail or truck.

### Physical and Chemical Characteristics

**Physical State** (20°C, 1 atm): Solid.
**Solubility** (Water): 0.0022 g/100 mL (18°C); 0.0065 g/100 mL (100°C).
**Molecular Weight:** 197.4
**Vapour Pressure:** No information.
**Boiling Point:** 1 450°C (decomposes).

**Floatability** (Water): Sinks.
**Odour:** Odourless.
**Flash Point:** Not flammable.
**Vapour Density:** No information.
**Specific Gravity:** 4.3 (20°C).

**Colour:** White.
**Explosive Limits:** Not flammable.
**Melting Point:** Decomposes 1 450°C; Melts at 1 740°C (90 atm).

## HAZARD DATA

### Human Health

**Symptoms: Contact:** eyes and skin – redness and pain. **Ingestion:** excessive salivation, coughing, vomiting, diarrhea, difficulty breathing, convulsions, muscular paralysis. **Inhalation:** symptoms similar to ingestion.
**Toxicology:** Toxic by ingestion.
TLV® (air) 0.5 mg/m³ (as soluble Ba).    LC$_{50}$ – No information.    LD$_{50}$ – Oral: rat = 0.63 g/kg
Short-term Inhalation Limits – No information.    Delayed Toxicity – No information.    LDLo – Oral: human = 0.057 g/kg

### Fire

**Fire Extinguishing Agents:** Not combustible; most fire extinguishing agents may be used on fires involving barium carbonate.
**Behaviour in Fire:** No information.    **Burning Rate:** Not combustible.
**Ignition Temperature:** Not combustible.

### Reactivity

**With Water:** No reaction.
**With Common Materials:** No information.
**Stability:** Stable.

### Environment

**Water:** Prevent entry into water intakes and waterways. **BOD:** No information.
**Land-Air:** No information.
**Food Chain Concentration Potential:** No information.

## EMERGENCY MEASURES

**Special Hazards**

POISON.

**Immediate Responses**

Keep non-involved people away from spill site. Issue warning: "POISON". Stop or reduce discharge if possible. Avoid contact. Notify manufacturer. Notify environmental authorities.

**Protective Clothing and Equipment**

Respiratory protection - where dust is present use approved dust mask; in confined areas use self-contained breathing apparatus. Clothing - coveralls, or as necessary to avoid excessive skin contact. Protective gloves - general purpose. Safety goggles - tight fitting, or safety glasses.

**Fire and Explosion**

Not combustible; most fire extinguishing agents may be used on fires involving barium carbonate.

**First Aid**

Move victim out of spill site to fresh air. Call for medical assistance, but start first aid at once. Ingestion: give conscious victim plenty of water to drink and induce vomiting. Contact: eyes - irrigate with plenty of water; skin - wash with plenty of water. If medical assistance is not immediately available, transport victim to hospital, doctor or clinic.

## ENVIRONMENTAL PROTECTION MEASURES

**Response**

**Water**
1. Stop or reduce discharge if safe to do so.
2. Contact manufacturer or supplier for advice.
3. If possible, contain discharge by damming or water diversion.
4. Dredge or vacuum pump to remove contaminants, liquids and contaminated bottom sediments.
5. Notify environmental authorities to discuss disposal and cleanup of contaminated materials.

**Land-Air**
1. Stop or reduce discharge if safe to do so.
2. Contact manufacturer or supplier for advice.
3. Contain spill by diking with earth or other barrier.
4. Remove material by manual or mechanical means.
5. Recover undamaged containers.
6. Remove contaminated soil for disposal.
7. Notify environmental authorities to discuss disposal and cleanup of contaminated materials.

**Disposal**
1. Contact manufacturer or supplier for advice on disposal.
2. Contact environmental authorities for advice on disposal.

BARIUM CARBONATE   $BaCO_3$

# BARIUM SULFATE   BaSO$_4$

## IDENTIFICATION

| Common Synonyms | Observable Characteristics | UN No. 1564 Barium Compounds |
|---|---|---|
| BARITE<br>BLANC FIXE<br>BARYTES<br>BASOFOR | White or yellowish powder. Odourless. | **Manufacturers**<br>Mountain Minerals Ltd.,<br>Lethbridge, Alberta. |

**Transportation and Storage Information**

Shipping State: Solid.
Classification: Not required.
Inert Atmosphere: No requirement.
Venting: Open.

Label(s): Not required.
Storage Temperature: Ambient.

Grades or Purity: Technical, pulp and bleach.
Containers and Materials: Drums, barrels, paper sacks.

**Physical and Chemical Characteristics**

Physical State (20°C, 1 atm): Solid.
Solubility (Water): 0.00025 g/100 mL (25°C); 0.00034 g/100 mL (50°C); 0.00041 g/100 mL (100°C).
Molecular Weight: 233.4
Vapour Pressure: No information.
Boiling Point: >1 580°C.

Floatability (Water): Sinks.
Odour: Odourless.
Flash Point: Not flammable.
Vapour Density: No information.
Specific Gravity: 4.3 to 4.5 (20°C).

Colour: White to yellowish.
Explosive Limits: Not flammable.
Melting Point: 1 580°C.

## HAZARD DATA

**Human Health**

Symptoms: Contact (with dust): eyes and skin - redness and pain. Ingestion: excessive salivation, coughing, vomiting, diarrhea, difficulty breathing, convulsions, muscular paralysis. Inhalation: of dust produces symptoms similar to ingestion.
Toxicology: Toxic by ingestion.
TLV® - (air) 0.5 mg/m$^3$ (as soluble Ba).
Short-term Inhalation Limits - No information.

LC$_{50}$ - No information.
Delayed Toxicity - No information.

LD$_{50}$ - No information.
TD$L_o$ - Intraperitoneal: rat = 0.2 g/kg

**Fire**

Fire Extinguishing Agents: Not combustible; most fire extinguishing agents may be used on fires involving barium sulfate.
Behaviour in Fire: No information.
Ignition Temperature: Not combustible.
Burning Rate: Not combustible.

**Reactivity**

With Water: No reaction.
With Common Materials: Can react explosively with aluminum when heated.
Stability: Stable.

**Environment**

Water: BOD: No information.
Land-Air: No information.
Food Chain Concentration Potential: No information.

## EMERGENCY MEASURES

**Special Hazards**

POISON.

**Immediate Responses**

Keep non-involved people away from spill site. Issue warning: "POISON". Stop discharge if possible. Avoid contact. Notify manufacturer. Notify environmental authorities.

**Protective Clothing and Equipment**

Respiratory protection - where dust is present, use approved dust mask. Clothing - coveralls, or as necessary to avoid excessive skin contact. Protective gloves - general purpose. Safety goggles - tight fitting, or safety glasses.

**Fire and Explosion**

Not combustible; most fire extinguishing agents may be used on fires involving barium sulfate.

**First Aid**

Move victim out of spill site to fresh air. Call for medical assistance, but start first aid at once. Ingestion: give conscious victim plenty of water to drink and induce vomiting. Contact: eyes - irrigate with plenty of water; skin - wash with plenty of water. If medical assistance is not immediately available, transport victim to hospital, doctor or clinic.

## ENVIRONMENTAL PROTECTION MEASURES

**Response**

Water
1. Stop or reduce discharge if safe to do so.
2. Contact manufacturer or supplier for advice.
3. If possible, contain discharge by damming or water diversion.
4. Dredge or vacuum pump to remove contaminants, liquids and contaminated bottom sediments.
5. Notify environmental authorities to discuss disposal and cleanup of contaminated materials.

Land-Air
1. Stop or reduce discharge if safe to do so.
2. Contact manufacturer or supplier for advice.
3. Contain spill by diking with earth or other barrier.
4. Remove material by manual or mechanical means.
5. Recover undamaged containers.
6. Remove contaminated soil for disposal.
7. Notify environmental authorities to discuss disposal and cleanup of contaminated materials.

**Disposal**
1. Contact manufacturer or supplier for advice on disposal.
2. Contact environmental authorities for advice on disposal.

BARIUM SULFATE  $BaSO_4$

# BENZENE $C_6H_6$

## IDENTIFICATION     UN No. 1114

### Common Synonyms
BENZOL
COAL NAPHTHA
BENZIN
PHENYLHYDRIDE

### Observable Characteristics
Colourless liquid. Aromatic odour.

### Manufacturers
Gulf Oil Canada Ltd., Montreal, Que.
Finachem Canada Ltd., Montreal, Quebec.
Shell Canada Ltd., Corunna, Ont.
Petrosar, Corunna, Ont.
Esso Chemical Canada, Sarnia, Ontario.

### Transportation and Storage Information
**Shipping State:** Liquid.
**Classification:** Flammable liquid.
**Inert Atmosphere:** No requirement.
**Venting:** Pressure-vacuum.
**Pump Type:** Gear or centrifugal, explosion-proof, grounded.
**Label(s):** Red label - FLAMMABLE LIQUID; Class 3.2, Group II.
**Storage Temperature:** Ambient.
**Hose Type:** Viton, polypropylene, Teflon, stainless steel.
**Grades or Purity:** Industrial, pure, 99+%; thiophene free, 99+%; nitration, 99+%; industrial, 90%, 85+%; reagent, 99+%.
**Containers and Materials:** Drums, tank cars, tank trucks; steel and stainless steel.

### Physical and Chemical Characteristics
**Physical State** (20°C, 1 atm): Liquid.
**Solubility** (Water): 0.18 g/100 mL (20°C).
**Molecular Weight:** 78.1
**Vapour Pressure:** 60 mm Hg (15°C); 76 mm Hg (20°C); 118 mm Hg (30°C).
**Boiling Point:** 80.1°C.
**Floatability** (Water): Floats.
**Odour:** Aromatic (0.16 to 4.7 ppm, odour threshold).
**Flash Point:** -11°C (c.c.).
**Vapour Density:** 2.8
**Specific Gravity:** 0.88 (20°C).
**Colour:** Colourless.
**Explosive Limits:** 1.3 to 7.1%.
**Melting Point:** 5.5°C.

## HAZARD DATA

### Human Health
**Symptoms:** Inhalation: irritation, headache, dizziness, dullness, nausea, euphoria, respiratory paralysis, unconsciousness. Contact: skin - skin irritation; prolonged skin exposure, same symptoms as inhalation. Ingestion: sore throat, nausea, vomiting, headache, unconsciousness.
**Toxicology:** Toxic upon inhalation, ingestion and contact.
TLV®- (inhalation, skin); 10 ppm; 30 mg/m³.    $LC_{50}$ - Inhalation: rat = 10 000 ppm/7 h    $LD_{50}$ - Oral: rodent = 3.8 g/kg
Short-term Inhalation Limits - 25 ppm; 75 mg/m³    Delayed Toxicity - Suspected carcinogen.
(15 min).

### Fire
**Fire Extinguishing Agents:** Use dry chemical, foam or carbon dioxide. Water may be ineffective, but may be used to cool fire-exposed containers.
**Behaviour in Fire:** Flashback may occur along vapour trail.
**Ignition Temperature:** 562°C.     **Burning Rate:** 6.0 mm/min.

### Reactivity
**With Water:** No reaction.
**With Common Materials:** Can react with oxidizing materials. Reacts vigorously with chlorine, chromates and perchlorates.
**Stability:** Stable.

### Environment
**Water:** Prevent entry into water intakes and waterways. Fish toxicity: 5 ppm/6 h/minnow/lethal/distilled water; 20 ppm/24 h/sunfish/TLm/tapwater; 386 ppm/96 h/mosquito fish/TLm/freshwater; Aquatic toxicity rating = 10 to 100 ppm/96 h/TLm/freshwater; BOD: 1.2 lb/lb, 10 days.
**Land-Air:** No information.
**Food Chain Concentration Potential:** None.

## EMERGENCY MEASURES

### Special Hazards
FLAMMABLE.

### Immediate Responses
Keep non-involved people away from spill site. Issue warning: "FLAMMABLE". Avoid contact. Evacuate downwind. CALL FIRE DEPARTMENT. Eliminate all sources of ignition. Notify manufacturer. Stop or reduce discharge, if this can be done without risk. Contain spill by diking to prevent runoff. In fire, stay upwind and use water spray to control vapours. Notify environmental authorities.

### Protective Clothing and Equipment
Respiratory protection - self-contained breathing apparatus and totally encapsulated protective clothing. Boots - high, synthetic rubber (pants worn outside boots). Gloves - synthetic rubber or suitable plastic.

### Fire and Explosion
Extinguish with dry chemical, foam or carbon dioxide. Water may be ineffective but may be used to cool fire-exposed containers.

### First Aid
Move victim out of spill area to fresh air. Call for medical assistance, but start first aid at once. Inhalation: if victim is not breathing, give artificial respiration; if breathing is laboured, give oxygen. Contact: skin - flush with plenty of water; use soap if available and remove contaminated clothing; eyes - irrigate with plenty of water. Ingestion: do not induce vomiting. Keep victim warm and quiet. If medical assistance is not immediately available, transport victim to hospital, doctor or clinic.

## ENVIRONMENTAL PROTECTION MEASURES

### Response

**Water**
1. Stop or reduce discharge if safe to do so.
2. Contact manufacturer or supplier for advice.
3. If possible, contain discharge by booming.
4. If floating, skim and remove.
5. Notify environmental authorities to discuss disposal and cleanup of contaminated materials.

**Land-Air**
1. Stop or reduce discharge if safe to do so.
2. Contact manufacturer or supplier for advice.
3. Contain spill by diking with earth or other barrier.
4. Remove material with vacuum pumps or vacuum equipment and place in appropriate containers.
5. Recover undamaged containers.
6. Absorb residual liquid on natural or synthetic sorbents.
7. Notify environmental authorities to discuss disposal and cleanup of contaminated materials.

### Disposal
1. Contact manufacturer or supplier for advice on disposal.
2. Contact environmental authorities for advice on disposal.
3. Incinerate (approval of environmental authorities required).

BENZENE $C_6H_6$

# BENZOIC ACID   $C_6H_5COOH$

## IDENTIFICATION

**UN No. 9094**

### Common Synonyms

BENZENECARBOXYLIC ACID
CARBOXYBENZENE
PHENYLFORMIC ACID
PHENYLCARBOXYLIC ACID

### Observable Characteristics

Colourless to white crystals, powder or flake. Faint, pleasant, aromatic odour.

### Manufacturers

Dow Chemical Canada Inc., Delta, B.C.

### Transportation and Storage Information

**Shipping State:** Solid.
**Classification:** Not regulated.
**Inert Atmosphere:** No requirement.
**Venting:** Open.

**Label(s):** None.
**Storage Temperature:** Ambient.

**Grades or Purity:** Technical.
**Containers and Materials:** Paper bags, barrels, drums, tank trucks, tank cars; stainless steel.

### Physical and Chemical Characteristics

**Physical State** (20°C, 1 atm): Solid.
**Solubility** (Water): 0.29 g/100 mL (25°C).
**Molecular Weight:** 122.1
**Vapour Pressure:** 1 mm Hg (96°C); 6.5 mm Hg (121°)C.
**Boiling Point:** 249.2°C.

**Floatability** (Water): Sinks.
**Odour:** Faint, pleasant, slightly aromatic odour.
**Flash Point:** 121°C (c.c.).
**Vapour Density:** 4.2
**Specific Gravity:** 1.27 (15°C); 1.32 (28°C).

**Colour:** Clear to white.
**Explosive Limits:** No information.
**Melting Point:** 121.7 to 122.4°C; (sublimes at 100°C).

## HAZARD DATA

### Human Health

**Symptoms:** Inhalation (dust): irritation of respiratory tract, nose, throat. Contact: irritation of skin and corrosive and irritating to eyes. Ingestion: nausea, gastrointestinal problems.
**Toxicology:** Moderately toxic by contact, inhalation and ingestion.
**TLV®-** No information.
**Short-term Inhalation Limits -** No information.

$LD_{50}$ - Oral: rat = 2.53 g/kg
Oral: mouse = 2.37 g/kg

$LC_{50}$ - No information.
Delayed Toxicity - None known.

### Fire

**Fire Extinguishing Agents:** Dry powder, chemical foam, water fog, carbon dioxide.
**Behaviour in Fire:** No information.
**Ignition Temperature:** 570 to 574°C.

**Burning Rate:** No information.

### Reactivity

**With Water:** No reaction.
**With Common Materials:** Can react with oxidizing materials.
**Stability:** Stable.

### Environment

**Water:** Prevent entry into water intakes and waterways. Harmful to aquatic life. Fish toxicity: 200 ppm/7 h/goldfish/lethal/freshwater; 500 ppm/1 h/sunfish/lethal/freshwater; 180 mg/L/96 h/mosquito fish/TLm/freshwater; BOD: 125 to 165%, 5 days.
**Land-Air:** No information.
**Food Chain Concentration Potential:** None.

## EMERGENCY MEASURES

**Special Hazards**

**Immediate Responses**

Keep non-involved people away from spill site. If there is a fire, call Fire Department. Avoid contact and inhalation (dust and fumes). Fight fire from upwind. If water is used, dike area to prevent runoff. Notify supplier. Notify environmental authorities.

**Protective Clothing and Equipment**

In fire, Respiratory protection – self-contained breathing apparatus and suit. Otherwise, dust respirator. Goggles – (mono), tight fitting. Gloves – rubber or plastic. Coveralls.

**Fire and Explosion**

Use water fog, dry chemical, foam or carbon dioxide to extinguish.

**First Aid**

Move victim out of spill area to fresh air. Call for medical assistance, but start first aid at once. Contact: remove contaminated clothing, wash eyes and skin with plenty of warm water. Ingestion: give plenty of water to conscious victim to drink. Keep victim warm and quiet. If medical assistance is not immediately available, transport victim to hospital, doctor, or clinic.

## ENVIRONMENTAL PROTECTION MEASURES

**Response**

**Water**
1. Stop or reduce discharge if safe to do so.
2. Contact manufacturer or supplier for advice.
3. If possible, contain discharge by damming or water diversion.
4. Dredge or vacuum pump to remove contaminants, liquids and contaminated bottom sediments.
5. Notify environmental authorities to discuss disposal and cleanup of contaminated materials.

**Land-Air**
1. Stop or reduce discharge if safe to do so.
2. Contact manufacturer or supplier for advice.
3. Contain spill by diking with earth or other barrier.
4. Remove material by manual or mechanical means.
5. Recover undamaged containers.
6. Notify environmental authorities to discuss disposal and cleanup of contaminated materials.

**Disposal**
1. Contact manufacturer or supplier for advice on disposal.
2. Contact environmental authorities for advice on disposal.
3. Incinerate (approval of environmental authorities required).

BENZOIC ACID  $C_6H_5COOH$

# BISPHENOL A  $(CH_3)_2C(C_6H_4OH)_2$

## IDENTIFICATION

### Common Synonyms
4-4'-ISOPROPYLIDENEDIPHENOL
p-p'-ISOPROPYLIDENEDIPHENOL
2,2-bis (4-hydroxyphenyl) propane

### Observable Characteristics
White crystals or flakes.
Weak phenolic odour.

### Manufacturers
Gulf Canada Ltd., Montreal, Quebec.

### Transportation and Storage Information
**Shipping State:** Solid.
**Classification:** None.
**Inert Atmosphere:** No requirement.
**Venting:** Open.
**Label(s):** Not regulated.
**Storage Temperature:** Ambient.
**Grades or Purity:** Commercial; high purity.
**Containers and Materials:** Bags and drums.

### Physical and Chemical Characteristics
**Physical State (20°C, 1 atm):** Solid.
**Solubility (Water):** Insoluble.
**Molecular Weight:** 228.3
**Vapour Pressure:** 5.2 mm Hg (220°C).
**Boiling Point:** 220°C (4 mm Hg).
**Floatability (Water):** Sinks.
**Odour:** Very weak phenolic.
**Flash Point:** 79°C (c.c.)
**Vapour Density:** 7.9
**Specific Gravity:** 1.2 (25°C).
**Colour:** White.
**Explosive Limits:** No information.
**Melting Point:** 153°C.

## HAZARD DATA

### Human Health
**Symptoms: Inhalation:** dust is irritating to respiratory passages, coughing and sneezing. **Contact: eyes** - dust is irritating. **Ingestion:** nausea and vomiting.
**Toxicology:** Moderately toxic by inhalation and ingestion.
**TLV -** No information. **LC$_{50}$ -** No information. **LD$_{50}$ - Oral:** rat = 4.04 g/kg
**Short-term Inhalation Limits -** No information. **Delayed Toxicity -** No information.

### Fire
**Fire Extinguishing Agents:** Use foam, dry chemical, carbon dioxide or water.
**Behaviour in Fire:** No information. **Burning Rate:** No information.
**Ignition Temperature:** No information.

### Reactivity
**With Water:** No reaction.
**With Common Materials:** No reactions known.
**Stability:** Stable.

### Environment
**Water:** Prevent entry into water intakes and waterways. Hazardous to aquatic life. Little information available on the hazards of this material.
**Land-Air:** No information.
**Food Chain Concentration Potential:** No information.

## EMERGENCY MEASURES

**Special Hazards**
COMBUSTIBLE.

**Immediate Responses**
Keep non-involved people away from spill site. Call Fire Department. Notify manufacturer for advice. Notify environmental authorities.

**Protective Clothing and Equipment**
Suitable dust respirator and protective clothing as required.

**Fire and Explosion**
Use foam, dry chemical, carbon dioxide or water to extinguish fires.

**First Aid**
Move victim out of spill site to fresh air. Call for medical assistance, but start first aid at once. <u>Inhalation</u>: if breathing has stopped give artificial respiration; if laboured, give oxygen. <u>Contact</u>: skin and eyes - flush with large amounts of water. <u>Ingestion</u>: give water to conscious victim to drink. If medical help is not immediately available, transport victim to hospital, doctor or clinic.

## ENVIRONMENTAL PROTECTION MEASURES

**Response**

**Water**
1. Stop or reduce discharge if safe to do so.
2. Contact manufacturer or supplier for advice.
3. If possible, contain discharge by damming or water diversion.
4. Dredge or vacuum pump to remove contaminants, liquids and contaminated bottom sediments.
5. Notify environmental authorities to discuss disposal and cleanup of contaminated materials.

**Land-Air**
1. Stop or reduce discharge if safe to do so.
2. Contact manufacturer or supplier for advice.
3. Remove material by manual or mechanical means.
4. Recover undamaged containers.
5. Notify environmental authorities to discuss disposal and cleanup of contaminated materials.

**Disposal**
1. Contact manufacturer or supplier for advice on disposal.
2. Contact environmental authorities for advice on disposal.
3. Incinerate (approval of environmental authorities required).

BISPHENOL A   $(CH_3)_2C(C_6H_4OH)_2$

# BORAX   $Na_2B_4O_7 \cdot 10H_2O$

## IDENTIFICATION

**Common Synonyms**
SODIUM BORATE, decahydrate
SODIUM TETRABORATE, decahydrate

**Observable Characteristics**
White crystals or powder.
Odourless.

**Manufacturers**

No Canadian manufacturers.
Canadian suppliers:
Bate Chemical, Toronto, Ontario.
Canada Colours and Chemicals, Toronto, Ontario.
Lawrason, SF, London, Ontario.
Van Waters and Rogers, Vancouver, B.C.

Canadian Industries Ltd., Toronto, Ontario.
Harrisons and Crosfield, Toronto, Ontario.
Pigment and Chemical, Montreal, Quebec.
St. Lawrence Chemical, Montreal, Quebec.

Originating from:
U.S. Borax, USA

Kerr-McGee, USA

### Transportation and Storage Information

**Shipping State:** Solid.
**Classification:** None.
**Inert Atmosphere:** No requirement.
**Venting:** Open.

**Label(s):** Not regulated.
**Storage Temperature:** Ambient.

**Grades or Purity:** Refined and technical.
**Containers and Materials:** Barrels, bags.

### Physical and Chemical Characteristics

**Physical State (20°C, 1 atm):** Solid.
**Solubility (Water):** 2.0 g/100 mL (0°C); 170 g/100 mL (100°C).
**Molecular Weight:** 381.4
**Vapour Pressure:** No information.
**Boiling Point:** Loses 10 $H_2O$ at 320°C.

**Floatability (Water):** Sinks and mixes.
**Odour:** None.
**Flash Point:** Not flammable.
**Vapour Density:** No information.
**Specific Gravity:** 1.7 (20°C).

**Colour:** White.
**Explosive Limits:** Not flammable.
**Melting Point:** Loses 8 $H_2O$ (60°C); after which it is metaborate, which melts at 75°C.

## HAZARD DATA

### Human Health

**Symptoms:** Contact: skin and eyes - redness and pain. Ingestion: nausea, vomiting and diarrhea.
**Toxicology:** Relatively nontoxic.
**TLV** - 5 mg/m³ (dust).                    $LC_{50}$ - No information.                    $LD_{50}$ - Oral: rat = 2.7 g/kg
**Short-term Inhalation Limits** - No information.   **Delayed Toxicity** - No information.   $LD_{Lo}$ - Oral: human = 0.71 g/kg

### Fire

**Fire Extinguishing Agents:** Not combustible. Most fire extinguishing agents may be used on fires involving borax.
**Behaviour in Fire:** Not combustible.     **Burning Rate:** Not combustible.
**Ignition Temperature:** Not combustible.

### Reactivity

**With Water:** No reaction; soluble.
**With Common Materials:** Reacts with zirconium.
**Stability:** Stable.

### Environment

**Water:** Prevent entry into water intakes and waterways. Harmful to aquatic life. Aquatic toxicity rating = 100 to 1 000 ppm/96 h/TLm/freshwater; 8 200 ppm/48 h/mosquito fish/TLm/freshwater; BOD: no information.
**Land-Air:** No information.
**Food Chain Concentration Potential:** No information.

## EMERGENCY MEASURES

**Special Hazards**

**Immediate Responses**

Keep non-involved people away from spill site. Dike area to prevent runoff from rainwater or water application. Contact manufacturer. Notify environmental authorities.

**Protective Clothing and Equipment**

Suitable dust respirator and protective clothing as required.

**Fire and Explosion**

Not combustible. Most fire extinguishing agents may be used on fires involving borax.

**First Aid**

Move victim out of spill site to fresh air. Call for medical assistance, but start first aid at once. <u>Inhalation:</u> if victim is not breathing, give artificial respiration; if laboured, give oxygen. <u>Contact:</u> eyes and skin - irrigate with plenty of water. <u>Ingestion:</u> give water to conscious victim to drink. If medical help is not immediately available, transport victim to doctor, clinic or hospital.

## ENVIRONMENTAL PROTECTION MEASURES

**Response**

**Water**
1. Stop or reduce discharge if safe to do so.
2. Contact manufacturer or supplier for advice.
3. If possible, contain discharge by damming or water diversion.
4. Dredge or vacuum pump to remove contaminants, liquids and contaminated bottom sediments.
5. Notify environmental authorities to discuss disposal and cleanup of contaminated materials.

**Land-Air**
1. Stop or reduce discharge if safe to do so.
2. Contact manufacturer or supplier for advice.
3. Dike to prevent runoff from rainwater or water application.
4. Remove material by manual or mechanical means.
5. Recover undamaged containers.
6. Notify environmental authorities to discuss disposal and cleanup of contaminated materials.

**Disposal**
1. Contact manufacturer or supplier for advice on disposal.
2. Contact environmental authorities for advice on disposal.

BORAX $Na_2B_4O_7 \cdot 10H_2O$

# BORIC ACID  $H_3BO_3$

## IDENTIFICATION

| Common Synonyms | Observable Characteristics | Manufacturers | |
|---|---|---|---|
| ORTHOBORIC ACID<br>BORACIC ACID | Colourless or white crystals or powder. Odourless. | No Canadian manufacturers.<br>Canadian suppliers:<br>Canada Colours and Chemicals, Toronto, Ont.<br>S.F. Lawrason and Co., London, Ont.<br>McArthur Chemical.<br><br>Canadian Industries Ltd., Toronto, Ont.<br>Harrisons and Crosfield, Toronto, Ont.<br>Pigment and Chemical, Montreal, Que.<br>St. Lawrence Chemical, Montreal, Que. | Originating from:<br>US Borax, USA<br><br><br><br>Kerr - McGee, USA |

### Transportation and Storage Information

**Shipping State:** Solid.
**Classification:** None.
**Inert Atmosphere:** No requirement.
**Venting:** Open.

**Label(s):** Not regulated.
**Storage Temperature:** Ambient.

**Grades or Purity:** Technical, 99.9%.
**Containers and Materials:** Cans, kegs, drums.

### Physical and Chemical Characteristics

**Physical State (20°C, 1 atm):** Solid.
**Solubility (Water):** 6.4 g/100 mL (20°C); 27.6 g/100mL (100°C).
**Molecular Weight:** 61.8
**Vapour Pressure:** 15 mm Hg (21°C); 46 mm Hg (38°C).
**Boiling Point:** Loses 1.5 $H_2O$ at 300°C. Converts to $HBO_2$ (169°C).

**Floatability (Water):** Sinks and mixes.
**Odour:** Odourless.
**Flash Point:** Not flammable.
**Vapour Density:** No information.
**Specific Gravity:** 1.4 (20°C).

**Colour:** Colourless to white.
**Explosive Limits:** Not flammable.
**Melting Point:** Loses 1.5 $H_2O$ at 300°C. Converts to $HBO_2$ at 169°C.

## HAZARD DATA

### Human Health

**Symptoms:** Toxic by skin absorption. **Contact:** skin - rash, blistering, vomiting, diarrhea, convulsions, anemia; eyes - redness and pain. **Ingestion:** vomiting and diarrhea, lethargy, convulsions, jaundice, cyanosis, collapse.
**Toxicology:** Highly toxic by oral route; moderate by other routes.
**TLV** - No information. **$LC_{50}$** - No information. **$LD_{50}$** - Oral: rat = 2.7 g/kg
**Short-term Inhalation Limits** - No information. **Delayed Toxicity** - No information. **$LD_{Lo}$** - Oral: human = 0.214 g/kg

### Fire

**Fire Extinguishing Agents:** Not combustible. Most fire extinguishing agents may be used on fires involving boric acid.
**Behaviour in Fire:** Not combustible.
**Ignition Temperature:** Not combustible. **Burning Rate:** Not combustible.

### Reactivity

**With Water:** No reaction; moderately soluble.
**With Common Materials:** Reacts with potassium and acetic anhydride.
**Stability:** Stable.

### Environment

**Water:** Prevent entry into water intakes and waterways. Harmful to aquatic life. **Fish toxicity:** 1 800 ppm/24 h/mosquito fish/TLm/freshwater; BOD: None.
**Land-Air:** No information.
**Food Chain Concentration Potential:** None.

## EMERGENCY MEASURES

**Special Hazards**

**Immediate Responses**

Keep non-involved people away from spill site. Dike area to prevent runoff from rainwater or water application. Contact manufacturer. Notify environmental authorities.

**Protective Clothing and Equipment**

Suitable dust respirator and protective clothing as required.

**Fire and Explosion**

Not combustible. Most fire extinguishing agents may be used on fires involving boric acid.

**First Aid**

Move victim out of spill site to fresh air. Call for medical assistance, but start first aid at once. <u>Inhalation</u>: if breathing has stopped, give artificial respiration; if laboured, give oxygen. <u>Contact</u>: eyes and skin – irrigate with large amounts of water. <u>Ingestion</u>: give water to conscious victim to drink. If medical help is not immediately available, transport victim to doctor, clinic or hospital.

## ENVIRONMENTAL PROTECTION MEASURES

**Response**

**Water**
1. Stop or reduce discharge if safe to do so.
2. Contact manufacturer or supplier for advice.
3. If possible, contain discharge by damming or water diversion.
4. Dredge or vacuum pump to remove contaminants, liquids and contaminated bottom sediments.
5. Notify environmental authorities to discuss disposal and cleanup of contaminated materials.

**Land-Air**
1. Stop or reduce discharge if safe to do so.
2. Contact manufacturer or supplier for advice.
3. Dike to prevent runoff from rainwater or water application.
4. Remove material by manual or mechanical means.
5. Recover undamaged containers.
6. Notify environmental authorities to discuss disposal and cleanup of contaminated materials.

**Disposal**
1. Contact manufacturer or supplier for advice on disposal.
2. Contact environmental authorities for advice on disposal.

BORIC ACID  $H_3BO_3$

# 1,3-BUTADIENE  $CH_2=CHCH=CH_2$

## IDENTIFICATION

UN No. 1010

| Common Synonyms | Observable Characteristics | Manufacturers | |
|---|---|---|---|
| BIETHYLENE<br>DIVINYL<br>VINYLETHYLENE<br>BIVINYL | Colourless gas. Mildly aromatic odour. | Canadian manufacturer.<br>Polysar, Sarnia, Ontario.<br>Canadian supplier:<br>Dow Chemical Canada Inc., Sarnia, Ontario.<br>Monsanto Canada, Sarnia, Ontario.<br>Polysar, Sarnia, Ontario. | Originating from:<br>Dow Chemical, USA<br>Monsanto, USA<br>Northern Petrochem, USA |

### Transportation and Storage Information

**Shipping State:** Liquid (compressed gas).
**Classification:** Flammable gas.
**Inert Atmosphere:** No requirement.
**Venting:** Safety-relief.
**Pump Type:** Centrifugal, gear, etc. Explosion-proof.

**Label(s):** Red label - FLAMMABLE GAS; Class 2.1.
**Storage Temperature:** Ambient.
**Hose Type:** Rubber, Hypalon, Viton, flexible stainless steel.

**Grades or Purity:** Commercial, 98% (inhibited).
**Containers and Materials:** Cylinders, tank cars, tank trucks.

### Physical and Chemical Characteristics

**Physical State (20°C, 1 atm):** Gas.
**Solubility (Water):** 0.074 g/100 mL (20°C).
**Molecular Weight:** 54.1
**Vapour Pressure:** 900 mm Hg (0°C); 1 500 mm Hg (14°C).
**Boiling Point:** -4.4°C.

**Floatability (Water):** Floats and boils.
**Odour:** Mildly aromatic (4 mg/m$^3$, odour threshold).
**Flash Point:** -76°C (c.c.).
**Vapour Density:** 1.9 (20°C).
**Specific Gravity:** (liquid) 0.62 (20°C).

**Colour:** Colourless.
**Explosive Limits:** 2 to 12 %.
**Melting Point:** -108.9°C.

## HAZARD DATA

### Human Health

**Symptoms:** Inhalation: coughing, dizziness, dullness, unconsciousness. Contact: skin - frostbite; burns.
**Toxicology:** Moderately toxic by inhalation.
TLV®- Inhalation 1 000 ppm; 2 200 mg/m$^3$.
Short-term Inhalation Limits - 1 250 ppm; 2 750 mg/m$^3$ (15 min).
Inhalation - rat = 2 280 ppm/4 h
LC$_{50}$ - No information.
Delayed Toxicity - Suspected carcinogen.
LD$_{50}$ - Oral: rat = 5.48 g/kg

### Fire

**Fire Extinguishing Agents:** Stop flow of gas before attempting to extinguish fire. Use dry chemical, carbon dioxide, foam or water spray. Use water spray to cool fire-exposed containers and protect men effecting shut off.
**Behaviour in Fire:** At elevated temperatures (as in fire) polymerization may occur. If this takes place in a container, violent rupture may occur.
**Ignition Temperature:** 420°C.
**Burning Rate:** 8.0 mm/min.

### Reactivity

**With Water:** No reaction.
**With Common Materials:** May polymerize. Reacts with phenol and crotonaldehyde. May form explosive peroxides when exposed to air. Can react with oxidizing agents.
**Stability:** Stable if inhibited.

### Environment

**Water:** Prevent entry into water intakes and waterways. Aquatic toxicity rating = 10 to 100 ppm/96 h/TLm/freshwater; 71.5 mg/L/24 h/pin perch/TLm/freshwater; BOD: No information.
**Land-Air:** No information.
**Food Chain Concentration Potential:** No information.

## EMERGENCY MEASURES

### Special Hazards
FLAMMABLE. DANGER OF VIOLENT POLYMERIZATION.

### Immediate Responses
Keep non-involved people away from spill site. Issue warning: "FLAMMABLE". Call Fire Department. Eliminate all sources of ignition. Contact manufacturer for advice. Dike to prevent runoff. Stop or reduce discharge, if this can be done without risk. Notify environmental authorities.

### Protective Clothing and Equipment
In fires; Respiratory protection - self-contained breathing apparatus and standard fire protection suits. Gloves - rubber. Outer protective clothing - as required. Acid suit or coveralls. Boots - high, rubber (pants worn outside boots).

### Fire and Explosion
Stop flow of gas before attempting to extinguish fires. For small fires, use dry chemical or carbon dioxide. In large fires, use foam or water spray. In fire, polymerization may occur. If this takes place in a container, violent rupture may occur. Cool tanks with straight streams of water. Beware of tank explosion. Use unmanned monitor nozzle. Apply from side. Keep clear of tank heads. If in doubt, withdraw from area and let fire burn.

### First Aid
Move victim out of spill area to fresh air. Call for medical assistance, but start first aid at once. Contact: eyes - (liquid or vapour) irrigate immediately with plenty of water for at least 15 minutes; skin - flush skin with plenty of water for at least 15 minutes; at the same time, remove contaminated clothing. Inhalation: if breathing has stopped, give artificial respiration; if breathing is laboured, give oxygen. If medical assistance is not immediately available, transport victim to hospital, doctor or clinic.

## ENVIRONMENTAL PROTECTION MEASURES

### Response

**Water**
1. Stop or reduce discharge if safe to do so.
2. Contact manufacturer or supplier for advice.
3. Notify environmental authorities to discuss disposal and cleanup of contaminated materials.

**Land-Air**
1. Stop or reduce discharge if safe to do so.
2. Contact manufacturer or supplier for advice.
3. Notify environmental authorities to discuss disposal and cleanup of contaminated materials.

### Disposal
1. Contact manufacturer or supplier for advice on disposal.
2. Contact environmental authorities for advice on disposal.
3. Incinerate (approval of environmental authorities required).

1,3-BUTADIENE $CH_2=CHCH=CH_2$

# BUTANE n-C$_4$H$_{10}$

## IDENTIFICATION

UN No. 1011

### Common Synonyms
BUTYL HYDRIDE
n-BUTANE
LPG (see also propane)

### Observable Characteristics
Colourless. Mild natural gas-like odour.

### Manufacturers
Dome Petroleum Limited, Calgary, Alta.
Pacific Petroleums, Ltd., Calgary, Alta.
Mobil Oil Canada Ltd, Calgary, Alta.
Home Oil Co. Ltd., Calgary, Alta.

### Transportation and Storage Information
**Shipping State:** Liquid (compressed gas).
**Classification:** Flammable gas.
**Inert Atmosphere:** No requirement.
**Venting:** Safety relief.
**Pump Type:** Rotary, LPG, compressor.

**Label(s):** Red label - FLAMMABLE GAS; Class 2.1.
**Storage Temperature:** Ambient.
**Hose Type:** Special LPG type, reinforced high pressure.

**Grades or Purity:** Pure 99.4%; commercial; technical (97.6%).
**Containers and Materials:** Cylinders, tank cars, tank trucks; steel.

### Physical and Chemical Characteristics
**Physical State** (20°C, 1 atm): Gas.
**Solubility** (Water): 0.003 g/100 mL (15°C); 0.0008 g/100 mL (20°C); 0.0021 g/100 mL (38°C).
**Molecular Weight:** 58.1
**Vapour Pressure:** 1 823 mm Hg (25°C).
**Boiling Point:** -1°C.

**Floatability** (Water): Floats and boils.
**Odour:** Mild, natural gas-like (~5 000 ppm, odour threshold.
**Flash Point:** -60°C (c.c.).
**Vapour Density:** 2.0
**Specific Gravity:** (liquid) 0.60 (-1°C).

**Colour:** Colourless.
**Explosive Limits:** 1.6 to 8.5%.
**Melting Point:** -138°C.

## HAZARD DATA

### Human Health
**Symptoms: Contact:** skin - no effect from gas; liquid causes "frostbite"; eyes - gas has no effect but liquid splashed in eyes causes stinging, watering, inflammation and cloudiness. **Inhalation:** asphyxia, irregular breathing, headache, fatigue, mental confusion, nausea and vomiting, loss of consciousness.
**Toxicology:** An asphyxiant.
TLV - (inhalation) 800 ppm; 1 900 mg/m$^3$
Short-term Inhalation Limits - No information.

LC$_{50}$ - Inhalation: rat = 658 g/m$^3$/4 h
Delayed Toxicity - No information.

LD$_{50}$ - No information.

### Fire
**Fire Extinguishing Agents:** Do not attempt to extinguish fire until leak has been shut off. For small fires use dry chemical. Let large fire burn. Use water spray to cool tanks exposed to fire and protect fire fighters.
**Behaviour in Fire:** When exposed to heat and flame, containers may rupture. Flash back may occur along vapour trail.
**Ignition Temperature:** 360°C.   **Burning Rate:** 7.9 mm/min.

### Reactivity
**With Water:** No reaction.
**With Common Materials:** May react with oxidizing agents.
**Stability:** Stable.

### Environment
**Water:** Prevent entry into water intakes and waterways. Toxic to aquatic life in high concentrations; which may not be achievable because of solubility. Aquatic toxicity rating >1 000 ppm/96 h/TLm/freshwater.
**Land-Air:** No information.
**Food Chain Concentration Potential:** No information.

## EMERGENCY MEASURES

**Special Hazards**

FLAMMABLE. Containers may rupture violently.

**Immediate Responses**

Keep non-involved people away from spill site. Issue warning: "FLAMMABLE". Call Fire Department. Eliminate all ignition sources. Use only non-sparking tools. Contact supplier or producer for guidance and assistance. Stay upwind. Use water spray to control vapour. Let fire burn. Do not attempt to extinguish fire until leak has been shut off. Notify environmental authorities.

**Protective Clothing and Equipment**

In fires, Respiratory protection - self-contained breathing apparatus and standard fire protection suit. Gloves - rubber. Outer protective clothing - as required. Acid suit or coveralls. Boots - high, rubber (pants worn outside boots).

**Fire and Explosion**

Do not put out fire until leak has been shut off. Use water spray to cool fire-exposed containers and to protect fire fighters. Beware of tank explosion.

**First Aid**

Move victim out of spill area to fresh air. Call for medical assistance, but start first aid at once. Inhalation: if breathing has stopped, give artificial respiration. If breathing is laboured, give oxygen. Contact: if exposed to liquefied butane, remove clothing, irrigate eyes and flush skin with water. Treat as for frostbite; do not rub affected areas. If medical assistance is not immediately available, transport victim to hospital, doctor or clinic.

## ENVIRONMENTAL PROTECTION MEASURES

**Response**

**Water**
1. Stop or reduce discharge if safe to do so.
2. Contact manufacturer or supplier for advice.
3. Notify environmental authorities to discuss disposal and cleanup of contaminated materials.

**Land-Air**
1. Stop or reduce discharge if safe to do so.
2. Contact manufacturer or supplier for advice.
3. Notify environmental authorities to discuss disposal and cleanup of contaminated materials.

**Disposal**
1. Contact manufacturer or supplier for advice on disposal.
2. Contact environmental authorities for advice on disposal.
3. Incinerate (approval of environmental authorities required).

BUTANE  n-$C_4H_{10}$

# BUTYL ALCOHOL  $C_4H_9OH$

## IDENTIFICATION

UN No. UN No. 1120

### Common Synonyms

n-BUTYL ALCOHOL $CH_3(CH_2)_2CH_2OH$
1-butanol, 1-hydroxybutane, n-propylcarbinol
TERT-BUTYL ALCOHOL $(CH_3)_3COH$
2-methyl-2-propanol, t-butyl alcohol, tert-butanol, trimethylcarbinol
SEC-BUTYL ALCOHOL $CH_3CH_2CH$ OH $CH_3$
2-butanol, s-butyl alcohol, methylethylcarbinol

### Observable Characteristics

Colourless liquids or crystals.
Strong alcohol-like odour for n-butanol.
Camphor-like odour for tert-butanol.

### Manufacturers

BASF Canada Ltd., Laval, Quebec.

### Transportation and Storage Information

**Shipping State:** Liquid or solid.
**Classification:** Flammable liquids.
**Inert Atmosphere:** No requirement.
**Venting:** Open (flame arrester); or pressure-vacuum.
**Label(s):** Red label - FLAMMABLE LIQUID; Class 3.2, Group I, II or III.
**Storage Temperature:** Ambient.
**Hose Type:** Natural rubber, polyethylene, butyl.
**Pump Type:** Standard flammable liquid types.
**Grades or Purity:** 99+% technical.
**Containers and Materials:** Cans, drums, tank trucks; steel, stainless steel and aluminum.

### Physical and Chemical Characteristics

**Physical State** (20°C, 1 atm): Liquid (n and sec); solid (tert).
**Solubility (Water):** n: 7.7 g/100 mL (20°C); sec: 12.5 g/100 mL (20°C); tert: soluble in all proportions.
**Molecular Weight:** 74.1
**Vapour Pressure:** n: 4.4 mm Hg (20°C), 6.5 mm Hg (25°C); 10 mm Hg (30°C); sec: 12 mm Hg (20°C), 24 mm Hg (30°C); tert: 31 mm Hg (20°C), 42 mm Hg (25°C), 56 mm Hg (30°C).
**Floatability (Water):** Floats and mixes.
**Odour:** Sharp alcohol odour, pungent, strong, non-residual. Tert. is camphor-like (0.3 to 2.0 ppm (n); 0.12 to 0.56 ppm (sec) odour threshold).
**Flash Point:** n: 35°C (c.c.); sec: 23.9°C (c.c.); tert: 11.0°C (c.c.).
**Vapour Density:** 2.6
**Specific Gravity:** (liquid) 0.81 (20°C) (n); 0.81 (20°C) (sec); 0.79 (20°C) (tert).
**Boiling Point:** 117.7°C (n); 99.5°C (sec); 83°C (tert).
**Colour:** Colourless.
**Explosive Limits:** 1.4 to 11.2% (n); 1.7 to 9.8% (sec); 2.4 to 8.0% (tert).
**Melting Point:** -89.3°C (n); -89°C (sec); 25.7°C (tert).

## HAZARD DATA

### Human Health

**Symptoms: Contact:** skin - dermatitis; eyes - redness and burns. **Inhalation:** headache, dizziness, nausea and narcosis. **Ingestion:** abdominal pains; nausea, dullness.
**Toxicology:** Moderately toxic by exposure and inhalation; n-butyl alcohol can be absorbed by the skin giving symptoms similar to inhalation. TLV*- (skin) 50 ppm, 150 mg/m³ (n); LC$_{50}$ - No information. LD$_{50}$ - Oral: rat = 0.79 g/kg (n)
100 ppm, 305 mg/m³ (sec); Delayed Toxicity - No information. LD$_{50}$ - Oral: rat = 6.48 g/kg (sec).
100 ppm, 300 mg/m³ (tert). LD$_{50}$ - Oral: rat = 3.5 g/kg (tert).
Short-term Inhalation Limits - 150 ppm; 455 mg/m³ (15 min) (sec); 150 ppm; 450 mg/m³ (15 min) (tert).

### Fire

**Fire Extinguishing Agents:** Use carbon dioxide, dry chemical or alcohol foam. Water may be ineffective. **Burning Rate:** 3.2 mm/min (n); 3.1 mm/min (sec); 3.4 mm/min (tert).
**Behaviour in Fire:** Flashback may occur along vapour trail.
**Ignition Temperature:** 343°C (n); 405°C (sec); 478°C (tert).

### Reactivity

**With Water:** No reaction; soluble.
**With Common Materials:** May react with oxidizers.
**Stability:** Stable.

### Environment

**Water:** Prevent entry into water intakes and waterways. Fish toxicities: (n) = 1 400 mg/L/24 h/creek chub/LC100/freshwater; (sec) = 4 300 mg/L/24 h/goldfish/LD$_{50}$/freshwater; (tert) = 6 000 mg/L/24 h/goldfish/LC$_{100}$/freshwater; Aquatic toxicity rating >1 000 ppm/96 h/TLm/freshwater; BOD: (n) 150%, 5 days, (sec) 187%, 5 days; (tert) 0%, 5 days.
**Land-Air:** No information.
**Food Chain Concentration Potential:** None.

## EMERGENCY MEASURES

**Special Hazards**
FLAMMABLE.

**Immediate Responses**
Keep non-involved people away from spill site. Issue warning: "FLAMMABLE". Call fire department. Eliminate all sources of ignition. Stop or reduce discharge, if safe to do so. Work from upwind and use water spray to control vapours. Dike to contain spill. Notify manufacturer or supplier. Notify environmental authorities.

**Protective Clothing and Equipment**
Respiratory protection - self-contained breathing apparatus. Gloves - rubber or plastic. Coveralls - rubber or plastic. Suit, rubber or plastic. Gastight suit. Boots - rubber, high (pants worn outside boots).

**Fire and Explosion**
Use dry chemical, carbon dioxide or alcohol foam to extinguish. Water may be ineffective on fire, but may be used to cool fire-exposed containers.

**First Aid**
Move victim out of spill area to fresh air. Call for medical assistance, but start first aid at once. Inhalation: give artificial respiration if breathing has stopped and oxygen if breathing is laboured. Contact: remove contaminated clothing and irrigate eyes and affected skin thoroughly with plenty of warm water. Ingestion: give milk or water to conscious victim to drink and induce vomiting. If medical assistance is not immediately available, transport victim to hospital, clinic or doctor.

## ENVIRONMENTAL PROTECTION MEASURES

**Response**

Water
1. Stop or reduce discharge if safe to do so.
2. Contact manufacturer or supplier for advice.
3. If possible, contain discharge by damming or water diversion.
4. Notify environmental authorities to discuss disposal and cleanup of contaminated materials.

Land-Air
1. Stop or reduce discharge if safe to do so.
2. Contact manufacturer or supplier for advice.
3. Contain spill by diking with earth or other barrier.
4. Remove material with pumps or vacuum equipment and place in appropriate containers.
6. Absorb residual liquid on natural or synthetic sorbents.
7. Notify environmental authorities to discuss disposal and cleanup of contaminated materials.

**Disposal**
1. Contact manufacturer or supplier for advice on disposal.
2. Contact environmental authorities for advice on disposal.
3. Incinerate (approval of environmental authorities required).

BUTYL ALCOHOL $C_4H_9OH$

# BUTYLENE  $CH_3CH_2CH = CH_2$

## IDENTIFICATION

UN No. 1012

### Common Synonyms
LPG 1-BUTENE, α-BUTYLENE, ETHYLETHYLENE, cis-2-butene, trans-2-butene are also available and have similar properties to 1-butene.

### Observable Characteristics
Colourless gas. Slightly aromatic odour.

### Manufacturers
Polysar, Sarnia, Ontario.
Finachem Canada, Montreal, Quebec.
Esso Chemical Canada, Dartmouth, Nova Scotia.

### Transportation and Storage Information
**Shipping State:** Liquid (compressed gas).
**Classification:** Flammable gas.
**Inter Atmosphere:** No requirement.
**Venting:** Safety relief.
**Pump Type:** Steel, stainless steel.

**Label(s):** Red and white label - FLAMMABLE GAS; Class 2.1.
**Storage Temperature:** Ambient.
**Hose Type:** No information.

**Grades or Purity:** Technical 95%; CP 99.0%.
**Containers and Materials:** Cylinders, tanks; steel.

### Physical and Chemical Characteristics
**Physical State** (20°C, 1 atm): Gas.
**Solubility:** Insoluble.
**Molecular Weight:** 56.1
**Vapour Pressure:** 430 mm Hg (-20°C); 960 mm Hg (0°C); 1976 mm Hg (21°C).
**Boiling Point:** -6.3°C.

**Floatability** (Water): Floats and boils.
**Odour:** Slightly aromatic (69 ppb, odour threshold).
**Flash Point:** -80°C (c.c.).
**Vapour Density:** 1.9
**Specific Gravity:** 0.68 (-40°C).

**Colour:** Colourless.
**Explosive Limits:** 1.6 to 10%.
**Melting Point:** -185°C.

## HAZARD DATA

### Human Health
**Symptoms: Contact:** skin - liquid causes frosbite. **Inhalation:** rapid respiration, gasping for air, fatigue, nausea and vomiting.
**Toxicology:** Asphyxiant.
**TLV®** - No information.  **LC$_{50}$** - No information.  **LD$_{50}$** - No information.
**Short-term Inhalation Limits** - No information.  **Delayed Toxicity** - No information.

### Fire
**Fire Extinguishing Agents:** Stop flow of gas before attempting to extinguish fire. Most fire extinguishing agents may be used. Cool fire-exposed containers with water.
**Behaviour in Fire:** Flash back may occur along vapour trail.  **Burning Rate:** 8.8 mm/min.
**Ignition Temperature:** 385°C.

### Reactivity
**With Water:** No reaction.
**With Common Materials:** Can react with oxidizing materials.
**Stability:** Stable.

### Environment
**Water:** Prevent entry into water intakes and waterways. Toxicity to aquatic life unknown.
**Land-Air:** No information.
**Food Chain Concentration Potential:** None.

## EMERGENCY MEASURES

**Special Hazards**
FLAMMABLE.

**Immediate Responses**
Keep non-involved people away from spill site. Issue warning: "FLAMMABLE". CALL FIRE DEPARTMENT. Shut off flow of gas if it can be done without risk. Notify manufacturer. Notify environmental authorities.

**Protective Clothing and Equipment**
In fires and confined spaces Respiratory protection – self-contained breathing apparatus; otherwise, protective outer clothing as required.

**Fire and Explosion**
Stop flow of gas before attempting to extinguish fire. Most fire extinguishing agents may be used. Water may be used to cool fire-exposed containers. Flash back may occur along vapour trail.

**First Aid**
Move victim out of spill site to fresh air. Call for medical assistance, but start first aid at once. Contact: skin – treat as for frostbite, do not rub affected areas; flush affected areas with water; eyes – irrigate with water. Inhalation: if breathing has stopped give artificial respiration; if laboured, give oxygen. If medical assistance is not immediately available, transport victim to doctor, clinic or hospital.

## ENVIRONMENTAL PROTECTION MEASURES

**Response**

**Water**
1. Stop or reduce discharge if safe to do so.
2. Contact manufacturer or supplier for advice.
3. Notify environmental authorities to discuss disposal and cleanup of contaminated materials.

**Land-Air**
1. Stop or reduce discharge if safe to do so.
2. Contact manufacturer or supplier for advice.
3. Notify environmental authorities to discuss disposal and cleanup of contaminated materials.

**Disposal**
1. Contact manufacturer or supplier for advice on disposal.
2. Contact environmental authorities for advice on disposal.
3. Contents of contaminated cylinders may be incinerated (with approval of environmental authorities).

BUTYLENE   $CH_3CH_2CH=CH_2$

# BUTYRALDEHYDE (NORMAL)   CH3(CH2)2CHO

## IDENTIFICATION

UN No. 1129

### Common Synonyms
n-BUTYRALDEHYDE
BUTYRIC ALDEHYDE
n-BUTANAL

### Observable Characteristics
Colourless liquid. Pungent odour.

### Manufacturers
Canadian Manufacturer:
BASF Canada Ltd.,
Laval, Quebec.

### Transportation and Storage Information
**Shipping State:** Liquid.
**Classification:** Flammable liquid.
**Inert Atmosphere:** No requirement.
**Venting:** Pressure-vacuum.
**Pump Type:** No information.

**Label(s):** Red Label - FLAMMABLE LIQUID;
Class 3.2, Group II.
**Storage Temperature:** Ambient.
**Hose Type:** No information.

**Grades or Purity:** Water-saturated, 97%;
dry, 99.5%.
**Containers and Materials:** Drums, tank cars, tank trucks, steel.

### Physical and Chemical Characteristics
**Physical State (20°C, 1 atm):** Liquid.
**Solubility (Water):** 3.7 g/100 mL (0°C); 7.1 m/100 mL (20°C).
**Molecular Weight:** 72
**Vapour Pressure:** 71 mm Hg (20°C).
**Boiling Point:** 75 to 76°C.

**Floatability (Water):** Floats and mixes.
**Odour:** Pungent (0.005 to 0.048 ppm, odour threshold).
**Flash Point:** -6.6 (c.c.)
**Vapour Density:** 2.5
**Specific Gravity:** 0.81 (20°C).

**Colour:** Colourless.
**Explosive Limits:** 1.9 to 12.5%.
**Melting Point:** -96 to -99°C.

## HAZARD DATA

### Human Health
**Symptoms: Contact:** skin - burning sensation and blisters; eyes - watering, burning sensation. **Inhalation:** irritation of mucous membranes, difficulty breathing, headache, cyanosis, nausea, muscular weakness. **Ingestion:** burning sensation, difficulty breathing, nausea and vomiting, convulsions.
**Toxicology:** Moderately toxic by dermal route.
**TLV -** No information.    $LC_{50}$ - Inhalation: rat = 60 000 ppm/1/2 h    $LD_{50}$ - Oral: rat = 2.49 g/kg
**Short-term Inhalation Limits -** No information.    **Delayed Toxicity -** No information.

### Fire
**Fire Extinguishing Agents:** Use foam, carbon dioxide or dry chemical.  Water may be ineffective but may be used to cool fire-exposed tanks or disperse vapours.
**Behaviour in Fire:** Flash back may occur along vapour trail.  Fire may be difficult to control due to ease of reignition.
**Ignition Temperature:** 230°C.    **Burning Rate:** 4.4 mm/min.

### Reactivity
**With Water:** No reaction, slightly soluble.
**With Common Materials:** Reacts vigorously with chlorosulfonic acid, nitric acid, sulfuric acid and oleum. Can react with oxidizing materials.
**Stability:** Stable.

### Environment
**Water:** Prevent entry into water intakes and waterways.  Aquatic toxicity rating = 1 to 10 ppm/96 h/TLm/freshwater; BOD: 106%, 5 days.
**Land-Air:** No information.
**Food Chain Concentration Potential:** None.

## EMERGENCY MEASURES

**Special Hazards**

FLAMMABLE.

**Immediate Responses**

Keep non-involved people away from spill site. Issue warning: "FLAMMABLE". CALL FIRE DEPARTMENT. Eliminate all sources of ignition. Call manufacturer or supplier for advice. Stop or reduce discharge, if this can be done without risk. Avoid contact and inhalation. Contain spill area by diking. Notify environmental authorities.

**Protective Clothing and Equipment**

Respiratory protection - self contained breathing apparatus. Gloves - rubber or plastic. Boots - high, rubber (pants worn outside boots). Outerwear - as required, coveralls, acid suit.

**Fire and Explosion**

Use foam, dry chemical or carbon dioxide. Water may be ineffective but may be used to cool fire-exposed containers and to knock down vapours. Flash back may occur along vapour trail.

**First Aid**

Move victim out of spill site to fresh air. Call for medical assistance, but start first aid at once. Inhalation: if breathing has stopped, give artificial respiration; if laboured, give oxygen. Contact: eyes - immediately irrigate with water for at least 15 min; skin: flush with plenty of water and remove contaminated clothing. Ingestion: rinse mouth with water, keep calm and warm. If medical assistance is not immediately available, transport victim to hospital, doctor or clinic.

## ENVIRONMENTAL PROTECTION MEASURES

**Response**

**Water**
1. Stop or reduce discharge if safe to do so.
2. Contact manufacturer or supplier for advice.
3. If possible, contain discharge by damming or water diversion.
4. Dredge or vacuum pump to remove contaminants, liquids and contaminated bottom sediments.

**Land-Air**
1. Stop or reduce discharge if safe to do so.
2. Contact manufacturer or supplier for advice.
3. Contain spill by diking with earth or other barrier.
4. Remove material with pumps or vacuum equipment and place in appropriate containers.
5. Recover undamaged containers.
6. Absorb residual liquid on natural or synthetic sorbents.
7. Notify environmental authorities to discuss disposal and cleanup of contaminated materials.

**Disposal**
1. Contact manufacturer or supplier for advice on disposal.
2. Contact environmental authorities for advice on disposal.
3. Incinerate (approval of environmental authorities required).

BUTYRALDEHYDE (NORMAL)  $CH_3(CH_2)_2CHO$

# CALCIUM CARBIDE   $CaC_2$

## IDENTIFICATION

**UN No. 1402**

### Common Synonyms
CARBIDE
ACETYLENOGEN
CALCIUM ACETYLIDE

### Observable Characteristics
Grey to black powder or lumps. Odourless when dry; weak acetylene (garlic-like) odour when damp.

### Manufacturers
Cyanamid of Canada Ltd., Niagara Falls, Ontario.
Gulf Oil Canada Ltd., Shawinigan, Quebec.

### Transportation and Storage Information
**Shipping State:** Solid.
**Classification:** Dangerous when wet.
**Inert Atmosphere:** No requirement (dry).
**Venting:** Closed.

**Label(s):** Blue label - DANGEROUS WHEN WET; Class 4.3, Group II.
**Storage Temperature:** Ambient.

**Grades or Purity:** According to granule size.
**Containers and Materials:** Airtight and watertight metal containers.

### Physical and Chemical Characteristics
**Physical State (20°C, 1 atm):** Solid.
**Solubility (Water):** Decomposes in water.
**Molecular Weight:** 64.1
**Vapour Pressure:** No information.
**Boiling Point:** >2 300°C (melts at 2 300°C).

**Floatability (Water):** Sinks and reacts to form acetylene gas and calcium hydroxide.
**Odour:** Acetylene (garlic-like).
**Flash Point:** -17.8°C (c.c.) (as acetylene).
**Vapour Density:** 0.91 (as acetylene).
**Specific Gravity:** 2.22 (18°C).

**Colour:** Grey to black.
**Explosive Limits:** 2.5 to 100%.
**Melting Point:** 2 300°C.

## HAZARD DATA

### Human Health
**Symptoms: Contact:** skin and eyes - burns, ulceration, damage to mucous membranes. **Inhalation (acetylene):** burning sensation in nose and throat, difficulty breathing, coughing, chemical bronchitis. **Ingestion:** painful swallowing, mouth cavity turns white, vomiting, stomach cramps, rapid breathing, diarrhea, unconsciousness.
**Toxicology:** Moderate toxicity upon contact.
**TLV** - No information.
**Short-term Inhalation Limits** - No information.

$LC_{50}$ - No information.
**Delayed Toxicity** - No information.

$LD_{50}$ - No information.

### Fire
**Fire Extinguishing Agents:** Use a suitable dry powder. Carbon dioxide is ineffective. Do not use water, vaporizing liquid or foam.
**Behaviour in Fire:** Not flammable in dry state but produces acetylene gas on contact with water or moisture. Will generate sufficient heat on contact with water to ignite acetylene formed.
**Ignition Temperature:** 305°C (as acetylene).
**Burning Rate:** No information (as acetylene).

### Reactivity
**With Water:** Reacts exothermically to form acetylene gas and calcium hydroxide.
**With Common Materials:** Reacts with selenium, hydrogen chloride, magnesium, silver nitrate, stannous chloride, and sulfur.
**Stability:** Stable in absence of moisture.

### Environment
**Water:** Prevent entry into water intakes and waterways. Toxicity to aquatic life is unknown.
**Land-Air:** No information.
**Food Chain Concentration Potential:** None.

## EMERGENCY MEASURES

**Special Hazards**

DANGEROUS WHEN WET. Hazardous acetylene gas generated on contact with water.

**Immediate Responses**

Keep non-involved people away from spill site. Issue warning: "DANGEROUS WHEN WET". Call Fire Department (Caution - no water or foam). Contact manufacturer for advice. If there is any possibility of acetylene gas being generated, proceed as follows: Eliminate all sources of ignition. Use only spark-proof tools. Stop or reduce discharge, if this can be done without risk. Contain spill by diking, particularly if water is present. Allow acetylene gas to disperse - but evacuate area downwind as explosion hazard may exist. Notify environmental authorities.

**Protective Clothing and Equipment**

Respiratory protection - self-contained breathing apparatus. Gloves - cotton or rubber work gloves. Clothing - as required, coveralls, etc.

**Fire and Explosion**

Use a suitable dry powder. Carbon dioxide is ineffective. Do not use water, vaporizing liquids or foam.

**First Aid**

Move victim out of spill site to fresh air. Call for medical assistance, but start first aid at once. Contact: eyes - irrigate with water for at least 15 minutes; skin - remove contaminated clothing and flush affected areas with plenty of water. Inhalation: if breathing has stopped, give artificial respiration; if laboured, give oxygen. If medical assistance is not immediately available, transport victim to hospital, doctor or clinic.

## ENVIRONMENTAL PROTECTION MEASURES

**Response**

**Water**
1. Stop or reduce discharge if safe to do so.
2. Contact manufacturer or supplier for advice.
3. Notify environmental authorities to discuss disposal and cleanup of contaminated materials.

**Land-Air**
1. Stop or reduce discharge if safe to do so.
2. Contact manufacturer or supplier for advice.
3. Make sure that material does not contact water moist materials.
4. Notify environmental authorities to discuss disposal and cleanup of contaminated materials.

**Disposal**
1. Contact manufacturer or supplier for advice on disposal.
2. Contact environmental authorities for advice on disposal.

CALCIUM CARBIDE    $CaC_2$

# CALCIUM CARBONATE   $CaCO_3$

## IDENTIFICATION

| Common Synonyms | Observable Characteristics | Manufacturers |
|---|---|---|
| CHALK<br>AGRICULTURAL LIMESTONE<br>MARBLE | White powder or crystals. Odourless. | WR Barnes, Perth, Ontario.<br>Industrial Fillers, St. Armand, Quebec. |

### Transportation and Storage Information

**Shipping State:** Solid.
**Classification:** None.
**Inert Atmosphere:** No requirements.
**Venting:** Open.

**Label(s):** Not regulated.
**Storage Temperature:** Ambient.

**Grades or Purity:** Variable. Commercial $CaCO_3$ has various proportions of magnesium oxide, silicon dioxide and a number of other compounds.
**Containers and Materials:** Multiwall paper bags; bulk in carlots, trucklots.

### Physical and Chemical Characteristics

**Physical State (20°C, 1 atm):** Solid.
**Solubility (Water):** 0.0015 g/100 mL (25°C); 0.0019 g/100 mL (75°C).
**Molecular Weight:** 100.1
**Vapour Pressure:** No information.
**Boiling Point:** Decomposes 825°C.

**Floatability (Water):** Sinks.
**Odour:** Odourless.
**Flash Point:** Not flammable.
**Vapour Density:** No information.
**Specific Gravity:** 2.7 to 2.95 (20°C).

**Colour:** White.
**Explosive Limits:** Not flammable.
**Melting Point:** Turns to calcite 520°C; decomposes 825°C.

## HAZARD DATA

### Human Health

**Symptoms:** Inhalation: sneezing and slight nose irritation. Can be irritating to respiratory tract. Ingestion: no symptoms. Contact: skin - no symptoms; eyes - watering, irritation.
**Toxicology:** Relatively nontoxic.
TLV - (dust) 30 mppcf or 10 mg/3 (total dust); 5 mg/m$^3$ (respirable dust).
Short-term Inhalation Limits - 20 mg/m$^3$ (dust).

$LC_{50}$ - None.
Delayed Toxicity - No information.

$LD_{50}$ - None.

### Fire

**Fire Extinguishing Agents:** Not combustible. Nost fire extinguishing agents may be used on fires involving calcium carbonate.
**Behaviour in Fire:** Not combustible. Releases carbon dioxide upon decomposition between 600 and 825°C.
**Ignition Temperature:** Not combustible.   **Burning Rate:** Not combustible.

### Reactivity

**With Water:** No reaction.
**With Common Materials:** Reacts violently with fluorine.
**Stability:** Stable.

### Environment

**Water:** Prevent entry into water intakes and waterways. Aquatic toxicity rating = >1 000 ppm/96 h/TLm/freshwater.
**Land-Air:** No information.
**Food Chain Concentration Potential:** None.

## EMERGENCY MEASURES

**Special Hazards**

**Immediate Responses**

Keep non-involved people away from spill site. Stop or reduce discharge if safe to do so. Notify manufacturer. Notify environmental authorities.

**Protective Clothing and Equipment**

Dust respirator may be required under certain circumstances. Outer protective clothing as required.

**Fire and Explosion**

Not combustible. Most fire extinguishing agents may be used on fires involving calcium carbonate.

**First Aid**

Move victim out of spill site to fresh air. Call for medical assistance, but start first aid at once. Inhalation: have victim blow nose to remove dust. Ingestion: induce vomiting in conscious victim. Contact: skin and eyes - remove contaminated clothing and flush affected areas with water. If necessary, transport victim to hospital, doctor or clinic.

## ENVIRONMENTAL PROTECTION MEASURES

**Response**

**Water**
1. Stop or reduce discharge if safe to do so.
2. Contact manufacturer or supplier for advice.
3. If possible, contain discharge by damming or water diversion.
4. Dredge or vacuum pump to remove contaminants, liquids and contaminated bottom sediments.
5. Notify environmental authorities to discuss disposal and cleanup of contaminated materials.

**Land-Air**
1. Stop or reduce discharge if safe to do so.
2. Contact manufacturer or supplier for advice.
3. Remove material by manual or mechanical means.
4. Notify environmental authorities to discuss disposal and cleanup of contaminated materials.

**Disposal**
1. Contact manufacturer or supplier for advice on disposal.
2. Contact environmental authorities for advice on disposal.
3. May be dumped in a municipal landfill (approval of environmental authorities required).

CALCIUM CARBONATE $CaCO_3$

# CALCIUM CHLORIDE  $CaCl_2 \cdot xH_2O$ (x = 0,1,2,6)

## IDENTIFICATION

### Common Synonyms
CALCIUM CHLORIDE, ANHYDROUS ($CaCl_2$)
CALCIUM CHLORIDE MONOHYDRATE ($CaCl_2 \cdot H_2O$)
CALCIUM CHLORIDE DIHYDRATE ($CaCl_2 \cdot 2H_2O$)
CALCIUM CHLORIDE HEXAHYDRATE ($CaCl_2 \cdot 6H_2O$)

### Observable Characteristics
Colourless to white crystals, flake, or solutions. Odourless.

### Manufacturers
Allied Chemical
Amherstburg, Ontario.
Canadian suppliers:
Allied Chemical, Mississauga, Ont.
Dow Chemical Canada Inc, Sarnia, Ont.

Originating from:
Allied Chemical, USA
Dow Chemical, USA

### Transportation and Storage Information
**Shipping State:** Solid or liquid (aqueous solution).
**Classification:** None.
**Inert Atmosphere:** No requirement.
**Venting:** Open.

**Label(s):** Not regulated.
**Storage Temperature:** Ambient.
**Hose Type:** Teflon, Viton, neoprene.
**Pump Type:** Centrifugal or displacement; stainless steel.

**Grades or Purity:** Technical, solutions from 30 to 95%.
**Containers and Materials:** Bags, drums; bulk by truck or train.

### Physical and Chemical Characteristics
**Physical State** (20°C, 1 atm): Solid.
**Solubility (Water):** Anhydrous 74.5 g/100 mL (20°C), 159 g/100 mL (100°C); monohydrate 76.8 g/100 mL (0°C), 249 g/100 mL (100°C); dihydrate 97.7 g/100 mL (0°C), 326 g/100 mL (60°C); hexahydrate 279 g/100 mL (0°C), 536 g/100 mL (20°C).
**Molecular Weight:** Anhydrous 111.0; monohydrate 129.0; dihydrate 147.0; hexahydrate 219.1.
**Vapour Pressure:** 9 mm Hg (21°C) (35.5% solution).

**Floatability (Water):** Sinks and mixes.
**Odour:** Odourless.
**Flash Point:** Not flammable.
**Vapour Density:** No information.
**Specific Gravity:** Anhydrous 2.15 (25°C); dihydrate 0.84 (25°C); hexahydrate 1.7 (25°C).
**Boiling Point:** Anhydrous >1 600°C; hexahydrate loses 4$H_2O$ at 30°C; loses 6 $H_2O$ at 200°C.

**Colour:** Colourless to white.
**Explosive Limits:** Not flammable.
**Melting Point:** Anhydrous, 782°C; monohydrate, 260°C; hexahydrate, 29.9°C.

## HAZARD DATA

### Human Health
**Symptoms:** Inhalation: irritation of nose and throat. Ingestion: irritation of mouth and stomach, nausea and vomiting. Dust can cause eye irritation and possible transient corneal injury. Skin contact with dry solid causes mild irritation; strong solutions - marked irritation, even superficial burns.
**Toxicology:** Relatively nontoxic.
**TLV®-** No information.           **$LC_{50}$** - No information.           **$LD_{50}$ - Oral:** rat = 1 g/kg
**Short-term Inhalation Limits -** No information.           **Delayed Toxicity -** None.

### Fire
**Fire Extinguishing Agents:** Not combustible. Most fire extinguishing agents may be used on fires involving calcium chloride.
**Behaviour in Fire:** Not combustible.
**Ignition Temperature:** Not combustible.           **Burning Rate:** Not combustible.

### Reactivity
**With Water:** No reaction; soluble.
**With Common Materials:** Reacts violently with (boric oxide and calcium oxide).
**Stability:** Stable.

### Environment
**Water:** Prevent entry into water intakes and waterways. Fish toxicity: 10 650 ppm/96 h/sunfish/TLm/freshwater; 13 400 ppm/96 h/mosquito fish/TLm/freshwater; BOD: No information.
**Land-Air:** No information.
**Food Chain Concentration Potential:** None.

## EMERGENCY MEASURES

**Special Hazards**

**Immediate Responses**

Keep non-involved people away from spill site. Stop or reduce discharge if safe to do so. Notify manufacturer. Notify environmental authorities.

**Protective Clothing and Equipment**

Dust respirator may be required under certain conditions. Rubber, neoprene and vinyl protective clothing as required.

**Fire and Explosion**

Not combustible. Most fire extinguishing agents may be used on fires involving calcium chloride.

**First Aid**

Move victim out of spill site to fresh air. Call for medical assistance, but start first aid at once. **Inhalation:** give artificial respiration if breathing has stopped; oxygen, if breathing is laboured. **Ingestion:** give water to conscious victim to drink and rinse mouth. <u>Contact</u>: skin and eyes - remove contaminated clothing and flush affected areas with water. If medical assistance is not immediately available, transport victim to doctor, clinic or hospital.

## ENVIRONMENTAL PROTECTION MEASURES

**Response**

**Water**
1. Stop or reduce discharge if safe to do so.
2. Contact manufacturer or supplier for advice.
3. Notify environmental authorities to discuss cleanup and disposal of contaminated materials.

**Land-Air**
1. Stop or reduce discharge if safe to do so.
2. Contact manufacturer or supplier for advice.
3. Dike to prevent runoff from rainwater or water application.
4. Remove material by manual or mechanical means.
5. Notify environmental authorities to discuss disposal and cleanup of contaminated materials.

**Disposal**
1. Contact manufacturer or supplier for advice on disposal.
2. Contact environmental authorities for advice on disposal.

CALCIUM CHLORIDE   $CaCl_2 \cdot xH_2O$ (x = 0,1,2,6)

# CALCIUM CYANIDE  $Ca(CN)_2$

## IDENTIFICATION

UN No. 1575

### Common Synonyms

CYANOGAS
CYANIDE OF CALCIUM

### Observable Characteristics

White, to grey, to black crystals or powder. Odourless when dry, faint almond-like odour when wet.

### Manufacturers

Cyanamid Canada Ltd., Niagara Falls, Ontario.

### Transportation and Storage Information

**Shipping State:** Solid.
**Classification:** Poison.
**Inert Atmosphere:** No requirement.
**Venting:** Closed.

**Label(s):** White label – POISON; Class 6.1, Group I.
**Storage Temperature:** Ambient.

**Grades or Purity:** No information.
**Containers and Materials:** Drums; steel.

### Physical and Chemical Characteristics

**Physical State** (20°C, 1 atm): Solid.
**Solubility** (Water): Reacts to form HCN gas.
**Molecular Weight:** 92.1
**Vapour Pressure:** No information.
**Boiling Point:** Decomposes >350°C.

**Floatability** (Water): Reacts to form HCN gas.
**Odour:** Odourless (dry); almond-like (wet).
**Flash Point:** Not flammable.
**Vapour Density:** No information.
**Specific Gravity:** 2.4 (20°C).

**Colour:** White, to grey, to black.
**Explosive Limits:** Not flammable.
**Melting Point:** Decomposes >350°C.

## HAZARD DATA

### Human Health

**Symptoms: Inhalation (HCN gas):** breathing cyanide gas or dust causes headache, nausea, vomiting and profound weakness, wheezing, convulsions. Contact and Ingestion: similar symptoms as inhalation. High concentrations are rapidly fatal.
**Toxicology:** Highly toxic by inhalation and ingestion.
TLV® – as CN (skin) 5 mg/m³.    $LC_{50}$ – No information.    $LD_{50}$ – Oral: rat = 0.4 g/kg
Short-term Inhalation Limits – No information.    Delayed Toxicity – No information.

### Fire

**Fire Extinguishing Agents:** DO NOT USE WATER OR CARBON DIOXIDE. Use dry chemical, sand or earth. If possible, allow fire from wet cyanide to burn out.
**Behaviour in Fire:** At high temperatures decomposes to cyanides.
**Ignition Temperature:** Not combustible.    **Burning Rate:** Not combustible.

### Reactivity

**With Water:** Reacts to form HCN gas.
**With Common Materials:** Reacts with acids producing hydrogen cyanide and acetylene.
**Stability:** Stable if dry.

### Environment

**Water:** Prevent entry into water intakes and waterways. Fish toxicity: 0.21 ppm/96 h/sunfish/TLm/freshwater; >25 ppm/48 h/cockle/$LC_{50}$/saltwater.
**Land-Air:** No information.
**Food Chain Concentration Potential:** No information.

## EMERGENCY MEASURES

### Special Hazards
POISON. Toxic and flammable HCN gas produced on contact with water.

### Immediate Responses
Keep non-involved people away from spill site. Issue warning: "POISON". Call Fire Department. Eliminate all sources of ignition (for HCN gas). Contact supplier for guidance. Avoid contact and inhalation. Do not attempt to extinguish fire with water. Evacuate people downwind from spill or fire. If water is present, dike to contain runoff. Notify environmental authorities.

### Protective Clothing and Equipment
Respiratory protection - self-contained breathing apparatus and totally encapsulated protective clothing.

### Fire and Explosion
DO NOT USE WATER OR CARBON DIOXIDE. Use dry chemical, sand or earth. If possible, allow fire from wet cyanide to burn out.

### First Aid
Move victim out of spill site to fresh air. Call for medical assistance, but start first aid at once. Inhalation: if victim is conscious, simply keep warm and quiet. If victim is not breathing, initiate artificial respiration (not by mouth-to-mouth). If breathing is laboured, give oxygen. Contact: skin and eyes - remove contaminated clothing and flush affected areas with large amounts of water. Ingestion: induce vomiting. If medical assistance is not immediately available, transport victim to hospital, clinic, or doctor.

## ENVIRONMENTAL PROTECTION MEASURES

### Response

**Water**
1. Stop or reduce discharge if safe to do so.
2. Contact manufacturer or supplier for advice.
3. Notify environmental authorities to discuss cleanup and disposal of contaminated materials.

**Land-Air**
1. Stop or reduce discharge if safe to do so.
2. Contact manufacturer or supplier for advice.
3. Dike to prevent runoff from rainwater or water application.
4. Remove material by manual or mechanical means.
5. Recover undamaged containers.
6. Remove contaminated soil for disposal.
7. Notify environmental authorities to discuss disposal and cleanup of contaminated materials.

### Disposal
1. Contact manufacturer or supplier for advice on disposal.
2. Contact environmental authorities for advice on disposal.

CALCIUM CYANIDE   $Ca(CN)_2$

# CALCIUM HYDROXIDE  Ca(OH)$_2$

## IDENTIFICATION

**UN No. 9098**

### Common Synonyms
LIME
CALCIUM HYDRATE
HYDRATED LIME
LIME HYDRATE
SLAKED LIME
CAUSTIC LIME
AGRICULTURAL LIME

### Observable Characteristics
White or greyish-white powder or lumps.

### Manufacturers
Domtar Chemicals, Beachville, Ontario.
Joliette, Quebec.
BeachviLime, Beachville, Ontario.
Algoma Steel, Sault-Ste-Marie, Ontario.

### Transportation and Storage Information
**Shipping State:** Solid.
**Classification:** None.
**Inert Atmosphere:** No requirement.
**Venting:** Open.
**Label(s):** None. Class 9,2, Group III.
**Storage Temperature:** Ambient.
**Grades or Purity:** Agricultural, 65 to 71%; industrial 70 to 73%; chemical, 71 to 73% (may contain magnesium hydroxide, magnesium oxide, silicon dioxide and others in trace amounts).
**Containers and Materials:** Multiwall paper bags; bulk by truck or tank; steel.

### Physical and Chemical Characteristics
**Physical State** (20°C, 1 atm): Solid.
**Solubility** (Water): 0.185 g/100 mL (0°C); 0.077 g/100 mL (100°C).
**Molecular Weight:** 74.1
**Vapour Pressure:** No information.
**Boiling Point:** Decomposes >580°C.
**Floatability** (Water): Sinks.
**Odour:** Odourless.
**Flash Point:** Not flammable.
**Vapour Density:** No information.
**Specific Gravity:** 2.08–2.34 (20°C).
**Colour:** White or greyish-white.
**Explosive Limits:** Not flammable.
**Melting Point:** Decomposes 580°C (loses H$_2$O).

## HAZARD DATA

### Human Health
**Symptoms: Contact:** skin - burning sensation and inflammation; eyes - pain and watering. **Inhalation:** irritation of respiratory tract, difficulty breathing, coughing, sneezing. **Ingestion:** burning sensation, pain, stomach cramps.
**Toxicology:**
TLV® (inhalation) 5 mg/m$^3$ (dust).
Short-term Inhalation Limits - No information.
LC$_{50}$ - No information.
Delayed Toxicity - None known.
LD$_{50}$ - Oral: rat = 7.34 g/kg

### Fire
**Fire Extinguishing Agents:** Not combustible. Most fire extinguishing agents may be used on fires involving calcium hydroxide.
**Behaviour in Fire:** Not combustible.
**Ignition Temperature:** Not combustible.
**Burning Rate:** Not combustible.

### Reactivity
**With Water:** No reaction.
**With Common Materials:** Reacts violently with phosphorus, maleic anhydride, nitromethane, nitropropane and nitroparaffins.
**Stability:** Stable.

### Environment
**Water:** Prevent entry into water intakes and waterways, toxic to aquatic life. Fish toxicity: 92 ppm/7 h/trout/toxic/freshwater; Aquatic toxicity rating = 10 to 1 000 ppm/96 h/TLm/freshwater; 240 ppm/24 h/mosquito fish/TLm/freshwater; 160 ppm/96 h/mosquito fish/TLm/freshwater; BOD: None.
**Land-Air:** Frequently used in agriculture to neutralize acidic soils.
**Food Chain Concentration Potential:** No information.

## EMERGENCY MEASURES

**Special Hazards**

**Immediate Responses**

Keep non-involved people away from spill site. Stop discharge. Notify supplier. Notify environmental authorities.

**Protective Clothing and Equipment**

If dust level is high, use suitable dust respirators.

**Fire and Explosion**

Not combustible. Most fire extinguishing agents may be used on fires involving calcium hydroxide.

**First Aid**

Move victim out of spill area to fresh air. Call for medical assistance, but start first aid at once. <u>Inhalation</u>: (dust) make victim blow nose. <u>Contact</u>: skin - remove contaminated clothing and flush affected areas with water. <u>Ingestion</u>: give water to conscious victim to rinse mouth. If medical assistance is not immediately available, transport victim to hospital, doctor or clinic.

## ENVIRONMENTAL PROTECTION MEASURES

**Response**

**Water**
1. Stop or reduce discharge if safe to do so.
2. Contact manufacturer or supplier for advice.
3. If possible, contain discharge by damming or water diversion.
4. Dredge or vacuum pump to remove contaminants, liquids and contaminated bottom sediments.
5. Notify environmental authorities to discuss disposal and cleanup of contaminated materials.

**Land-Air**
1. Stop or reduce discharge if safe to do so.
2. Contact manufacturer or supplier for advice.
3. Contain spill by diking with earth or other barrier.
4. Remove material by manual or mechanical means.
5. Absorb residual liquid on natural or synthetic sorbents.
6. Notify environmental authorities to discuss disposal and cleanup of contaminated materials.

**Disposal**
1. Contact manufacturer or supplier for advice on disposal.
2. Contact environmental authorities for advice on disposal.
3. Apply to farmer's fields (approval of environmental authorities required).
4. May be buried in municipal landfill sites (approval of environmental authorities required).

CALCIUM HYDROXIDE   $Ca(OH)_2$

# CALCIUM HYPOCHLORITE   $Ca(OCl)_2$   $CaCl(ClO) \cdot 4H_2O$ (hydrate)

## IDENTIFICATION

**Common Synonyms**

ANHYDROUS $Ca(OCl)_2$
Calcium Oxychloride

HYDRATE $CaCl(ClO) \cdot 4H_2O$
Chloride of Lime
Bleaching Powder
Lime Chloride

**Observable Characteristics**

White powder or crystals. Strong chlorine odour.

**UN No. 2880 hydrate**
1748 dry >38% Cl
2208 dry 10 to 39% Cl

**Manufacturers**

Canadian manufacturer:
CIL Industries Ltd., Shawinigan, Quebec.
Selected U.S. manufacturers:
Olin Corporation, Stamford, Conn.
Pennwalt Corp., Indchem Division, Philadelphia, Pa.
Canadian suppliers:
Canadian Industries Limited,
General Chemicals Division, Toronto, Ont.
Pennwalt of Canada Ltd., Oakville, Ont.
Standard Chemical Ltd., Montreal, Que.

## Transportation and Storage Information

**Shipping State:** Solid.
**Classification:** Oxidizing material.
**Inert Atmosphere:** No requirement.
**Venting:** Open or closed.

**Label(s):** Yellow label - OXIDIZER; Class 5.2, Group III.
**Storage Temperature:** Ambient.

**Grades or Purity:** Anhydrous - Commercial 70 %; high purity 99.2%; hydrate 35 to 37% active chlorine or technical.
**Containers and Materials:** Cans, drums; steel, plastic.

## Physical and Chemical Characteristics

**Physical State** (20°C, 1 atm): Solid.
**Solubility (Water):** Decomposes.
**Molecular Weight:** 143 anhydrous; 199 hydrate (varies).
**Vapour Pressure:** No information.
**Boiling Point:** Decomposes >100°C.

**Floatability (Water):** Sinks and decomposes.
**Odour:** Chlorine odour.
**Flash Point:** Not flammable.
**Vapour Density:** No information.
**Specific Gravity:** 2.35 (20°C) (anhydrous); hydrate is similar, but varies.

**Colour:** White.
**Explosive Limits:** Not flammable.
**Melting Point:** Decomposes >100°C.

## HAZARD DATA

### Human Health

**Symptoms: Contact:** skin - itching, burning, sensation, inflammation; eyes - stinging, watering, inflammation. **Ingestion:** burning sensation in mouth and throat, stomach cramps, nausea, vomiting, weakness, shock, convulsion, coma. **Inhalation:** irritation of nose and eyes, coughing, difficulty breathing, cyanosis.
**Toxicology:** Highly toxic by ingestion and inhalation (as chlorine).
TLV® 1 ppm; 3 mg/m³ (as chlorine).    $LC_{50}$ - No information.    $LD_{50}$ - Oral: rat = 0.85 g/kg (anhydrous).
Short-term Inhalation Limits - 3 ppm; 9 mg/m³    Delayed Toxicity - No information.
(as chlorine) (15 min).

### Fire

**Fire Extinguishing Agents:** Not combustible. Most fire extinguishing agents (preferably water spray) may be used on fires involving calcium hypochlorite.
**Behaviour in Fire:** Not combustible, but evolves $O_2$ and $Cl_2$ at high temperatures. Readily ignites organic materials when in contact.
**Ignition Temperature:** Not combustible.    **Burning Rate:** Not combustible.

### Reactivity

**With Water:** Reacts with water to produce chlorine.
**With Common Materials:** Readily oxidizes combustible and organic substances. Reacts violently with carbon tetrachloride and amines.
**Stability:** Stable when dry and not exposed to heat or organic materials.

### Environment

**Water:** Prevent entry into water intakes and waterways; harmful to aquatic life in very low concentrations. Fish toxicity: 0.5 ppm/tns/trout/killed/freshwater; Aquatic toxicity rating = 1 to 10 ppm/96 h/TLm/freshwater; BOD: No information.
**Land-Air:** No information.
**Food Chain Concentration Potential:** No information.

## EMERGENCY MEASURES

### Special Hazards
OXIDIZER. Releases chlorine upon decomposition by heat or on contact with water.

### Immediate Responses
Keep non-involved people away from spill site. Issue warning: "OXIDIZER". Call Fire Department. Avoid contact and inhalation. Contact supplier or manufacturer. Stop discharge, if this can be done without risk. Move undamaged containers out of spill or fire area if this can be done without risk. Dike to prevent runoff. Notify environmental authorities.

### Protective Clothing and Equipment
Respiratory protection - in the case of a fire or in enclosed spaces, self-contained breathing apparatus. Otherwise, chemical goggles - (mono), tight fitting. Rubber gloves. Protective outerwear - suitable for the situation. If high rubber boots are worn, pants should be outside boots.

### Fire and Explosion
Not combustible. Most fire extinguishing agents (preferably water spray) may be used on fires involving calcium hypochlorite.

### First Aid
Move victim out of spill site to fresh air. Call for medical assistance, but start first aid at once. Contact: eyes - irrigate immediately with plenty of water for at least 15 minutes; skin - remove contaminated clothing and flood affected skin with water for at least 15 minutes. Inhalation: if breathing has stopped, give artificial respiration; if laboured, give oxygen. Ingestion: wash out mouth thoroughly with water. Give conscious victim plenty of water to drink. If medical assistance is not immediately available, transport victim to hospital, doctor or clinic.

## ENVIRONMENTAL PROTECTION MEASURES

### Response

**Water**
1. Stop or reduce discharge if safe to do so.
2. Contact manufacturer or supplier for advice.
3. If possible, contain discharge by damming or water diversion.
4. Dredge or vacuum pump to remove contaminants, liquids and contaminated bottom sediments.
5. Notify environmental authorities to discuss disposal and cleanup of contaminated materials.

**Land-Air**
1. Stop or reduce discharge if safe to do so.
2. Contact manufacturer or supplier for advice.
3. Contain spill by diking with earth or other barrier.
4. Remove material by manual or mechanical means.
5. Recover undamaged containers.
6. Notify environmental authorities to discuss disposal and cleanup of contaminated materials.

### Disposal
1. Contact manufacturer or supplier for advice on disposal.
2. Contact environmental authorities for advice on disposal.

CALCIUM HYPOCHLORITE   $Ca(OCl)_2$   $CaCl(ClO) \cdot 4H_2O$ (hydrate)

CALCIUM OXIDE  CaO

## IDENTIFICATION

UN No. 1910

### Common Synonyms
BURNT LIME
CALX
FLUXING LIME
LIME
QUICKLIME
UNSLAKED LIME
PEBBLE LIME

### Observable Characteristics
White or greyish-white powder or lumps. Odourless.

### Manufacturers
Domtar, Beachville, Ontario.
Joliette, Québec.
BeachviLime, Beachville, Ontario.
Algoma Steel, Sault Ste-Marie, Ontario.

### Transportation and Storage Information
**Shipping State:** Solid.
**Classification:** Corrosive.
**Inert Atmosphere:** No requirement.
**Venting:** Open.

**Label(s):** White and black label - CORROSIVE; Class 8, Group III.
**Storage Temperature:** Ambient.

**Grades or Purity:** Technical; agriculture; construction. Typically contains magnesium oxide; silicon dioxide, aluminum oxide, and other trace minerals.
**Containers and Materials:** Multiwall paper bags, barrels; bulklots by truck or train.

### Physical and Chemical Characteristics
**Physical State** (20°C, 1 atm): Solid.
**Solubility (Water):** Reacts to form Ca(OH)$_2$
**Molecular Weight:** 56.1
**Vapour Pressure:** No information.
**Boiling Point:** 2 850°C.

**Floatability (Water):** Reacts to form Ca(OH)$_2$.
**Odour:** Odourless.
**Flash Point:** Not flammable.
**Vapour Density:** No information.
**Specific Gravity:** 3.3 to 3.4 (20°C).

**Colour:** White or greyish-white.
**Explosive Limits:** Not flammable.
**Melting Point:** 2 570°C.

## HAZARD DATA

### Human Health
**Symptoms:** Contact: skin - pain, ulceration, shock; eyes - pain, watering, burns, ulceration of eyes, perforation, blindness. Inhalation: burning sensation in nose and throat, difficulty breathing, coughing, chemical bronchitis. Ingestion: burning sensation in mouth, throat, stomach, pain, stomach cramps, vomiting, diarrhea, unconsciousness.
**Toxicology:** Toxic by all routes.
TLV® 2 mg/m$^3$ (dust).
Short-term Inhalation Limits - No information.

LC$_{50}$ - No information.
Delayed Toxicity - None known.

LD$_{50}$ - No information.

### Fire
**Fire Extinguishing Agents:** Not combustible. Most fire extinguishing agents, except water, may be used in fires involving calcium oxide. Use water only in flooding amounts.
**Behaviour in Fire:** Not combustible.
**Ignition Temperature:** Not combustible.

**Burning Rate:** Not combustible.

### Reactivity
**With Water:** Reacts exothermically to form calcium hydroxide.
**With Common Materials:** May oxidize organic materials. Reacts violently with boron trifluoride, chlorine trifluoride, fluorine, hydrofluoric acid and phosphorous pentoxide.
**Stability:** Stable.

### Environment
**Water:** Prevent entry into water intakes and waterways. Fish toxicity: 92 ppm/7 h/(trout)/toxic/freshwater; 240 ppm/24 h/mosquito fish/TLm/freshwater; BOD: None.
**Land-Air:** Frequently used in agriculture to neutralize acidic soils.
**Food Chain Concentration Potential:** No information.

## EMERGENCY MEASURES

**Special Hazards**

CORROSIVE.

**Immediate Responses**

Keep non-involved people away from spill site. Issue warning "CORROSIVE". Stop discharge, if possible. Exercise caution with water application. Notify manufacturer or supplier. Notify environmental authorities.

**Protective Clothing and Equipment**

Dusty conditions, suitable dust mask. Gloves - work gloves with gauntlets. Coveralls. Boots - safety or high rubber (pants worn outside boots).

**Fire and Explosion**

Not combustible. Most fire extinguishing agents may be used on fires involving calcium oxide. If water is used, use flooding amounts.

**First Aid**

Move victim out of spill site to fresh air. Call for medical assistance, but start first aid at once. Inhalation: (dust) make victim blow nose. Contact: skin - remove contaminated clothing and flush affected areas with water; eyes - immediately flush with plenty of water. If medical assistance is not immediately available, transport victim to hospital, doctor or clinic.

## ENVIRONMENTAL PROTECTION MEASURES

**Response**

**Water**
1. Stop or reduce discharge if safe to do so.
2. Contact manufacturer or supplier for advice.
3. Notify environmental authorities to discuss disposal and cleanup of contaminated materials.

**Land-Air**
1. Stop or reduce discharge if safe to do so.
2. Contact manufacturer or supplier for advice.
3. Contain spill by diking with earth or other barrier.
4. Remove material by manual or mechanical means.
5. Recover undamaged containers.
6. Notify environmental authorities to discuss disposal and cleanup of contaminated materials.

**Disposal**
1. Contact manufacturer or supplier for advice on disposal.
2. Contact environmental authorities for advice on disposal.
3. Dump in a municipal landfill site (approval of environmental authorities required).

CALCIUM OXIDE    CaO

# CALCIUM PHOSPHATE

## IDENTIFICATION

### Common Synonyms

MONOBASIC CALCIUM PHOSHATE $(Ca(H_2PO_4)_2 2H_2O$
calcium biphosphate
acid calcium phosphate
calcium phosphate, primary
monocalcium phosphate

DIBASIC CALCIUM PHOSPHATE $CaHPO_4 \cdot 2H_2O$
dicalcium orthophosphate
bicalcium phosphate
secondary calcium phosphate

TRIBASIC CALCIUM PHOSPHATE $Ca_3(PO_4)_2$
bone ash    tricalcium phosphate
calcium orthophosphate    precipitated calcium phosphate
tricalcium orthophosphate    tertiary calcium phosphate

### Observable Characteristics

Colourless to white powder or crystals. Odourless.

### Manufacturers

International Minerals and Chemical Co. Port Maitland, Ont.
Cyanamid, Niagara Falls, Ont.

Canadian suppliers:
Agricultural Chemical London, Ont.
Erco Industries, Toronto, Ont.
Harrison's and Crosfield, Toronto, Ont.
Occidental Petroleum

Originating from:
Agrico, USA
Stauffer, USA

Occidental Petroleum, USA

### Transportation and Storage Information

**Shipping State:** Solid.
**Classification:** None.
**Inert Atmosphere:** No requirement.
**Venting:** Open.

**Label(s):** Not regulated.
**Storage Temperature:** Ambient.

**Grades or Purity:** Technical, by percentage of P (phosphorus).
**Containers and Materials:** Bags, barrels; bulklot by truck or train.

### Physical and Chemical Characteristics

**Physical State (20°C, 1 atm):** Solid.
**Solubility (Water):** monobasic 1.8 g/100 mL (20°C); dibasic: 0.03 g/100 mL (20°), 0.075 g/100 mL (100°C); tribasic: 0.002 g/100 mL (20°C).
**Molecular Weight:** 252.1 (mono); 172.1 (di); 310.2 (tri).
**Vapour Pressure:** No information.
**Boiling Point:** monobasic: decomposes 203°C. dibasic and tribasic: greater than melting point.

**Floatability (Water):** Sinks.
**Odour:** Odourless.
**Flash Point:** Not flammable.
**Vapour Density:** No information.
**Specific Gravity:** monobasic 2.2; dibasic 2.3; tribasic 3.1 (20°C).

**Colour:** Colourless to white.
**Explosive Limits:** Not flammable.
**Melting Point:** monobasic and dibasic lose $H_2O$ (109°C); tribasic: 1 670°C.

## HAZARD DATA

### Human Health

**Symptoms: Contact:** skin and eyes – redness, irritation. **Inhalation:** irritation of upper respiratory tract. **Ingestion:** may cause nausea, vomiting, stomach cramps, diarrhea.
**Toxicology:** Relatively nontoxic by all routes.
TLV - No information.  $LC_{50}$ - No information.  $LD_{50}$ - No information.
Short-term Inhalation Limits - No information.  Delayed Toxicity - No information.

### Fire

**Fire Extinguishing Agents:** Not combustible. Most fire extinguishing agents may be used on fires involving calcium phosphate.
**Behaviour in Fire:** Not combustible, but releases toxic $PO_x$ fumes at high temperatures.
**Ignition Temperature:** Not combustible.  **Burning Rate:** Not combustible.

### Reactivity

**With Water:** No reaction.
**With Common Materials:** Reacts violently with magnesium.
**Stability:** Stable.

### Environment

**Water:** Prevent entry into water intakes and waterways. Effects on aquatic life have not been determined.
**Land-Air:** No information.
**Food Chain Concentration Potential:** No information.

## EMERGENCY MEASURES

### Special Hazards

### Immediate Responses
Keep non-involved people away from spill site. Notify manufacturer. Notify environmental authorities.

### Protective Clothing and Equipment
Protective outer clothing as required.

### Fire and Explosion
Not combustible. Most fire extinguishing agents may be used on fires involving calcium phosphate.

### First Aid
Move victim out of spill site to fresh air. Call for medical assistance, but start first aid at once. Inhalation: give artificial respiration if breathing has stopped; if laboured, give oxygen. Ingestion: give water to conscious victim to drink. Contact: skin – remove contaminated clothing and flush affected areas with water; eyes – flush immediately with water. If medical assistance is not immediately available, transport victim to doctor, clinic or hospital.

## ENVIRONMENTAL PROTECTION MEASURES

### Response

**Water**
1. Stop or reduce discharge if safe to do so.
2. Contact manufacturer or supplier for advice.
3. If possible, contain discharge by damming or water diversion.
4. Dredge or vacuum pump to remove contaminants, liquids and contaminated bottom sediments.
5. Notify environmental authorities to discuss disposal and cleanup of contaminated materials.

**Land-Air**
1. Stop or reduce discharge if safe to do so.
2. Contact manufacturer or supplier for advice.
3. Contain spill by diking with earth or other barrier.
4. Remove material by manual or mechanical means.
5. Recover undamaged containers.
6. Notify environmental authorities to discuss disposal and cleanup of contaminated materials.

### Disposal
1. Contact manufacturer or supplier for advice on disposal.
2. Contact environmental authorities for advice on disposal.
3. Dump in municipal landfill site (approval of environmental authorities required).

CALCIUM PHOSPHATE

# CAPROLACTAM  CH$_2$(CH$_2$)$_4$NHCO

## IDENTIFICATION

**Common Synonyms**
AMINOCAPROIC LACTAM
2-KETOHEXAMETHYLENIMINE
2-OXOHEXAMETHYLENEIMINE
EPSILON-CAPROLACTAM
HEXAHYDRO-2H-AZEPINE-2-ONE

**Observable Characteristics**
White crystals or flakes.
Mild odour.

**Manufacturers**
Re-refined: Dow Badische Canada Ltd., Arnprior, Ontario.

Canadian Supplier:  Originating from:
Badische Canada, Arnprior, Ont.  Badische, USA
Firestone Textiles, Woodstock, Ont.  Firestone, USA

### Transportation and Storage Information

**Shipping State:** Solid.
**Classification:** None.
**Inert Atmosphere:** No requirement.
**Venting:** Open.

**Label(s):** Not regulated.
**Storage Temperature:** Ambient.

**Grades or Purity:** 99% flakes.
**Containers and Materials:** Multiwall paper bags, fibre drums.

### Physical and Chemical Characteristics

**Physical State** (20°C, 1 atm): Solid.
**Solubility** (Water): 82 g/100 mL (20°C).
**Molecular Weight:** 113.2
**Vapour Pressure:** 0.001 mm Hg (20°C); 0.0035 mm Hg (30°C).
**Boiling Point:** 268°C.

**Floatability** (Water): Sinks and mixes.
**Odour:** Mild (63 ppb, odour threshold).
**Flash Point:** 110°C (c.c.); 125°C (o.c.).
**Vapour Density:** 3.9
**Specific Gravity:** 1.1 (20°C).

**Colour:** White.
**Explosive Limits:** 1.4 to 8%.
**Melting Point:** 69 to 71°C.

## HAZARD DATA

### Human Health

**Symptoms:** Contact: eyes and skin - irritation. Inhalation: coughing or mild irritation. Ingestion: nausea.
**Toxicology:** Moderately toxic by contact and ingestion. Toxic by inhalation.
TLV® 1 mg/m$^3$; (dust) 5 ppm, 20 mg/m$^3$ (vapour). TC$_{Lo}$ - Inhalation: human = 100 ppm   LD$_{50}$ - Oral: rat = 2.14 g/kg
Short-term Inhalation Limits – 3 mg/m$^3$ (dust)   Delayed Toxicity - No information.
15 min; 10 ppm, 40 mg/m$^3$ (vapour) (15 min).

### Fire

**Fire Extinguishing Agents:** Use water, dry chemical, foam, or carbon dioxide.
**Behaviour in Fire:** Emits toxic NO$_x$ fumes when heated to decomposition.
**Ignition Temperature:** 375°C.  **Burning Rate:** 2.4 mm/min.

### Reactivity

**With Water:** No reaction; soluble.
**With Common Materials:** No information.
**Stability:** Stable.

### Environment

**Water:** Prevent entry into water intakes and waterways. Fish toxicity: 5 g/L/18 h/catfish/killed/freshwater; BOD: 60%, 20 days.
**Land-Air:** No information.
**Food Chain Concentration Potential:** No information.

## EMERGENCY MEASURES

**Special Hazards**

**Immediate Responses**

Keep non-involved people away from spill site. If there is a fire call Fire Department. Avoid contact and inhalation. Dike to prevent runoff. Notify manufacturer. Notify environmental authorities.

**Protective Clothing and Equipment**

Respiratory protection - self-contained breathing apparatus. Other protective clothing as required.

**Fire and Explosion**

Use water, dry chemical, foam or carbon dioxide.

**First Aid**

Move victim out of spill site to fresh air. Call for medical assistance, but start first aid at once. <u>Contact</u>: skin and eyes - remove contaminated clothing, flush affected areas with water. <u>Inhalation</u>: if breathing has stopped give artificial respiration; if laboured, give oxygen. <u>Ingestion</u>: give conscious victim water to drink. If medical assistance is not immediately available, transport victim to doctor, clinic or hospital.

## ENVIRONMENTAL PROTECTION MEASURES

**Response**

**Water**
1. Stop or reduce discharge if safe to do so.
2. Contact manufacturer or supplier for advice.
3. If possible, contain discharge by damming or water diversion.
4. Notify environmental authorities to discuss disposal and cleanup of contaminated materials.

**Land-Air**
1. Stop or reduce discharge if safe to do so.
2. Contact manufacturer or supplier for advice.
3. Dike to prevent runoff from rainwater or water application.
4. Remove material by manual or mechanical means.
5. Recover undamaged containers.
6. Notify environmental authorities to discuss disposal and cleanup of contaminated materials.

1. Contact manufacturer or supplier for advice on disposal.
2. Contact environmental authorities for advice on disposal.

CAPROLACTAM   $CH_2(CH_2)_4NHCO$

# CARBARYL  $C_{10}H_7OOCNHCH_3$

## IDENTIFICATION

**UN No.** 2757
**Danger Group According to Percentage of Active Substance**
**Group III** — solid 80 to 100%
— liquid 20 to 100%

### Common Synonyms
1-NAPHTHYL-N-METHYLCARBAMATE
**Common Trade Names**
SEVIN
(A general purpose insecticide).

### Observable Characteristics
White to grey powder or solution. Weak odour.

### Manufacturers
Ciba-Geigy, Cambridge Ont.
Chipman Chemical Ltd., Stoney Creek, Ontario.
Interprovincial Cooperatives, Saskatoon, Sask.

### Transportation and Storage Information
**Shipping State:** Solid or liquid (formulation).
**Classification:** None.
**Inert Atmosphere:** No requirement.
**Venting:** Open.
**Pump Type:** No information.

**Label(s):** Not regulated.
**Storage Temperature:** Ambient.
**Hose Type:** No information.

**Grades or Purity:** Various purities as described below.
**Containers and Materials:** Glass bottles, metal cans and drums.

### Physical and Chemical Characteristics
**Physical State (20°C, 1 atm):** Solid (technical).
**Solubility (Water):** All types insoluble, WP, EC and SU are dispersible; 0.012 g/100 mL (30°C) (technical).
**Molecular Weight:** 201.2
**Vapour Pressure:** 0.002 mm Hg (40°C) technical.
**Boiling Point:** No information.

**Floatability (Water):** Technical sinks; EC, WP and SU are dispersible in water. DU and SN may float.
**Odour:** Weak.
**Flash Point:** PP products flammable.
**Vapour Density:** No information.
**Specific Gravity:** 1.23 (20°C) (technical); DU 0.46 to 0.58 (20°C).

**Colour:** White to grey.
**Explosive Limits:** PP products flammable.
**Melting Point:** 142°C (technical).

## HAZARD DATA

### Human Health
**Symptoms:** **Inhalation:** irritation of eyes, nose and throat. **Contact:** skin - irritation; eyes - watering. **Ingestion:** salivation, sweating, abdominal cramps, nausea, vomiting, diarrhea, cyanosis, tremors and convulsions.
**Toxicology:** Highly toxic by inhalation and ingestion.
TLV® (inhalation) 5 mg/m³
Short-term Inhalation Limits - (inhalation) - 10 mg/m³ (15 min).

$LC_{50}$ - No information.
Delayed Toxicity - Liver damage.

$LD_{50}$ - Oral: rat = 0.250 g/kg
$LD_{50}$ - Inhalation: rat = 0.721 g/kg

### Fire
**Fire Extinguishing Agents:** Use carbon dioxide or dry chemical to extinguish.
**Behaviour in Fire:** Only PP is flammable, toxic fumes are released in fires.
**Ignition Temperature:** No information.
**Burning Rate:** No information.

### Reactivity
**With Water:** No reaction, EC, WP and SU dispersible in water.
**With Common Materials:** May react with oxidizing agents.
**Stability:** Stable.

### Environment
**Water:** Prevent entry into water intakes and waterways. Aquatic toxicity rating = 1 to 10 ppm/96 h/TLm/freshwater; 1.47 mg/L/$LC_{50}$/rainbow trout/96 h/freshwater; 13.0 mg/L/$LC_{50}$/fathead minnow/96 h/freshwater; 0.038 mg/L/$LC_{50}$/grass shrimp/24h/saltwater.
**Land-Air:** $LD_{50}$ - Oral: Chicken = 0.197 g/kg, $LD_{50}$ - Oral: wild bird = 0.056 g/kg
**Food Chain Concentration Potential:** Unknown.

## EMERGENCY MEASURES

**Special Hazards**

POISON.

**Immediate Responses**

Keep non-involved people away from spill site. Stop or reduce discharge if safe to do so. Notify manufacturer or supplier. Dike to contain material or water runoff. Notify environmental authorities.

**Protective Clothing and Equipment**

In fires or confined spaces - Respiratory Protection - self-contained breathing apparatus and totally encapsulated suit. Otherwise, approved pesticide respirator and impervious outer clothing.

**Fire and Explosion**

Use carbon dioxide, foam or dry chemical to extinguish. Releases toxic fumes in fires.

**First Aid**

Move victim out of spill site to fresh air. Call for medical assistance, but start first aid at once. Inhalation: if breathing has stopped, give artificial respiration (not mouth-to-mouth method); if laboured, give oxygen. Contact: skin - remove contaminated clothing and flush affected areas with plenty of water; eyes - irrigate with plenty of water. Ingestion: give water to conscious victim and induce vomiting; in the case of petroleum distillates, do not induce vomiting for fear of aspiration and chemical pneumonia. If medical assistance is not immediately available, transport victim to hospital, doctor or clinic.

## ENVIRONMENTAL PROTECTION MEASURES

**Response**

**Water**

1. Stop or reduce discharge if safe to do so.
2. Contact manufacturer or supplier for advice.

Floats        Sinks or mixes
3. If possible contain       3. If possible contain discharge
discharge by booming.         by damming or water diversions.
4. If floating, skim and     4. Dredge or vacuum pump to remove
remove.                       contaminants, liquids and con-
                              taminated bottom sediments.
5. Notify environmental authorities to discuss disposal and cleanup of contaminated materials.

**Land-Air**

1. Stop or reduce discharge if safe to do so.
2. Contact manufacturer or supplier for advice.
3. Contain spill by diking with earth or other barrier.
4. If liquid, remove material with pumps or vacuum equipment and place in appropriate containers.
5. If solid, remove material by manual or mechanical means.
6. Recover undamaged containers.
7. Absorb residual liquid on natural or synthetic sorbents.
8. Remove contaminated soil for disposal.
9. Notify environmental authorities to discuss cleanup and disposal of contaminated materials.

**Disposal**

1. Contact manufacturer or supplier for advice on disposal.
2. Contact environmental authorities for advice on disposal.

**Available Formulations**

**Technical Grade:** Purity: typically 99% carbaryl.
Properties: solid, sinks in water, immiscible with water.

**Formulations:**

| Type: | Purity: | Properties: |
|---|---|---|
| DU - dust | - typically 5% carbaryl | - insoluble in water, floats |
| EC - emulsifiable concentrate | - typically 5% carbaryl | - dispersible in water |
| WP - wettable powder | - typically 5% carbaryl | - combustible, dispersible in water |
| SU - suspension | - typically 38% carbaryl | - miscible in water |
| SN - solution | - typically 25% carbaryl in petroleum distillates | - combustible |
| PP - pressurized product | - typically 1% carbaryl | - flammable, insoluble in water. |

Other Possible Ingredients Found in Formulations: Zineb, Folpet, Sulfur, Malathion, pyrethrins, Methoxychlor, rotenone, captan.

**CARBARYL  $C_{10}H_7OOCNHCH_3$**

# CARBOFURAN $C_{12}H_{15}NO_3$

## IDENTIFICATION

**Common Synonyms**
2,3-DIHYDRO-2,2-DIMETHYL-7-BENZO-FURANYL METHYLCARBAMATE

**Common Trade Names**
FURADAN
(A systemic insecticide.)

**Observable Characteristics**
White solid. Odourless.

**UN No.** 2757
**Danger Group According to Percentage of Active Substance**
Group II    >10 to 100%
Group III   solid 1 to 10%
            liquid >0 to 10%

**Manufacturers**
FMC of Canada, Regina, Sask.
Chipman Chemicals, Hamilton, Ont.

### Transportation and Storage Information

**Shipping State:** Solid or liquid (formulation).
**Classification:** None.
**Inert Atmosphere:** No requirement.
**Venting:** Open.
**Pump Type:** No information.

**Label(s):** Not regulated.
**Storage Temperature:** Ambient.
**Hose Type:** No information.

**Grades or Purity:** Various, as described below.
**Containers and Materials:** Glass bottles; cans, drums; steel.

### Physical and Chemical Characteristics

**Physical State** (20°C, 1 atm): Solid.
**Solubility** (Water): 0.07 g/100 mL (20°C)
**Molecular Weight:** 221.3
**Vapour Pressure:** 0.00002 mm Hg (33°C).
**Boiling Point:** No information.

**Floatability** (Water): Sinks.
**Odour:** Odourless.
**Flash Point:** Not flammable.
**Vapour Density:** No information.
**Specific Gravity:** 1.18 (20°C) (technical).

**Colour:** White.
**Explosive Limits:** Not flammable.
**Melting Point:** 150 to 153°C (technical).

## HAZARD DATA

### Human Health

**Symptoms:** Inhalation, Ingestion and Contact: headaches, dizziness, salivation, tearing, nausea, vomiting, sweating, abdominal cramps, diarrhea. Absorbed by skin.

**Toxicology:** Highly toxic by all routes.
**TLV®** (inhalation) 0.1 mg/m$^3$
**Short-term Inhalation Limits** - No information.

$LC_{50}$ - Inhalation: rat = 85 mg/m$^3$
$LC_{50}$ - Inhalation: guinea pig = 43 mg/m$^3$/4 h
**Delayed Toxicity** - No information.

$LD_{50}$ - Oral: human = 0.011 g/kg
$LD_{50}$ - Oral: rat = 0.0053 g/kg

### Fire

**Fire Extinguishing Agents:** Use dry chemical or carbon dioxide to extinguish.
**Behaviour in Fire:** Only PP flammable. Toxic fumes are released in fires.
**Ignition Temperature:** No information.
**Burning Rate:** No information.

### Reactivity

**With Water:** No reaction.
**With Common Materials:** No information.
**Stability:** Stable.

### Environment

**Water:** Prevent entry into water intakes and waterways. Fish toxicity = 0.64 mg/L/96 h/bluegill/$LC_{50}$/freshwater; 0.0046 mg/L/48 h/pink shrimp/$LC_{50}$/saltwater; 0.28 mg/L/96 h/rainbow trout/$LC_{50}$/freshwater.
**Land-Air:** Oral: wild bird = 0.00042 g/kg; $LD_{50}$ - skin: wild bird = 0.10 g/kg/$LD_{50}$. $LD_{50}$ - Oral: chicken = 0.006 g/kg; $LD_{50}$ - Oral: duck = 0.000415 g/kg.
**Food Chain Concentration Potential:** No information.

## EMERGENCY MEASURES

**Special Hazards**
POISON.

**Immediate Responses**
Keep non-involved people away from spill site. Stop or reduce discharge if safe to do so. Notify manufacturer or supplier. Dike to contain material or water runoff. Notify environmental authorities.

**Protective Clothing and Equipment**
In fires or confined spaces - Respiratory Protection - self-contained breathing apparatus and totally encapsulated suit. Otherwise, approved pesticide respirator and impervious outer clothing.

**Fire and Explosion**
Use carbon dioxide, foam or dry chemical to extinguish. Releases toxic fumes in fires.

**First Aid**
Move victim out of spill site to fresh air. Call for medical assistance, but start first aid at once. Inhalation: if breathing has stopped, give artificial respiration (not mouth-to-mouth method); if laboured, give oxygen. Contact: skin - remove contaminated clothing and flush affected areas with plenty of water; eyes - irrigate with plenty of water. Ingestion: give water to conscious victim and induce vomiting; in the case of petroleum distillates, do not induce vomiting for fear of aspiration and chemical pneumonia. If medical assistance is not immediately available, transport victim to hospital, doctor or clinic.

## ENVIRONMENTAL PROTECTION MEASURES

**Response**

**Water**
1. Stop or reduce discharge if safe to do so.
2. Contact manufacturer or supplier for advice.

Floats    Sinks or mixes
3. If possible contain discharge by damming or water diversions.
4. If floating, skim and remove.
4. Dredge or vacuum pump to remove contaminants, liquids and contaminated bottom sediments.

5. Notify environmental authorities to discuss disposal and cleanup of contaminated materials.

**Land-Air**
1. Stop or reduce discharge if safe to do so.
2. Contact manufacturer or supplier for advice.
3. Contain spill by diking with earth or other barrier.
4. If liquid, remove material with pumps or vacuum equipment and place in appropriate containers.
5. If solid, remove material by manual or mechanical means.
6. Recover undamaged containers.
7. Absorb residual liquid on natural or synthetic sorbents.
8. Remove contaminated soil for disposal.
9. Notify environmental authorities to discuss cleanup and disposal of contaminated materials.

**Disposal**
1. Contact manufacturer or supplier for advice on disposal.
2. Contact environmental authorities for advice on disposal.

**Available Formulations**

**Technical Grade:** Purity: typically 99%
Properties: sinks in water, combustible.

**Formulations:**

Type:
GR - granular
SU - suspension

Purity:
- typically 10%
- typically 47%

Properties:
- sinks in water
- disperses in water

**CARBOFURAN** $C_{12}H_{15}NO_3$

# CARBON DIOXIDE  $CO_2$

## IDENTIFICATION

UN No. 1013

### Common Synonyms
CARBONIC ACID GAS
CARBONIC ANHYDRIDE
DRY ICE

### Observable Characteristics
Colourless liquid or gas, or white solid. Odourless.

### Manufacturers
Cominco, Calgary, Alta.
Canadian Fertilizers, Medicine Hat, Alta.
Liquid Carbonic Canada, Montreal, Quebec.

### Transportation and Storage Information
**Shipping State:** Liquid (compressed gas); solid (frozen liquid).
**Classification:** Nonpoisonous, nonflammable gas.
**Inert Atmosphere:** No requirement.
**Venting:** Liquid - safety relief; solid - open.
**Pump type:** No information.

**Label(s):** Green and white label - NONPOISONOUS; NONFLAMMABLE gas.
**Storage Temperature:** Ambient.
**Hose Type:** No information.

**Grades or Purity:** Commercial, 99.5%.
**Containers and Materials:** Liquid - cylinders, tank trucks; steel. Solid - food containers.

### Physical and Chemical Characteristics
**Physical State** (20°C, 1 atm): Gas.
**Solubility** (Water): 0.35 g/100 mL (0°C); 0.15 g/100 mL (25°C); 0.097 (40°C); 0.058 (60°C).
**Molecular Weight:** 44.0
**Vapour Pressure:** 569.1 mm Hg (-82°C).
**Boiling Point:** Sublimes, -79°C.

**Floatability** (Water): Sinks.
**Odour:** Odourless.
**Flash Point:** Not flammable.
**Vapour Density:** 1.5
**Specific Gravity:** 1.56 (-79°C) solid; 1.1 (-37°C) liquid.

**Colour:** Colourless (gas); solid is white.
**Explosive Limits:** Not flammable.
**Melting Point:** -78.5°C (sublimes).

## HAZARD DATA

### Human Health
**Symptoms:** Inhalation: increased respiration rate, headache, dizziness, drowsiness, shortness of breath, weakness and ringing in the ears. Contact: skin and eyes - solid can cause cold contact burns. Liquid or cold gas can cause freezing injury to skin or eyes similar to a burn.
**Toxicology:** An asphyxiant.
TLV*- 5 000 ppm; 9 000 mg/m³.
Short-term Inhalation Limits - 15 000 ppm; 27 000 mg/m³ (15 min).

$LC_{Lo}$ - Inhalation: rat = 657 190 ppm (15 min).
$LC_{Lo}$ - Inhalation: human = 100 000 ppm (1 min).
Delayed Toxicity - None.

$LD_{50}$ - No information.

### Fire
**Fire Extinguishing Agents:** Not combustible. Use water to cool fire-exposed containers.
**Behaviour in Fire:** Will not support fires and is frequently used as a fire fighting agent. Tanks may rupture in fires.
**Ignition Temperature:** Not combustible.
**Burning Rate:** Not combustible.

### Reactivity
**With Water:** No reaction.
**With Common Materials:** Reacts violently with diethyl magnesium, lithium, potassium, sodium and titanium.
**Stability:** Stable.

### Environment
**Water:** Prevent entry into water intakes and waterways. Fish toxicity: 100 to 200 mg/L/tns/various organisms/$LC_{50}$/freshwater; BOD: None.
**Land-Air:** Waterfowl - Inhalation 5 to 8%; no effect.
**Food Chain Concentration Potential:** None.

## EMERGENCY MEASURES

**Special Hazards**

**Immediate Responses**

Keep non-involved people away from spill site. Avoid contact and inhalation. Notify supplier. Notify environmental authorities.

**Protective Clothing and Equipment**

Respiratory protection - self-contained breathing apparatus. Protective outer clothing as required.

**Fire and Explosion**

Not combustible.

**First Aid**

Move victim out of spill site to fresh air. Call for medical assistance, but start first aid at once. <u>Inhalation</u>: if breathing has stopped, give artificial respiration; if laboured, give oxygen. <u>Contact</u>: skin (frostbite) do not rub affected areas, treat as a burn. If medical assistance is not immediately available, transport victim to doctor, clinic or hospital.

## ENVIRONMENTAL PROTECTION MEASURES

**Response**

**Water**
1. Stop or reduce discharge if safe to do so.
2. Contact manufacturer or supplier for advice.
3. Notify environmental authorities to discuss disposal and cleanup of contaminated materials.

**Land-Air**
1. Stop or reduce discharge if safe to do so.
2. Contact manufacturer or supplier for advice.
3. Recover undamaged containers.
6. Notify environmental authorities to discuss disposal and cleanup of contaminated materials.

**Disposal**
1. Contact manufacturer or supplier for advice on disposal.
2. Contact environmental authorities for advice on disposal.
3. Allow to disperse to atmosphere in a controlled manner (environmental authorities approval required).

CARBON DIOXIDE $CO_2$

# CARBON DISULFIDE  $CS_2$

## IDENTIFICATION

UN No. 1131

### Common Synonyms
CARBON DISULFIDE

### Observable Characteristics
Clear, colourless to yellow liquid. Disagreeable, pungent, decaying cabbage odour (almost odourless when pure).

### Manufacturers
Cornwall Chemicals Ltd., Cornwall, Ontario.
Thio Pet Chemicals Limited, Fort Saskatchewan, Alta.

### Transportation and Storage Information

**Shipping State:** Liquid.
**Classification:** Flammable; poison.
**Inert Atmosphere:** Nitrogen or other inert gas.
**Venting:** Pressure-vacuum (flame arrester).
**Pump Type:** Centrifugal, gear, etc; steel or stainless steel.

**Label(s):** Red label - FLAMMABLE LIQUID; Class 3.1, Group I. White label - POISON; Class 6.1, Group I.
**Storage Temperature:** Ambient.
**Hose Type:** Polyethylene, polypropylene flexible, steel, stainless steel.

**Grades or Purity:** Technical grade, 99.95% min.; commercial, 99.9%.
**Containers and Materials:** Drums, tank cars, tank trucks; steel, stainless steel.

### Physical and Chemical Characteristics

**Physical State** (20°C, 1 atm): Liquid.
**Solubility** (Water): 0.23 g/100 mL (20°C).
**Molecular Weight:** 76.1
**Vapour Pressure:** 200 mm Hg (10°C); 260 mm Hg (20°); 430 mm Hg (30°C).
**Boiling Point:** 46.3°C.

**Floatability** (Water): Sinks.
**Odour:** Disagreeable, cabbage, pungent. (0.1 to 0.21 ppm, odour threshold).
**Flash Point:** -30°C, (c.c.).
**Vapour Density:** 2.6
**Specific Gravity:** 1.26 (20°C).

**Colour:** Colourless to faint yellow.
**Explosive Limits:** 1.3 to 50%.
**Melting Point:** -111°C.

## HAZARD DATA

### Human Health

**Symptoms: Ingestion:** vomiting, headache, cyanosis, respiratory depression, convulsions, unconsciousness. **Inhalation:** vapours are rapidly absorbed, headache, nausea and dizziness, followed by restlessness, depression, nausea, vomiting, blurred vision and unconsciousness. **Contact:** (eyes and skin) liquid rapidly absorbed by skin producing symptoms similar to inhalation. Causes redness, burning, cracking and peeling.
**Toxicology:** Highly toxic by all routes.
TLV® - (skin) 10 ppm; 30 mg/m³.  LC$_{Lo}$ - **Inhalation:** human = 4 000 ppm (30 min).  LD$_{Lo}$ - **Oral:** human = 0.014 g/kg
Short-term Inhalation Limits - No information.  Delayed Toxicity - No information.

### Fire

**Fire Extinguishing Agents:** Use dry chemical, carbon dioxide or inert gases. Foam is ineffective. Water may be ineffective but may be used to cool fire-exposed containers.
**Behaviour in Fire:** Flashback may occur along vapour trail. Burning $CS_2$ releases toxic $SO_x$ fumes.
**Ignition Temperature:** 90°C.  **Burning Rate:** 2.7 mm/min.

### Reactivity

**With Water:** No reaction.
**With Common Materials:** Reacts violently with aluminum, chlorine, azides, fluorine, zinc, potassium and chlorine.
**Stability:** Stable.

### Environment

**Water:** Prevent entry into water intakes and waterways. Harmful to aquatic life in low concentrations. Fish toxicity: 135 ppm/48 h/mosquito fish/ TLm/freshwater; Aquatic toxicity rating = 100 to 1 000 ppm/96 h/TLm/freshwater; BOD: No information.
**Land-Air:** No information.
**Food Chain Concentration Potential:** No information.

## EMERGENCY MEASURES

**Special Hazards**

FLAMMABLE. POISON.

**Immediate Responses**

Keep non-involved people away from spill site. Issue warning: "FLAMMABLE; POISON". CALL FIRE DEPARTMENT. Eliminate all ignition sources. Stop or reduce discharge if safe to do so. Contact manufacturer for assistance. Avoid contact or inhalation. Dike to prevent runoff. Notify environmental authorities.

**Protective Clothing and Equipment**

Respiratory protection - self-contained breathing apparatus, and totally encapsulated suit.

**Fire and Explosion**

Use dry chemical, carbon dioxide or inert gases to extinguish. Foam is ineffective. Water is ineffective but may be used to cool fire-exposed containers. Flashback may occur along vapour trail. Burning $CS_2$ releases toxic $SO_x$ fumes.

**First Aid**

Move victim out of spill area to fresh air. Call for medical assistance, but start first aid at once. <u>Contact</u>: irrigate eyes with plenty of water; skin - immediately flush affected areas with plenty of water; remove contaminated clothing. <u>Inhalation</u>: if breathing has stopped, give artificial respiration (not mouth-to-mouth method), if breathing is laboured give oxygen. <u>Ingestion</u>: if victim is conscious, induce vomiting and give water to drink. Repeat until vomitus is clear. If medical assistance is not immediately available, transport victim to hospital, doctor or clinic.

## ENVIRONMENTAL PROTECTION MEASURES

**Response**

**Water**
1. Stop or reduce discharge if safe to do so.
2. Contact manufacturer or supplier for advice.
3. If possible, contain discharge by damming or water diversion.
4. Dredge or vacuum pump to remove contaminants, liquids and contaminated bottom sediments.
5. Notify environmental authorities to discuss disposal and cleanup of contaminated materials.

**Land-Air**
1. Stop or reduce discharge if safe to do so.
2. Contact manufacturer or supplier for advice.
3. Contain spill by diking with earth or other barrier.
4. Remove material with pumps or vacuum equipment and place in appropriate containers.
5. Recover undamaged containers.
6. Absorb residual liquid on natural or synthetic sorbents.
7. Notify environmental authorities to discuss disposal and cleanup of contaminated materials.

**Disposal**

1. Contact manufacturer or supplier for advice on disposal.
2. Contact environmental authorities for advice on disposal.

CARBON DISULFIDE $CS_2$

# CARBON MONOXIDE   CO

## IDENTIFICATION

UN No. 1016

**Common Synonyms**
MONOXIDE

**Observable Characteristics**
Colourless gas or liquid. Odourless.

**Manufacturers**
No Canadian manufacturer.
Canadian supplier:
Liquid Carbonic Canada Ltd.,
Montreal, Quebec,
Scarborough, Ontario,
Vancouver, B.C.,
Regina, Sask.

### Transportation and Storage Information

**Shipping State:** Liquid (compressed gas).
**Classification:** Flammable gas; poison.
**Inert Atmosphere:** No requirement.
**Venting:** Pressure-relief.
**Pump Type:** No requirement.

**Label(s):** Red label - FLAMMABLE GAS; Class 3. White label - POISONOUS GAS; Class 2.3.
**Storage Temperature:** Ambient.
**Hose Type:** No information.

**Grades or Purity:** Commercial, 98%. Also generated in various percentages from carbon combustion.
**Containers and Materials:** Cylinders, special trucks; steel.

### Physical and Chemical Characteristics

**Physical State** (20°C, 1 atm): Gas.
**Solubility** (Water): 0.0044 g/100 mL (0°C); 0.0019 g/100 mL (60°C).
**Molecular Weight:** 28
**Vapour Pressure:** 760 mm Hg (-191°C); 4 406 mm Hg (20°C).
**Boiling Point:** -191.5°C.

**Floatability** (Water): Floats and boils.
**Odour:** Odourless.
**Flash Point:** No information.
**Vapour Density:** 0.97 (0°C).
**Specific Gravity:** 0.791 (-191.5°C) liquid; 0.97 (20°C) gas.

**Colour:** Colourless.
**Explosive Limits:** 12.5 to 74%.
**Melting Point:** -199 to -207°C.

## HAZARD DATA

### Human Health

**Symptoms:** Inhalation: headache, nausea, irritability, increased respiration rate, chest pain, confusion, impaired judgement, unconsciousness. Contact: skin - liquid causes frostbite.
**Toxicology:** Highly toxic by inhalation.
TLV® (inhalation) 50 ppm; 55 mg/m³.    LC$_{50}$ - Inhalation: rat = 1 807 ppm (4 h).    LD$_{50}$ - No information.
Short-term Inhalation Limits - 400 ppm; 440 mg/m³  Delayed Toxicity - No information.
(15 min).

### Fire

**Fire Extinguishing Agents:** Do not put out fire until leak has been shut off. Use water, carbon dioxide, dry chemical to extinguish.
**Behaviour in Fire:** No information.    **Burning Rate:** No information.
**Ignition Temperature:** 609°C.

### Reactivity

**With Water:** No reaction.
**With Common Materials:** Reacts violently with bromine trifluoride, chlorine trifluoride, (lithium and water), oxygen, oxygen difluoride, (potassium and oxygen), silver oxide, (sodium and ammonia).
**Stability:** Stable.

### Environment

**Water:** Prevent entry into water intakes and waterways. Fish toxicity: 1.5 ppm/1 to 6 h/minnows and sunfish/killed/freshwater; BOD: No information.
**Land-Air:** No information.
**Food Chain Concentration Potential:** No information.

## EMERGENCY MEASURES

**Special Hazards**
POISON. FLAMMABLE.

**Immediate Responses**
Keep non-involved people away and upwind from spill site. Issue warning: "POISON; FLAMMABLE". CALL FIRE DEPARTMENT. Extinguish all ignition sources. Notify manufacturer. Stop or reduce discharge if safe to do so. Notify environmental authorities.

**Protective Clothing and Equipment**
Respiratory protection - self-contained breathing apparatus. Protective outer clothing as required.

**Fire and Explosion**
Do not put out fire until leak has been shut off, allow escaping carbon monoxide to burn. Fight fire from safe distance. Use water, carbon dioxide or dry chemical to extinguish. Use water to cool fire-exposed, adjacent containers.

**First Aid**
Move victim out of spill site to fresh air. Call for medical assistance, but start first aid at once. **Inhalation:** administer oxygen if available; use artificial respiration (not mouth-to-mouth method) if necessary. If medical assistance is not immediately available, transport victim to hospital, doctor or clinic.

## ENVIRONMENTAL PROTECTION MEASURES

**Response**

**Water**
1. Stop or reduce discharge if safe to do so.
2. Contact manufacturer or supplier for advice.
3. Notify environmental authorities to discuss disposal and cleanup of contaminated materials.

**Land-Air**
1. Stop or reduce discharge if safe to do so.
2. Contact manufacturer or supplier for advice.
3. Notify environmental authorities to discuss disposal and cleanup of contaminated materials.

**Disposal**
1. Contact manufacturer or supplier for advice on disposal.
2. Contact environmental authorities for advice on disposal.

CARBON MONOXIDE   CO

# CARBON TETRACHLORIDE  $CCl_4$

## IDENTIFICATION

UN No. 1846

### Common Synonyms

CARBON TET
TETRACHLOROMETHANE
PERCHLOROMETHANE
BENZINOFORM
NECATORINA

### Observable Characteristics

Clear, colourless, liquid. Strong, ethereal odour somewhat resembling chloroform.

### Manufacturers

Cornwall Chemicals Ltd., Cornwall, Ontario.
Dow Chemical Canada Inc., Sarnia, Ontario.

### Transportation and Storage Information

**Shipping State:** Liquid.
**Classification:** Poison.
**Inert Atmosphere:** No requirement.
**Venting:** Pressure-vacuum.
**Pump Type:** Standard types.

**Label(s):** White label - POISON; Class 6.1, Group II.
**Storage Temperature:** Ambient.
**Hose Type:** Cross-linked polyethylene, Viton, Buna-N flexisteel, Teflon, stainless.

**Grades or Purity:** Commercial, technical.
**Containers and Materials:** Cans, drums, tank cars, tank trucks; steel, stainless steel.

### Physical and Chemical Characteristics

**Physical State** (20°C, 1 atm): Liquid.
**Solubility** (Water): 0.12 g/100 mL (25°C); 0.08 g/100 mL (20°C).
**Molecular Weight:** 153.8
**Vapour Pressure:** 56 mm Hg (10°C); 90 mm Hg (20°C); 137 mm Hg (30°C).
**Boiling Point:** 76.7°C.

**Floatability** (Water): Sinks.
**Odour:** Strong ethereal odour (like chloroform). (21 to 200 ppm, odour threshold).
**Flash Point:** Not flammable.
**Vapour Density:** 5.5
**Specific Gravity:** 1.59 (25°C).

**Colour:** Colourless.
**Explosive Limits:** Not flammable.
**Melting Point:** -23.0°C.

## HAZARD DATA

### Human Health

**Symptoms: Contact:** skin - readily absorbed by skin. Symptoms for inhalation, ingestion and skin absorption are: abdominal pains, nausea, vomiting, confusion, weakness, unconsciousness.
**Toxicology:** Highly toxic by inhalation, ingestion or skin absorption.
TLV® - (skin) 5 ppm; 30 mg/m³  LC$_{50}$ - Inhalation: mouse = 9 526 ppm/8 h  LD$_{50}$ - Oral: rat = 2.8 g/kg
Short-term Inhalation Limits - 20 ppm; 125 mg/m³ (15 min).
**Delayed Toxicity** - Delayed effects may include severe damage to liver and kidneys, but may not become evident until 1 to 10 days after exposure. $CCl_4$ is a suspected carcinogen.

### Fire

**Fire Extinguishing Agents:** Not combustible. Most fire extinguishing agents may be used on fires involving carbon tetrachloride.
**Behaviour in Fire:** Decomposes at high temperatures (in fire) to give phosgene, chlorine and hydrogen chloride.
**Ignition Temperature:** Not combustible.  **Burning Rate:** Not combustible.

### Reactivity

**With Water:** No reaction.
**With Common Materials:** Reacts violently with allyl alcohol, aluminum and its alloys, barium, bromine trifluoride, calcium hypochlorite, diborane, dimethylformamide and fluorine, lithium, magnesium, liquid oxygen, potassium, sodium and strong alkalis.
**Stability:** Stable.

### Environment

**Water:** Prevent entry into water intakes and waterways. Aquatic toxicity rating = 10 to 100 ppm/96 h/TLm/freshwater; BOD: None.
**Land-Air:** No information.
**Food Chain Concentration Potential:** Possibility of concentration in food chain.

## EMERGENCY MEASURES

**Special Hazards**

POISON.

**Immediate Responses**

Keep non-involved people away from spill site. Issue warning; "POISON". Avoid contact and inhalation. Stop or reduce discharge, if this can be done without risk. Contain spill by diking to prevent entry into water intakes or courses. Notify manufacturer. Notify environmental authorities.

**Protective Clothing and Equipment**

Respiratory protection - self-contained breathing apparatus; totally encapsulated chemical protection suit.

**Fire and Explosion**

Not combustible. Most fire extinguishing agents may be used on fires involving carbon tetrachloride. Decomposes at high temperatures evolving phosgene, hydrogen chloride and chlorine gases.

**First Aid**

Move victim out of spill area to fresh air. Call for medical assistance, but start first aid at once. Inhalation: if breathing has stopped, give artificial respiration; if laboured, give oxygen. Ingestion: give plenty of water to conscious victim to drink. Keep warm and quiet. Contact: irrigate eyes and flush skin with plenty of water. Remove contaminated clothing at same time. If medical assistance is not immediately available. transport victim to hospital, doctor or clinic.

## ENVIRONMENTAL PROTECTION MEASURES

**Response**

Water

1. Stop or reduce discharge if safe to do so.
2. Contact manufacturer or supplier for advice.
3. If possible, contain discharge by damming or water diversion.
4. Dredge or vacuum pump to remove contaminants, liquids and contaminated bottom sediments.

Land-Air

1. Stop or reduce discharge if safe to do so.
2. Contact manufacturer or supplier for advice.
3. If possible, contain discharge by damming or water diversion.
4. Remove material with pumps or vacuum equipment and place in appropriate containers.
5. Recover undamaged containers.
6. Absorb residual liquid on natural or synthetic sorbents.
7. Remove contaminated soil for disposal.
8. Notify environmental authorities to discuss disposal and cleanup of contaminated materials.

**Disposal**

1. Contact manufacturer or supplier for advice on disposal.
2. Contact environmental authorities for advice on disposal.

CARBON TETRACHLORIDE   $CCl_4$

# CHLORDANE $C_{10}H_6Cl_8$

## IDENTIFICATION

UN No. 2761
Danger Group According to Percentage of Active Substance
Group III     solid 55 to 100%
                 liquid 10 to 100%

### Common Synonyms

1,2,4,5,6,7,8,8-OCTACHLOR-2,3,3a,4,7,7a-HEXAHYDRO-4,7, METHANOINDANE

### Observable Characteristics

White to grey powder, granules or grey to brown liquids. Solid, odourless. Liquids, chlorine-like odour.

### Manufacturers

Chipman, Incorporated, Stoney Creek, Ontario.
Interprovincial Co-ops Ltd., Saskatoon, Saskatchewan.
Velsicol Corp., Mississauga, Ontario.
Chevron Chemical, Burlington, Ontario.

(Chlordane is a pesticide commonly used to control ground insects such as ants and grubs.)

### Transportation and Storage Information

**Shipping State:** Solid or liquid (formulation).
**Classification:** None.
**Inert Atmosphere:** No requirement.
**Venting:** Open; SN (flame arrester).
**Pump Type:** No information.

**Label(s):** Not regulated.
**Storage Temperature:** Ambient.
**Hose Type:** No information.

**Grades or Purity:** Various, as described below.
**Containers and Materials:** Glass bottles, cans, drums; aluminum clad or enamel-lined.

### Physical and Chemical Characteristics

**Physical State** (20°C, 1 atm): Liquid (technical).
**Solubility (Water):** Technical-insoluble, EC and WP are dispersable.
**Molecular Weight:** 409.8
**Vapour Pressure:** 0.0001 mm Hg (20°C) (technical).
**Boiling Point:** 175°C (technical).

**Floatability (Water):** All except SN sink.
**Odour:** Odourless.
**Flash Point:** SN 56-90°C (c.c.).
**Vapour Density:** No information.
**Specific Gravity:** 1.59 to 1.63 (25°C) (technical).

**Colour:** Solid-white to grey. Liquid-grey to brown.
**Explosive Limits:** Only SN may be flammable.
**Melting Point:** No information.

## HAZARD DATA

### Human Health

**Symptoms:** Inhalation, Ingestion or Skin Contact: blurred vision, confusion, abdominal pain, nausea, vomiting, diarrhea, hyperexcitability, tremors, convulsions, followed by CNS depression which may terminate in respiratory failure.
**Toxicology:** Highly toxic by ingestion. Moderately toxic by skin contact.
TLV® (skin) 0.5 mg/m³     $LC_{50}$ - Inhalation: cat = 100 mg/m³/4 h     $LD_{50}$ - Oral: rabbit = 0.10 g/kg
Short-term Inhalation Limits - (skin) 2.0 mg/m³     Delayed Toxicity - Suspected carcinogen.     $LD_{50}$ - Oral: human = 0.04 g/kg

### Fire

**Fire Extinguishing Agents:** Foam, carbon dioxide, dry chemical.
**Behaviour in Fire:** Releases toxic fumes.
**Ignition Temperature:** No information.     **Burning Rate:** No information.

### Reactivity

**With Water:** No reaction.
**With Common Materials:** No information.
**Stability:** Stable.

### Environment

**Water:** Prevent entry into water intakes and waterways. Fish toxicity: 0.0082 to 0.097 mg/L/96 h/rainbow trout/$LC_{50}$/freshwater; 0.058 mg/L/24 h/bluegill/$LC_{50}$/freshwater; 0.062 to 0.214 mg/L/96 h/fathead minnow/$LC_{50}$/freshwater.
**Land-Air:** $LD_{50}$ - Oral: chicken = 0.22 g/kg; $LD_{50}$ - Oral: mallard = 800-850 mg/L (5 day); $LD_{50}$ - Oral: pheasant = 400-500 mg/L (5 day).
**Food Chain Concentration Potential:** Possible; chlordane is known to be present in the environment.

## EMERGENCY MEASURES

**Special Hazards**
POISON.

**Immediate Responses**
Keep non-involved people away from spill site. Stop or reduce discharge if safe to do so. Notify manufacturer or supplier. Dike to contain material or water runoff. Notify environmental authorities.

**Protective Clothing and Equipment**
In fires or confined spaces - Respiratory Protection - self-contained breathing apparatus and totally encapsulated suit. Otherwise, approved pesticide respirator and impervious outer clothing.

**Fire and Explosion**
Use carbon dioxide, foam or dry chemical to extinguish. Releases toxic fumes in fires.

**First Aid**
Move victim out of spill site to fresh air. Call for medical assistance, but start first aid at once. Inhalation: if breathing has stopped, give artificial respiration (not mouth-to-mouth method); if laboured, give oxygen. Contact: skin - remove contaminated clothing and flush affected areas with plenty of water; eyes - irrigate with plenty of water. Ingestion: give water to conscious victim and induce vomiting; in the case of petroleum distillates, do not induce vomiting for fear of aspiration and chemical pneumonia. If medical assistance is not immediately available, transport victim to hospital, doctor or clinic.

## ENVIRONMENTAL PROTECTION MEASURES

**Response**

**Water**
1. Stop or reduce discharge if safe to do so.
2. Contact manufacturer or supplier for advice.

Floats | Sinks or mixes
3. If possible contain discharge by booming.
4. If floating, skim and remove.

3. If possible contain discharge by damming or water diversions.
4. Dredge or vacuum pump to remove contaminants, liquids and con-taminated bottom sediments.

5. Notify environmental authorities to discuss disposal and cleanup of contaminated materials.

**Land-Air**
1. Stop or reduce discharge if safe to do so.
2. Contact manufacturer or supplier for advice.
3. Contain spill by diking with earth or other barrier.
4. If liquid, remove material with pumps or vacuum equipment and place in appropriate containers.
5. If solid, remove material by manual or mechanical means.
6. Recover undamaged containers.
7. Absorb residual liquid on natural or synthetic sorbents.
8. Remove contaminated soil for disposal.
9. Notify environmental authorities to discuss cleanup and disposal of contaminated materials.

**Disposal**
1. Contact manufacturer or supplier for advice on disposal.
2. Contact environmental authorities for advice on disposal.

**Available Formulations**

**Technical Grade:** Purity: 60 to 75%
Properties: sinks in water.

**Formulations:**

Type:
Du - dust
EC - emulsifiable concentrate
GR - granular
SN - solution
WP - wettable powder

Purity:
- typically 5%, remainder inerts
- typically 75%
- typically 5 to 25%, remainder inerts
- typically 2% in deodorized kerosene
- typically 40%

Properties:
- low combustibility
- dispersible in water
- sinks in water, low combustibility
- flammable to combustible, floats on water
- dispersable in water

CHLORDANE    $C_{10}H_6Cl_8$

# CHLORINE  Cl₂

## IDENTIFICATION

UN No. 1017

### Common Synonyms
None.

### Observable Characteristics
Greenish-yellow gas or clear, amber-coloured liquid under pressure. Pungent, bleach-like, irritating odour.

### Manufacturers
C.I.L. Inc., Cornwall, Ont., Dalhousie, N.B., Becancour, Que. Dow Chemical Canada Inc., Sarnia, Ont., Fort Saskatchewan, Alta. Canadian Occidental Petroleum Ltd., Vancouver, B.C., Nanaimo, B.C. F.M.C., Squamish, B.C.

### Transportation and Storage Information

**Shipping State:** Liquid (compressed gas).
**Classification:** Poisonous gas.
**Inert Atmosphere:** No requirement.
**Venting:** Safety relief.
**Pump Type:** Normally transferred by its own pressure.

**Label(s):** White label - POISONOUS GAS; Class 2.3.
**Storage Temperature:** Ambient.
**Hose Type:** Rigid piping (steel, special grade); special flexible steel or stainless steel, copper tubing.

**Grades or Purity:** Commercial, technical, pure (99.9%).
**Containers and Materials:** Cylinders, tank cars, tank trucks, ton containers; steel.

### Physical and Chemical Characteristics

**Physical State** (20°C, 1 atm): Gas.
**Solubility** (Water): 1.46 g/100 mL (0°C); 0.57 g/100 mL (30°C).
**Molecular Weight:** 70.9
**Vapour Pressure:** 2 749 mm Hg (0°C); 6 780 mm Hg (31°C).
**Boiling Point:** -34.1°C.

**Floatability** (Water): Liquid sinks and boils.
**Odour:** Characteristic, pungent, irritating (0.3 to 3.5 ppm, odour threshold).
**Flash Point:** Not flammable.
**Vapour Density:** (Gas) 2.5 g/L (20°C, 1 atm)
**Specific Gravity:** 1.57 (-40°C) (liquid).

**Colour:** Greenish-yellow.
**Explosive Limits:** Not flammable.
**Melting Point:** -101°C.

## HAZARD DATA

### Human Health

**Symptoms: Inhalation:** irritation of mucous membranes, watering of eyes, nasal discharge, sneezing, coughing, difficulty breathing, headache, nausea, muscular weakness and pulmonary edema. **Ingestion:** pain and burning sensation, thirst, abdominal cramps, nausea and vomiting, difficulty breathing, shock, convulsions and coma.
**Contact:** skin - burning sensation, inflammation and blisters; eyes - burning, watering, loss of sight.
**Toxicology:** High toxicity by inhalation.
TLV®- (inhalation) 1 ppm; 3 mg/m³.
Short-term Inhalation Limits - 3 ppm; 9 mg/m³ (15 min).

LC₅₀ - Inhalation:  rat = 293 ppm/1 h
LC₅₀ - Inhalation:  mouse = 137 ppm/1 h
Delayed Toxicity - None known.

LD₅₀ - No information.

### Fire

**Fire Extinguishing Agents:** Not combustible in air; use water with caution in fires involving chlorine as chlorine in water at some concentrations is very corrosive. Use water to keep fire-exposed containers cool.
**Behaviour in Fire:** Most combustible materials burn in chlorine as they do in oxygen. May combine with H₂O or steam to produce toxic and corrosive HCl fumes.
**Ignition Temperature:** Not combustible.
**Burning Rate:** Not combustible.

### Reactivity

**With Water:** No reaction; slightly soluble. May react to produce toxic and corrosive HCl fumes.
**With Common Materials:** Can react with: turpentine, ether, ammonia gas, hydrocarbons, hydrogen and powdered metals, polypropylene, rubber, sulfamic acid, acetaldehyde, acetylene, alcohols, alkyl phosphenes, aluminum, benzene, boron, brass, calcium, carbon, carbon disulfide, copper, ethane, ethylene, glycerol, iron, magnesium, manganese, several mercuric compounds, phosphorus, PCBs, silicone, sodium and zinc.
**Stability:** Stable, within the limits of the foregoing.

### Environment

**Water:** Prevent entry into water intakes and waterways. Harmful to aquatic life in very low concentrations. Fish toxicity: 0.08 ppm/168 h/trout/TLm/freshwater; 0.7 to 0.15 ppm/96 h/fathead minnow/LC₅₀/freshwater; 0.5 mg/L/72 h/Daphnia magna/LC₅₀/freshwater; Aquatic toxicity rating = <1 ppm/96 h/TLm/freshwater; BOD: None.
**Land-Air:** Lethal to plant life.

## EMERGENCY MEASURES

**Special Hazards**

CORROSIVE, POISON. HIGH CHEMICAL REACTIVITY. Vapour cloud collects in low lying areas.

**Immediate Responses**

Keep non-involved people away from spill site. Issue warning: "CORROSIVE; POISON". Call Fire Department. Evacuate from downwind. Avoid contact and inhalation. Stop or control leak if safe to do so. Contact manufacturer for advice. Notify environmental authorities.

**Protective Clothing and Equipment**

Respiratory protection - self-contained breathing apparatus and totally encapsulated suit. Gloves - rubber. Boots rubber (pants worn outside boots).

**Fire and Explosion**

Not combustible. In fires involving chlorine, use water with caution, but water should not be applied directly to a chlorine leak. Use water to cool fire-exposed containers.

**First Aid**

Move victim out of spill site to fresh air. Call for medical assistance, but start first aid at once. Inhalation: if breathing has stopped, give artificial respiration (not mouth-to-mouth method); if laboured, give oxygen. Contact: eyes (for liquid or gas) - irrigate with large amount of water for at least 30 min; skin - wash with plenty of water and soap for at least 15 minutes, and remove contaminated clothing; if both inhalation and contact have occurred, first aid for inhalation should be given first. Ingestion: rinse mouth of conscious victim liberally with water and give water to drink. Keep warm and quiet. If medical assistance is not immediately available, transport victim to hospital, doctor or clinic.

## ENVIRONMENTAL PROTECTION MEASURES

**Response**

**Water**
1. Stop or reduce discharge if safe to do so.
2. Contact manufacturer or supplier for advice.
3. Notify environmental authorities to discuss disposal and cleanup of contaminated materials.

**Land-Air**
2. Stop or reduce discharge if safe to do so.
2. Contact manufacturer or supplier for advice.
3. Dike to prevent runoff from rainwater or water application.
4. Recover undamaged containers.
5. Notify environmental authorities to discuss disposal and cleanup of contaminated materials.

**Disposal**
1. Contact manufacturer or supplier for advice on disposal.
2. Contact environmental authorities for advice on disposal.

CHLORINE $Cl_2$

# CHLOROBENZENE $C_6H_5Cl$

## IDENTIFICATION

UN No. 1134

### Common Synonyms

MONOCHLOROBENZENE
PHENYL CHLORIDE
BENZENE CHLORIDE
MCB
CHLOROBENZOL

### Observable Characteristics

Clear, colourless liquid.
Almond-like odour.

### Manufacturers

No Canadian manufacturers.
Canadian supplier:
Dow Chemical Canada Inc.,
Sarnia, Ontario.

Originating from:
Dow Chemical,
Midland, Michigan, USA

### Transportation and Storage Information

**Shipping State:** Liquid.
**Classification:** Flammable liquid.
**Inert Atmosphere:** No requirement.
**Venting:** Pressure-vacuum.
**Pump Type:** Centrifugal or positive displacement; carbon or stainless steel.

**Label(s):** Red label - FLAMMABLE LIQUID; Class 3.2, Group II.
**Storage Temperature:** Ambient.
**Hose Type:** Cross-linked polyethylene, Viton, Teflon, braided, flexible stainless steel or bronze.

**Grades or Purity:** Technical, 99.5%.
**Containers and Materials:** Drums, tank cars, tank trucks; carbon steel, stainless steel. Not aluminum.

### Physical and Chemical Characteristics

**Physical State** (20°C, 1 atm): Liquid.
**Solubility** (Water): 0.05 g/100 mL (20°C); 0.0488 g/100 mL (30°C).
**Molecular Weight:** 112.6
**Vapour Pressure:** 8.8 mm Hg (20°C); 11.8 mm Hg (25°C); 15 mm Hg (30°C).
**Boiling Point:** 132°C.

**Floatability** (Water): Sinks.
**Odour:** Almond-like (0.21 to 0.22 ppm, odour threshold).
**Flash Point:** 29°C (c.c.).
**Vapour Density:** 3.9
**Specific Gravity:** 1.1 (20°C).

**Colour:** Colourless.
**Explosive Limits:** 1.3 to 7.1%.
**Melting Point:** -45°C.

## HAZARD DATA

### Human Health

**Symptoms:** Contact: skin - irritation, mild burns, may be absorbed; eyes - irritation, watering, inflammation. Inhalation: irritation of eyes, nose and throat, coughing, dizziness, headache, extreme drowsiness, loss of consciousness, coma. Ingestion: irritation of lips and mouth, nausea, vomiting, diarrhea, drowsiness, loss of consciousness, narcosis and shock.
**Toxicology:** Moderately toxic by inhalation, ingestion and contact.
TLV®- (inhalation) 75 ppm; 350 mg/m³.    LC50 - No information.    LD50 - Oral: rat = 2.9 g/kg
Short-term Inhalation Limits - No information.    Delayed Toxicity - Long-term exposure, suspected liver and kidney damage.

### Fire

**Fire Extinguishing Agents:** Carbon dioxide, dry chemical or foam. Water spray may be used to control large fires. Layer of water may be used to blanket the fire.
**Behaviour in Fire:** Flashback may occur along vapour trail. Combustion products may be toxic.
**Ignition Temperature:** 593 to 638°C.    **Burning Rate:** 4.6 mm/min.

### Reactivity

**With Water:** No reaction.
**With Common Materials:** Can react vigorously with oxidizing compounds. Reacts with dimethyl sulfoxide and silver perchlorate. Degrades aluminum, rubber and plastic. Hazardous polymerization will not occur.
**Stability:** Stable.

### Environment

**Water:** Prevent entry into water intakes and waterways. Harmful to aquatic life in low concentrations. Fish toxicity: 20 ppm/96 h/bluegill/TLm/freshwater; Aquatic toxicity rating = 1 to 100 ppm/96 h/TLm/freshwater; 29 to 30 mg/L/48 h/fathead minnow/TLm/freshwater; BOD: 0.3 lb./lb., 5 days.
**Land-Air:** No information.
**Food Chain Concentration Potential:** No information.

## EMERGENCY MEASURES

**Special Hazards**

FLAMMABLE.

**Immediate Responses**

Keep non-involved people away from spill site. Issue warning: "FLAMMABLE". Call Fire Department. Eliminate all ignition sources. Avoid contact and inhalation. Stay upwind and use water spray to control vapours. Stop or reduce discharge, if safe to do so. Contact manufacturer. Dike or dam spill to prevent runoff. Notify environmental authorities.

**Protective Clothing and Equipment**

Respiratory protection - self-contained breathing apparatus. Acid suit - (jacket and pants) rubber or plastic, or coveralls. Boots - high, rubber (pants worn outside boots). Gloves - rubber or plastic.

**Fire and Explosion**

Use dry chemical, foam or carbon dioxide. Water spray may be used to control small fires. Layer of water may be used to blanket the fire. Flashback may occur along vapour trail.

**First Aid**

Move victim out of spill area to fresh air. Call for medical assistance, but start first aid at once. Inhalation: give artificial respiration if breathing has stopped; give oxygen if breathing is laboured. Contact: remove contaminated clothing; wash eyes and affected skin with plenty of warm water for at least 15 minutes. Ingestion: give warm water to conscious victim to drink. Keep warm and quiet. If medical assistance is not immediately available, transport victim to hospital, doctor or clinic.

## ENVIRONMENTAL PROTECTION MEASURES

**Response**

Water
1. Stop or reduce discharge if safe to do so.
2. Contact manufacturer or supplier for advice.
3. If possible, contain discharge by damming or water diversion.
4. Dredge or vacuum pump to remove contaminants, liquids and contaminated bottom sediments.
5. Notify environmental authorities to discuss disposal and cleanup of contaminated materials.

Land-Air
1. Stop or reduce discharge if safe to do so.
2. Contact manufacturer or supplier for advice.
3. Contain spill by diking with earth or other barrier.
4. Remove material with pumps or vacuum equipment and place in appropriate containers.
5. Recover undamaged containers.
6. Remove contaminated soil for disposal.
7. Notify environmental authorities to discuss disposal and cleanup of contaminated materials.

**Disposal**

1. Contact manufacturer or supplier for advice on disposal.
2. Contact environmental authorities for advice on disposal.

CHLOROBENZENE    $C_6H_5Cl$

# CHLOROFORM CHCl$_3$

## IDENTIFICATION

UN No. 1888

### Common Synonyms
TRICHLOROMETHANE
METHENYL TRICHLORIDE

### Observable Characteristics
Clear, colourless, liquid.
Characteristic ethereal odour.

### Manufacturers

**Canadian Supplier:**

Allied Chemical, Mississauga, Ont.
Dow Chemical Canada, Inc., Sarnia, Ont.
Du Pont Canada, Montreal, Que.
Mallinckrodt Canada, Pte Claire, Que.

**Originating from:**

Allied Chemical, USA
Dow Chemical, USA
West Germany
Mallinckrodt, USA

### Transportation and Storage Information

**Shipping State:** Liquid.
**Classification:** Poison.
**Inert Atmosphere:** No requirement.
**Venting:** Open.
**Pump Type:** Gear or centrifugal; steel or stainless steel.

**Label(s):** White label - POISON; Class 6.1, Group II.
**Storage Temperature:** Ambient.
**Hose Type:** Flexible stainless steel, Teflon, Viton.

**Grades or Purity:** Technical.
**Containers and Materials:** Drums, tank trucks, tank cars; steel, stainless steel.

### Physical and Chemical Characteristics

**Physical State** (20°C, 1 atm)**:** Liquid.
**Solubility** (Water)**:** 1.0 g/100 mL (15°C); 0.8 g/100 mL (20°C); 0.93 g/100 mL (25°C).
**Molecular Weight:** 119.4
**Vapour Pressure:** 160 mm Hg (20°C); 245 mm Hg (30°C).
**Boiling Point:** 61.2°C.

**Floatability** (Water)**:** Sinks.
**Odour:** Characteristic ethereal (205 to 307 ppm, odour threshold).
**Flash Point:** Not flammable.
**Vapour Density:** 4.1
**Specific Gravity:** 1.49 (20°C).

**Colour:** Colourless.
**Explosive Limits:** Not flammable.
**Melting Point:** -63.5°C.

## HAZARD DATA

### Human Health

**Symptoms:** <u>Inhalation:</u> of high concentrations may result in narcosis, shock and coma. <u>Contact:</u> irritating to skin and mucous membranes. <u>Ingestion:</u> followed by severe burning in mouth and throat, pain in chest and abdomen and vomiting, loss of consciousness.
**Toxicology:** Moderate by inhalation and ingestion.
TLV® (inhalation) 10 ppm; 50 mg/m$^3$.
Short-term Inhalation Limits - 50 ppm; 225 mg/m$^3$ (15 min).

LC$_{50}$ - Inhalation: mouse = 28 g/m$^3$
TC$_{10}$ - Inhalation: human = 5 000 mg/m$^3$/7 min.
Delayed Toxicity - Suspected carcinogen.
Causes liver and kidney damage.

LD$_{50}$ - Oral: rat = 0.8 g/kg
LD$_{Lo}$ - Oral: human = 0.14 g/kg

### Fire

**Fire Extinguishing Agents:** Not combustible. Most fire extinguishing agents may be used on fires involving chloroform.
**Behaviour in Fire:** Decomposes at high temperatures (as in fires) to give phosgene and hydrochloric gases.
**Ignition Temperature:** Not combustible.
**Burning Rate:** Not combustible.

### Reactivity

**With Water:** No reaction.
**With Common Materials:** Reacts violently with acetone and a base, aluminum, lithium, magnesium, nitrogen tetroxide, potassium, (potassium hydroxide and methanol) and sodium. Dissolves rubber and plastics.
**Stability:** Stable.

### Environment

**Water:** Prevent entry into water intakes and waterways. Fish toxicity: critical concentration = 10 mg/L; Aquatic toxicity rating = 10 to 100 ppm/96 h/TLm/freshwater; BOD: 0.08%, 5 days.
**Land-Air:** No information.
**Food Chain Concentration Potential:** None.

## EMERGENCY MEASURES

### Special Hazards
POISON.

### Immediate Responses
Keep non-involved people away from spill site. Issue warning: "POISON". Avoid contact and inhalation. Stop or reduce discharge, if this can be done without risk. Contain spill by diking to prevent entry into water intakes or courses. Notify manufacturer. Notify environmental authorities.

### Protective Clothing and Equipment
Respiratory protection - self-contained breathing apparatus. Gloves - rubber or vinyl. Boots - rubber (pants worn outside boots). Outerwear - coveralls, aprons, suit - impervious material.

### Fire and Explosion
Not combustible. Most fire extinguishing agents may be used on fires involving chloroform. Chloroform decomposes at high temperatures (as in fires) to evolve phosgene and hydrogen chloride.

### First Aid
Move victim out of spill area to fresh air. Call for medical assistance, but start first aid at once. Inhalation: if breathing has stopped, give artificial respiration (not mouth-to-mouth method); if laboured, give oxygen. Ingestion: keep warm and quiet. Treat as severe emergency. Contact: remove contaminated clothing and irrigate eyes and flush skin with plenty of water for at least 15 min. If medical assistance is not immediately available, transport victim to hospital, doctor or clinic.

## ENVIRONMENTAL PROTECTION MEASURES

### Response

**Water**
1. Stop or reduce discharge if safe to do so.
2. Contact manufacturer or supplier for advice.
3. If possible, contain discharge by damming or water diversion.
4. Dredge or vacuum pump to remove contaminants, liquids and contaminated bottom sediments.
5. Notify environmental authorities to discuss disposal and cleanup of contaminated materials.

**Land-Air**
1. Stop or reduce discharge if safe to do so.
2. Contact manufacturer or supplier for advice.
3. Contain spill by diking with earth or other barrier.
4. Remove material with pumps or vacuum equipment and place in appropriate containers.
5. Recover undamaged containers.
6. Absorb residual liquid on natural or synthetic sorbents.
7. Notify environmental authorities to discuss disposal and cleanup of contaminated materials.

### Disposal
1. Contact manufacturer or supplier for advice on disposal.
2. Contact environmental authorities for advice on disposal.

CHLOROFORM   $CHCl_3$

# CHLOROSULFONIC ACID   $ClSO_3H$ or $(ClSO_2OH)$

## IDENTIFICATION   UN No. 1754

### Common Synonyms
SULFURIC CHLOROHYDRIN
CHLORSULFONIC ACID
CHLOROSULFURIC ACID.

### Observable Characteristics
Colourless to pale yellow liquid.
Sharp, acrid, choking odour.
Fumes when exposed to air.

### Manufacturers
No Canadian manufacturers.
Canadian Supplier:
Du Pont Canada,
Montreal, Que., Toronto, Ont.
and branches across Canada.

Originating from:
E.I. Du Pont de Nemours and
Co. Inc., Wilmington, Del., USA

### Transportation and Storage Information
**Shipping State:** Liquid.
**Classification:** Corrosive liquid.
**Inert Atmosphere:** No requirement.
**Venting:** Pressure-vacuum.
**Pump Type:** Centrifugal, gear; steel; stainless steel.
**Label(s):** Black and white label - CORROSIVE; Class 8, Group I.
**Storage Temperature:** Ambient.
**Hose Type:** Teflon, flexible steel, stainless steel.
**Grades or Purity:** Technical (commercial).
**Containers and Materials:** Carboys, drums, tank cars, tank trucks.

### Physical and Chemical Characteristics
**Physical State** (20°C, 1 atm): Liquid.
**Solubility** (Water): Reacts to form $H_2SO_4$ and $HCl$.
**Molecular Weight:** 116.5
**Vapour Pressure:** 0.8 mm Hg (30°C); 1 mm Hg (32°C).
**Boiling Point:** 151 to 158°C.
**Floatability** (Water): Reacts violently, decomposes into sulfuric and hydrochloric acids.
**Odour:** Sharp, acrid, penetrating. (1 to 5 ppm, odour threshold).
**Flash Point:** Not flammable.
**Vapour Density:** 4.0
**Specific Gravity:** 1.77 (20°C).
**Colour:** Colourless to pale yellow.
**Explosive Limits:** Not flammable.
**Melting Point:** -80°C.

## HAZARD DATA

### Human Health
**Symptoms:** Causes severe burns. **Contact:** skin - concentrated solutions are rapidly destructive to body tissues; eyes - rapidly causes severe damage, which may be followed by total loss of sight. **Inhalation:** vapour or mist will cause damage to the upper respiratory tract and even to lung tissue, and loss of consciousness. **Ingestion:** destructive to mouth, throat and stomach, pain and burning sensation.
**Toxicology:** Highly toxic by ingestion, inhalation and contact.
**TLV®** - (inhalation) No information.   $LC_{50}$ - No information.   $LD_{50}$ - No information.
**Short-term Inhalation Limits** - No information.   **Delayed Toxicity** - No information.

### Fire
**Fire Extinguishing Agents:** Not combustible. Most fire extinguishing agents may be used on fires involving chlorosulfonic acid. Water may be used to cool fire-exposed containers. Do not let water contact chlorosulfonic acid.
**Behaviour in Fire:** Not combustible. Can cause ignition by contact with combustible material.
**Ignition Temperature:** Not combustible.   **Burning Rate:** Not combustible.

### Reactivity
**With Water:** Reacts violently to produce sulfuric and hydrochloric acids.
**With Common Materials:** Reacts violently with acetic acid, acetic anhydride, acetonitrile, acrolein, acrylic acid, acrylonitrile, allyl alcohol, allyl chloride, ammonium hydroxide, aniline, butyraldehyde, creosote, cresol, cumene, diisobutylene, epichlorohydrin, ethylacetate, ethylacrylate, ethylenediamine, ethyleneglycol, hydrochloric acid, hydrofluoric acid, hydrogen peroxide, metal powders, methylethyl ketone, nitric acid, phosphorus, propylene oxide, pyridine, sodium hydroxide, sulfuric acid, styrene, vinyl acetate and organic matter.
**Stability:** Stable, within the limits of the foregoing.

### Environment
**Water:** Prevent entry into water intakes and waterways. Harmful to aquatic life. Fish toxicity: 282 ppm/96 h/mosquito fish/TLm/freshwater; 100 to 300 ppm/48 h/shrimp/$LC_{50}$/saltwater; Aquatic toxicity rating = 10 to 100 ppm/96 h/TLm/freshwater; BOD: None.
**Land-Air:** No information.
**Food Chain Concentration Potential:** No information.

## EMERGENCY MEASURES

**Special Hazards**

CORROSIVE. VIOLENT REACTION IN CONTACT WITH WATER. REACTS WITH MANY MATERIALS.

**Immediate Responses**

Keep non-involved people away from spill site. Issue warning: "CORROSIVE". Call Fire Department. Call manufacturer or supplier for assistance. Stop or reduce discharge, if this can be done without risk. Use water spray to control vapour. Do not allow water to contact material. Contain spill by diking to prevent runoff. Notify environmental authorities.

**Protective Clothing and Equipment**

Respiratory protection - use self-contained breathing apparatus. Acid suit - complete, rubber with hood or totally encapsulated suit. Boots - rubber, pants worn outside boots. Gloves - gauntlet, rubber.

**Fire and Explosion**

Not combustible. Most fire extinguishing agents may be used on fires involving chlorosulfonic acid. Water may be used to cool fire-exposed containers. Do not let water contact chlorosulfonic acid.

**First Aid**

Move victim out of spill area to fresh air. Call for medical assistance, but start first aid at once. Contact: flush eyes and skin with plenty of water for at least 15 minutes, while removing contaminated clothing. Inhalation: if breathing has stopped, give artificial respiration (not mouth-to-mouth method), if laboured give oxygen. Ingestion: give conscious victim large amounts of water to drink. Do not induce vomiting. If medical assistance is not immediately available, transport victim to hospital, doctor or clinic.

## ENVIRONMENTAL PROTECTION MEASURES

**Response**

**Water**
1. Stop or reduce discharge if safe to do so.
2. Contact manufacturer or supplier for advice.
3. Notify environmental authorities to discuss disposal and cleanup of contaminated materials.

**Land-Air**
1. Stop or reduce discharge if safe to do so.
2. Contact manufacturer or supplier for advice.
3. Contain spill by diking with earth or other barrier.
4. Remove material with pumps or vacuum equipment and place in appropriate containers.
5. Recover undamaged containers. Caution must be taken to make sure water does not contact material during recovery procedures.
6. Notify environmental authorities to discuss disposal and cleanup of contaminated materials.

**Disposal**
1. Contact manufacturer or supplier for advice on disposal.
2. Contact environmental authorities for advice on disposal.

CHLOROSULFONIC ACID   $ClSO_3H$ or $(ClSO_2OH)$

# CHROMIC ANHYDRIDE  $CrO_3$

## IDENTIFICATION

UN No. 1463 Solid
1755 Solution

### Common Synonyms
CHROMIC ACID
CHROMIC ACID, SOLID
CHROMIUM TRIOXIDE
CHROMIUM OXIDE
CHROMIUM VI OXIDE

### Observable Characteristics
Dark red, crystals or powder. Odourless.

### Manufacturers
No Canadian manufacturer.
Selected US manufacturer:
Essex Chemical Company,
Clifton, NJ, USA

Canadian supplier:
Canada Chrome and
Chemicals Ltd.,
Toronto, Ont.
Van Waters and Rogers,
Vancouver, B.C.

Originating from:
British Crome, UK
Diamond Shamrock, USA

### Transportation and Storage Information
**Shipping State:** Solid.
**Classification:** Corrosive; Oxidizer.
**Inert Atmosphere:** No requirement.
**Venting:** No requirement.

**Label(s):** Yellow label - OXIDIZER; Class 5.1, Group II. Black and white label - CORROSIVE; Class 8, Group II.
**Storage Temperature:** Ambient.

**Grades or Purity:** Technical flake, 99.75% $CrO_3$.
**Containers and Materials:** Drums; steel.

### Physical and Chemical Characteristics
**Physical State** (20°C, 1 atm): Solid.
**Solubility** (Water): 61.7 g/100 mL (0°C); 67.5 g/100 mL (100°C).
**Molecular Weight:** 100
**Vapour Pressure:** No information.
**Boiling Point:** Decomposes at 250°C.

**Floatability** (Water): Sinks and dissolves.
**Odour:** Odourless.
**Flash Point:** Not flammable.
**Vapour Density:** No information.
**Specific Gravity:** 2.70 (20°C).

**Colour:** Dark red.
**Explosive Limits:** Not flammable.
**Melting Point:** 196°C.

## HAZARD DATA

### Human Health
**Symptoms:** Contact: skin - irritation, inflammation and tissue destruction; eyes - irritation, inflammation, risk of serious lesions. Ingestion: burning sensation in mouth and throat, stomach cramps, nausea and vomiting, diarrhea, shock, coma. Inhalation: irritation of nose and eyes, difficulty breathing and sneezing.
**Toxicology:** Very toxic by inhalation; moderately toxic by ingestion and contact.
TLV® (inhalation) 0.05 mg/m³ (as Cr VI).
$LC_{50}$ - No information.
$LD_{50}$ - No information.
$LD_{Lo}$ - Subcutaneous: dog = 0.3 g/kg
Short-term Inhalation Limits - No information.
Delayed Toxicity - Suspected carcinogen.

### Fire
**Fire Extinguishing Agents:** Not combustible. Use water on fires involving chromic anhydride.
**Behaviour in Fire:** Chromic acid is a very powerful oxidizer and can ignite organic materials. In fire, containers may rupture.
**Ignition Temperature:** Not combustible.
**Burning Rate:** Not combustible.

### Reactivity
**With Water:** No reaction; soluble.
**With Common Materials:** Chromic acid is a powerful oxidizing agent and may ignite organic materials. Reacts violently with acetic acid, acetic anhydride, acetone, aluminum, ammonia, anthracene, benzene, camphor, diethyl ether, alcohol, glycerol, hydrocarbons, hydrogen sulfide, methanol, naphthalene, phosphorus, potassium, pyridine, sodium, sulfur and turpentine.
**Stability:** Stable.

### Environment
**Water:** Prevent entry into water intakes and waterways. Fish toxicity: 52 ppm/96 h/goldfish/died/freshwater; 0.01 ppm/48 h/Daphnia magna/TLm/freshwater; BOD: None.
**Land-Air:** No information.
**Food Chain Concentration Potential:** None.

## EMERGENCY MEASURES

**Special Hazards**

OXIDIZER, POISON.

**Immediate Responses**

Keep non-involved people away from spill site. Issue warning: "OXIDIZER, POISON". Avoid contact and inhalation. Stop or reduce discharge, if this can be done without risk. Dike spill to prevent runoff. Notify supplier and environmental authorities.

**Protective Clothing and Equipment**

In fires or confined spaces Respiratory protection - self-contained breathing apparatus. Otherwise goggles - (mono), tight fitting. Gloves - rubber, plastic coated. Dust respirators - with suitable filters for protection against dusts and mists. Protective clothing - coveralls, aprons, suits. Rubber safety shoes.

**Fire and Explosion**

Not combustible. Use water on fires involving chromic anhydride. Chromic anhydride is a very powerful oxidizing agent and may ignite organic materials. In fire, containers may rupture.

**First Aid**

Move victim out of spill site to fresh air. Call for medical assistance, but start first aid at once. Contact: eyes - irrigate with plenty of water; skin - flush affected areas with large volumes of water; at the same time, remove contaminated clothing. Ingestion: if victim is conscious, rinse mouth with water and give water to drink. Inhalation: if breathing has stopped give artificial respiration (not mouth-to-mouth method); if breathing is laboured, give oxygen. If medical assistance is not immediately available, transport victim to hospital, doctor or clinic.

## ENVIRONMENTAL PROTECTION MEASURES

**Response**

**Water**
1. Stop or reduce discharge if safe to do so.
2. Contact manufacturer or supplier for advice.
3. If possible, contain discharge by damming or water diversion.
4. Dredge or vacuum pump to remove contaminants, liquids and contaminated bottom sediments.
5. Notify environmental authorities to discuss disposal and cleanup of contaminated materials.

**Land-Air**
1. Stop or reduce discharge if safe to do so.
2. Contact manufacturer or supplier for advice.
3. Contain spill by diking with earth or other barrier.
4. Remove material by manual or mechanical means.
5. Recover undamaged containers.
6. Remove contaminated soil for disposal.
7. Notify environmental authorities to discuss disposal and cleanup of contaminated materials.

**Disposal**

1. Contact manufacturer or supplier for advice on disposal.
2. Contact environmental authorities for advice on disposal.

CHROMIC ANHYDRIDE   $CrO_3$

## COBALT  Co
## IDENTIFICATION

| Common Synonyms | Observable Characteristics | Manufacturers |
|---|---|---|
| COBALT METAL (nonradioactive). | Silvery, bluish-white shining metal. Black powder. Odourless. | Shell Canada Ltd., Toronto, Ontario. |

**Transportation and Storage Information**

Shipping State: Solid.
Classification: None.
Inert Atmosphere: No requirement.
Venting: No requirement.

Label(s): Not regulated.
Storage Temperature: Ambient.

Grades or Purity: 95 to 99.9%.
Containers and Materials: Kegs, drums.

**Physical and Chemical Characteristics**

Physical State (20°C, 1 atm): Solid.
Solubility (Water): Insoluble.
Molecular Weight: 58.9
Vapour Pressure: No information.
Boiling Point: 2 900 to 3 500°C.

Floatability (Water): Sinks.
Odour: Odourless.
Flash Point: Not flammable.
Vapour Density: No information.
Specific Gravity: 8.9 (20°C).

Colour: Silver-grey or black.
Explosive Limits: Not flammable.
Melting Point: 1 493°C.

## HAZARD DATA

### Human Health

**Symptoms:** Inhalation: shortness of breath, irritation of respiratory tract; asthma-like symptoms. Ingestion: pain, vomiting, diarrhea, convulsions. Contact: skin-dermatitis, irritation; eyes - conjunctivitis.
**Toxicology:** Highly toxic by inhalation and ingestion.
TLV® (inhalation) 0.05 mg/m$^3$ (as metal dust and fume).
Short-term Inhalation Limits - 0.1 mg/m$^3$ (15 min) (as metal dust and fume).

$LC_{50}$ - No information.
Delayed Toxicity - Some evidence of damage to heart, thyroid and pancreas.

$LD_{50}$ - No information.
$LD_{Lo}$ - Oral: rat = 1.5 g/kg

### Fire

**Fire Extinguishing Agents:** Solid metal is not combustible. Powder is combustible; use specially formulated dry-type agents to extinguish fires involving cobalt (dry sand, dry dolomite, dry graphite or sodium chloride).
**Behaviour in Fire:** Powdered cobalt may be pyrophoric in air.
**Ignition Temperature:** 370°C (dust); 760°C (cloud).   **Burning Rate:** No information.

### Reactivity

**With Water:** No reaction.
**With Common Materials:** Powdered cobalt reacts violently with acetylene, air and ammonium nitrate.
**Stability:** Stable.

### Environment

**Water:** Prevent entry into water intakes and waterways. Fish toxicity: 50.0 mg/L/tns/Daphnia magna/$LC_{100}$/freshwater; 1.0 mg/L/rainbow trout/not harmful/freshwater.
**Land-Air:** No information.
**Food Chain Concentration Potential:** No information.

## EMERGENCY MEASURES

**Special Hazards**

POWDERED COBALT IS PYROPHORIC.

**Immediate Responses**

Keep non-involved people away from spill site. Stop or reduce discharge if safe to do so. Avoid contact or inhalation of dust or fumes. Notify manufacturer or supplier. Notify environmental authorities.

**Protective Clothing and Equipment**

In fires, or enclosed spaces, Respiratory protection - self-contained breathing apparatus; otherwise, Gloves - rubber. Safety goggles or glasses. Coveralls.

**Fire and Explosion**

Solid metal is not combustible. Powder is combustible and pyrophoric. Use specially formulated dry-type agents to extinguish fires involving cobalt. (dry sand, dry dolomite, dry graphite or sodium chloride).

**First Aid**

Move victim out of spill area to fresh air. Call for medical assistance. but start first aid at once. Inhalation: if breathing has stopped, give artificial respiration; if laboured, give oxygen. Ingestion: give milk or water to conscious victim to drink, induce vomiting. Contact: skin - remove contaminated clothing and wash affected areas thoroughly with plenty of water; eyes - irrigate with plenty of water. If medical assistance is not immediately available, transport victim to hospital, doctor or clinic.

## ENVIRONMENTAL PROTECTION MEASURES

**Response**

**Water**
1. Stop or reduce discharge if safe to do so.
2. Contact manufacturer or supplier for advice.
3. If possible, contain discharge by damming or water diversion.
4. Dredge or vacuum pump to remove contaminants, liquids and contaminated bottom sediments.
5. Notify environmental authorities to discuss disposal and cleanup of contaminated materials.

**Land-Air**
1. Stop or reduce discharge if safe to do so.
2. Contact manufacturer or supplier for advice.
3. Contain spill by diking with earth or other barrier.
4. Remove material by manual or mechanical means.
5. Recover undamaged containers.
6. Remove contaminated soil for disposal.
7. Notify environmental authorities to discuss disposal and cleanup of contaminated materials.

**Disposal**
1. Contact manufacturer or supplier for advice on disposal.
2. Contact environmental authorities for advice on disposal.

COBALT   Co

# COPPER  Cu

## IDENTIFICATION

| Common Synonyms | Observable Characteristics | Manufacturers |
|---|---|---|
| COPPER METAL | Reddish, metallic solid or powder. Odourless. | Canada Metal Co. Ltd., Toronto, Ontario. Gaspé Copper Mines Ltd., Murdochville, Que. |

### Transportation and Storage Information

**Shipping State:** Solid.
**Classification:** Not regulated.
**Inert Atmosphere:** No requirement.
**Venting:** No requirement.
**Label(s):** Not regulated.
**Storage Temperature:** Ambient.
**Grades or Purity:** 99.9%+.
**Containers and Materials:** Kegs, drum, bulk by rail or truck; steel.

### Physical and Chemical Characteristics

**Physical State** (20°C, 1 atm): Solid.
**Solubility** (Water): 0.00002g/100 mL (30°C).
**Molecular Weight:** 63.6
**Vapour Pressure:** No information.
**Boiling Point:** 2 560 to 2 600°C.
**Floatability** (Water): Sinks.
**Odour:** Odourless.
**Flash Point:** Not flammable.
**Vapour Density:** No information.
**Specific Gravity:** 8.9 (20°C).
**Colour:** Reddish.
**Explosive Limits:** Dust can explode.
**Melting Point:** 1 083°C.

## HAZARD DATA

### Human Health

**Symptoms:** Inhalation: dust irritates the nose and upper respiratory tract. Ingestion: causes metallic taste, nausea, gastrointestinal problems.
**Toxicology:** Highly toxic by inhalation (dust and fumes); moderately toxic by ingestion.
TLV®- (inhalation) 0.2 mg/m$^3$ (copper fume); $LC_{50}$ - No information. $LD_{50}$ - Intraperitoneal: mouse = 0.0035 g/kg 1 mg/m$^3$ (dust, mist).
Delayed Toxicity - Possible kidney damage, jaundice,
Short-term Inhalation Limits - 2.0 mg/m$^3$ perforation of nasal septum.
(15 min) (dusts and mists).

### Fire

**Fire Extinguishing Agents:** Solid is not combustible. Powder is combustible. Use specially formulated dry-type agents to extinguish fires involving copper (dry sand, dry dolomite, dry graphite or sodium chloride).
**Behaviour in Fire:** Solid is not combustible. Powder is combustible and can explode.
**Ignition Temperature:** No information. **Burning Rate:** No information.

### Reactivity

**Water:** No reaction.
**With Common Materials:** Reacts violently with acetylene, ammonium nitrate, bromates, chlorates, iodates, chlorine, ethylene oxide, fluorine, hydrogen peroxide and hydrogen sulfide.
**Stability:** Stable.

### Environment

**Water:** Prevent entry into water intakes and waterways. Saltwater aquatic guideline: 0.7 µg/L/24 h/average; concentration should not exceed 8 µg/L at any time; human drinking water criterion = 1 µg/L; BOD: None.
**Land-Air:** No information.
**Food Chain Concentration Potential:** No information.

## EMERGENCY MEASURES

**Special Hazards**
Powdered copper is flammable. Dust is explosive.

**Immediate Responses**
Keep non-involved people away from spill site. In the case of powdered copper, call Fire Department. Stop or reduce discharge if safe to do so. Contact manufacturer. Notify environmental authorities.

**Protective Clothing and Equipment**
In fires, or enclosed spaces, Respiratory protection - self-contained breathing apparatus; otherwise, Gloves - rubber. Safety goggles or glasses. Coveralls.

**Fire and Explosion**
Powdered copper may ignite when heated.

**First Aid**
Move victim out of spill area to fresh air. Call for medical assistance, but start first aid at once. Inhalation: if breathing has stopped, give artificial respiration; if laboured, give oxygen. Ingestion: give milk or water to conscious victim to drink. Induce vomiting. Contact: skin - remove contaminated clothing and wash affected areas throughly with plenty of water; eyes - irrigate with plenty of water. If medical assistance is not immediately available, transport victim to hospital, doctor or clinic.

## ENVIRONMENTAL PROTECTION MEASURES

**Response**

**Water**
1. Stop or reduce discharge if safe to do so.
2. Contact manufacturer or supplier for advice.
3. If possible, contain discharge by damming or water diversion.
4. Dredge or vacuum pump to remove contaminants, liquids and contaminated bottom sediments.
5. Notify environmental authorities to discuss disposal and cleanup of contaminated materials.

**Land-Air**
1. Stop or reduce discharge if safe to do so.
2. Contact manufacturer or supplier for advice.
3. Contain spill by diking with earth or other barrier.
4. Remove material by manual or mechanical means.
5. Recover undamaged containers.
6. Remove contaminated soil for disposal.
7. Notify environmental authorities to discuss disposal and cleanup of contaminated materials.

**Disposal**
1. Contact manufacturer or supplier for advice on disposal.
2. Contact environmental authorities for advice on disposal.

COPPER   Cu

# COPPER CHLORIDE  $CuCl_2$

## IDENTIFICATION

UN No. 2802

### Common Synonyms

CUPRIC CHLORIDE $CuCl_2$
COPPER (II) CHLORIDE (dihydrate) $CuCl_2.2H_2O$

### Observable Characteristics

Anhydrous: Brownish-yellow powder. Odourless.
Dihydrate: Green crystals. Odourless.

### Manufacturers

No Canadian Manufacturers:
US Manufacturers:
Diamond Shamrock Corp.
Cleveland, Ohio.
Mallinckrodt Inc.
St. Louis, Missouri.

### Transportation and Storage Information

**Shipping State:** Solid.
**Classification:** Corrosive.
**Inert Atmosphere:** No requirement.
**Venting:** No requirement.

**Label(s):** Black and white label – CORROSIVE; Class 8; Group III.
**Storage Temperature:** Ambient.

**Grades or Purity:** Technical.
**Containers and Materials:** Fibre drums, barrels.

### Physical and Chemical Characteristics

**Physical State** (20°C, 1 atm): Solid.
**Solubility** (Water): 70.6 g/100 mL (0°C); 107.9 g/100 mL (100°C) (anhydrous); 110.4 g/100 mL (0°C), 192.4 g/100 mL (100°C) (dihydrate).
**Molecular Weight:** 134.4 (anhydrous); 170.5 (dihydrate).
**Vapour Pressure:** No information.
**Boiling Point:** Anhydrous decomposes to CuCl at 993°C. Dihydrate loses $2H_2O$ at 100°C.

**Floatability** (Water): Sinks and mixes.
**Odour:** Odourless.
**Flash Point:** Not flammable.
**Vapour Density:** No information.
**Specific Gravity:** 3.39 (anhydrous); 2.54 (dihydrate) (20°C).

**Colour:** Brownish-yellow or green.
**Explosive Limits:** Not flammable.
**Melting Point:** 620°C (anhydrous); dihydrate loses $2H_2O$ at 100°C.

## HAZARD DATA

### Human Health

**Symptoms:** Inhalation: coughing, sneezing, irritation of nose and throat. Ingestion: burning sensation, stomach cramps, nausea, vomiting, diarrhea, convulsions, coma. Contact: skin - itching and inflammation; eyes - stinging and inflammation.
**Toxicology:** Highly toxic by ingestion and inhalation.
TLV® - 1 mg/m³ (as Cu dusts and mists), 0.2 mg/m³ (copper fume). Short-term Inhalation Limits - 2 mg/m³ (15 min)(as Cu dust and mists).
$LC_{50}$ - No information.
Delayed Toxicity - No information.

$LD_{50}$ - Oral: rat = 0.14 g/kg
$LD_{50}$ - Oral: mouse = 0.19 g/kg
$LD_{50}$ - Oral: guinea pig = 0.03 g/kg

### Fire

**Fire Extinguishing Agents:** Not combustible. Most fire extinguishing agents may be used on fires involving copper chloride.
**Behaviour in Fire:** No information.
**Ignition Temperature:** Not combustible.
**Burning Rate:** Not combustible.

### Reactivity

**With Water:** No reaction; soluble.
**With Common Materials:** Can react violently with potassium and sodium.
**Stability:** Stable.

### Environment

**Water:** Prevent entry into water intakes and waterways. Fish toxicity: critical concentration = 3.3 mg/L; 0.044 ppm/3 weeks/<u>Daphnia magna</u>/$LC_{50}$/freshwater; 0.009 ppm (as Cu)/tns/goldfish/rapid death/freshwater; BOD: none.
**Land-Air:** No information.
**Food Chain Concentration Potential:** No information.

## EMERGENCY MEASURES

**Special Hazards**

CORROSIVE.

**Immediate Responses**

Keep non-involved people away from spill site. Issue warning: "CORROSIVE". Dike spill area, particularly if there is any danger of water runoff. Avoid inhalation of dust and fumes. Lightly wet down dry spillage, particularly if there is any dust drift by wind. Notify supplier or manufacturer. Notify environmental authorities.

**Protective Clothing and Equipment**

In fires or confined spaces Respiratory protection - self-contained breathing apparatus. Other protective equipment as required.

**Fire and Explosion**

Not combustible. Most fire extinguishing agents may be used on fires involving copper chloride.

**First Aid**

Move victim out of spill area to fresh air. Call for medical assistance, but start first aid at once. Contact: skin - remove contaminated clothing and flush affected areas with plenty of water; eyes - irrigate with plenty of water. Ingestion: conscious victim should be given water to drink. Keep warm and quiet. If medical assistance is not immediately available, transport victim to hospital, doctor or clinic.

## ENVIRONMENTAL PROTECTION MEASURES

**Response**

**Water**
1. Stop or reduce discharge if safe to do so.
2. Contact manufacturer or supplier for advice.
3. If possible, contain discharge by damming or water diversion.
4. Dredge or vacuum pump to remove contaminants, liquids and contaminated bottom sediments.
5. Notify environmental authorities to discuss disposal and cleanup of contaminated materials.

**Land-Air**
1. Stop or reduce discharge if safe to do so.
2. Contact manufacturer or supplier for advice.
3. Dike to prevent runoff from rainwater or water application.
4. Remove material by manual or mechanical means.
5. Notify environmental authorities to discuss disposal and cleanup of contaminated materials.

**Disposal**
1. Contact manufacturer or supplier for advice on disposal.
2. Contact environmental authorities for advice on disposal.

COPPER CHLORIDE $CuCl_2$

# COPPER NAPHTHENATE

## IDENTIFICATION

### Common Synonyms
CUPRENOL
WITTOX-C
COPPER UVERSOL
NAPHTHENIC ACID, COPPER SALT
Copper naphthenate is derived from naphthenic acid which is a mixture of cyclic aliphatic acids.

### Observable Characteristics
Dark green liquid (solution) or green-blue solid. Solution smells of the solvent, typically gasoline. mineral oil distillates.

### Manufacturers
Dussek Bros. (Canada) Ltd., Belleville, Ontario.
Nuodex Canada Limited, Toronto, Ontario.

### Transportation and Storage Information
**Shipping State:** Liquid (solution).
**Classification:** None.
**Inert Atmosphere:** No requirement.
**Venting:** Pressure-vacuum or open (flame arrester).
**Pump Type:** No information.
**Label(s):** Not regulated.
**Storage Temperature:** Ambient.
**Hose Type:** No information.
**Grades or Purity:** 6, 8 or 11.5% copper.
**Containers and Materials:** Drums; steel.

### Physical and Chemical Characteristics
**Physical State (20°C, 1 atm):** Solid.
**Solubility (Water):** Insoluble.
**Molecular Weight:** Variable.
**Vapour Pressure:** No information.
**Boiling Point:** 154 to 202°C.
**Floatability (Water):** Sinks or floats depending on specific gravity.
**Odour:** Solution smells of the solvent.
**Flash Point:** 38°C (c.c. typical).
**Vapour Density:** No information.
**Specific Gravity:** 0.93 to 1.06 (20°C).
**Colour:** Green-blue (solid) or dark green liquid
**Explosive Limits:** 0.8 to 5.0% (mineral spirits).
**Melting Point:** No information.

## HAZARD DATA

### Human Health
**Symptoms: Inhalation:** concentrated vapour of solvent may cause headache, and in severe cases, uncoordination. **Ingestion:** burning pain in mouth and throat, causes gastrointestinal irritation, vomiting, convulsions and collapse. **Contact:** may cause skin or eye irritation.
**Toxicology:** Highly toxic by ingestion or inhalation.
TLV® (inhalation) - 0.2 mg/m$^3$ (copper fume);
1 mg/m$^3$ (copper dust, mist).
Short-term Inhalation Limits - 2.0 mg/m$^3$ (15 min) (copper dust and mist).
LC$_{50}$ - No information.
Delayed Toxicity - No information.
LD$_{50}$ - No information.
LD$_{Lo}$ - Oral: mouse = 0.11 g/kg

### Fire
**Fire Extinguishing Agents:** Use foam, carbon dioxide and dry chemical. Water may be used to cool fire-exposed containers.
**Behaviour in Fire:** No information.
**Ignition Temperature:** 282°C (mineral spirits).
**Burning Rate:** 4 mm/min (mineral spirits).

### Reactivity
**With Water:** No reaction.
**With Common Materials:** Can react with oxidizing materials.
**Stability:** Stable.

### Environment
**Water:** Prevent entry into water intakes and waterways. Harmful to aquatic life. Aquatic toxicity rating = 2.0 ppm/72 h/blue-green algae/100% kill/freshwater; BOD: 8%, 5 days.
**Land-Air:** No information.
**Food Chain Concentration Potential:** No information.

## EMERGENCY MEASURES

**Special Hazards**

FLAMMABLE.

**Immediate Responses**

Keep non-involved people away from spill site. Call Fire Department. Avoid contact and inhalation. Stop or reduce discharge, if safe to do so. Dike to prevent runoff. Notify manufacturer. Notify environmental authorities.

**Protective Clothing and Equipment**

In fires or confined spaces Respiratory protection - self-contained breathing apparatus. Otherwise, goggles, respirator, rubber gloves, coveralls.

**Fire and Explosion**

Use foam, carbon dioxide or dry chemical to extinguish fires. Water may be used to cool fire-exposed containers.

**First Aid**

Move victim out of spill area to fresh air. Call for medical assistance, but start first aid at once. **Contact:** remove contaminated clothing and wash eyes and affected skin thoroughly with plenty of water. **Ingestion:** give water or milk to conscious victim to drink. Allow vomiting to occur. If medical assistance is not immediately available, transport victim to doctor, clinic or hospital.

## ENVIRONMENTAL PROTECTION MEASURES

**Response**

**Water**

Floating
1. Stop or reduce discharge if safe to do so.
2. Contact manufacturer or supplier for advice.
3. If possible, contain discharge by booming.
4. If floating, skim and remove.
5. Notify environmental authorities to discuss disposal and cleanup of contaminated materials.

Sinking
1. Stop or reduce discharge if safe to do so.
2. Contact manufacturer or supplier for advice.
3. If possible, contain discharge by damming or water diversion.
4. Dredge or vacuum pump to remove contaminants, liquids and contaminated bottom sediments.
5. Notify environmental authorities to discuss disposal and cleanup of contaminated materials.

**Land-Air**
1. Stop or reduce discharge if safe to do so.
2. Contact manufacturer or supplier for advice.
3. Contain spill by diking with earth or other barrier.
4. Remove material with pumps or vacuum equipment and place in appropriate containers.
5. Recover undamaged containers.
6. Absorb residual liquid on natural or synthetic sorbents.
7. Notify environmental authorities to discuss disposal and cleanup of contaminated materials.

**Disposal**

1. Contact manufacturer or supplier for advice on disposal.
2. Contact environmental authorities for advice on disposal.

## COPPER NAPHTHENATE

# COPPER SULFATE   $CuSO_4 \cdot 5H_2O$

## IDENTIFICATION

### Common Synonyms
BLUESTONE
CUPRIC SULFATE
BLUE VITRIOL
ROMAN VITRIOL
COPPER 2 SULFATE PENTAHYDRATE
COPPER SULFATE PENTAHYDRATE

### Observable Characteristics
Whitish-blue to greenish-blue powder or crystals. Odourless.

### Manufacturers
Canadian Copper Refiners Ltd., Montreal, Que.
Cominco Limited, Kimberley, B.C.
Canadian Metafina, Vancouver, B.C.

### Transportation and Storage Information
**Shipping State:** Solid.
**Classification:** Not regulated.
**Inert Atmosphere:** No requirement.
**Venting:** Open.

**Label(s):** None. Class 9.2, Group I.
**Storage Temperature:** Ambient.

**Grades or Purity:** Technical.
**Containers and Materials:** Multiwall paper and poly bags, drums; steel.

### Physical and Chemical Characteristics
**Physical State** (20°C, 1 atm): Solid.
**Solubility** (Water): 31.6 g/100 mL (0°C); 203.3 g/100 mL (100°C).
**Molecular Weight:** 249.7
**Vapour Pressure:** No information.
**Boiling Point:** Loses $4H_2O$ at 110°C; loses $5H_2O$ at 150°C.

**Floatability** (Water): Sinks and mixes.
**Odour:** Odourless.
**Flash Point:** Not flammable.
**Vapour Density:** No information.
**Specific Gravity:** 2.28 (20°C).

**Colour:** Whitish-blue to greenish-blue.
**Explosive Limits:** Not flammable.
**Melting Point:** Loses $4H_2O$ at 110°C; loses $5H_2O$ at 150°C.

## HAZARD DATA

### Human Health
**Symptoms:** _Contact_: irritation of eyes and skin, as well as congestion of nasal mucous membranes. _Ingestion_: vomiting, gastric pain, hemorrhagic gastritis, diarrhea, convulsions and collapse. _Inhalation_: sore throat, coughing, shortness of breath.
**Toxicology:** Moderately toxic by ingestion and inhalation.
TLV® (inhalation) 0.2 mg/m³ (as fume Cu);  $LC_{50}$ - No information.   $LD_{50}$ - Oral: rat= 0.3 g/kg
1 mg/m³ (dust, mist Cu).   Delayed Toxicity - No information.
Short-term Inhalation Limits - 2 mg/m³ for 15 min (as Cu, dust and mist).

### Fire
**Fire Extinguishing Agents:** Not combustible. Most fire extinguishing agents may be used on fires involving copper sulfate.
**Behaviour in Fire:** Not combustible. When heated above 400°C can release toxic $SO_x$ fumes. Closed containers may rupture when heated above 110°C due to loss of water vapour and expansion.
**Ignition Temperature:** Not combustible.   **Burning Rate:** Not combustible.

### Reactivity
**With Water:** No reaction, soluble.
**With Common Materials:** Can react with hydroxylamine and magnesium.
**Stability:** Stable.

### Environment
**Water:** Prevent entry into water intakes and waterways. Harmful to aquatic life in low concentrations. Fish toxicity: 0.15 mg/L/48 h/rainbow trout/TLm/freshwater; 3.8 ppm/24 h/rainbow trout/TLm/freshwater; 1.0 mg/L/24 h/_Daphnia magna_/LC100/freshwater; BOD: None.
**Land-Air:** $LD_{Lo}$ - wild bird = 0.3 g/kg
**Food Chain Concentration Potential:** No information.

# EMERGENCY MEASURES

## Special Hazards

### Immediate Responses

Keep non-involved people away from spill site. Avoid contact and inhalation. Stop or reduce discharge if safe to do so. Contain spill by diking. Notify manufacturer or supplier. Notify environmental authorities.

### Protective Clothing and Equipment

In fires and confined spaces - Respiratory protection - self-contained breathing apparatus. Otherwise, Gloves - rubber or plastic. Goggles - (mono), tight fitting. Rubber boots (pants worn outside boots).

### Fire and Explosion

Not combustible. Most fire extinguishing agents may be use on fires involving copper sulfate.

### First Aid

Move victim out of spill area to fresh air. Call for medical assistance, but start first aid at once. Contact: remove contaminated clothing, wash eyes and affected skin with plenty of water. Keep warm and quiet. Ingestion: give milk or water to conscious victim to drink. Induce vomiting. If medical assistance is not immediately available, transport victim to hospital, doctor or clinic.

# ENVIRONMENTAL PROTECTION MEASURES

## Response

**Water**
1. Stop or reduce discharge if safe to do so.
2. Contact manufacturer or supplier for advice.
3. If possible, contain discharge by damming or water diversion.
4. Dredge or vacuum pump to remove contaminants, liquids and contaminated bottom sediments.
5. Notify environmental authorities to discuss disposal and cleanup of contaminated materials.

**Land-Air**
1. Stop or reduce discharge if safe to do so.
2. Contact manufacturer or supplier for advice.
3. Dike to prevent runoff from rainwater or water application.
4. Remove material by manual or mechanical means.
5. Recover undamaged containers.
6. Notify environmental authorities to discuss disposal and cleanup of contaminated materials.

## Disposal

1. Contact manufacturer or supplier for advice on disposal.
2. Contact environmental authorities for advice on disposal.

COPPER SULFATE   $CuSO_4 \cdot 5H_2O$

# CRESOL  CH₃C₆H₄OH

## IDENTIFICATION  UN No. 2076

### Common Synonyms
HYDROXYMETHYLBENZENE
CRESYLIC ACID
HYDROXYTOLUENE
METHYL PHENOL
ORTHO-CRESOL; o-CRESOL,2-CRESOL) Cresol (cresylic acid)
META-CRESOL; m-CRESOL,3-CRESOL ) is usually a mixture
PARA-CRESOL; p-CRESOL,4-CRESOL ) of these forms.

### Observable Characteristics
Colourless to yellow or pink to yellow-brown liquid (sometimes solid). Tarry or phenol-like odour.

### Manufacturers
Canadian supplier:
Domtar Chemicals Group,
Tar and Chemicals Division,
Toronto, Ont.

### Transportation and Storage Information
**Shipping State:** Liquid or solid.
**Classification:** Poison.
**Inert Atmosphere:** No requirement.
**Venting:** Open.
**Pump Type:** Centrifugal; steel, stainless steel.

**Label(s):** White label - Poison; Class 6.1, Group II.
**Storage Temperature:** Ambient.
**Hose Type:** Polyethylene, butyl, Viton.
**Containers and Materials:** Drums, tank cars, tank trucks; steel, stainless steel.

**Grades or Purity:** USP liquid (mixed isomers). Phenol-cresol mixtures: Ortho-cresol; 80 to 98% (with phenol). Meta-cresol; 60 to 98% (with other cresols and xylenol). Meta-para-cresol; containing cresol and xylenols. Para-cresol 92 to 98% containing meta-cresol.

### Physical and Chemical Characteristics
**Physical State (20°C, 1 atm):** Liquid (sometimes a solid).
**Solubility (Water):** (o) 3.1 g/100 mL (40°C), 5.6 g/100 mL (100°C); (m) 2.4 g/100 mL (20°C), 5.8 g/100 mL (100°C); (p) 2.4 g/100 mL (40°C) 5.3 g/100 mL (100°C).
**Molecular Weight:** 108.1 (pure o, m, or p).

**Floatability (Water):** Sinks.
**Odour:** Tarry or phenol-like (0.001 to 27 ppm, odour threshold); (o) 0.26 ppm, (m) 0.27 ppm, (p) 0.2 ppm.
**Flash Point:** (o) 81°C; (m or p) 86°C(c.c.).
**Vapour Density:** (o,m,p) 3.7
**Vapour Pressure:** (o) 0.24 mm Hg (25°C), 5 mm Hg (64°C); (m) 0.04 mm Hg (20°C), 0.12 mm Hg (30°C), 5 mm Hg (76°C); (p) 0.04 mm Hg (20°C), 1 mm Hg (53°C).

**Colour:** Colourless to yellow or pink to yellow-brown.
**Explosive Limits:** 1.4% (LEL, o); 1.1% (LEL, m, p).
**Melting Point:** (o) 31°C; (m) 12°C; (p) 34.8°C.
**Specific Gravity:** 1.04 (20°C)(o,m,p).
**Boiling Point:** (o) 191°C; (m) 202°C; (p) 202°C.

## HAZARD DATA

### Human Health
**Symptoms:** Contact: skin - readily absorbed producing burns and symptoms similar to inhalation. Inhalation: results in muscular weakness, headache, dizziness, dimness of vision, ringing in ears, irregular and rapid respiration, weak pulse, mental confusion, muscular twitching, loss of consciousness. Ingestion: in addition to the above there can also be nausea, with or without vomiting, and severe abdominal pains.
**Toxicology:** Highly toxic by ingestion. Moderately toxic by contact and inhalation. Readily absorbed through skin.
TLV®- (skin) 5 ppm; 22 mg/m³.         LC₅₀ - No information.         LD₅₀ - Oral: rat = 1.45 g/kg (cresol mixture).
Short-term Inhalation Limits - No information.    Delayed Toxicity - Possible damage to liver    LD₅₀ - Oral: rat = 0.12 g/kg (ortho).
                                                  and kidney.                                   LD₅₀ - Oral: rat = 0.24 g/kg (meta).
                                                                                                LD₅₀ - Oral: rat = 0.21 g/kg (para).

### Fire
**Fire Extinguishing Agents:** Use water spray, dry chemical, foam or carbon dioxide. Use water spray to disperse vapours and keep fire-exposed containers cool.
**Behaviour in Fire:** Emits highly toxic fumes in fires.
**Ignition Temperature:** 599°C (o-cresol); 558°C (m- or p-cresol)                **Burning Rate:** No information.

### Reactivity
**With Water:** No reaction; slightly soluble.
**With Common Materials:** Can react vigorously with oxidizing materials. Reacts with chlorosulfonic acid, nitric acid and oleum.
**Stability:** Stable.

### Environment
**Water:** Prevent entry into water intakes and waterways. Harmful to aquatic life in low concentrations. Fish toxicity: 24 ppm/96 h/bluegill/TLm/freshwater; Aquatic toxicity rating = 1 to 10 ppm/96 h/TLm/freshwater; (ortho) 22.2 to 20.8 mg/L/48 h/bluegill/TLm/freshwater; (meta) 24 mg/L/48 h/mosquito fish/TLm/freshwater; BOD: o-cresol: 164%, 5 days; m-cresol:170%, 5 days; p-cresol: 144%, 5 days.
**Land-Air:** No information.

## EMERGENCY MEASURES

### Special Hazards
POISON.

### Immediate Responses
Keep non-involved people away from spill site. Issue warning: "POISON". Call Fire Department. Avoid contact and inhalation. Contact manufacturer or supplier for advice. Stop or reduce discharge, if safe to do so. Dike to contain toxic runoff. Notify environmental authorities.

### Protective Clothing and Equipment
Respiratory protection - self-contained breathing apparatus and totally encapsulated suit. Gloves - rubber. Boots - high, rubber - pants worn outside boots.

### Fire and Explosion
Use water spray, dry chemical, foam or carbon dioxide. Use water spray to disperse vapours and keep fire-exposed containers cool.

### First Aid
Move victim out of spill area to fresh air. Call for medical assistance, but start first aid at once. Contact: skin - remove contaminated clothing, flush affected areas with plenty of water; eyes - irrigate with plenty of water. Ingestion: give water to conscious victim to drink. Do not induce vomiting. Inhalation: oxygen is recommended for respiratory distress. Treat as an emergency. If medical assistance is not immediately available, transport victim to hospital, doctor or clinic.

## ENVIRONMENTAL PROTECTION MEASURES

### Response

**Water**
1. Stop or reduce discharge if safe to do so.
2. Contact manufacturer or supplier for advice.
3. If possible, contain discharge by damming or water diversion.
4. Dredge or vacuum pump to remove contaminants, liquids and contaminated bottom sediments.
5. Notify environmental authorities to discuss disposal and cleanup of contaminated materials.

**Land-Air**
1. Stop or reduce discharge if safe to do so.
2. Contact manufacturer or supplier for advice.
3. Dike to prevent runoff from rainwater or water application.
4. Remove material with pumps or vacuum equipment and place in appropriate containers.
5. If material is a solid, remove by manual or mechanical means.
6. Recover undamaged containers.
7. Absorb residual liquid on natural or synthetic sorbents.
8. Notify environmental authorities to discuss disposal and cleanup of contaminated materials.

### Disposal
1. Contact manufacturer or supplier for advice on disposal.
2. Contact environmental authorities for advice on disposal.
3. Incinerate (approval of environmental authorities required).

CRESOL   $CH_3C_6H_4OH$

# CRYOLITE  AlF$_3$·2NaF
## IDENTIFICATION

| Common Synonyms | Observable Characteristics | Manufacturers |
|---|---|---|
| GREENLAND SPAR<br>ICETONE<br>KRYOLITH<br>SODIUM ALUMINUM FLUORIDE<br>ALUMINUM SODIUM FLUORIDE<br>SODIUM FLUOALUMINATE | Colourless to white, sometimes reddish or brown to black, solid. Odourless. | Aluminum Company of Canada Ltd., Montreal, Quebec.<br>Alcan Smelters and Chemicals Ltd., Jonquière, Quebec. |

### Transportation and Storage Information

**Shipping State:** Solid.
**Classification:** None.
**Inert Atmosphere:** No requirement.
**Venting:** Open.

**Label(s):** Not regulated.
**Storage Temperature:** Ambient.

**Grades or Purity:** Commercial, technical.
**Containers and Materials:** Drums, bulk by rail or truck; steel.

### Physical and Chemical Characteristics

**Physical State (20°C, 1 atm):** Solid.
**Solubility (Water):** 3 g/100 mL (0°C); 4 g/100 mL (25°C); 9 g/100 mL (90°C).
**Molecular Weight:** 210
**Vapour Pressure:** No information.
**Boiling Point:** No information.

**Floatability (Water):** Sinks.
**Odour:** Odourless.
**Flash Point:** Not flammable.
**Vapour Density:** No information.
**Specific Gravity:** 2.95

**Colour:** Colourless to white, sometimes reddish or brown to black.
**Explosive Limits:** Not flammable.
**Melting Point:** 1 000°C.

## HAZARD DATA

### Human Health

**Symptoms: Contact:** skin - irritation and dermatitis. **Inhalation** (dust): coughing, chills, tightness in chest, cyanosis. **Ingestion:** salivation, nausea, vomiting, diarrhea, abdominal pain, convulsions.
**Toxicology:** Highly toxic by ingestion and skin absorption.
TLV® (inhalation) 2.5 mg/m$^3$ (as Fluoride, F).    LC$_{50}$ - No information.    LD$_{50}$ - Oral: rat = 0.2 g/kg
Short-term Inhalation Limits - No information.    Delayed Toxicity - may result in long-term    LD$_{50}$ - Intraperitoneal: rat = 0.059 g/kg
                                                    bone damage especiallly for repeated exposure.

### Fire

**Fire Extinguishing Agents:** Not combustible. Most fire extinguishing agents may be used on fires involving cryolite.
**Behaviour in Fire:** No information.
**Ignition Temperature:** Not combustible.    **Burning Rate:** Not combustible.

### Reactivity

**With Water:** No reaction.
**With Common Materials:** No information.
**Stability:** Stable.

### Environment

**Water:** Prevent entry into water intakes and waterways. **Fish toxicity:** 5 ppm/Daphnia magna/static bioassay; BOD: None.
**Land-Air:** No information.
**Food Chain Concentration Potential:** No information.

## EMERGENCY MEASURES

### Special Hazards

### Immediate Responses
Keep non-involved people away from spill site. Stop or reduce discharge, if this can be done without risk. Contain spill by diking, particularly if there is any water runoff. Notify manufacturer. Notify environmental authorities.

### Protective Clothing and Equipment
Respiratory protection - dust respirator with suitable filter for dust protection. Goggles - (mono) type, tight fitting. Gloves - rubber, neoprene or PVC coated. Outer clothing, as required.

### Fire and Explosion
Not combustible. Most fire extinguishing agents may be used in fires involving cryolite.

### First Aid
Move victim out of spill site to fresh air. Call for medical assistance but start first aid at once. Inhalation: if breathing has stopped, give artificial respiration; if laboured, give oxygen. Ingestion: give plenty water or milk to conscious victim to drink. Keep warm and quiet. Contact: skin and eyes - remove contaminated clothing and flush affected areas with plenty of water. If medical assistance is not immediately available, transport victim to hospital, doctor or clinic.

## ENVIRONMENTAL PROTECTION MEASURES

### Response

**Water**
1. Stop or reduce discharge if safe to do so.
2. Contact manufacturer or supplier for advice.
3. If possible, contain discharge by damming or water diversion.
4. Dredge or vacuum pump to remove contaminants, liquids and contaminated bottom sediments.
5. Notify environmental authorities to discuss disposal and cleanup of contaminated materials.

**Land-Air**
1. Stop or reduce discharge if safe to do so.
2. Contact manufacturer or supplier for advice.
3. Dike to prevent runoff from rainwater or water application.
4. Remove material by manual or mechanical means.
5. Recover undamaged containers.
6. Notify environmental authorities to discuss disposal and cleanup of contaminated materials.

### Disposal
1. Contact manufacturer or supplier for advice on disposal.
2. Contact environmental authorities for advice on disposal.

CRYOLITE    $AlF_3 \cdot 3NaF$

# CUMENE   $C_6H_5CH(CH_3)_2$

## IDENTIFICATION

UN No. 1918

### Common Synonyms

CUMOL
ISOPROPYLBENZENE
2-PHENYLPROPANE

### Observable Characteristics

Clear, colourless liquid. Sharp aromatic odour.

### Manufacturers

Gulf Oil Canada Limited, Montreal, Quebec.

### Transportation and Storage Information

**Shipping State:** Liquid.
**Classification:** Flammable liquid.
**Inert Atmosphere:** No requirement.
**Venting:** Open (flame arrester).
**Pump Type:** Centrifugal, gear, explosion-proof, grounded.

**Label(s):** Red label - FLAMMABLE LIQUID; Class 3.2, Group II.
**Storage Temperature:** Ambient.
**Hose Type:** Viton, polypropylene, Teflon.

**Grades or Purity:** Technical.
**Containers and Materials:** Drums, tank cars, tank trucks; steel.

### Physical and Chemical Characteristics

**Physical State** (20°C, 1 atm): Liquid.
**Solubility** (Water): 0.005 g/100 mL (20°C).
**Molecular Weight:** 120.2
**Vapour Pressure:** 3.2 mm Hg (20°C); 10 mm Hg (38°C).
**Boiling Point:** 152.0 to 152.7°C.

**Floatability** (Water): Floats.
**Odour:** Sharp aromatic, (0.008 to 0.047 ppm, odour threshold).
**Flash Point:** 36 to 44°C (c.c.).
**Vapour Density:** 4.1
**Specific Gravity:** 0.86 (20°C).

**Colour:** Colourless.
**Explosive Limits:** 0.9 to 6.5%.
**Melting Point:** -96.1°C.

## HAZARD DATA

### Human Health

**Symptoms:** Inhalation: irritation of respiratory tract, sore throat, coughing, shortness of breath, nausea. Ingestion: abdominal pain and vomiting. Contact: skin - redness and irritation. Also absorbed by skin to produce symptoms similar to inhalation.
**Toxicology:** Moderately toxic by inhalation, ingestion and contact.
TLV® - (skin) 50 ppm; 245 mg/m³.    $LC_{50}$ - Inhalation: rat = 8 000 ppm/4 h    $LD_{50}$ - Oral: rat = 1.4 g/kg
Short-term Inhalation Limits - (skin) 75 ppm;    Delayed Toxicity - No information.
365 mg/m³ (15 min).

### Fire

**Fire Extinguishing Agents:** Use water spray, dry chemical, foam or carbon dioxide. Use water to keep fire-exposed containers cool. Water spray may be ineffective in fighting fire.
**Behaviour in Fire:** Flashback may occur along vapour trail.
**Ignition Temperature:** 425°C.    **Burning Rate:** 5 mm/min.

### Reactivity

**With Water:** No reaction.
**With Common Materials:** Can react with oxidizing materials. Reacts violently with nitric acid, oleum, chlorosulfonic acid.
**Stability:** Stable.

### Environment

**Water:** Prevent entry into water intakes and waterways. Fish toxicity: >110 ppm/24 h/brine shrimp/TLm/saltwater; BOD: 40%, 5 days (freshwater).
**Land-Air:** No information.
**Food Chain Concentration Potential:** No information.

## EMERGENCY MEASURES

**Special Hazards**
FLAMMABLE.

**Immediate Responses**
Keep non-involved people away from spill site. Issue warning: "FLAMMABLE". CALL FIRE DEPARTMENT. Eliminate all sources of ignition. Call manufacturer or supplier for guidance. Contain spill by diking to prevent runoff. Stop or reduce discharge, if safe to do so. Notify environmental authorities.

**Protective Clothing and Equipment**
Respiratory protection - self-contained breathing apparatus, and totally encapsulated suit. Gloves - butyl or PVA. Boots - high, rubber.

**Fire and Explosion**
Use water spray, dry chemical, foam or carbon dioxide, to extinguish. Use water to keep fire-exposed containers cool. Flashback may occur along vapour trail.

**First Aid**
Move victim out of spill area to fresh air. Call for medical assistance, but start first aid at once. Inhalation: give artificial respiration if necessary. Contact: eyes - flush with plenty of water; skin - remove contaminated clothing and flush affected areas with water. Ingestion: Do not induce vomiting. If medical assistance is not immediately available, transport victim to hospital, doctor or clinic.

## ENVIRONMENTAL PROTECTION MEASURES

**Response**

**Water**
1. Stop or reduce discharge if safe to do so.
2. Contact manufacturer or supplier for advice.
3. If possible, contain discharge by booming.
4. If floating, skim and remove.
5. Notify environmental authorities to discuss disposal and cleanup of contaminated materials.

**Land-Air**
1. Stop or reduce discharge if safe to do so.
2. Contact manufacturer or supplier for advice.
3. Contain spill by diking with earth or other barrier.
4. Remove material with pumps or vacuum equipment and place in appropriate containers.
5. Recover undamaged containers.
6. Absorb residual liquid on natural or synthetic sorbents.
7. Notify environmental authorities to discuss disposal and cleanup of contaminated materials.

**Disposal**
1. Contact manufacturer or supplier for advice on disposal.
2. Contact environmental authorities for advice on disposal.
3. Incinerate (approval of environmental authorities required).

CUMENE  $C_6H_5CH(CH_3)_2$

# CUMENE HYDROPEROXIDE  $C_6H_5C(CH_3)_2OOH$

## IDENTIFICATION

UN No. 2116

### Common Synonyms

α,α-DIMETHYLBENZYL HYDROPEROXIDE
CUMYL HYDROPEROXIDE
ISOPROPYLBENZENE HYDROPEROXIDE

### Observable Characteristics

Colourless to pale yellow liquid.

### Manufacturers

Gulf Oil Canada Limited, Montreal, Quebec.

### Transportation and Storage Information

**Shipping State:** Liquid.
**Classification:** Organic peroxide.
**Inert Atmosphere:** No requirement.
**Venting:** Open (flame arrester).
**Pump Type:** Standard types.

**Label(s):** Yellow label - ORGANIC PEROXIDE; Class 5.2, Group I.
**Storage Temperature:** Ambient.
**Hose Type:** Standard types.

**Grades or Purity:** 77 to 85%; 90% cut with cumene.
**Containers and Materials:** Drums, tank cars, tank trucks; steel.

### Physical and Chemical Characteristics

**Physical State** (20°C, 1 atm): Liquid.
**Solubility** (Water): 1.5 g/100 mL (20°C).
**Molecular Weight:** 152.2
**Vapour Pressure:** 25 mm Hg at 100°C.
**Boiling Point:** 153°C (decomposes).

**Floatability** (Water): Floats or sinks depending on specific gravity.
**Odour:** No information.
**Flash Point:** 79°C (c.c.); 88°C (o.c.).
**Vapour Density:** 5.3
**Specific Gravity:** (70%) 1.00; (90%) 1.05 (20°C).

**Colour:** Colourless to pale yellow.
**Explosive Limits:** 0.9 to 6.5% (as cumene).
**Melting Point:** -10°C.

## HAZARD DATA

### Human Health

**Symptoms: Inhalation:** sore throat, coughing, shortness of breath, laboured breathing. **Contact:** skin and eyes - redness and pain, burning sensation. **Ingestion:** abdominal pain, vomiting, diarrhea.
**Toxicology:** Highly toxic by inhalation, skin absorption and ingestion.
**TLV®** - No information.   **LC$_{50}$ - Inhalation:** rat = 220 ppm/4 h    **LD$_{50}$ - Oral:** rat = 0.4 g/kg
**Short-term Inhalation Limits** - No information.   **Delayed Toxicity** - No information.

### Fire

**Fire Extinguishing Agents:** Use dry chemical, foam, carbon dioxide or water spray.
**Behaviour in fire:** At concentrations of 91% and above, decomposes violently at about 150°C. At temperatures greater than 153°C also decomposes with some violence.
**Ignition Temperature:** Decomposes above 153°C.   **Burning Rate:** Decomposes.

### Reactivity

**With Water:** No reaction.
**With Common Materials:** Strong oxidizing agent, may react with organic materials.
**Stability:** Stable; however, compositions 91% and above decompose.

### Environment

**Water:** Prevent entry into water intakes and waterways; BOD: No information.
**Land-Air:** No information.
**Food Chain Concentration Potential:** No information.

## EMERGENCY MEASURES

**Special Hazards**
ORGANIC PEROXIDE.

**Immediate Responses**
Keep non-involved people away from spill site. Issue warning: "ORGANIC PEROXIDE". CALL FIRE DEPARTMENT. Eliminate all sources of ignition. Call manufacturer. Dike to prevent runoff. Stop or reduce discharge, if this can be done without risk. Notify environmental authorities.

**Protective Clothing and Equipment**
Respiratory protection - self contained breathing apparatus. Gloves. Boots. Clothing - acid suit, coveralls (neoprene, natural rubber, butyl) (pants worn outside boots).

**Fire and Explosion**
Use dry chemical, foam, carbon dioxide, water spray to extinguish. Decomposes with some violence at temperatures greater than 153°C.

**First Aid**
Move victim out of spill area to fresh air. Call for medical assistance, but start first aid at once. Inhalation: if breathing has stopped, give artificial respiration; if laboured, give oxygen. Contact: eyes - irrigate thoroughly with water for at least 15 minutes; skin - flush thoroughly with water and soap for at least 15 minutes. Ingestion: give plenty of water to conscious victim to drink. If medical assistance is not immediately available, transport victim to hospital, doctor or clinic.

## ENVIRONMENTAL PROTECTION MEASURES

**Response**

**Water**

Floats
1. Stop or reduce discharge if safe to do so.
2. Contact manufacturer or supplier for advice.
3. If possible, contain discharge by booming.
4. If floating, skim and remove.
5. Notify environmental authorities to discuss disposal and cleanup of contaminated materials.

Sinks
1. Stop or reduce discharge if safe to do so.
2. Contact manufacturer or supplier for advice.
3. If possible, contain discharge by damming or water diversion.
4. Dredge or vacuum pump to remove contaminants, liquids and contaminated bottom sediments.
5. Notify environmental authorities to discuss disposal and cleanup of contaminated materials.

**Land-Air**
1. Stop or reduce discharge if safe to do so.
2. Contact manufacturer or supplier for advice.
3. Dike to prevent runoff from rainwater or water application.
4. Remove material with pumps or vacuum equipment and place in appropriate containers.
5. Recover undamaged containers.
6. Absorb residual liquid on natural or synthetic sorbents.
7. Notify environmental authorities to discuss disposal and cleanup of contaminated materials.

**Disposal**
1. Contact manufacturer or supplier for advice on disposal.
2. Contact environmental authorities for advice on disposal.
3. Incinerate (approval of environmental authorities required).

CUMENE HYDROPEROXIDE   $C_6H_5C(CH_3)_2OOH$

# CYANIDES

## IDENTIFICATION

UN No. 1587 Copper cyanide CU(CN)$_2$
1680 Potassium cyanide KCN
1713 Zinc cyanide Zn(CN)$_2$

**Common Synonyms**
COPPER CYANIDE
- CUPRICIN
- CUPRIC CYANIDE
- CUPROUS CYANIDE
(see also sodium cyanide)

**Observable Characteristics**
KCN - white crystals.
Zn(CN)$_2$ - colourless to white powder.
CU(CN)$_2$ - white to yellow-green powder.
All have distinct hydrocyanic acid odour (almond-like).

**Manufacturers**
No Canadian manufacturers
Canadian suppliers:
Canadian Industries Ltd., Toronto, Ontario.
Du Pont Canada Ltd., Montreal, Que.; Toronto, Ont.

Originating from:

E.I. Du Pont de Nemours Inc., Wilmington, Delaware; Niagara Falls, NY.

### Transportation and Storage Information

**Shipping State:** Solid.
**Classification:** Poisonous solid.
**Inert Atmosphere:** No requirement.
**Venting:** Sealed containers.

**Label(s):** White lavel - POISON; Class 6.1, Group I, II or III.
**Storage Temperature:** Ambient.

**Grades or Purity:** Technical: Potassium cyanide: 97% KCN; Copper cyanide: 99% CU(CN)$_2$; Zinc cyanide: 95.3% Zn(CN)$_2$.
**Containers and Materials:** Cans, drums; steel.

### Physical and Chemical Characteristics

**Physical State** (20°C, 1 atm): Solid.
**Solubility** (Water): (Cu) insoluble;
(K) 50 g/100 mL (25°C); 100 g/100 mL (100°C);
(Zn) 0.005 g /100 mL (20°C).
**Molecular Weight:** (Cu) 115.6; (K) 65.1; (Zn) 117.4.
**Vapour Pressure:** No information.
**Boiling Point:** Cu (decomposes in air);
K (no information); Zn (decomposes 800°C).

**Floatability** (Water): KCN sinks and mixes. Zn and Cu cyanides sink.
**Odour:** Distinct HCN odour (almond-like).
**Flash Point:** Not flammable.
**Vapour Density:** No information.
**Specific Gravity:** (Cu) 2.92 (20°C); (K) 1.52 (16°C); (Zn) 1.85 (20°C0.

**Colour:** White to yello-green crystals or powders.
**Explosive Limits:** Not flammable.
**Melting Point:** Cu (decomposes in air); K, 634.5°C; Zn (decomposes 800°C).

## HAZARD DATA

### Human Health

**Symptoms:** Inhalation, Ingestion and Contact (skin): dizziness, rapid respiration, vomiting, flushing, headache, drowsiness, rapid pulse and unconsciousness.
**Toxicology:** Highly toxic by inhalation, ingestion and skin contact.
TLV® (skin) 5 mg/m$^3$ (as CN$^-$).    LC$_{50}$ - No information    LD$_{Lo}$ - Intraperitoneal: rat  0.05 g/kg (Cu)
short-term Inhalation Limits - No information.    Delayed Toxicity - No information.    LD$_{50}$ - Oral: rat = 0.01 g/kg (K)
     LD$_{Lo}$ - Intrapertional: rat = 0.1 g/kg (Zn)

### Fire

**Fire Extinguishing Agents:** Not combustible. In fires involving cyanides, most fire extinguishing agents may be used, except water.
**Behaviour in Fire:** Not combustible.
**Ignition Temperature:** Not combustible.    **Burning Rate:** Not combustible.

### Reactivity

**With Water:** In contact with water or moist air, low concentration of HCN gas is released.
**With Common Materials:** Contact with acids or weak alkalis liberates poisonous and flammable HCN gas. Reacts violently with chlorates, fluorine, magnesium, nitrates, nitric acid and nitrites.
**Stability:** Stable.

### Environment

**Water:** Prevent entry into water intakes and waterways. Harmful to aquatic life in very low concentrations. Fish toxicity: 0.16 ppm/48 h/bluegill/TLm/ freshwater; BOD: 0%, 7 days (theoretical) (K).
**Land-Air:** No information.
**Food Chain Concentration Potential:** None.

## EMERGENCY MEASURES

| | |
|---|---|
| **Special Hazards** | POISON. |
| **Immediate Responses** | Keep non-involved people away from spill site. Issue warning: "POISON". Avoid contact and inhalation. Contact supplier or manufacturer for guidance. Stop or reduce discharge, if this can be done without risk. Dike to prevent runoff from rainwater or water application. Notify environmental authorities. |
| **Protective Clothing and Equipment** | Respiratory protection - self-contained breathing apparatus and totally encapsulated suit. Gloves - cotton for dry product, rubber for solutions. Boots, coveralls - as required. |
| **Fire and Explosion** | Not combustible. In fires involvng cyanides, most fire extinguishing agents may be used, except water. |
| **First Aid** | Move victim out of spill area to fresh air. Call for medical assistance, but start first aid at once. Contact: immediately flush skin and eyes with plenty of water, while removing contaminated clothes. Keep warm and quiet. Inhalation: give artificial respiration or oxygen if breathing has stopped or is laboured (do not use mouth-to-mouth technique). Ingestion: if victim is conscious, induce vomiting and repeat until vomitus is clear. Give oxygen. If medical assistance is not immediately available, transport victim to hospital, doctor or clinic. |

## ENVIRONMENTAL PROTECTION MEASURES

| | |
|---|---|
| **Response** | |
| **Water** | **Land-Air** |
| 1. Stop or reduce discharge if safe to do so. | 1. Stop or reduce discharge if safe to do so. |
| 2. Contact manufacturer or supplier for advice. | 2. Contact manufacturer or supplier for advice. |
| 3. If possible, contain discharge by damming or water diversion. | 3. Dike to prevent runoff from rainwater or water application. |
| 4. Dredge or vacuum pump to remove contaminants, liquids and contaminated bottom sediments. | 4. Remove material by manual or mechanical means. |
| 5. Notify environmental authorities to discuss disposal and cleanup of contaminated materials. | 5. Recover undamaged containers. |
| | 6. Remove contaminated soil for disposal. |
| | 7. Notify environmental authorities to discuss disposal and cleanup of contaminated materials. |
| **Disposal** | |
| 1. Contact manufacturer or supplier for advice on disposal. | |
| 2. Contact environmental authorities for advice on disposal. | |

CYANIDES

# CYCLOHEXANE (CH$_2$)$_6$

## IDENTIFICATION

| Common Synonyms | Observable Characteristics | UN No. 1145 Manufacturers |
|---|---|---|
| HEXAMETHYLENE<br>HEXAHYDROBENZENE<br>HEXANAPHTHENE | Clear, colourless liquid. Aromatic odour. | Gulf Oil Canada Limited, Montreal, Quebec. |

### Transportation and Storage Information

**Shipping State:** Liquid.
**Classification:** Flammable liquid.
**Inert Atmosphere:** No requirement.
**Venting:** Open (flame arrester) or pressure-vacuum.
**Pump Type:** Centrifugal, gear, explosion-proof motors.

**Label(s):** Red label - FLAMMABLE; Class 3.1, Group II.
**Storage Temperature:** Ambient.
**Hose Type:** Reinforced antistatic rubber, neoprene.

**Grades or Purity:** Commercial 85 to 98%; technical.
**Containers and Materials:** Cans, drums, tank cars, tank trucks; steel.

### Physical and Chemical Characteristics

**Physical State** (20°C, 1 atm): Liquid.
**Solubility** (Water): 0.0055 g/100 mL (20°C); 0.0045 g/100 mL (15°C).
**Molecular Weight:** 84.2
**Vapour Pressure:** 77 mm Hg (20°C); 120 mm Hg (30°C).
**Boiling Point:** 80.7°C.

**Floatability** (Water): Floats.
**Odour:** Aromatic (0.04 to 0.41 ppm, odour threshold).
**Flash Point:** -20°C (c.c.).
**Vapour Density:** 2.9
**Specific Gravity:** (liquid) 0.78 (20°C).

**Colour:** Colourless.
**Explosive Limits:** 1.3 to 8.4%
**Melting Point:** 6.3 to 6.6°C.

## HAZARD DATA

### Human Health

**Symptoms:** Inhalation: headache, nausea, vomiting, dizziness, unconsciousness, convulsions. Ingestion: dizziness, fatigue, unconsciousness, coma. Contact: skin - dryness and irritation; eyes - stinging, watering and inflammation.
**Toxicology:** Moderately toxic by inhalation and ingestion.
TLV® (inhalation) 300 ppm; 1 050 mg/m$^3$.     LC$_{50}$ - No information.     LD$_{50}$ - Oral: rat = 29.8 g/kg
Short-term Inhalation Limits - 375 ppm;     Delayed Toxicity - No information.     LD$_{50}$ - Oral: mouse = 1.3 g/kg
1 300 mg/m$^3$ (15 min).

### Fire

**Fire Extinguishing Agents:** Use foam, dry chemical or carbon dioxide. Water may be ineffective. Cool fire-exposed containers with water.
**Behaviour in Fire:** No information.     **Burning Rate:** 6.9 mm/min.
**Ignition Temperature:** 245°C.

### Reactivity

**With Water:** No reaction.
**With Common Materials:** Can react with strong oxidizing agents. Reacts violently with nitrogen dioxide.
**Stability:** Stable.

### Environment

**Water:** Prevent entry into water intakes and waterways. Fish toxicity: 15 500 ppm/24 h/mosquito fish/TLm/freshwater; 32 to 43 mg/L/48 h/fathead minnow/TLm/freshwater; 34 to 43 mg/L/48 h/bluegill/TLm/freshwater; BOD: 238%, 25 days.
**Land-Air:** No information.
**Food Chain Concentration Potential:** None.

## EMERGENCY MEASURES

### Special Hazards
FLAMMABLE.

### Immediate Responses
Keep non-involved people away from spill site. Issue warning: "FLAMMABLE". CALL FIRE DEPARTMENT. Eliminate all sources of ignition. Use only spark-proof tools. Call manufacturer or supplier for guidance. Dike to prevent runoff. Stop or reduce discharge, if this can be done without risk. Notify environmental authorities.

### Protective Clothing and Equipment
Respiratory protection - self-contained breathing apparatus. Gloves - butyl, neoprene, or PVC. Outer protective clothing - as required. Acid suit, coveralls, etc. Boots - high, rubber (pants worn outside boots).

### Fire and Explosion
Use foam, dry chemical or carbon dioxide to extinguish. Water may be ineffective. Cool fire-exposed containers with water spray.

### First Aid
Move victim out of spill area to fresh air. Call for medical assistance, but start first aid at once. Inhalation: administer artificial respiration, if necessary. Contact: eyes - irrigate with plenty of water; skin - flush with plenty of water, while removing contaminated clothing. Ingestion: Do not induce vomiting. If medical assistance is not immediately available, transport victim to hospital, doctor or clinic.

## ENVIRONMENTAL PROTECTION MEASURES

### Response

**Water**
1. Stop or reduce discharge if safe to do so.
2. Contact manufacturer or supplier for advice.
3. If possible contain discharge by booming.
4. If floating, skim and remove.
5. Notify environmental authorities to discuss disposal and cleanup of contaminated materials.

**Land-Air**
1. Stop or reduce discharge if safe to do so.
2. Contact manufacturer or supplier for advice.
3. Contain spill by diking with earth or other barrier.
4. Remove material with pumps or vacuum equipment and place in appropriate containers.
5. Recover undamaged containers.
6. Absorb residual liquid on natural or synthetic sorbents.
7. Notify environmental authorities to discuss disposal or cleanup of contaminated materials.

### Disposal
1. Contact manufacturer or supplier for advice on disposal.
2. Contact environmental authorities for advice on disposal.
3. Incinerate (approval of environmental authorities required).

CYCLOHEXANE $(CH_2)_6$

# 2,4-D $C_8H_6Cl_2O_3$ (acid)

## IDENTIFICATION

| Common Synonyms | Observable Characteristics | Manufacturers |
|---|---|---|
| (2,4-DICHLOROPHENOXY) ACETIC ACID<br>2,4-D AMINE<br>2,4-D ESTER | White to brown or grey solids (flakes), brown liquids. Characteristic. | Ciba-Geigy Canada Ltd., Cambridge, Ontario.<br>Chipman Inc., Stoney Creek, Ontario.<br>Allied Chemical Services, Calgary, Alberta.<br>Interprovincial Co-ops Ltd., Saskatoon, Saskatchewan.<br>Dow Chemical Canada Inc., Sarnia, Ontario. (from Midland, MI). |

(A commonly used herbicide for the control of broadleaf weeds.)

### Transportation and Storage Information

**Shipping State:** Solid or liquid (formulation).
**Classification:** Miscellaneous Dangerous Goods, Class 9.1, 9.2.
**Inert Atmosphere:** No requirement.
**Venting:** Open.

**Pump Type:** No information.
**Label(s):** Not required.
**Storage Temperature:** Ambient.
**Hose Type:** No information.

**Grades or Purity:** Various, as described below.
**Containers and Materials:** Glass bottles; jugs, pails, cans, drums; steel. Not aluminum.

### Physical and Chemical Characteristics

**Physical State** (20°C, 1 atm): Solid (technical acid), liquid (technical ester or amine or solutions).
**Solubility** (Water): 0.09 g/100 mL (25°C) technical acid; technical butoxyethanol ester - insoluble; 300 g/100 mL (20°C) technical amine; 333 g/100 mL isoctyl ester.
**Molecular Weight:** 221.0 (acid)
**Vapour Pressure:** 0.4 mm Hg (160°C) technical acid.

**Floatability** (Water): Sinks; EC will disperse, amine is soluble.
**Odour:** Characteristic (3 ppm, odour threshold).
**Flash Point:** Variable.
**Vapour Density:** No information.
**Specific Gravity:** 1.57 (30°C) technical acid.
**Boiling Point:** 160°C (0.4 mm Hg) technical acid; 870°C technical amine, decomposes; 156-162°C (1.5 mm Hg) butoxyethanol ester.

**Colour:** White to brown or grey.
**Explosive Limits:** PP products can form explosive mixtures in air.
**Melting Point:** 135 to 138°C (technical acid); 85 to 87°C (technical amine).

## HAZARD DATA

### Human Health

**Symptoms: Ingestion:** irritation of gastrointestinal tract, flushing of skin, nausea, vomiting, CNS depression, muscular twitching, muscular weakness. **Inhalation:** respiratory tract irritation, asthma. **Contact:** skin - irritation and allergic reaction.
**Toxicology:** Highly toxic by ingestion. Moderately toxic by inhalation and skin contact.
**TLV®** - No information.
**Short-term Inhalation Limits** - No information.

$LC_{50}$ - No information.
$TC_{Lo}$ - Inhalation: Human = 1 mg/m³ (butylester)
Delayed Toxicity - Suspected carcinogen.

$LD_{50}$ - Oral: human = 0.05 g/kg (acid)
$LD_{50}$ - Oral: rat = 0.375 g/kg (acid)
$LD_{50}$ - Oral: rat = 0.666 g/kg (sodium salt)

### Fire

**Fire Extinguishing Agents:** Use foam, carbon dioxide or dry chemical.
**Behaviour in Fire:** Releases toxic fumes.
**Ignition Temperature:** No information.

**Burning Rate:** No information.

### Reactivity

**With Water:** No reaction.

**With Common Materials:** No information.

**Stability:** Stable.

### Environment

**Water:** Prevent entry into water. Toxicity rating = 1 to 10 ppm/96 h/TLm/freshwater; Fish toxicities: 5.6 mg/L/48 h/$LC_{50}$/Daphnia magna/freshwater (ester); 2.2 mg/L/48 h/$LC_{50}$/rainbow trout/freshwater (acid); 1.1 mg/L/48 h/bluegill/$LC_{50}$/freshwater (ester); 5.0 mg/L/48 h/killifish/$LC_{50}$/saltwater.
**Land-Air:** $LD_{50}$ - Oral: Chicken = 0.54 g/kg (acid); $LD_{50}$ >5000 ppm/duck/5 days (ester), mule deer = 0.5 g/kg (acid), pheasant = 0.47 g/kg (acid).
**Food Chain Concentration Potential:** No information.

## EMERGENCY MEASURES

**Special Hazards**
POISON.

**Immediate Responses**
Keep non-involved people away from spill site. Stop or reduce discharge if safe to do so. Notify manufacturer or supplier. Dike to contain material or water runoff. Notify environmental authorities.

**Protective Clothing and Equipment**
In fires or confined spaces - Respiratory protection - self-contained breathing apparatus and totally encapsulated suit. Otherwise, approved pesticide respirator and impervious outer clothing.

**Fire and Explosion**
Use carbon dioxide, foam or dry chemical to extinguish. Releases toxic fumes in fires.

**First Aid**
Move victim out of spill area to fresh air. Call for medical assistance, but start first aid at once. Inhalation: if breathing has stopped, give artificial respiration; if laboured, give oxygen. Contact: skin - remove contaminated clothing and flush affected areas with plenty of water; eyes - irrigate with plenty of water. Ingestion: give water to conscious victim to drink and induce vomiting; in the case of petroleum distillates, do not induce vomiting for fear of aspiration and chemical pneumonia. If medical assistance is not immediately available, transport victim to hospital, doctor, or clinic.

## ENVIRONMENTAL PROTECTION MEASURES

**Response**

**Water**
1. Stop or reduce discharge if safe to do so.
2. Contact manufacturer or supplier for advice.

Floats         Sinks or mixes
3. If possible contain discharge by booming.
3. If possible contain discharge by damming or water diversions.
4. If floating, skim and remove.
4. Dredge or vacuum pump to remove contaminants, liquids and con-taminated bottom sediments.
5. Notify environmental authorities to discuss disposal and cleanup of contaminated materials.

**Land-Air**
1. Stop or reduce discharge if safe to do so.
2. Contact manufacturer or supplier for advice.
3. Contain spill by diking with earth or other barrier.
4. If liquid, remove material with pumps or vacuum equipment and place in appropriate containers.
5. If solid, remove material by manual or mechanical means.
6. Recover undamaged containers.
7. Adsorb residual liquid on natural or synthetic sorbents.
8. Remove contaminated soil for disposal.
9. Notify environmental authorities to discuss cleanup and disposal of contaminated materials.

**Disposal**
1. Contact manufacturer or supplier for advice on disposal.
2. Contact environmental authorities for advice on disposal.

| Formulations (typical parities) | acid | dimethylamine salt | mixed butyl esters | isooctyl esters | butoxyethyl ester | propylene glycol butyl ester | diethanolamine salt | other amine salts | sodium salt |
|---|---|---|---|---|---|---|---|---|---|
| EC - emulsifiable concentrate - dispersible in water | 5% | 0.1 to 0.8% | 50 to 80% | 5 to 60% | 40 to 50% | 50% | - | - | - |
| GR - granular - low combustibility | - | - | - | - | - | - | - | 5% | - |
| PP - pressurized product - flammable | - | 0.2% | - | - | 20% | - | 0.3% | - | - |
| PE - pellet - low combustibility | 3 to 5% | - | - | - | - | - | - | - | - |
| SN - solution | 60% | 10 to 60% | - | - | - | - | 1 to 50% | 1 to 20% | - |
| SO - solid - combustibility varies from low to high | 15% | - | - | - | - | - | - | - | - |
| TA - tablet - low combustibility | - | - | - | - | - | - | - | - | 35% |

Other possible ingredients found in formulations: atrazine, bromacil, dicamba, mecoprop, MCPA, trifluralin, borax, prometon.

2,4-D  $C_8H_6Cl_2O_3$ (acid)

# DICAMBA  HOOC(C$_2$)$_2$C$_6$H$_2$Cl(OCH$_3$)

## IDENTIFICATION

**Common Synonyms**
3,6-DICHLORO-O-ANISIC ACID
3,6-DICHLORO-2-METHOXYBENZOIC ACID
**Common Trade Names**
BANVEL DYCLEER
(A herbicide for broadleaf plants.)

**Observable Characteristics**
White, grey to brown solid or brownish liquid. Odourless.

UN No. 2769
Danger Group According to Percentage of Active Substance
Group III      liquid 50 to 100%

**Manufacturers**
Ciba-Geigy, Cambridge, Ont.
Chipman Chemicals, Hamilton, Ont.
Velsicol Chemical Corporation, Mississauga, Ont.

### Transportation and Storage Information

**Shipping State:** Solid or liquid (formulation).
**Classification:** None.
**Inert Atmosphere:** No requirement.
**Venting:** Open.
**Pump Type:** No information.

**Label(s):** Not regulated.
**Storage Temperature:** Ambient.
**Hose Type:** No information.

**Grades or Purity:** Various, as described below.
**Containers and Materials:** Glass bottles; cans, drums; steel.

### Physical and Chemical Characteristics

**Physical State** (20°C, 1 atm): Solid.
**Solubility** (Water): 0.45 g/100 mL (Solid), SN 72 g/100 mL; EC is dispersible in water.
**Molecular Weight:** 221.1
**Vapour Pressure:** 0.000034 mm Hg (25°C).
**Boiling Point:** No information.

**Floatability** (Water): Sinks; SN soluble, EC dispersible in water.
**Odour:** Odourless.
**Flash Point:** Only PP flammable.
**Vapour Density:** 7.64
**Specific Gravity:** 1.57 (20°C) technical.

**Colour:** White, grey to brown.
**Explosive Limits:** Only PP products are flammable.
**Melting Point:** 114 to 116°C (technical).

## HAZARD DATA

### Human Health

**Symptoms: Inhalation:** dizziness, weakness, headache, nausea, vomiting and difficulty breathing. **Contact: eyes** - extremely irritating. **Ingestion:** symptoms similar to inhalation.
**Toxicology:** Moderately toxic by ingestion.
**TLV®** - No information.
**Short-term Inhalation Limits** - No information.

**LC$_{50}$** - No information.
**Delayed Toxicity** - No information.

**LD$_{50}$ - Oral:** rat = 1.04 g/kg
**LD$_{50}$ - Oral:** Guinea pig = 3.00 g/kg

### Fire

**Fire Extinguishing Agents:** Use carbon dioxide, foam or dry chemical.
**Behaviour in Fire:** Releases toxic fumes in fires.
**Ignition Temperature:** No information.

**Burning Rate:** No information.

### Reactivity

**With Water:** No reaction.
**With Common Materials:** No information.
**Stability:** Stable.

### Environment

**Water:** Prevent entry into water intakes or waterways. Fish toxicity: 40 mg/L/48 h/bluegill/TLm/freshwater; 35 mg/L/48 h/rainbow trout/LC$_{50}$/freshwater.
**Land-Air:** No information.
**Food Chain Concentration Potential:** No information.

## EMERGENCY MEASURES

**Special Hazards**
POISON.

**Immediate Responses**
Keep non-involved people away from spill site. Stop or reduce discharge if safe to do so. Notify manufacturer or supplier. Dike to contain material or water runoff. Notify environmental authorities.

**Protective Clothing and Equipment**
In fires or confined spaces - Respiratory Protection - self-contained breathing apparatus and totally encapsulated suit. Otherwise, approved pesticide respirator and impervious outer clothing.

**Fire and Explosion**
Use carbon dioxide, foam or dry chemical to extinguish. Releases toxic fumes in fires.

**First Aid**
Move victim out of spill site to fresh air. Call for medical assistance, but start first aid at once. Inhalation: if breathing has stopped, give artificial respiration (not mouth-to-mouth method); if laboured, give oxygen. Contact: skin - remove contaminated clothing and flush affected areas with plenty of water; eyes - irrigate with plenty of water. Ingestion: give water to conscious victim and induce vomiting; in the case of petroleum distillates, do not induce vomiting for fear of aspiration and chemical pneumonia. If medical assistance is not immediately available transport victim to hospital, doctor or clinic.

## ENVIRONMENTAL PROTECTION MEASURES

**Response**

**Water**
1. Stop or reduce discharge if safe to do so.
2. Contact manufacturer or supplier for advice.

Floats     Sinks or mixes
3. If possible contain discharge by booming.
4. If floating, skim and remove.

3. If possible contain discharge by damming or water diversions.
4. Dredge or vacuum pump to remove contaminants, liquids and contaminated bottom sediments.

5. Notify environmental authorities to discuss disposal and cleanup of contaminated materials.

**Land-Air**
1. Stop or reduce discharge if safe to do so.
2. Contact manufacturer or supplier for advice.
3. Contain spill by diking with earth or other barrier.
4. If liquid, remove material with pumps or vacuum equipment and place in appropriate containers.
5. If solid, remove material by manual or mechanical means.
6. Recover undamaged containers.
7. Absorb residual liquid on natural or synthetic sorbents.
8. Remove contaminated soil for disposal.
9. Notify environmental authorities to discuss cleanup and disposal of contaminated materials.

**Disposal**
1. Contact manufacturer or supplier for advice on disposal.
2. Contact environmental authorities for advice on disposal.

**Available Formulations**
**Technical Grade:** Purity: typically 98%
Properties: combustible solid, slightly soluble in water

Formulations:

| Type: | Purity: | Properties: |
|---|---|---|
| EC - emulsifiable concentrate | - typically 1% | - dispersible in water |
| GR - granular | - typically 1%, remainder inerts | |
| PE - pellet | - typically 1%, remainder inerts | |
| SN - solution | - typically 15% in water | - not combustible, miscible with water (dimethylammonium salt) |
| PP - pressurized product | - typically 0.1% | - flammable |

Other Possible Ingredients Found in Formulations: MCPA; 2,4-D; Mecoprop; bromacil.

DICAMBA    $HOOC(C_2)_2C_6H_2Cl(OCH_3)$

# 1,2-DICHLOROBENZENE $C_6H_4Cl_2$

## IDENTIFICATION

| Common Synonyms | Observable Characteristics | UN No. 1591 |
|---|---|---|
| ORTHO-DICHLOROBENZENE<br>o-DICHLOROBENZENE<br>DCB<br>DOWTHERM-E | Colourless liquid. Pleasant, aromatic odour. | **Manufacturers**<br>Record Chemical Co. Inc.,<br>Montreal, Quebec. |

### Transportation and Storage Information

**Shipping State:** Liquid.
**Classification:** Poison.
**Inert Atmosphere:** No requirement.
**Venting:** Open.
**Pump Type:** Standard types. Steel or stainless steel.

**Label(s):** White label - POISON; Class 6.1; Group III.
**Storage Temperature:** Ambient.
**Hose Type:** Polyethylene, Viton, stainless steel, Teflon.

**Grades or Purity:** Technical, 99.5%; technical, 85% ortho, 15% para-; technical 80% ortho, 17% para-; 2% meta-. Pure >99.5% ortho, <0.5% para.
**Containers and Materials:** Drums, tank cars, tank trucks; steel; stainless steel.

### Physical and Chemical Characteristics

**Physical State** (20°C, 1 atm): Liquid.
**Solubility** (Water): 0.010 g/100 mL (20°C); 0.015 g/100 mL (25°C).
**Molecular Weight:** 147.0
**Vapour Pressure:** 1 mm Hg (20°C); 1.9 mm Hg (30°C).
**Boiling Point:** 180.5°C.

**Floatability** (Water): Sinks.
**Odour:** Pleasant, aromatic, (4.0 to 50.0 ppm, odour threshold).
**Flash Point:** 66°C (c.c.).
**Vapour Density:** 5.1
**Specific Gravity:** 1.31 (20°C).

**Colour:** Colourless.
**Explosive Limits:** 2.2 to 9.2%.
**Melting Point:** -16.7 to -18°C.

## HAZARD DATA

### Human Health

**Symptoms:** Inhalation: coughing, faintness, trembling, coma. Contact: skin and eyes - redness and burning sensation. Ingestion: abdominal pain, vomiting, shock.
**Toxicology:** Moderately toxic by inhalation and ingestion.
TLV® (inhalation) 50 ppm, 300 mg/m³.          $LC_{50}$ - No information.                         $LD_{50}$ - Oral:  rat = 0.5 g/kg
Short-term Inhalation Limits - No information.   $LC_{Lo}$ - Inhalation: rat = 821 ppm/7 h
                                                  Delayed Toxicity - Possible long-term effects of liver and kidney damage.

### Fire

**Fire Extinguishing Agents:** Use water spray, foam, dry chemical, or carbon dioxide. Use water to keep fire-exposed containers cool.
**Behaviour in Fire:** Toxic gases (HCl and other chlorine-containing compounds) may be given off at elevated temperatures.
**Ignition Temperature:** 648°C.                **Burning Rate:** 1.3 mm/min.

### Reactivity

**With Water:** No reaction.
**With Common Materials:** May react with oxidizing materials. Can react violently with finely divided aluminum.
**Stability:** Stable.

### Environment

**Water:** Prevent entry to water intakes and waterways. Fish toxicity: 13 ppm/tns/marine plankton/no growth/saltwater; harmful to <u>Chlorella</u> sp. (alga) at 18 mg/L; BOD: less than 0.1%, 1/8 day (theoretical).
**Land-Air:** No information.
**Food Chain Concentration Potential:** No information.

## EMERGENCY MEASURES

**Special Hazards**
POISON.

**Immediate Responses**
Keep non-involved people away from spill site. Issue warning: "POISON". Call Fire Department. Avoid contact and inhalation. Notify manufacturer. Stop or reduce discharge, if this can be done without risk. Dike spill to contain runoff. Notify environmental authorities.

**Protective Clothing and Equipment**
Respiratory protection - self-contained breathing apparatus. Gloves and apron or coveralls, plastic coated. Boots - high, rubber (pants worn outside boots).

**Fire and Explosion**
Use water spray, foam, dry chemical or carbon dioxide to extinguish. Water may be used to keep fire-exposed containers cool. Toxic gases (HCl and other chlorine-containing compounds) may be given off at elevated temperatures.

**First Aid**
Move victim out of spill site to fresh air. Call for medical assistance, but start first aid at once. Contact: remove contaminated clothing and wash eyes and skin with plenty of warm water. Ingestion: Give water to conscious victim to drink and induce vomiting. Inhalation: apply artificial respiration if breathing has stopped (not mouth-to-mouth method); oxygen if breathing is laboured. If medical assistance is not immediately available, transport victim to hospital, doctor or clinic.

## ENVIRONMENTAL PROTECTION MEASURES

**Response**

**Water**
1. Stop or reduce discharge if safe to do so.
2. Contact manufacturer or supplier for advice.
3. If possible, contain discharge by damming or water diversion.
4. Dredge or vacuum pump to remove contaminants, liquids and contaminated bottom sediments.
5. Notify environmental authorities to discuss disposal and cleanup of contaminated materials.

**Land-Air**
1. Stop or reduce discharge if safe to do so.
2. Contact manufacturer or supplier for advice.
3. Contain spill by diking with earth or other barrier.
4. Remove material with pumps or vacuum equipment and place in appropriate containers.
5. Recover undamaged containers.
6. Absorb residual liquids on natural or synthetic sorbents.
7. Notify environmental authorities to discuss disposal and cleanup of contaminated materials.

**Disposal**
1. Contact manufacturer or supplier for advice on disposal.
2. Contact environmental authorities for advice on disposal.

1,2-DICHLOROBENZENE   $C_6H_4Cl_2$

# DIMETHYL ETHER  $CH_3OCH_3$

## IDENTIFICATION  UN No. 1033

| Common Synonyms | Observable Characteristics | Manufacturers |
|---|---|---|
| METHYL ETHER<br>METHYL OXIDE<br>WOOD ETHER<br>WOOD NAPHTHA | Colourless gas. Pleasant, ethereal odour. | No Canadian manufacturer.<br>Canadian supplier:<br>Union Carbide Canada Limited,<br>Pointe aux Trembles, Quebec. | Originating from:<br>Union Carbide Corporation,<br>Chemicals and Plastics,<br>New York, NY. |

### Transportation and Storage Information

**Shipping State:** Liquid (compressed gas).
**Classification:** Flammable gas.
**Inert Atmosphere:** No requirement
**Venting:** Safety-relief.
**Pump Type:** No information.

**Label(s):** Red label - FLAMMABLE GAS; Class 2.1.
**Storage Temperature:** Ambient.
**Hose Type:** No information.

**Grades or Purity:** Technical, 99.5%.
**Containers and Materials:** Cylinders; steel.

### Physical and Chemical Characteristics

**Physical State** (20°C, 1 atm): Gas.
**Solubility** (Water): 7 g/100 mL (18°C).
**Molecular Weight:** 46.1
**Vapour Pressure:** 1 915 mm Hg (0°C); 3 745 mm Hg (20°C); 5 692 mm Hg (34°C).
**Boiling Point:** -24.8°C.

**Floatability** (Water): Floats.
**Odour:** Ethereal.
**Flash Point:** -41.1°C (c.c.).
**Vapour Density:** 1.6
**Specific Gravity:** 0.67 (20°C).

**Colour:** Colourless.
**Explosive Limits:** 3.4 to 27.0%.
**Melting Point:** -141.5°C.

## HAZARD DATA

### Human Health

**Symptoms:** Inhalation or Ingestion: headache, dizziness and rapid loss of consciousness, respiratory paralysis, anaesthesia. Contact: skin or eyes - liquid may cause frostbite.
**Toxicology:** Moderately toxic by ingestion or inhalation.
**TLV•** - (inhalation) No information.   $LC_{50}$ - Inhalation: mouse = 386 ppm/15 min.   $LD_{50}$ - No information.
**Short-term Inhalation Limits** - No information.   **Delayed Toxicity** - No information.

### Fire

**Fire Extinguishing Agents:** Stop flow of gas before attempting to put out fire. Most fire extinguishing agents may be used on fires involving dimethyl ether. Water spray may be used to cool fire-exposed containers and protect people effecting shutoff.
**Behaviour in Fire:** Flashback may occur along vapour trail.
**Ignition Temperature:** 350°C.   **Burning Rate:** 6.6 mm/min.

### Reactivity

**With Water:** No reaction.
**With Common Materials:** Reacts violently with aluminum and lithium aluminum hydrides. "Burns" in a chlorine atmosphere.
**Stability:** Stable.

### Environment

**Water:** Prevent entry into water intakes and waterways. Information on aquatic toxicities not available.
**Land-Air:** No information.
**Food Chain Concentration Potential:** No information.

## EMERGENCY MEASURES

**Special Hazards**

FLAMMABLE.

**Immediate Responses**

Keep non-involved people away from spill site. Issue warnings: "FLAMMABLE". Call Fire Department. Eliminate all sources of ignition. Use only spark-proof tools. Avoid contact and inhalation. Notify manufacturer or supplier for advice. Stop or reduce discharge if this can be done without risk. Contain Spill area by diking to prevent runoff. Notify environmental authorities.

**Protective Clothing and Equipment**

Respiratory protection - use self-contained breathing apparatus. Gloves - rubber. Boots - rubber (pants worn outside boots). Outerwear - as required: acid "slicker" suit, coveralls.

**Fire and Explosion**

Stop flow of gas before attempting to put out fire. Most fire extinguishing agents may be used on fires involving dimethyl ether. Water may be used to cool fire-exposed containers and protect men effecting shutoff. Flashback may occur along vapour trail.

**First Aid**

Move victim out of spill area to fresh air. Call for medical assistance, but start first aid at once. Inhalation: if breathing has stopped, give artificial respiration; if laboured, give oxygen. Contact: eyes - irrigate with water for at least 15 minutes; skin - flush with plenty of water; at same time, remove contaminated clothing. Treat as for frostbite. If medical assistance is not immediately available, transport victim to hospital, doctor or clinic.

## ENVIRONMENTAL PROTECTION MEASURES

**Response**

**Water**
1. Stop or reduce discharge if safe to do so.
2. Contact manufacturer or supplier for advice.
3. Notify environmental authorities to discuss disposal and cleanup of contaminated materials.

**Land-Air**
1. Stop or reduce discharge if safe to do so.
2. Contact manufacturer or supplier for advice.
3. Contain spill by diking with earth or other barrier.
4. Recover undamaged containers.
5. Notify environmental authorities to discuss disposal and cleanup of contaminated materials.

**Disposal**
1. Contact manufacturer or supplier for advice on disposal.
2. Contact environmental authorities for advice on disposal.
3. Incinerate (approval of environmental authorities required).

DIMETHYL ETHER $CH_3OCH_3$

# DIMETHYL TEREPHTHALATE  $C_6H_4(COOCH_3)_2$

## IDENTIFICATION

### Common Synonyms
DMT, DMP
TEREPHTHALIC ACID, DIMETHYL ESTER
1,4-BENZENEDICARBOXYLIC ACID,
DIMETHYL ESTER
DIMETHYL-1,4-BENZENEDICARBOXYLATE
DIMETHYLPHTHALATE

### Observable Characteristics
Colourless crystals. Odourless.

### Manufacturers
No Canadian manufacturers.
Canadian suppliers:
Eastman Chemicals,
Millhaven Fibres, Millhaven, Ont.

Originating from:
Eastman, USA
Hercules, USA

### Transportation and Storage Information
**Label(s):** None. Not regulated.
**Shipping State(s):** Solid.
**Classification:** None.
**Storage Temperature:** Ambient.
**Inert Atmosphere:** No requirement.
**Venting:** Open.

**Grades or Purity:** Technical 99.9%.
**Containers and Materials:** Bottles, drums; glass or steel.

### Physical and Chemical Characteristics
**Physical State** (20°C, 1 atm): Solid.
**Solubility** (Water): 0.33 g/100 mL (80°C).
**Molecular Weight:** 194.2
**Vapour Pressure:** 16 mm Hg (100°C); 140 mm Hg (150°C).
**Boiling Point:** >300°C (sublimes).

**Floatability** (Water): Sinks.
**Odour:** Odourless.
**Flash Point:** 146°C (c.c.).
**Vapour Density:** 6.7
**Specific Gravity:** 1.04 at 20°C.

**Colour:** Colourless.
**Explosive Limits:** 0.9% at 180°C (LEL) (dust).
**Melting Point:** 140°C.

## HAZARD DATA

### Human Health
**Symptoms:** Contact: eyes and skin - irritation and redness. Inhalation: irritation of nasal passages, sore throat, coughing. Ingestion: abdominal pain, nausea and vomiting.
**Toxicology:** Moderate toxicity by ingestion and contact. Low toxicity by inhalation.
TLV® - 5 mg/m³.    $LC_{50}$ - No information.    $LD_{50}$ - Oral: rat = 4.39 g/kg
Short-term Inhalation Limits - 10 mg/m³    Delayed Toxicity - Suspected carcinogen.
(15 min).

### Fire
**Fire Extinguishing Agents:** Use dry chemical or carbon dioxide. Water or foam may cause frothing.
**Behaviour in Fire:** No information.
**Ignition Temperature:** (dust) 490°C.    **Burning Rate:** No information.

### Reactivity
**With Water:** No reaction.
**With Common Materials:** May react with strong oxidizers (nitrates, alkalis, acids).
**Stability:** Stable.

### Environment
**Water:** Prevent entry into water intakes and waterways. Toxicity to aquatic life unknown.
**Land-Air:** No information.
**Food Chain Concentration Potential:** No information.

## EMERGENCY MEASURES

**Special Hazards**

**Immediate Responses**

Keep non-involved people away from spill site. Stop or reduce discharge if safe to do so. Dike to prevent runoff from rainwater or water application. Notify manufacturer. Notify environmental authorities.

**Protective Clothing and Equipment**

In fires or confined spaces, Respiratory protection - self-contained breathing apparatus; otherwise, protective clothing as required.

**Fire and Explosion**

Use carbon dioxide or dry chemical to extinguish. Water or foam may cause frothing.

**First Aid**

Move victim out of spill site to fresh air. Call for medical assistance, but start first aid at once. Contact: eyes and skin - remove contaminated clothing and flush affected areas with plenty of water. Inhalation: give artificial respiration if necessary. Ingestion: give plenty of water to conscious victim to drink. If medical assistance is not immediately available, transport victim to doctor, clinic or hospital.

## ENVIRONMENTAL PROTECTION MEASURES

**Response**

**Water**
1. Stop or reduce discharge if safe to do so.
2. Contact manufacturer or supplier for advice.
3. If possible, contain spill by damming or water diversion.
4. Dredge or vacuum pump to remove contaminants, liquids and contaminated bottom sediments.
5. Notify environmental authorities to discuss disposal and cleanup of contaminated materials.

**Land-Air**
1. Stop or reduce discharge if safe to do so.
2. Contact manufacturer or supplier for advice.
3. Contain spill by diking with earth or other barrier.
4. Remove material by manual or mechanical means.
5. Recover undamaged containers.
6. Notify environmental authorities to discuss disposal and cleanup of contaminated materials.

**Disposal**
1. Contact manufacturer or supplier for advice on disposal.
2. Contact environmental authorities for advice on disposal.
3. Incinerate (approval of environmental authorities required).

DIMETHYL TEREPHTHALATE   $C_6H_4(COOCH_3)_2$

# DINOSEB  $(C_4H_9)C_6H_2(NO_2)_2OH$

## IDENTIFICATION

### Common Synonyms
2-SEC-BUTYL-4,6-DINITROPHENOL
**Common Trade Names**
DYTOP SINOX DINOCAP PREMERGE

(A herbicide used for pre-emergent and post-emergent weed control.)

### Observable Characteristics
Technical, orange to brown solid or brownish liquid; pungent odour.

UN No. 2779
Danger Group According to Percentage of Active Substance
Group II   >40 to 100%
Group III  solid 5 to 40%
           liquid 5 to 40%

**Manufacturers**
Dow Chemical Canada, Inc., Sarnia, Ontario (from Midland, MI).
FMC of Canada, Hamilton, Ontario.
Niagara Chemical, Burlington, Ontario.

### Transportation and Storage Information
**Shipping State:** Solid or liquid (formulation).
**Classification:** Poison, substituted nitrophenol pesticide, MAS.
**Inert Atmosphere:** No requirement.
**Venting:** Open.

**Label(s):** White label - POISON; Class 6.1.
**Storage Temperature:** Ambient.
**Hose Type:** Seamless stainless steel, Teflon, Viton, neoprene, cross-linked polyethylene.
**Pump Type:** Centrifugal or positive displacement, stainless steel.

**Grades or Purity:** Various as described below.
**Containers and Materials:** Glass bottles, cans, drums, tank trucks, tank cars, intermodel tanks; steel.

### Physical and Chemical Characteristics
**Physical State** (20°C, 1 atm): Solid (technical).
**Solubility** (Water): 0.0052 g/100 mL (20°C); EC is dispersible in water.
**Molecular Weight:** 240.2
**Vapour Pressure:** 1 mm Hg (151°C) technical.
**Boiling Point:** >300°C technical.

**Floatability** (Water): Sinks, SN floats.
**Odour:** Pungent.
**Flash Point:** 177°C technical; 15 to 30°C SN.
**Vapour Density:** No information.
**Specific Gravity:** 1.26 (45°C) technical.

**Colour:** Orange to brown.
**Explosive Limits:** SN may be explosive.
**Melting Point:** 32 to 44°C (technical).

## HAZARD DATA

### Human Health
**Symptoms:** Inhalation, Ingestion or Contact (skin): high fever, thirst, nausea, vomiting, excessive perspiration and difficulty breathing. Symptoms may later progress to cyanosis, muscular tremors and coma.
**Toxicology:** Very highly toxic by ingestion. Moderately toxic by skin contact.
**TLV®** - No information.    LC$_{50}$ - No information.                              LD$_{50}$ - Oral: rat = 0.025 g/kg
**Short-term Inhalation Limits** - No information.   Delayed Toxicity - Possible kidney/liver damage.   LD$_{50}$ - Skin: rat = 0.080 g/kg

### Fire
**Fire Extinguishing Agents:** Use foam, carbon dioxide or dry chemical.
**Behaviour in Fire:** Releases toxic fumes. Exothermic decomposition can occur resulting in an explosion
**Ignition Temperature:** No information.   **Burning Rate:** No information.

### Reactivity
**With Water:** No reaction.
**With Common Materials:** No information.
**Stability:** Normally stable, may undergo exothermic decomposition on heating.

### Environment
**Water:** Prevent entry into water intakes and waterways. Very toxic to aquatic life. Fish toxicity = 0.038 to 0.051/mg/L/96 h/lake trout/LC$_{50}$/freshwater; 0.056 to 0.081 mg/L/96 h/cutthroat trout/LC$_{50}$/freshwater.
**Land-Air:** LD$_{50}$ - Oral: chicken = 0.04 g/kg; wild bird = 0.007 g/kg.
**Food Chain Concentration Potential:** No information.

## EMERGENCY MEASURES

### Special Hazards
POISON.

### Immediate Responses
Keep non-involved people away from spill site. Stop or reduce discharge if safe to do so. Notify manufacturer or supplier. Dike to contain material or water runoff. Notify environmental authorities.

### Protective Clothing and Equipment
In fires or confined spaces - Respiratory protection - self-contained breathing apparatus and totally encapsulated suit. Otherwise, approved pesticide respirator and impervious outer clothing.

### Fire and Explosion
Use carbon dioxide, foam or dry chemical to extinguish. Releases toxic fumes in fires.

### First Aid
Move victim out of spill site to fresh air. Call for medical assistance, but start first aid at once. Inhalation: if breathing has stopped, give artificial respiration; if laboured, give oxygen. Contact: skin - remove contaminated clothing and flush affected areas with plenty of water; eye⁻ - irrigate with plenty of water. Ingestion: give water to conscious victim to drink and induce vomiting; in the case of petroleum distillates, do not induce vomiting for fear of aspiration and chemical pneumonia. If medical assistance is not immediately available, transport victim to hospital, doctor, or clinic.

## ENVIRONMENTAL PROTECTION MEASURES

### Response

**Water**
1. Stop or reduce discharge if safe to do so.
2. Contact manufacturer or supplier for advice.

Floats     Sinks or mixes
3. If possible contain discharge by booming.
3. If possible contain discharge by damming or water diversions.
4. If floating, skim and remove.
4. Dredge or vacuum pump to remove contaminants, liquids and contaminated bottom sediments.
5. Notify environmental authorities to discuss disposal and cleanup of contaminated materials.

**Land-Air**
1. Stop or reduce discharge if safe to do so.
2. Contact manufacturer or supplier for advice.
3. Contain spill by diking with earth or other barrier.
4. If liquid, remove material with pumps or vacuum equipment and place in appropriate containers.
5. If solid, remove material by manual or mechanical means.
6. Recover undamaged containers.
7. Adsorb residual liquid on natural or synthetic sorbents.
8. Remove contaminated soil for disposal.
9. Notify environmental authorities to discuss cleanup and disposal of contaminated materials.

### Disposal
1. Contact manufacturer or supplier for advice on disposal.
2. Contact environmental authorities for advice on disposal.

### Available Formulations
**Technical Grade:** Purity: 95 to 98%
Properties: Combustible.

**Formulations:**

Type:
EC - emulsifiable concentrate
SN - solution

Purity:
- typically 30 or 60%
- typically 35% in petroleum distillates

Properties:
- dispersible in water
- flammable, floats on water.

DINOSEB    $(C_4H_9)C_6H_2(NO_2)_2OH$

# DIPHENYL AMINE  ($C_6H_5)_2NH$

## IDENTIFICATION

| Common Synonyms | Observable Characteristics | Manufacturers |
|---|---|---|
| ANILINOBENZENE<br>N-PHENYLANILINE<br>DPA, DFA | Colourless to greyish crystals. Floral odour. | No Canadian manufacturers.<br>Canadian suppliers:<br>Bayer (Canada) Inc., Pointe Claire, Quebec.<br>Cyanamid Canada Inc., Toronto, Ontario. |

### Transportation and Storage Information

**Shipping State(s):** Solid.
**Classification:** Not regulated.
**Inert Atmosphere:** No requirement.
**Venting:** Open.

**Label(s):** None.
**Storage Temperature:** Ambient.

**Grades or Purity:** Technical.
**Containers and Materials:** Polyethylene-lined paper bags, fibre, tank cars, tank trucks.

### Physical and Chemical Characteristics

**Physical State** (20°C, 1 atm): Solid.
**Solubility** (Water): Insoluble.
**Molecular Weight:** 169.2
**Vapour Pressure:** 1 mm Hg (108.3°C).
**Boiling Point:** 302°C.

**Floatability** (Water): Sinks.
**Odour:** Floral.
**Flash Point:** 153°C (c.c.).
**Vapour Density:** 5.8
**Specific Gravity:** 1.16 (20°C).

**Colour:** Colourless to greyish.
**Explosive Limits:** No information.
**Melting Point:** 52.8°C.

## HAZARD DATA

### Human Health

**Symptoms: Contact:** eyes - irritation and watering. **Inhalation:** irritation of nose, headache, coughing, nausea. **Ingestion:** slight irritation of stomach, nausea and vomiting, diarrhea and general fatigue.
**Toxicology:** Moderately toxic by ingestion.
TLV® 10 mg/m³.  $LC_{50}$ - No information.  $LD_{50}$ - Oral: guinea pig = 0.3 g/kg
**Short-term Inhalation Limits** - 20 mg/m³ (15 min).  **Delayed Toxicity** - Suspected teratogen.

### Fire

**Fire Extinguishing Agents:** Use dry chemical or carbon dioxide. Water or foam may cause frothing.
**Behaviour in Fire:** When heated to decomposition, emits highly toxic fumes.
**Ignition Temperature:** 634°C.  **Burning Rate:** No information.

### Reactivity

**With Water:** No reaction.
**With Common Materials:** Can react with oxidizing agents. Reacts violently with melamines.
**Stability:** Stable.

### Environment

**Water:** Prevent entry into water intakes and waterways. Toxicity to aquatic life unknown.
**Land-Air:** No information.
**Food Chain Concentration Potential:** No information.

## EMERGENCY MEASURES

**Special Hazards**

**Immediate Responses**

Keep non-involved people away from spill site. Stop or reduce discharge if safe to do so. Dike to prevent runoff from rainwater or water application. Notify manufacturer. Notify environmental authorities.

**Protective Clothing and Equipment**

In fires or confined spaces, Respiratory protection - self-contained breathing apparatus; otherwise, protective clothing as required.

**Fire and Explosion**

Use dry chemical or carbon dioxide to extinguish. Water or foam may cause frothing.

**First Aid**

Move victim out of spill site to fresh air. Call for medical assistance, but start first aid at once. Inhalation: give artificial respiration if necessary. Contact: eyes and skin - remove contaminated clothing and flush affected areas with plenty of water. Ingestion: give plenty of water to conscious victim to drink. If medical assistance is not immediately available, transport victim to doctor, hospital or clinic.

## ENVIRONMENTAL PROTECTION MEASURES

**Response**

**Water**
1. Stop or reduce discharge if safe to do so.
2. Contact manufacturer or supplier for advice.
3. If possible, contain spill by damming or water diversion.
4. Dredge or vacuum pump to remove contaminants, liquids and contaminated bottom sediments.
5. Notify environmental authorities to discuss disposal and cleanup of contaminated materials.

**Land-Air**
1. Stop or reduce discharge if safe to do so.
2. Contact manufacturer or supplier for advice.
3. Contain spill by diking with earth or other barrier.
4. Remove material by manual or mechanical means.
5. Recover undammaged containers.
6. Notify environmental authorities to discuss disposal and cleanup of contaminated materials.

**Disposal**
1. Contact manufacturer or supplier for advice on disposal.
2. Contact environmental authorities for advice on disposal.

DIPHENYL AMINE   $(C_6H_5)_2NH$

# DIPHENYLMETHANE-4,4'-DIISOCYANATE  (P-OCNC$_6$H$_4$)$_2$CH$_2$

UN No. 2489

## IDENTIFICATION

### Common Synonyms
MDI
DIPHENYLMETHANE DIISOCYANATE
METHYLENE-DIPARAPHENYLENE ISOCYANATE
METHYLENE-BIS (PHENYL ISOCYANATE)

### Observable Characteristics
White to light yellow. Crystals or solid.

### Manufacturers
No Canadian manufacturers.
Canadian suppliers:  Originating from:
Bayer Canada,
Mississauga, Ont.
Du Pont Canada, Montreal.  E.I. Du Pont de Nemours, USA

### Transportation and Storage Information
**Shipping State:** Solid.
**Classification:** None.
**Inert Atmosphere:** No requirement.
**Venting:** Pressure-vacuum.
**Label(s):** Not regulated.
**Storage Temperature:** Ambient.

**Grades or Purity:** Technical; 91 to 99%.
**Containers and Materials:** Drums.

### Physical and Chemical Characteristics
**Physical State (20°C, 1 atm):** Solid.
**Solubility (Water):** Reacts slowly.
**Molecular Weight:** 250.3
**Vapour Pressure:** 0.00001 mm Hg (25°C).
**Boiling Point:** 314°C.

**Floatability (Water):** Sinks and reacts slowly.
**Odour:** No information.
**Flash Point:** 196°C (c.c.); 202°C (o.c.).
**Vapour Density:** 8.6 to 8.7
**Specific Gravity:** 1.2 (20°C).

**Colour:** White to light yellow.
**Explosive Limits:** No information.
**Melting Point:** 37 to 41°C.

## HAZARD DATA

### Human Health
**Symptoms:** Inhalation: sore throat, coughing, laboured breathing. Ingestion: severe irritation, vomiting and abdominal spasm. Contact: skin and eyes - redness, swelling, irritation.
**Toxicology:** Highly toxic by inhalation and ingestion.
TLV - 0.02 ppm; 0.2 mg/m$^3$.   LC$_{50}$ - No information.   LD$_{50}$ - No information.
Short-term Inhalation Limits - No information.   Delayed Toxicity - No information.

### Fire
**Fire Extinguishing Agents:** Use carbon dioxide and dry powders. No water or water-containing agents should be used.
**Behaviour in Fire:** Upon heating or combustion, decomposes yielding toxic gases such as hydrogen cyanide (HCN), carbon monoxide (CO) and NO$_x$.
**Ignition Temperature:** Decomposes.   **Burning Rate:** No information.

### Reactivity
**With Water:** Slowly reacts forming carbon dioxide and other products.
**With Common Materials:** Reacts violently with acids, bases, alcohols and amines.
**Stability:** Stable (within the limits of the foregoing).

### Environment
**Water:** Prevent entry into water intakes and waterways. Toxic to aquatic life.
**Land-Air:** No information.
**Food Chain Concentration Potential:** No information.

## EMERGENCY MEASURES

**Special Hazards**

Toxic in low concentrations.

**Immediate Responses**

Keep non-involved people away from spill site. Stop or reduce discharge if safe to do so. Notify manufacturer. Notify environmental authorities.

**Protective Clothing and Equipment**

Respiratory protection - self-contained breathing apparatus and totally encapsulated suit.

**Fire and Explosion**

Use carbon dioxide and dry powders to extinguish. No water or water-containing agents should be used. Upon heating yields toxic gases such as HCN, CO and $NO_x$.

**First Aid**

Move victim out of spill site to fresh air. Call for medical assistance, but start first aid at once. Contact: eyes - irrigate with plenty of water; skin - remove contaminated clothing and flush affected areas with plenty of water. Inhalation: give artificial respiration, if breathing has stopped; oxygen, if breathing is laboured. Ingestion: give plenty of water to conscious victim to drink. If medical assistance is not immediately available, transport victim to doctor, clinic or hospital.

## ENVIRONMENTAL PROTECTION MEASURES

**Response**

**Water**
1. Stop or reduce discharge if safe to do so.
2. Contact manufacturer or supplier for advice.
3. If possible, contain spill by damming or water diversion.
4. Dredge or vacuum pump to remove contaminants, liquids and contaminated bottom sediments.
5. Notify environmental authorities to discuss disposal and cleanup of contaminated materials.

**Land-Air**
1. Stop or reduce discharge if safe to do so.
2. Contact manufacturer or supplier for advice.
3. Contain spill by diking with earth or other barrier.
4. Remove material by manual or mechanical means.
5. Recover undamaged containers.
6. Notify environmental authorities to discuss disposal and cleanup of contaminated materials.

**Disposal**
1. Contact manufacturer or supplier for advice on disposal.
2. Contact environmental authorities for advice on disposal.

DIPHENYLMETHANE-4,4'-DIISOCYANATE  $(p-OCNC_6H_4)_2CH_2$

# EPICHLOROHYDRIN   O·CH$_2$·CH·CH$_2$Cl

## IDENTIFICATION

UN No. 2023

### Common Synonyms

ECH
CHLOROPROPYLENE OXIDE
CHLOROMETHYLOXIRANE
1-CHLORO-2,3-EPOXYPROPANE
3-CHLORO-1,2-PROPYLENE OXIDE
EPI
EPICHLORHYDRIN

### Observable Characteristics

Colourless liquid. Ethereal odour.

### Manufacturers

No Canadian manufacturers.
Canadian suppliers:
Dow Chemical Canada Inc., Sarnia, Ontario
Shell Canada Limited, Toronto, Ontario

Originating from: Dow Chemical, Freeport, TX

### Transportation and Storage Information

**Shipping State:** Liquid.
**Classification:** Poison.
**Inert Atmosphere:** Recommended for storage.
**Venting:** Pressure-vacuum.
**Pump Type:** Standard types (grounded). Not copper or copper alloys.

**Label(s):** White label - POISON; Class 6.1, Group II.
**Storage Temperature:** Ambient
**Hose Type:** Tite flex, Teflon; stainless steel. Allchem (cross-linked polyethylene). Not copper or copper alloys.

**Grades or Purity:** 99%.
**Containers and Materials:** Drums, tank cars, tank trucks; steel or stainless steel.

### Physical and Chemical Characteristics

**Physical State** (20°C, 1 atm): Liquid.
**Solubility** (Water): 6 g/100 mL (20°C).
**Molecular Weight:** 92.5
**Vapour Pressure:** 12 mm Hg (20°C); 22 mm Hg (30°C).
**Boiling Point:** 115 to 118°C.

**Floatability** (Water): Reacts mildly; sinks and mixes slightly.
**Odour:** Ethereal (0.08 to 100 ppm; odour threshold).
**Flash Point:** 32-39°C (o.c.); 31-38°C (c.c.).
**Vapour Density:** 3.3
**Specific Gravity:** 1.18 (20°C).

**Colour:** Colourless.
**Explosive Limits:** 3.8 to 21%.
**Melting Point:** -25°C.

## HAZARD DATA

### Human Health

**Symptoms: Contact:** eyes - irritation, watering, burning; skin - readily absorbed; itching, irritation, inflammation, blisters and burning. **Ingestion:** irritation, pain in swallowing, stomach and abdominal pain, nausea and vomiting, diarrhea. **Inhalation:** irritation of mucous membranes, headache, nausea, cyanosis, dizziness, fatigue and diarrhea.
**Toxicology:** Highly toxic by ingestion. Moderately toxic by skin absorption.
TLV® - (skin) 2 ppm; 10 mg/m$^3$        LC$_{50}$ - No information.        LD$_{50}$ - Oral: rat = 0.09 g/kg
Short-term Inhalation Limits - 5 ppm; 20 mg/m$^3$        Delayed Toxicity - Suspected carcinogen.
(skin) (15 min).

### Fire

**Fire Extinguishing Agents:** Use water spray, alcohol foam, carbon dioxide or dry chemical. Use water to cool fire-exposed containers and disperse vapours.
**Behaviour in Fire:** In fires, emits highly toxic fumes of phosgene gas and hydrogen chloride. Flashback may occur along vapour trail.
**Ignition Temperature:** 411°C.

### Reactivity

**With Water:** Mild reaction; slightly soluble.
**With Common Materials:** Reacts violently with nitric acid, chlorosulfonic acid, ethylenediamine, ethyleneimine, oleum and sulfuric acid. May react with oxidizers, acids and bases.
**Stability:** Stable.

### Environment

**Water:** Prevent entry into water intakes and waterways. Harmful to aquatic life in low concentrations. Fish toxicity: 10 ppm/48 h/Daphia magna/ lethal concentration; aquatic toxicity rating = 10 to 100 mg/L/96 h/TLm/freshwater; 23 mg/L/24 h/goldfish/LD50/freshwater; BOD: 3 to 16%, 5 days.
**Land-Air:** No information.
**Food Chain Concentration Potential:** None.

## EMERGENCY MEASURES

**Special Hazards**

POISON.

**Immediate Responses**

Keep non-involved people away from spill site. Issue warning: "POISON". Call Fire Department. Avoid contact and inhalation. Evacuate from downwind. Contact manufacturer for guidance. Stop or reduce discharge if this can be done without risk. Dike to prevent runoff. Notify environmental authorities.

**Protective Clothing and Equipment**

Respiratory protection - self-contained breathing apparatus and totally encapsulated protective clothing. Destroy contaminated leather clothing after use.

**Fire and Explosion**

Use water spray, alcohol foam, carbon dioxide or dry chemical to extinguish. Water spray may be used to cool fire-exposed containers and disperse vapours. When heated to decomposition, emits highly toxic fumes of phosgene gas and hydrogen chloride. Flashback may occur along vapour tail.

**First Aid**

Move victim out of spill area to fresh air. Call for medical assistance, but start first aid at once. Inhalation: if breathing has stopped, give artificial respiration; if laboured, give oxygen. Contact: immediately irrigate eyes and flush with plenty of warm water for at least 15 minutes. Remove contaminated clothing while washing. Ingestion: give water to conscious victim to drink and induce vomiting. If medical assistance is not immediately available, transport victim to hospital, doctor or clinic.

## ENVIRONMENTAL PROTECTION MEASURES

**Response**

**Water**

1. Stop or reduce discharge if safe to do so.
2. Contact manufacturer or supplier for advice.
3. If possible, contain discharge by damming or water diversion.
4. Dredge or vacuum pump to remove contaminants, liquids and contaminated bottom sediments.
5. Notify environmental authorities to discuss disposal and cleanup of contaminated materials.

**Land-Air**

1. Stop or reduce discharge if safe to do so.
2. Contact manufacturer or supplier for advice.
3. Contain spill by diking with earth or other barrier.
4. Remove material with pumps or vacuum equipment and place in appropriate containers.
5. Adsorb residual liquid on natural or synthetic sorbents.
6. Remove contaminated soil for disposal.
7. Notify environmental authorities to discuss disposal and cleanup of contaminated materials.

**Disposal**

1. Contact manufacturer or supplier for advice on disposal.
2. Contact environmental authorities for advice on disposal.

EPICHLOROHYDRIN  O·CH$_2$·CH·CH$_2$Cl

# ETHANOLAMINE  $HOCH_2CH_2NH_2$

## IDENTIFICATION  UN No. 2491

### Common Synonyms
MEA
MONOETHANOLAMINE
2-AMINOETHANOL
β-AMINOETHYL ALCOHOL
ETHYLOLAMINE
COLAMINE
GLYCINOL

### Observable Characteristics
Colourless liquid. Mild ammoniacal odour. May be a solid under some atmospheric conditions.

### Manufacturers
Union Carbide, Montreal, Que.
Dow Chemical Canada Inc,
Fort Saskatchewan, Alta.
Canadian supplier:
Dow Chemical Canada Inc., Sarnia, Ont.
Union Carbide, Toronto, Ont.

Originating from:
Dow Chemical, USA
Union Carbide, USA

### Grades or Purity: Commercial, 99+%.
**Containers and Materials:** Cans, drums, tank cars; steel and stainless steel.

## Transportation and Storage Information

**Shipping State:** Liquid.
**Classification:** Corrosive.
**Inert Atmosphere:** No requirement.
**Venting:** Open.
**Pump Type:** Centrifugal or positive displacement, stainless steel.

**Label(s):** Black and white label - CORROSIVE; Class 8, Group III.
**Storage Temperature:** Ambient.
**Hose Type:** Stainless steel, Teflon.

## Physical and Chemical Characteristics

**Physical State** (20°C, 1 atm): Liquid.
**Solubility:** (Water): Miscible in all proportions.
**Molecular Weight:** 61.1
**Vapour Pressure:** 0.4 mm Hg (20°C); 6 mm Hg (60°C).
**Boiling Point:** 170 to 172°C.

**Floatability** (Water): Sinks in fresh water; floats in salt water.
**Odour:** Ammoniacal (3 to 4 ppm, odour threshold).
**Flash Point:** 85°C (c.c.).
**Vapour Density:** 2.1
**Specific Gravity:** 1.02 (20°C).

**Colour:** Colourless.
**Explosive Limits:** No information.
**Melting Point:** 10 to 11°C.

## HAZARD DATA

### Human Health
**Symptoms:** Contact: skin - irritation; eyes - irritation, possible corneal injury. Ingesiton: irritation, nausea and vomiting. Inhalation: irritation and coughing.
**Toxicology:** Moderately toxic by contact, inhalation and ingestion.
TLV® - (inhalation) 3 ppm; 8 mg/m³.    $LC_{50}$ - No information.    $LD_{50}$ - Oral: rat = 2.1 g/kg
Short-term Inhalation Limits - 6 ppm; 15 mg/m³    Delayed Toxicity - No information.
(15 min).

### Fire
**Fire Extinguishing Agents:** Use water spray, dry chemical, alcohol foam or carbon dioxide. Water and regular foam may cause excessive frothing. Water spray may be used to cool fire-exposed containers and disperse vapours.
**Behaviour in Fire:** No information.    **Burning Rate:** No information.
**Ignition Temperature:** 410°C.

### Reactivity
**With Water:** No reaction; soluble.
**With Common Materials:** Reacts violently with acetic acid, acetic anhydride, acrolein, acrylic acid, acrylonitrile, chlorosulfonic acid, epichlorohydrin, hydrochloric acid, hydrofluoric acid, nitric acid, oelum, sulfuric acid, vinyl acetate. May react with halocarbons, epoxides, oxidizing materials, brass, bronze, zinc and cooper.
**Stability:** Stable within limits of foregoing.

### Environment
**Water:** Prevent entry into water intakes and waterways. Aquatic toxicity rating = 100 to 1 000 ppm/96 h/TLm/freshwater; Fish toxicity: >5 000 mg/L/24 h/ goldfish/$LC_{50}$/freshwater; 7 100 ppm/48 h/shrimp/$LC_{50}$/saltwater; BOD: 78 to 93%, 5 days.
**Land-Air:** No information.
**Food Chain Concentration Potential:** None.

## EMERGENCY MEASURES

### Special Hazards
CORROSIVE.

### Immediate Responses
Keep non-involved people away from spill site. Issue warning: "CORROSIVE". Call Fire Department. Notify manufacturer for advice. Shut off leak, if safe to do so. Dike to prevent runoff from rainwater or water application. Notify environmental authorities.

### Protective Clothing and Equipment
Respiratory protection - self-contained breathing apparatus and totally encapsulated protective clothing.

### Fire and Explosion
Use water spray, dry chemical, alcohol foam or carbon dioxide to extinguish. Water spray may be used to cool fire-exposed containers and disperse vapours.

### First Aid
Move victim out of spill area to fresh air. Call for medical assistance, but start first aid at once. Inhalation: if breathing has stopped, give artificial respiration; if laboured, give oxygen. Contact: flush skin and eyes with plenty of water for at least 30 minutes; remove contaminated clothing. Ingestion: give water to conscious victim to drink; induce vomiting. If medical assistance is not immediately available, transport victim to hospital, doctor or clinic.

## ENVIRONMENTAL PROTECTION MEASURES

### Response

**Water**

Sinks
1. Stop or reduce discharge if safe to do so.
2. Contact manufacturer or supplier for advice.
3. If possible, contain discharge by damming or water diversion.
4. Dredge or vacuum pump to remove contaminants, liquids and contaminated bottom sediments.
5. Notify environmental authorities to discuss disposal and cleanup of contaminated materials.

Floats
1. Stop or reduce discharge if safe to do so.
2. Contact manufacturer or supplier for advice.
3. If possible, contain discharge by booming.
4. If floating, skim and remove.
5. Notify environmental authorities to discuss disposal and cleanup of contaminated materials.

**Land-Air**

Liquid
1. Stop or reduce discharge if safe to do so.
2. Contact manufacturer or supplier for advice.
3. Contain spill by diking with earth or other barrier.
4. Remove material with pumps or vacuum equipment and place in appropriate containers.
5. Recover undamaged containers.
6. Adsorb residual liquid on natural or synthetic sorbents.
7. Notify environmental authorities to discuss disposal and cleanup of contaminated materials.

Solid
1. Stop or reduce discharge if safe to do so.
2. Contact manufacturer or supplier for advice.
3. Dike to prevent runoff from rainwater or water application.
4. Remove material by manual or mechanical means.
5. Recover undamaged containers.
6. Notify environmental authorities to discuss disposal and cleanup of contaminated materials.

### Disposal
1. Contact manufacturer or supplier for advice on disposal.
2. Contact environmental authorities for advice on disposal.

ETHANOLAMINE   $HOCH_2CH_2NH_2$

# ETHYL ACETATE   $CH_3COOC_2H_5$

UN No. 1173

## IDENTIFICATION

### Common Synonyms
VINEGAR NAPHTHA
ACETIC ETHER
ACETIC ESTER
ACETIC ACID ETHYL ESTER
ETHYL ETHANOATE

### Observable Characteristics
Colourless liquid. Fragrant, fruity odour.

### Manufacturers
Canadian manufacturer.
Caledon Laboratories Ltd.
Georgetown, Ont.
Canadian suppliers:
Bate Chemical, Toronto, Ont.
Celanese Canada, Edmonton, Alta.
Stanchem, Montreal, Quebec.
Eastman Chemical, Toronto, Ont.

Originating from:
Union Carbide, USA
Celanese, USA
Eastman, USA

### Transportation and Storage Information
**Shipping State:** Liquid.
**Classification:** Flammable liquid.
**Inert Atmosphere:** No requirement.
**Venting:** Open (flame arrester) or pressure-vacuum.
**Pump Type:** Gear, centrifugal, flammable liquid types.

**Label(s):** Red label - FLAMMABLE LIQUID; Class 3.2, Group II.
**Storage Temperature:** Ambient.
**Hose Type:** Polyethylene, polypropylene, butyl, Hypalon.

**Grades or Purity:** Commercial 85 to 88%.
**Containers and Materials:** Drums, tank cars, tank trucks; steel, stainless steel.

### Physical and Chemical Characteristics
**Physical State** (20°C, 1 atm): Liquid.
**Solubility** (Water): 7.9 to 8.6 g/100 mL (20°C); 7.4 g/100 mL (35°C).
**Molecular Weight:** 88.1
**Vapour Pressure:** 73 mm Hg (20°C); 115 mm Hg (30°C).
**Boiling Point:** 77°C.

**Floatability** (Water): Floats and mixes slightly.
**Odour:** Fragrant, fruity (6 to 70 ppm, odour threshold).
**Flash Point:** 10°C (o.c.); -4.4°C (c.c.)
**Vapour Density:** 3.04
**Specific Gravity:** 0.90 (20°C).

**Colour:** Colourless.
**Explosive Limits:** 2.0 to 11.5%.
**Melting Point:** -82 to -84°C.

## HAZARD DATA

### Human Health
**Symptoms: Contact:** skin - irritation and dermatitis; eyes - irritation. <u>Ingestion:</u> irritation, headache and nausea. <u>Inhalation:</u> sore throat, coughing, dizziness, drowsiness.
**Toxicology:** Moderately toxic by ingestion and inhalation.
TLV® - (inhalation) 400 ppm; 1 400 mg/m³.    $LC_{50}$ - Inhalation: rat = 1 600 ppm/8 h    $LD_{50}$ - Oral: rat = 11 g/kg
Short-term Inhalation Limits - No information.    Delayed Toxicity - May cause liver or kidney damage.

### Fire
**Fire Extinguishing Agents:** Use carbon dioxide, dry chemical or alcohol-type foam. Water may be ineffective, but may be used to cool fire-exposed containers.
**Behaviour in Fire:** Flashback may occur along vapour trail.
**Ignition Temperature:** 426°C.    **Burning Rate:** 3.7 mm/min

### Reactivity
**With Water:** No reaction; slightly soluble.
**With Common Materials:** Can react vigorously with oxidizing materials. Reacts violently with chlorosulfonic acid, oleum and potassium-t-butoxide.
**Stability:** Stable.

### Environment
**Water:** Prevent entry into water intakes and waterways. Aquatic toxicity rating = 100 to 1 000 ppm/96 h/TLm/freshwater; BOD: 15 to 36%, 5 days.
**Land-Air:** No information.
**Food Chain Concentration Potential:** None.

## EMERGENCY MEASURES

### Special Hazards
FLAMMABLE.

### Immediate Responses
Keep non-involved people away from spill site. Issue warning: "FLAMMABLE". Call Fire Department. Eliminate all ignition sources. Contact manufacturer or supplier for advice. Dike to prevent runoff from rainwater or water application. Stop or reduce discharge if this can be done without risk. Notify environmental authorities.

### Protective Clothing and Equipment
Respiratory protection - self-contained breathing apparatus in fires or enclosed spaces. Otherwise, Eye protection - goggles or face shield. Gloves - rubber or plastic. Outer clothing - suitable for situation, coveralls, etc. Boots - rubber.

### Fire and Explosion
Use carbon dioxide, dry chemical or alcohol-type foam to extinguish. Flashback may occur along vapour trail. Water may be used to cool fire-exposed containers.

### First Aid
Move victim out of spill area to fresh air. Call for medical assistance. but start first aid at once. Inhalation: if breathing has stopped, give artificial respiration; if breathing is laboured, give oxygen. Contact: eyes - irrigate with water for at least 15 minutes; skin - remove contaminated clothing; wash affected areas thoroughly with water. Ingestion: give milk or water to conscious victim to drink and induce vomiting. If medical assistance is not immediately available, transport victim to hospital, doctor or clinic.

## ENVIRONMENTAL PROTECTION MEASURES

### Response

**Water**
1. Stop or reduce discharge if safe to do so.
2. Contact manufacturer or supplier for advice.
3. If possible, contain discharge by booming.
4. If floating, skim and remove.
5. Notify environmental authorities to discuss disposal and cleanup of contaminated materials.

**Land-Air**
1. Stop or reduce discharge if safe to do so.
2. Contact manufacturer or supplier for advice.
3. Contain spill by diking with earth or other barrier.
4. Remove material with pumps or vacuum equipment and place in appropriate containers.
5. Recover undamaged containers.
6. Adsorb residual liquid on natural or synthetic sorbents.
7. Notify environmental authorities to discuss disposal and cleanup of contaminated materials.

### Disposal
1. Contact manufacturer or supplier for advice on disposal.
2. Contact environmental authorities for advice on disposal.
3. Incinerate (approval of environmental authorities required).

ETHYL ACETATE $CH_3COOC_2H_5$

# ETHYL ACRYLATE    CH$_2$:CHCOOC$_2$H$_5$    UN No. 1917

## IDENTIFICATION

| Common Synonyms | Observable Characteristics | Manufacturers |
|---|---|---|
| ACRYLIC ACID, ETHYL ESTER<br>ETHYL 2-PROPENOATE<br>2-PROPENOIC ETHYL ESTER | Colourless liquid. Sharp acrid odour. | No Canadian manufacturers.<br>Canadian supplier:<br>Celanese Canada, Montreal, Quebec. | Originating from:<br>Celanese Chemical Co.,<br>New York, NY. |

### Transportation and Storage Information

**Shipping State:** Liquid.
**Classification:** Flammable liquid.
**Inert Atmosphere:** No requirement.
**Venting:** Pressure-vacuum.
**Pump Type:** No information.

**Label(s):** Red and white label - FLAMMABLE LIQUID; Class 3.2, Group II.
**Storage Temperature:** Ambient.
**Hose Type:** No information.

**Grades or Purity:** 98.5 to 99.5%.
**Containers and Materials:** Drums, tank cars, tank trucks; steel, stainless steel.

### Physical and Chemical Characteristics

**Physical State (20°C, 1 atm):** Liquid.
**Solubility (Water):** 2 g/100 mL (20°C).
**Molecular Weight:** 100.1
**Vapour Pressure:** 29 mm Hg (20°C); 49 mm Hg (30°C).
**Boiling Point:** 99.1 to 99.6°C.

**Floatability (Water):** Floats and mixes slightly.
**Odour:** Sharp acid (0.1 to 0.47 ppb, odour threshold).
**Flash Point:** 10°C (o.c.).
**Vapour Density:** 3.5
**Specific Gravity:** 0.92 (20°C) (liquid).

**Colour:** Colourless.
**Explosive Limits:** 1.4 to 14%.
**Melting Point:** -71 to -75°C.

## HAZARD DATA

### Human Health

**Symptoms: Inhalation:** irritation of mucous membranes, thoracic congestion, coughing, cyanosis. **Ingestion:** irritation, pain in swallowing, nausea and vomiting, diarrhea. **Contact: skin** - readily absorbed, itching, irritation, inflammation and blistering; **eyes** - irritation and burning.
**Toxicology:** Moderately toxic by inhalation, ingestion and skin absorption.
TLV® - (skin) 5 ppm; 20 mg/m$^3$.       LC$_{50}$ - No information.            LD$_{50}$ - Oral: rat = 1.02 g/kg
Short-term Inhalation Limits - 25 ppm; 100 mg/m$^3$     LC$_{Lo}$ - Inhalation: rat = 1 000 ppm/4 h
(15 min).                   Delayed Toxicity - No information.

### Fire

**Fire Extinguishing Agents:** Use dry chemical, alcohol foam or carbon dioxide. Water may be ineffective but may be used to cool fire-exposed containers and knock down vapours.
**Behaviour in Fire:** At high temperatures, polymerization may take place stopping vents and causing rupture of containers. Flashback may occur along vapour trail.
**Ignition Temperature:** 372°C.       **Burning Rate:** 4.3 mm/min.

### Reactivity

**With Water:** No reaction; slightly soluble.
**With Common Materials:** Reacts vigorously with oxidizing materials. Reacts violently with chlorosulfonic acid.
**Stability:** May polymerize under certain conditions.

### Environment

**Water:** Prevent entry into water intakes and waterways. Harmful to aquatic life. Fish toxicity = 12 ppm/24 h/brine shrimp/TLm/saltwater; Aquatic toxicity rating = 100 to 1 000 ppm/96 h/TLm/freshwater; BOD: 11 to 66% (5 days).
**Land-Air:** No information.
**Food Chain Concentration Potential:** No information.

## EMERGENCY MEASURES

### Special Hazards
FLAMMABLE. May polymerize.

### Immediate Responses
Keep non-involved people away from spill site. Issue warning: "FLAMMABLE". Call Fire Department. Eliminate all sources of ignition. Call manufacturer for advice. Dike to prevent runoff from rainwater or water application. Notify environmental authorities.

### Protective Clothing and Equipment
Respiratory protection - self-contained breathing apparatus and totally encapsulated protective clothing.

### Fire and Explosion
Use dry chemical, alcohol foam or carbon dioxide to extinguish. Water may be ineffective, but may be used to cool fire-exposed containers and knock down vapours. At high temperatures, polymerization may take place stopping vents and causing rupture of containers. Flashback may occur along vapour trail.

### First Aid
Move victim from spill site to fresh air. Call for medical assistance, but start first aid at once. Inhalation: if breathing has stopped, give artificial respiration; if laboured, give oxygen. Contact: skin - remove contaminated clothing and flush affected areas with plenty of water; eyes - irrigate with plenty of water. Ingestion: give conscious victim plenty of water to drink and induce vomiting. If medical assistance is not immediately available, transport victim to doctor, clinic or hospital.

## ENVIRONMENTAL PROTECTION MEASURES

### Response

**Water**
1. Stop or reduce discharge if safe to do so.
2. Contact manufacturer or supplier for advice.
3. If possible, contain discharge by booming.
4. If floating, skim and remove.
5. Notify environmental authorities to discuss disposal and cleanup of contaminated materials.

**Land-Air**
1. Stop or reduce discharge if safe to do so.
2. Contact manufacturer or supplier for advice.
3. Contain spill by diking with earth or other barrier.
4. Remove material with pumps or vacuum equipment and place in appropriate containers.
5. Recover undamaged containers.
6. Adsorb residual liquid on natural or synthetic sorbents.
7. Notify environmental authorities to discuss disposal and cleanup of contaminated materials.

### Disposal
1. Contact manufacturer or supplier for advice on disposal.
2. Contact environmental authorities for advice on disposal.
3. Incinerate (approval of environmental authorities required).

ETHYL ACRYLATE   $CH_2{:}CHCOOC_2H_5$

# ETHYL ALCOHOL  $C_2H_5OH$

## IDENTIFICATION

### Common Synonyms
ALCOHOL
ETHANOL
GRAIN ALCOHOL
DENATURED ALCOHOL
ETHYL HYDRATE
ABSOLUTE ETHANOL
SPIRITS
ALCOHOL, DEHYDRATED

### Observable Characteristics
Clear, colourless liquid. Typical alcohol odour.

### Grades or Purity:
Anhydrous (100%); (95%) specially denatured; completely denatured.
**Containers and Materials:** Cans, drums, tank cars, tank trucks; steel, stainless steel, aluminum.

### Manufacturers
Ontario Paper Co. Ltd., Thorold, Ontario.
Commercial Alcohols, Montreal, Que.
Mohawk Oil, Minnedosa, Manitoba.
Consolidated Alcohols Ltd., Toronto, Ontario.

### Transportation and Storage Information
**Shipping State:** Liquid.
**Classification:** Flammable liquid.
**Inert Atmosphere:** No requirement.
**Venting:** Open with flame arrester or pressure vacuum.
**Pump Type:** Centrifugal, gear, etc.; steel, stainless steel.

**Label(s):** Red label - FLAMMABLE LIQUID; Class 3.2, Groups I, II or III.
**Storage Temperature:** Ambient.
**Hose Type:** Polyethylene, polypropylene, butyl, Hypalon, natural rubber, Viton.

### Physical and Chemical Characteristics
**Physical State** (20°C, 1 atm): Liquid.
**Solubility (Water):** Miscible in all proportions.
**Molecular Weight:** 46.1
**Vapour Pressure:** 43.9 mm Hg (20°C); 50 mm Hg (25°C); 75 mm Hg (30°C).
**Boiling Point:** 78.4°C.

**Floatability (Water):** Mixes.
**Odour:** Alcohol (1 to 10 ppm, odour threshold).
**Flash Point:** 13°C (c.c.) (pure); 17°C (c.c.) (95%+ water); 24°C (c.c.) (50%+ water).
**Vapour Density:** 1.6
**Specific Gravity:** 0.79 (20°C).

**Colour:** Colourless.
**Explosive Limits:** 3.3% to 19%.
**Melting Point:** -110 to -118°C.

## HAZARD DATA

### Human Health
**Symptoms:** Ingestion: visual impairment, muscular incoordination and slowing of reaction time, slurring of speech, nausea and vomiting. Bizarre symptoms (other than typical intoxication) can result from the denaturants often present in industrial ethyl alcohol. Contact: skin - drying and dermatitis; eyes - irritation and watering. Inhalation: irritation of nose and eyes.
**Toxicology:** Low toxicity through ingestion, inhalation and contact.
TLV® (inhalation) 1 000 ppm; 1 900 mg/m³.   $LC_{50}$ - No information.   $LD_{50}$ - Oral: rat = 14 g/kg
Short-term Inhalation Limits - No information.   Delayed Toxicity - None.   $TD_{Lo}$ - Oral: man = 0.05 g/kg

### Fire
**Fire Extinguishing Agents:** Use carbon dioxide, dry chemical, or alcohol foam. Water may be ineffective.
**Behaviour in Fire:** Flash back may occur along vapour trail.
**Ignition Temperature:** 363°C.   **Burning Rate:** 3.9 mm/min.

### Reactivity
**With Water:** No reaction, soluble.
**With Common Materials:** Can react vigorously with oxidizing materials. Reacts violently with acetyl chloride, chromates, hydrogen peroxide, nitric acid, perchlorates, (permanganates and sulfuric acid), and silver nitrate.
**Stability:** Stable.

### Environment
**Water:** Prevent entry into water intakes and waterways. Harmful to aquatic life. Fish toxicity: 13 000 mg/L/95 h/rainbow trout/$LC_{50}$/freshwater; 9 000 mg/L/24 h/creek chub/$LC_{100}$/freshwater; BOD: 93 to 167%, 5 days.
**Land-Air:** No information.
**Food Chain Concentration Potential:** None.

## EMERGENCY MEASURES

**Special Hazards**
FLAMMABLE.

**Immediate Responses**
Keep non-involved people away from spill site. Issue warning: "FLAMMABLE". CALL FIRE DEPARTMENT. Eliminate all ignition sources. Notify manufacturer or supplier. Dike to prevent runoff. Stop or reduce discharge if this can be done without risk. Notify environmental authorities.

**Protective Clothing and Equipment**
Eye protection - goggles or face shield. Gloves - rubber. Boots - rubber. Clothing - suitable for situation.

**Fire and Explosion**
Use dry chemical, alcohol foam or carbon dioxide to extinguish. Water may be ineffective on fire but may be used to cool fire-exposed containers.

**First Aid**
Move victim out of spill area to fresh air. Call for medical assistance, but start first aid at once. Contact: wash eyes and skin with water and remove contaminated clothing. Inhalation: if breathing has stopped, give artificial respiration; if laboured, give oxygen. Ingestion: give water to conscious victim to drink and induce vomiting. If medical attention is considered necessary, transport victim to hospital, doctor or clinic.

## ENVIRONMENTAL PROTECTION MEASURES

**Response**

**Water**
1. Stop or reduce discharge if safe to do so.
2. Contact manufacturer or supplier for advice.
3. Notify environmental authorities to discuss disposal and cleanup of contaminated materials.

**Land-Air**
1. Stop or reduce discharge if safe to do so.
2. Contact manufacturer or supplier for advice.
3. Dike to prevent runoff from rainwater or water application.
4. Remove material with pumps or vacuum equipment and place in appropriate containers.
5. Recover undamaged containers.
6. Absorb residual liquid on natural or synthetic sorbents.
7. Notify environmental authorities to discuss disposal and cleanup of contaminated materials.

**Disposal**
1. Contact manufacturer or supplier for advice on disposal.
2. Contact environmental authorities for advice on disposal.
3. Incinerate (approval of environmental authorities required).

ETHYL ALCOHOL $C_2H_5OH$

# ETHYLBENZENE $C_6H_5CH_2CH_3$

## IDENTIFICATION

**Common Synonyms**
PHENYLETHANE
EB
ETHYLBENZOL

**Observable Characteristics**
Clear, colourless liquid with an aromatic odour.

**UN No. 1175**

**Manufacturers**
Polysar, Sarnia, Ontario.
Dow Chemical Canada Inc., Sarnia, Ontario.

### Transportation and Storage Information

**Shipping State:** Liquid.
**Classification:** Flammable liquid.
**Inert Atmosphere:** No requirement.
**Venting:** Open (flame arrester) or pressure vacuum.
**Pump Type:** Gear or centrifugal, explosion-proof, grounded.

**Label(s):** Red and white label - FLAMMABLE LIQUID; Class 3.2, Group II.
**Storage Temperature:** Ambient.
**Hose Type:** Viton, polypropylene, Teflon, neoprene.

**Grades or Purity:** Pure grade, 99.5%; technical grade, 99.0%.
**Containers and Materials:** Drums, tank cars, tank trucks; steel, stainless steel.

### Physical and Chemical Characteristics

**Physical State** (20°C, 1 atm): Liquid.
**Solubility** (Water): 0.014 g/100 mL (15°C); 0.021 g/100 mL (25°C).
**Molecular Weight:** 106.2
**Vapour Pressure:** 7 mm Hg (20°C); 12 mm Hg (30°C).
**Boiling Point:** 136.2°C.

**Floatability** (Water): Floats.
**Odour:** Aromatic (140 ppm, odour threshold).
**Flash Point:** 15°C (c.c.).
**Vapour Density:** 3.7
**Specific Gravity:** 0.87 at 20°C.

**Colour:** Colourless.
**Explosive Limits:** 1.0 to 6.7%.
**Melting Point:** -95°C.

## HAZARD DATA

### Human Health

**Symptoms:** Inhalation: irritation of mucous membranes, headache, dizziness, narcosis and coma. Ingestion: symptoms similar to inhalation. Contact: skin - irritation, defatting and dermatitis; eyes - irritation and burning.
**Toxicology:** Moderately toxic by inhalation, contact and ingestion.
TLV® - (inhalation) 100 ppm; 435 mg/m$^3$. $LC_{50}$ - No information. $LD_{50}$ - Oral: rat = 3.5 g/kg
Short-term Inhalation Limits - 125 ppm; 545 mg/m$^3$ $LC_{Lo}$ - Inhalation: rat = 4 000 ppm/4 h
(15 min.) Delayed Toxicity - No information.

### Fire

**Fire Extinguishing Agents:** Use foam, dry chemical or carbon dioxide. Water may be ineffective but may be used to cool fire-exposed containers and knock down vapours.
**Behaviour in Fire:** Flashback may occur along vapour trail.
**Ignition Temperature:** 432°C.
**Burning Rate:** 5.8 mm/min.

### Reactivity

**With Water:** No reaction.
**With Common Materials:** Can react with oxidizing materials.
**Stability:** Stable.

### Environment

**Water:** Prevent entry into water intakes and waterways. Fish toxicity: 29 ppm/96 h/bluegill/TLm/freshwater; 32 to 35.1 ppm/48 h/bluegill/TLm/freshwater; 42.3 mg/L/48 h/fathead minnow/TLm/hard water; 14 mg/L/96 h/rainbow trout/$LC_{50}$/freshwater; BOD: 2.8% 5 days.
**Land-Air:** No information.
**Food Chain Concentration Potential:** None.

## EMERGENCY MEASURES

**Special Hazards**
FLAMMABLE.

**Immediate Responses**
Keep non-involved people away from spill site. Issue warnings: "FLAMMABLE". Call Fire Department. Eliminate all sources of ignition. Call manufacturer. Shut off leak if safe to do so. Dike to prevent runoff. Notify environmental authorities.

**Protective Clothing and Equipment**
Respiratory protection - self-contained breathing apparatus and totally encapsulated protective clothing.

**Fire and Explosion**
Use foam, dry chemical or carbon dioxide to extinguish. Water may be ineffective, but may be used to cool fire-exposed containers and knock down vapours. Flashback may occur along vapour trail. Do not extinguish fire until leak stopped.

**First Aid**
Move victim out of spill site to fresh air. Call for medical assistance, but start first aid at once. Inhalation: give artificial respiration if breathing has stopped; oxygen if laboured. Contact: skin - remove contaminated clothing and flush affected areas with plenty of water; eyes - irrigate with water. Ingestion: do not induce vomiting. If medical assistance is not immediately available, transport victim to doctor, hospital or clinic.

## ENVIRONMENTAL PROTECTION MEASURES

**Response**

**Water**
1. Stop or reduce discharge if safe to do so.
2. Contact manufacturer or supplier for advice.
3. If possible, contain discharge by booming.
4. If floating, skim and remove.
5. Notify environmental authorities to discuss disposal and cleanup of contaminated materials.

**Land-Air**
1. Stop or reduce discharge if safe to do so.
2. Contact manufacturer or supplier for advice.
3. Contain spill by diking with earth or other barrier.
4. Remove material with pumps or vacuum equipment and place in appropriate containers.
5. Recover undamaged containers.
6. Adsorb residual liquid on natural or synthetic sorbents.
7. Notify environmental authorities to discuss disposal and cleanup of contaminated materials.

**Disposal**
1. Contact manufacturer or supplier for advice on disposal.
2. Contact environmental authorities for advice on disposal.
3. Incinerate (approval of environmental authorities required).

ETHYLBENZENE $C_6H_5CH_2CH_3$

# ETHYL CHLORIDE  $C_2H_5Cl$

## IDENTIFICATION   UN No. 1037

### Common Synonyms
CHLOROETHANE
MONOCHLOROETHANE
HYDROCHLORIC ETHER

### Observable Characteristics
Colourless liquid or gas. Ethereal odour.

### Manufacturers
Ethyl Corporation of Canada Ltd., Corunna, Ontario.
Dow Chemical Canada Inc., Sarnia, Ontario
Originating from: Dow Chemical, Freeport, TX.

### Transportation and Storage Information
**Shipping State:** Liquid (compressed gas).
**Classification:** Flammable gas.
**Inert Atmosphere:** No requirement.
**Venting:** Safety relief.
**Pump Type:** Centrifugal, positive displacement or gear. Steel, stainless steel.

**Label(s):** Red label - FLAMMABLE GAS; Class 2.1.
**Storage Temperature:** Ambient.
**Hose Type:** Teflon, Viton A, flexible stainless steel.

**Grades or Purity:** Technical, 99.5 to 100%; USP, 100%.
**Containers and Materials:** Cylinders, tank cars, tank trucks (pressure vessels). Steel, stainless steel.

### Physical and Chemical Characteristics
**Physical State** (20°C, 1 atm): Gas.
**Solubility** (Water): Reacts to form HCl. 0.33 g/100 mL (0°C); 0.57 g/100 mL (20°C).
**Molecular Weight:** 64.5
**Vapour Pressure:** 457 mm Hg (0°C); 700 mm Hg (10°C); 1 000 mm Hg (20°C); 1 444 mm Hg (30°C).
**Boiling Point:** 12.3°C.

**Floatability** (Water): Floats and reacts.
**Odour:** Ethereal.
**Flash Point:** -43°C (o.c.); -50°C (c.c.).
**Vapour Density:** 2.2
**Specific Gravity:** 0.89 (25°C).

**Colour:** Colourless.
**Explosive Limits:** 3.8 to 15.4%.
**Melting Point:** -136 to -139°C.

## HAZARD DATA

### Human Health
**Symptoms:** Inhalation: dizziness, disorientation, incoordination, inhalation, narcosis, nausea or vomiting. Ingestion: similar to inhalation. Contact: skin - pain, irritation, defatting, frostbite; may be absorbed; eyes - pain and irritation, frostbite burn.
**Toxicology:** Moderately toxic by inhalation, ingestion and contact.
TLV® - (inhalation) 1 000 ppm; 2 600 mg/m³.      LC$_{50}$ - No information.      LD$_{50}$ - No information.
Short-term Inhalation Limits - 1 250 ppm;        LC$L_O$ - Inhalation: guinea pig = 4 000 ppm/45 min
3 250 mg/m³ (15 min).                            Delayed Toxicity - No information.

### Fire
**Fire Extinguishing Agents:** Stop or reduce discharge before attempting to extinguish fire. Use carbon dioxide or dry chemical to extinguish small fires. Water spray, fog, or foam may be necessary on large fires. Use water to cool fire-exposed containers.
**Behaviour in Fire:** Vessels may rupture. Decomposition from heat or fire produces toxic hydrochloric acid fumes and phosgene. Flashback may occur along vapour trail.
**Ignition Temperature:** 519°C.                **Burning Rate:** 3.8 mm/min.

### Reactivity
**With Water:** Reacts with water and steam to produce hydrochloric acid.
**With Common Materials:** Can react vigorously with oxidizing materials. Reacts violently with alkaline metals, magnesium, aluminum and zinc.
**Stability:** Stable.

### Environment
**Water:** Prevent entry into water intakes and waterways. Aquatic toxicity rating: >1 000 ppm/96 h/TLm/freshwater; BOD: No information.
**Land-Air:** No information.
**Food Chain Concentration Potential:** None.

## EMERGENCY MEASURES

### Special Hazards
FLAMMABLE.

### Immediate Responses
Keep non-involved people away from spill site. Issue warning: "FLAMMABLE". Call Fire Department. Eliminate all sources of ignition. Call manufacturer for advice. Avoid contact and inhalation. If material is burning, use water spray to knock down vapours and fumes; work from upwind. If water spray is used, dike area to contain toxic runoff. Stop or reduce discharge if this can be done without risk. Notify environmental authorities.

### Protective Clothing and Equipment
Respiratory protection - self-contained breathing apparatus and totally encapsulated suit. Boots - high, rubber (pants worn outside boots). Gloves - rubber or plastic.

### Fire and Explosion
Stop or reduce discharge before attempting to extinguish fire. Use carbon dioxide to extinguish. Use water to cool fire-exposed containers. Flashback may occur along vapour trail.

### First Aid
Move victim out of spill area to fresh air. Call for medical assistance, but start first aid at once. Inhalation: give artificial respiration if breathing has stopped; give oxygen if breathing is laboured. Contact: remove contaminated clothing. Wash eyes and affected skin with plenty of warm water for at least 15 minutes. Do not rub areas which appear to have been frostbitten. If medical assistance is not immediately available, transport victim to hospital, doctor or clinic.

## ENVIRONMENTAL PROTECTION MEASURES

### Response

**Water**

Control - favourable conditions
1. Stop or reduce discharge if safe to do so.
2. Contact manufacturer or supplier for advice.
3. If floating, skim and remove.
4. Notify environmental authorities to discuss disposal and cleanup of contaminated materials.

Control - not possible
1. Stop or reduce discharge if safe to do so.
2. Contact manufacturer or supplier for advice.
3. Notify environmental authorities to discuss disposal and cleanup of contaminated materials.

**Land-Air**

Control - favourable conditions
1. Stop or reduce discharge if safe to do so.
2. Contact manufacturer or supplier for advice.
3. Contain spill by diking with earth or other barrier.
4. Remove material with pumps or vacuum equipment and place in appropriate containers.
5. Adsorb residual liquid on natural or synthetic sorbents.
6. Remove contaminated soil for disposal.
7. Notify environmental authorities to discuss disposal and cleanup of contaminated materials.

Control - not possible
1. Stop or reduce discharge if safe to do so.
2. Contact manufacturer or supplier for advice.
3. Notify environmental authorities to discuss disposal and cleanup of contaminated materials.

### Disposal
1. Contact manufacturer or supplier for advice on disposal.
2. Contact environmental authorities for advice on disposal.

ETHYL CHLORIDE $C_2H_5Cl$

# ETHYLENE $C_2H_4$

## IDENTIFICATION

| Common Synonyms | Observable Characteristics | UN No. 1962 Manufacturers |
|---|---|---|
| ETHENE<br>OLEFIANT GAS | Colourless liquid or gas. Sweet odour. | Esso Chemicals Canada Ltd., Sarnia, Ontario.<br>Gulf Oil Canada Ltd., Varennes, Quebec.<br>Alberta Gas, Joffre, Alberta.<br>Petrosar Ltd., Corunna, Ontario.<br>Dow chemical Canada Inc., Sarnia, Ont.,<br>Fort Saskatchewan, Alta. |

### Transportation and Storage Information

| | | |
|---|---|---|
| **Shipping State:** Liquid (compressed gas).<br>**Classification:** Flammable gas.<br>**Inert Atmosphere:** No requirement.<br>**Venting:** Safety relief. | **Label(s):** Red label - FLAMMABLE GAS; Class 2.1.<br>**Storage Temperature:** Ambient (gas); -104°C (liquid).<br>**Hose Type:** Teflon bore 304 stainless steel (300-1 400 psig). | **Grades or Purity:** Technical, 98 to 99%; CP 99.5 to 100%.<br>**Containers and Materials:** Cylinders, tube trailers, tank trucks (under special permit), tank cars; steel. |

### Physical and Chemical Characteristics

| | | |
|---|---|---|
| **Physical State** (20°C, 1 atm): Gas.<br>**Solubility** (Water): 0.013 g/100 mL (20°C).<br>**Molecular Weight:** 28.1<br>**Vapour Pressure:** 30 400 mm Hg (0°C).<br>**Boiling Point:** -103.9°C. | **Floatability** (Water): Floats and boils.<br>**Odour:** Sweet (260 to 700 ppm, odour threshold).<br>**Flash Point:** -136°C approximately.<br>**Vapour Density:** 0.98 (0°C)<br>**Specific Gravity:** 0.57 (-104°C); 0.61 (0°C). | **Colour:** Colourless.<br>**Explosive Limits:** 2.7 to 36%.<br>**Melting Point:** -169.2°C. |

## HAZARD DATA

### Human Health

**Symptoms:** Inhalation: headache, drowsiness, dizziness, weakness, narcosis, nausea and vomiting, asphyxia. Contact: skin - liquid can cause frostbite; eyes - pain, burning.
**Toxicology:** Moderately toxic by inhalation.
**TLV®** - (inhalation) Asphyxiant.     $LC_{50}$ - Inhalation: mouse = 950 000 ppm.     $LD_{50}$ - No information.
**Short-term Inhalation Limits** - No information.     **Delayed Toxicity** - None known.

### Fire

**Fire Extinguishing Agents:** Stop flow of gas before attempting extinguishment. Use carbon dioxide, dry chemical or water fog. Use water to cool fire-exposed containers.
**Behaviour in Fire:** Flashback may occur along vapour trail.
**Ignition Temperature:** 450°C.     **Burning Rate:** 7.4 mm/min.

### Reactivity

**With Water:** No reaction.
**With Common Materials:** Can react vigorously with oxidizing materials. Reacts violently with aluminum chloride, bromotrichloromethane, carbon tetrachloride, chlorine, chlorine dioxide, nitrogen dioxide and ozone. May react with acids.
**Stability:** Stable.

### Environment

**Water:** Prevent entry into water intakes and waterways. Fish toxicity: 22 ppm/1 h/sunfish/killed/freshwater; Aquatic toxicity rating = 100 to 1 000 ppm/96 h/TLm/freshwater; BOD: No information.
**Land-Air:** Concentrations of 0.5 to 4 ppm cause leaf loss; 0.1 to 0.5 ppm causes growth retardation or inhibits flower opening.
**Food Chain Concentration Potential:** None.

## EMERGENCY MEASURES

**Special Hazards**
FLAMMABLE.

**Immediate Responses**
Keep non-involved people away. Issue warning: "FLAMMABLE". Call Fire Department. Eliminate sources of ignition. Avoid contact and inhalation. Work from upwind and use water spray to control vapour. Stop or reduce discharge if this can be done without risk. Notify manufacturer or supplier. Notify environmental authorities.

**Protective Clothing and Equipment**
Respiratory protection – self-contained breathing apparatus. Gloves – rubber or plastic. Acid suit – jacket and pants; rubber or plastic. Boots – high, rubber (pants worn outside boots).

**Fire and Explosion**
Stop flow of gas before attempting extinguishment. Use carbon dioxide, dry chemical or water fog to extinguish. Use water to cool fire-exposed containers. Flashback may occur along vapour trail.

**First Aid**
Move victim out of spill area to fresh air. Call for medical assistance, but start first aid at once. Inhalation: give artificial respiration if breathing has stopped; give oxygen if breathing is laboured. Contact: remove contaminated clothing and wash eyes and affected skin thoroughly with plenty of warm water. Do not rub affected areas. Keep victim warm and quiet. If medical assistance is not immediately available, transport victim to hospital, clinic or doctor.

## ENVIRONMENTAL PROTECTION MEASURES

**Response**

Water-Land-Air
1. Stop or reduce discharge if safe to do so.
2. Contact manufacturer or supplier for advice.
3. Notify environmental authorities to discuss disposal and cleanup of contaminated materials.

**Disposal**
1. Contact manufacturer or supplier for advice on disposal.
2. Contact environmental authorities for advice on disposal.
3. Incinerate (approval of environmental authorities required).

ETHYLENE $C_2H_4$

# ETHYLENEDIAMINE   $NH_2CH_2CH_2NH_2$

## IDENTIFICATION

**UN No. 1604**

**Common Synonyms**
1,2 DIAMINOETHANE
1,2 ETHANEDIAMINE

**Observable Characteristics**
Colourless liquid. Ammonia-like odour.

**Manufacturers**
No Canadian manufacturer.
Canadian supplier:
Dow Chemical Canada Inc.,
Sarnia, Ontario.
Originating from: Dow Chemical, Midland, MI.

### Transportation and Storage Information

**Shipping State:** Liquid.
**Classification:** Corrosive liquid.
**Inert Atmosphere:** No requirement.
**Venting:** Pressure-vacuum.
**Pump Type:** Centrifugal or positive displacement; stainless steel.

**Label(s):** Black and white label – CORROSIVE; Class 8, Group II.
**Storage Temperature:** Ambient.
**Hose Type:** Natural rubber, Viton, butyl, flexible stainless steel. No bronze, brass or copper alloy fittings.

**Grades or Purity:** Technical, 97+%.
**Containers and Materials:** Drums (tin-lined), tank trucks; stainless steel, aluminum.

### Physical and Chemical Characteristics

**Physical State** (20°C, 1 atm): Liquid.
**Solubility (Water):** Completely soluble.
**Molecular Weight:** 60.1
**Vapour Pressure:** 9 mm Hg (20°C); 16 mm Hg (30°C).
**Boiling Point:** 115-119°C.

**Floatability** (Water): Floats and mixes.
**Odour:** Ammonia-like odour (1.0 to 11.2 ppm, odour threshold).
**Flash Point:** 50-66°C (o.c.); 40-43°C (c.c.).
**Vapour Density:** 2.1
**Specific Gravity:** 0.90 (20°C).

**Colour:** Colourless.
**Explosive Limits:** 4.2 to 14.4% (2.5 to 12% at 100°C).
**Melting Point:** 11°C.

## HAZARD DATA

### Human Health

**Symptoms: Inhalation:** intense respiratory tract irritation, headache, dizziness, shortness of breath. **Contact:** burns to skin and eyes. **Ingestion:** irritation to digestive tract, nausea and vomiting.
**Toxicology:** Highly toxic by inhalation and ingestion; moderately toxic by skin contact.
TLV®– (inhalation) 10 ppm; 25 mg/m³.   $LC_{50}$ – Inhalation: rat = 4 000 ppm/8 h
Short-term Inhalation Limits – No information.   Delayed Toxicity – No information.

$LD_{50}$ – Oral: guinea pig = 0.47 g/kg
Oral: rat = 1.16 g/kg

### Fire

**Fire Extinguishing Agents:** Use carbon dioxide, dry chemical or alcohol foam. Water spray may be ineffective, but may be used to cool fire-exposed containers and knock down vapours.
**Behaviour in Fire:** Irritating $NO_x$ gases may be generated.
**Ignition Temperature:** 380-400°C.   **Burning Rate:** 2.2 mm/min.

### Reactivity

**With Water:** No reaction; soluble.
**With Common Materials:** May react with oxidizing materials. Reacts violently with acetic acid, acetic anhydride, acrolein, acrylic acid, acrylonitrile, allyl chloride, carbon disulfide, chlorosulfonic acid, epichlorohydrin, hydrochloric acid, nitric acid, oleum, silver perchlorate, sulfuric acid and vinyl acetate.
**Stability:** Stable.

### Environment

**Water:** Prevent entry into water intakes and waterways. Harmful to aquatic life in very low concentrations. Fish toxicity: 30 to 60 ppm/24 h/creek chub/killed/freshwater; Aquatic toxicity rating = 10 to 100 ppm/96 h/TLm/freshwater; BOD: 75%, 5 days (theoretical).
**Land-Air:** No information.
**Food Chain Concentration Potential:** None.

## EMERGENCY MEASURES

**Special Hazards**

CORROSIVE. FLAMMABLE.

**Immediate Responses**

Keep non-involved people away from spill site. Issue warnings: "CORROSIVE; FLAMMABLE". Call Fire Department. Avoid contact and inhalation. If water is used to control fire, dike area to prevent runoff. Notify supplier. Notify environmental authorities.

**Protective Clothing and Equipment**

Respiratory protection - self-contained breathing apparatus and totally encapsulated protection suit. Boots - high, rubber (pants worn outside boots). Gloves - rubber.

**Fire and Explosion**

Use dry chemical, alcohol foam or carbon dioxide to extinguish. Water may be used to cool fire-exposed containers and knock down vapours.

**First Aid**

Move victim out of spill area to fresh air. Call for medical assistance, but start first aid at once. Inhalation: give artificial respiration if breathing has stopped; give oxygen if breathing is laboured. Contact: remove contaminated clothing. Wash eyes and skin with plenty of warm water for at least 30 minutes. Ingestion: give milk or water to conscious victim only. Do not induce vomiting. If medical assistance is not immediately available, transport victim to hospital, doctor or clinic.

## ENVIRONMENTAL PROTECTION MEASURES

**Response**

Water
1. Stop or reduce discharge if safe to do so.
2. Contact manufacturer or supplier for advice.
3. If possible, contain discharge by damming or water diversion.
4. Notify environmental authorities to discuss disposal and cleanup of contaminated materials.

Land-Air
1. Stop or reduce discharge if safe to do so.
2. Contact manufacturer or supplier for advice.
3. Contain spill by diking with earth or other barrier.
4. Remove material by manual or mechanical means.
5. Recover undamaged containers.
6. Adsorb residual liquid on natural or synthetic sorbents.
7. Notify environmental authorities to discuss disposal and cleanup of contaminated materials.

**Disposal**

1. Contact manufacturer or supplier for advice on disposal.
2. Contact environmental authorities for advice on disposal.

ETHYLENEDIAMINE    $NH_2CH_2CH_2NH_2$

# ETHYLENE DIBROMIDE $BrCH_2CH_2Br$

## IDENTIFICATION

UN No. 1605

### Common Synonyms
EDB
1,2-DIBROMOETHANE
ETHYLENE BROMIDE

### Observable Characteristics
Colourless liquid. Sweet odour.

### Manufacturers
No Canadian manufacturers.
Canadian suppliers:
Basile Import/Export, St. Laurent, Quebec.
Chorney Chemical Co., Toronto, Ontario.
Dow Chemical Canada Inc.
Originating from:
Dow Chemical, Magnolia, AR.

### Transportation and Storage Information

**Shipping State:** Liquid.
**Classification:** Poisonous liquid.
**Inert Atmosphere:** No requirement.
**Venting:** Pressure-vacuum.
**Pump Type:** Centrifugal or positive displacement, steel or stainless steel.
**Label(s):** White label – POISON; Class 6.1, Group II.
**Storage Temperature:** Ambient
**Hose type:** stainless steel, Viton.
**Grades or Purity:** Commercial.
**Containers and Materials:** Drums; steel. Tanks trucks, tank cars; steel (not aluminum).

### Physical and Chemical Characteristics

**Physical State** (20°C, 1 atm): Liquid.
**Solubility** (Water): 0.431 g/100 mL (30°C)
**Molecular Weight:** 187.9
**Vapour Pressure:** 11 mm Hg (20°C); 17 mm Hg (30°C)
**Boiling Point:** 131 to 132°C.
**Floatability** (Water): Sinks.
**Odour:** Sweet (26 ppm, odour threshold).
**Flash Point:** Not flammable.
**Vapour Density:** 6.5
**Colour:** Colourless.
**Explosive Limits:** Not flammable.
**Melting Point:** 9 to 10°C

## HAZARD DATA

### Human Health

**Symptoms:** <u>Inhalation</u>: shortness of breath, coughing, irritation of respiratory tract, nausea, vomiting, drowsiness, internal injury. <u>Contact</u>: skin – readily absorbed, inflammation, blisters and symptoms similar to inhalation. <u>Ingestion</u>: symptoms similar to inhalation.
**Toxicology:** Highly toxic by all routes.
**TLV®** – No information; contact should not be made with ethylene dibromide.
**Short-term Inhalation Limits** – No information.
$LC_{50}$ – No information.
$LC_{Lo}$ – Inhalation: rat = 400 ppm/2 h
**Delayed Toxicity** – Suspected carcinogen.
$LD_{50}$ – Oral: rat = 0.117 g/kg

### Fire

**Fire Extinguishing Agents:** Not combustible. In fires involving ethylene bromide, use water spray, dry chemical, foam or carbon dioxide. Water may also be used to disperse vapours.
**Behaviour in Fire:** When heated to decomposition, emits highly toxic fumes of bromine and hydrogen bromide.
**Ignition Temperature:** Not combustible.
**Burning Rate:** Not combustible.

### Reactivity

**With Water:** No reaction.
**With Common Materials:** Reacts violently with metal powders of aluminum, zinc and magnesium. May react with strong oxidizers.
**Stability:** Stable.

### Environment

**Water:** Prevent entry into water intakes and waterways. Fish toxicity: 18 mg/L/48 h/bluegill/freshwater; BOD: No information.
**Land-Air:** No information.
**Food Chain Concentration Potential:** None.

## EMERGENCY MEASURES

**Special Hazards**

POISON.

**Immediate Responses**

Keep non-involved people away from spill site. Issue warning: "POISON". Avoid contact and inhalation. Notify manufacturer or supplier. Notify environmental authorities.

**Protective Clothing and Equipment**

Respiratory protection - self-contained breathing apparatus and totally encapsulated suit.

**Fire and Explosion**

Not combustible. In fires involving ethylene dibromide, use water spray, dry chemical, foam or carbon dioxide to extinguish. Water may also be used to knock down vapours. When heated to decomposition, emits highly toxic fumes of bromide.

**First Aid**

Move victim out of spill site to fresh air. Call for medical assistance, but start first aid at once. Inhalation: if breathing has stopped, give artificial respiration (not mouth-to-mouth method); if laboured, give oxygen. Contact: skin - remove contaminated clothing and flush affected areas with plenty of water; eyes - irrigate with plenty of water for at least 15 minutes. Ingestion: give water to conscious victim to drink and induce vomiting. If medical assistance is not immediately available, transport victim to hospital, doctor or clinic.

## ENVIRONMENTAL PROTECTION MEASURES

**Response**

**Water**
1. Stop or reduce discharge if safe to do so.
2. Contact manufacturer or supplier for advice.
3. If possible, contain spill by damming or water diversion.
4. Dredge or vacuum pump to remove contaminants, liquids and contaminated bottom sediments.
5. Notify environmental authorities to discuss disposal and cleanup of contaminated materials.

**Land-Air**
1. Stop or reduce discharge if safe to do so.
2. Contact manufacturer or supplier for advice.
3. Contain spill by diking with earth or other barrier.
4. Remove material with pumps or vacuum equipment and place in appropriate containers.
5. Recover undamaged containers.
6. Adsorb residual liquid on natural or synthetic sorbents.
7. Remove contaminated soil for disposal.
8. Notify environmental authorities to discuss disposal and cleanup of contaminated materials.

**Disposal**
1. Contact manufacturer or supplier for advice on disposal.
2. Contact environmental authorities for advice on disposal.

ETHYLENE DIBROMIDE   $BrCH_2CH_2Br$

# ETHYLENE DICHLORIDE  $CH_2Cl \cdot CH_2Cl$

## IDENTIFICATION

UN No. 1184

### Common Synonyms
1,2-DICHLOROETHANE
ETHYLENE CHLORIDE
DUTCH OIL OR LIQUID

### Observable Characteristics
Clear, colourless liquid. Sweet, chloroform-like odour.

### Manufacturers
Dow Chemical Canada Inc., Sarnia, Ontario, Fort Saskatchewan, Alberta.

### Transportation and Storage Information
**Shipping State:** Liquid.
**Classification:** Flammable liquid.
**Inert Atmosphere:** No requirement.
**Venting:** Pressure-vacuum.
**Pump Type:** Standard types for flammable liquids (centrifugal); steel or stainless steel.

**Label(s):** Red label - FLAMMABLE LIQUID; Class 3.2, Group II.
**Storage Temperature:** Ambient.
**Hose Type:** Flexible stainless steel, Teflon.

**Grades or Purity:** Commercial.
**Containers and Materials:** Drums, tank cars, trucks; steel, stainless steel.

### Physical and Chemical Characteristics
**Physical State** (20°C, 1 atm): Liquid.
**Solubility (Water):** 0.92 g/100 mL (0°C); 0.869 g/100 mL (20°C).
**Molecular Weight:** 99.0
**Vapour Pressure:** 40 mm Hg (10°C); 61 mm Hg (20°C); 105 mm Hg (30°C).
**Boiling Point:** 83.5°C.

**Floatability (Water):** Sinks.
**Odour:** Chloroform-like (6 to 40 ppm, odour threshold).
**Flash Point:** 18°C (o.c.); 13°C (c.c.).
**Vapour Density:** 3.4
**Specific Gravity:** 1.26 (20°C).

**Colour:** Colourless.
**Explosive Limits:** 6.2 to 16%.
**Melting Point:** -35.3 to -40°C.

## HAZARD DATA

### Human Health
**Symptoms:** Inhalation: irritation of eyes, nose and throat, mental confusion, dizziness, nausea and vomiting, diarrhea, coma; damage to liver, lungs and nervous system. Ingestion: irritation of the gastrointestinal tract, nausea and vomiting. Contact: skin - absorbed readily, burning, with symptoms similar to inhalation; eyes - irritation, clouding of cornea.
**Toxicology:** Moderately to highly toxic by inhalation, ingestion and contact.
TLV® - (inhalation) 10 ppm; 40 mg/m³.
Short-term Inhalation Limits - 15 ppm; 60 mg/m³ (15 min).
$LC_{50}$ - No information.
$LC_{Lo}$ - Inhalation: human - 4 000 ppm/1 h
Delayed Toxicity - Suspected carcinogen and damage to the liver.
$LD_{50}$ - Oral: rat = 0.68 g/kg
$LC_{Lo}$ - Oral: human = 0.81 g/kg

### Fire
**Fire Extinguishing Agents:** Use dry chemical, foam or carbon dioxide. Water may be ineffective, but may be used to cool fire-exposed containers and knock down vapours.
**Behaviour in Fire:** When heated to decomposition (or in fires) emits highly toxic fumes of phosgene, vinyl chloride and hydrochloric acid. Flashback may occur along vapour trail.
**Ignition Temperature:** 413°C.
**Burning Rate:** 1.6 mm/min.

### Reactivity
**With Water:** No reaction.
**With Common Materials:** Can react vigorously with oxidizing materials. Reacts violently with aluminum, ammonia and dimethylaminopropylamine.
**Stability:** Stable.

### Environment
**Water:** Prevent entry into water intakes and waterways; Fish toxicity: 225 mg/L/96 h/rainbow trout/$LC_{50}$/freshwater; 150 ppm/tns/pin perch/TLm/saltwater; Aquatic toxicity rating - 100-1 000 ppm; BOD: 0.002%, 5 days.
**Land-Air:** No information.
**Food Chain Concentration Potential:** None.

## EMERGENCY MEASURES

**Special Hazards**
FLAMMABLE.

**Immediate Responses**
Keep non-involved people away from spill site. Issue warning: "FLAMMABLE". Call Fire Department. Eliminate all sources of ignition. Avoid contact or inhalation. Contact manufacturer for guidance. Stop or reduce discharge, if this can be done without risk. Contain spill by diking. Notify environmental authorities.

**Protective Clothing and Equipment**
Respiratory protection - self-contained breathing apparatus and totally encapsulated suit.

**Fire and Explosion**
Use dry chemical, foam or carbon dioxide. Water may be used to cool fire-exposed containers and knock down vapours. Flashback may occur along vapour trail. In fires, emits highly toxic fumes of phosgene, vinyl chloride and hydrochloric acid.

**First Aid**
Move victim out of spill area to fresh air. Call for medical assistance, but start first aid at once. <u>Inhalation:</u> if breathing has stopped, give artificial respiration; if laboured, give oxygen. <u>Contact:</u> eyes - irrigate with plenty of water for at least 15 minutes; skin - remove contaminated clothing and flush affected areas with plenty of water. <u>Ingestion:</u> give water or milk to conscious victim to drink. Do not induce vomiting. Keep victim warm and quiet. If medical assistance is not immediately available, transport victim to hospital, doctor or clinic.

## ENVIRONMENTAL PROTECTION MEASURES

**Response**

**Water**
1. Stop or reduce discharge if safe to do so.
2. Contact manufacturer or supplier for advice.
3. If possible, contain discharge by damming or water diversion.
4. Dredge or vacuum pump to remove contaminants, liquids and contaminated bottom sediments.
5. Notify environmental authorities to discuss disposal and cleanup of contaminated materials.

**Land-Air**
1. Stop or reduce discharge if safe to do so.
2. Contact manufacturer or supplier for advice.
3. Contain spill by diking with earth or other barrier.
4. Remove material with vacuum pumps or vacuum equipment and place in appropriate containers.
5. Recover undamaged containers.
6. Adsorb residual liquid on natural or synthetic sorbents.
7. Notify environmental authorities to discuss disposal and cleanup of contaminated materials.

**Disposal**
1. Contact manufacturer or supplier for advice on disposal.
2. Contact environmental authorities for advice on disposal.

ETHYLENE DICHLORIDE    $CH_2Cl \cdot CH_2Cl$

# ETHYLENE GLYCOL $CH_2OHCH_2OH$

## IDENTIFICATION

### Common Synonyms
ANTIFREEZE
GLYCOL
MONOETHYLENE GLYCOL
1,2-ETHANEDIOL
ETHYLENE DIHYDRATE

### Observable Characteristics
Colourless liquid. Slight odour.

### Manufacturers
Dow Chemical Canada Inc., Fort Saskatchewan, Alberta, Sarnia, Ontario.
Union Carbide, Montreal, Quebec.

### Transportation and Storage Information
**Shipping State:** Liquid.
**Classification:** None.
**Inert Atmosphere:** No requirement.
**Venting:** Open (flame arrester).
**Pump Type:** Most types.
**Label(s):** Not regulated.
**Storage Temperature:** Ambient.
**Hose Type:** Most types.
**Grades or Purity:** Industrial.
**Containers and Materials:** Drums, tank cars, tank trucks; steel, stainless steel.

### Physical and Chemical Characteristics
**Physical State** (20°C, 1 atm): Liquid.
**Solubility (Water):** Soluble in all proportions.
**Molecular Weight:** 62.1
**Vapour Pressure:** 0.05 mm Hg (20°C); 0.2 mm Hg (30°C).
**Boiling Point:** 196 to 198°C.
**Floatability (Water):** Sinks and mixes.
**Odour:** Slight (0.08 to 25 ppm, odour threshold).
**Flash Point:** 116°C (o.c.); 111°C (c.c.).
**Vapour Density:** 2.1
**Specific Gravity:** 1.11 (20°C).
**Colour:** Colourless.
**Explosive Limits:** 3.2 to 15.3%.
**Melting Point:** -13°C.

## HAZARD DATA

### Human Health
**Symptoms:** Inhalation: intoxication, headache; prolonged inhalation may cause throat irritation and nervous system disorder. Ingestion: intoxication, headache, vomiting, cyanosis, unconsciousness with convulsions. Contact: skin - absorbed causing intoxication; eyes - irritation.
**Toxicology:** Moderately toxic by ingestion, contact and inhalation.
TLV® - 10 mg/m³ (particulate); 50 ppm; 125 mg/m³ (vapour).
Short-term Inhalation Limits - 20 mg/m³ (15 min) (particulate).
$LC_{50}$ - No information.
$LD_{50}$ - Oral: rat = 5.84 g/kg
Delayed Toxicity - Fatal kidney injury may result from ingestion.

### Fire
**Fire Extinguishing Agents:** Use water fog, alcohol foam, carbon dioxide, or dry chemical. Water or foam may cause frothing.
**Behaviour in Fire:** No information.
**Ignition Temperature:** 398°C.
**Burning Rate:** 1.0 mm/min.

### Reactivity
**With Water:** No reaction, soluble.
**With Common Materials:** Can react violently with chlorosulfonic acid, oleum and sulfuric acid. May react with strong oxidizing materials.
**Stability:** Stable.

### Environment
**Water:** Prevent entry into water intakes and waterways. Fish toxicity: >100 ppm/48 h/shrimp/$LC_{50}$/saltwater; Aquatic toxicity rating = 100 to 1 000 h/TLm/freshwater; >5 000 mg/L/24 h/goldfish/$LD_{50}$/freshwater; 41 000 mg/L/96 h/rainbow trout/$LC_{50}$/freshwater; BOD: 16 to 68%, 5 days.
**Land-Air:** No information.
**Food Chain Concentration Potential:** None.

## EMERGENCY MEASURES

**Special Hazards**

**Immediate Responses**

Keep non-involved people away from spill site. Dike to prevent runoff. Notify manufacturer for advice. Notify environmental authorities.

**Protective Clothing and Equipment**

Protective outer clothing as required.

**Fire and Explosion**

Use water fog, alcohol foam, carbon dioxide or dry chemical to extinguish. Water or foam may cause frothing.

**First Aid**

Move victim out of from spill site to fresh air. Call for medical assistance, but start first aid at once. <u>Inhalation</u>: if breathing has stopped give artificial respiration; if laboured, give oxygen. <u>Contact</u>: skin – remove contaminated clothing and wash affected areas with plenty of water; eyes – irrigate with water. <u>Ingestion</u>: give water to conscious victim to drink and induce vomiting. If medical assistance is not immediately available, transport victim to doctor, clinic or hospital.

## ENVIRONMENTAL PROTECTION MEASURES

**Response**

**Water**
1. Stop or reduce discharge if safe to do so.
2. Contact manufacturer or supplier for advice.
3. Notify environmental authorities to discuss disposal and cleanup of contaminated materials.

**Land-Air**
1. Stop or reduce discharge if safe to do so.
2. Contact manufacturer or supplier for advice.
3. Contain spill by diking with earth or other barrier.
4. Remove material with pumps or vacuum equipment and place in appropriate containers.
5. Recover undamaged containers.
6. Adsorb residual liquid on natural or synthetic sorbents.
7. Notify environmental authorities to discuss disposal and cleanup of contaminated materials.

**Disposal**
1. Contact manufacturer or supplier for advice on disposal.
2. Contact environmental authorities for advice on disposal.
3. Incinerate (approval of environmental authorities required).

ETHYLENE GLYCOL    $CH_2OHCH_2OH$

# ETHYLENEIMINE  $CH_2CH_2NH$

## IDENTIFICATION

**UN No. 1185**

### Common Synonyms
ETHYLENEIMINE (inhibited)
AZIRIDINE
DIMETHYLENIMINE
AZACYCLOPROPANE
DIHYDROAZIRINE

### Observable Characteristics
Colourless, oily liquid. Ammoniacal odour.

### Manufacturers
No Canadian or U.S. Manufacturers: Imported from Nippon Shokubai, Japan

### Transportation and Storage Information
**Shipping State:** Liquid.
**Classification:** Flammable liquid, poison.
**Inert Atmosphere:** Inerted.
**Venting:** Safety relief.
**Pump Type:** Centrifugal or positive displacement; stainless steel.

**Label(s):** Red label - FLAMMABLE LIQUID; Class 3.2, Group I. White label - POISON; Class 6.1, Group I.
**Storage Temperature:** Ambient.
**Hose Type:** Flexible, stainless steel.

**Grades or Purity:** 99.0%.
**Containers and Materials:** Drums (plastic-lined), tank cars (with inhibitor; NaOH).

### Physical and Chemical Characteristics
**Physical State (20°C, 1 atm):** Liquid.
**Solubility (Water):** Completely soluble.
**Molecular Weight:** 43.1
**Vapour Pressure:** 160 mm Hg (20°C); 250 mm Hg (30°C).
**Boiling Point:** 55 to 57°C.

**Floatability (Water):** Floats and mixes.
**Odour:** Ammoniacal (1.96 to 2 ppm, odour threshold).
**Flash Point:** -4°C (o.c.); -11°C (c.c.).
**Vapour Density:** 1.5
**Specific Gravity:** 0.83 (20°C).

**Colour:** Colourless.
**Explosive Limits:** 3.6 to 46%.
**Melting Point:** -71°C.

## HAZARD DATA

### Human Health

**Symptoms:** Inhalation: difficulty breathing, headache, dizziness, coughing, burning of respiratory tract and mucous membranes. Contact: eyes - burns; skin - severe vesicant-desicant can produce third degree burns, readily absorbed, producing symptoms similar to inhalation. Ingestion: nausea, vomiting and burning of mucous membranes.
**Toxicology:** Highly toxic by ingestion, inhalation or contact.
TLV® - (skin) 0.5 ppm; 1.0 mg/m³.   $LC_{50}$ - No information.   $LD_{50}$ - Oral: rat = 0.015 g/kg
Short-term Inhalation Limits - No information.   $LC_{Lo}$ - Inhalation: rat = 25 ppm/8 h
Delayed Toxicity - Suspected carcinogen.

### Fire

**Fire Extinguishing Agents:** Use alcohol foam, dry chemical or carbon dioxide. Water may be ineffective, but may be used to cool fire-exposed containers and knock down vapours.
**Behaviour in Fire:** Containers may rupture when exposed to heat and flame. At high temperatures (in fires), may polymerize with the evolution of heat. Flash-back may occur along vapour trail. Toxic nitric oxide fumes produced by burning.
**Ignition Temperature:** 320°C.   **Burning Rate:** No information.

### Reactivity

**With Water:** Mild reaction; soluble.
**With Common Materials:** In the presence of acidic type materials catalytically active metals or chloride ions, a violent exothermic reaction can occur. Reacts violently with acetic acid, acetic anhydride, acrolein, acrylic acid, allyl chloride, chlorine, carbon disulfide, chlorosulfonic acid, epichlorohydrin, hydrochloric acid, hydrofluoric acid, nitric acid, oleum, silver, sodium hypochlorite, sulfuric acid and vinyl acetate.
**Stability:** Unstable. May polymerize violently.

### Environment

**Water:** Prevent entry into water intakes and waterways. Fish toxicity: No information.
**Land-Air:** No information.
**Food Chain Concentration Potential:** None.

## EMERGENCY MEASURES

### Special Hazards
FLAMMABLE. POISON. UNSTABLE. CORROSIVE.

### Immediate Responses
Keep non-involved people away from spill site. Issue warnings: "FLAMMABLE; POISON". Call supplier or manufacturer. Avoid contact and inhalation. If material is burning, work from upwind. If water spray is used, dike area to contain toxic runoff. Stop or reduce discharge, if this can be done without risk. Notify environmental authorities.

### Protective Clothing and Equipment
Respiratory protection - self-contained breathing apparatus and totally encapsulated suit. Boots - rubber (pants worn outside boots). Gloves - rubber.

### Fire and Explosion
Use dry chemical, alcohol foam or carbon dioxide to extinguish. Water may be ineffective on fire, but may be used to cool fire-exposed containers. Fight fires from behind barrier. Flashback may occur along vapour trail. Toxic nitric oxide fumes produced by burning.

### First Aid
Move victim out of spill area to fresh air. Call for medical assistance, but start first aid at once. Inhalation: if breathing has stopped, give artificial respiration (not mouth-to-mouth method); if laboured, give oxygen. Contact: eyes - immediate and continuous irrigation with flowing water; skin - remove contaminated clothing and flush affected areas with plenty of water. Ingestion: do not induce vomiting; give large quantities of milk or water to conscious victim to drink. If medical assistance is not immediately available, transport victim to hospital, doctor or clinic.

## ENVIRONMENTAL PROTECTION MEASURES

### Response

**Water**
1. Stop or reduce discharge if safe to do so.
2. Contact manufacturer or supplier for advice.
3. Notify environmental authorities to discuss disposal and cleanup of contaminated materials.

**Land-Air**
1. Stop or reduce discharge if safe to do so. Caution: highly toxic; use non-sparking tools.
2. Contact manufacturer or supplier for advice.
3. Contain spill by diking with earth or other barrier.
4. Remove material with pumps or vacuum equipment and place in appropriate containers.
5. Adsorb residual liquid on natural or synthetic sorbents.
6. Remove contaminated soil for disposal.
7. Notify environmental authorities to discuss disposal and cleanup of contaminated materials.

### Disposal
1. Contact manufacturer or supplier for advice on disposal.
2. Contact environmental authorities for advice on disposal.

ETHYLENEIMINE   $CH_2CH_2NH$

# ETHYLENE OXIDE  $CH_2CH_2O$

## IDENTIFICATION

UN No. 1040

### Common Synonyms
OXIRANE
1,2-EPOXYETHANE
OXANE
DIMETHYLENE OXIDE

### Observable Characteristics
Colourless gas or liquid at room temperature. Ethereal odour.

### Manufacturers
Union Carbide Canada Ltd., Montreal, Quebec (sold in cylinders by Matheson). Dow Chemical Canada Inc., Sarnia, Ontario, Fort Saskatchewan, Alta.

### Transportation and Storage Information
**Shipping State:** Liquid (compressed gas).
**Classification:** Poisonous gas and flammable gas.
**Inert Atmosphere:** Inerted, with nitrogen.
**Venting:** Safety relief.
**Pump Type:** Centrifugal (not gear), explosion-proof.

**Label(s):** Red label - FLAMMABLE GAS; Class 2.1. White label - POISON; Class 2.3.
**Storage Temperature:** Ambient or refrigerated.
**Hose Type:** Carbon steel, flexible stainless steel, no natural rubber.

**Grades or Purity:** Technical, pure (99.7%).
**Containers and Materials:** Cylinders, tank cars; steel.

### Physical and Chemical Characteristics
**Physical State** (20°C, 1 atm): Gas.
**Solubility** (Water): Completely miscible.
**Molecular Weight:** 44.1
**Vapour Pressure:** 1 095 mm Hg (20°C); 1 596 mm Hg (30°C).
**Boiling Point:** 10.6°C.

**Floatability** (Water): Floats and mixes.
**Odour:** Ethereal (0.8 to 500 ppm, odour threshold).
**Flash Point:** <-18°C (c.c.); -29°C (o.c.).
**Vapour Density:** 1.5
**Specific Gravity:** 0.89 (7°C).

**Colour:** Colourless.
**Explosive Limits:** 3.6 to 100.0% (under certain conditions as low as 29%).
**Melting Point:** -111°C.

## HAZARD DATA

### Human Health
**Symptoms:** Inhalation: nose and throat irritation; nausea, vomiting, abdominal pain, difficulty breathing, coughing, dizziness. Contact: skin - burns, blistering, frostbite; eyes - burns. Ingestion: sore throat, nausea, vomiting, diarrhea, convulsions.
**Toxicology:** Highly toxic by ingestion and moderately toxic by inhalation.
TLV® - (inhalation) 1 ppm; 20 mg/m³.   $LC_{50}$ - Inhalation: rat = 1 462 ppm/4 h   $LD_{50}$ - Oral: rat = 0.072 g/kg
Short-term Inhalation Limits - No information.   $TC_{Lo}$ - Inhalation: human 12,500 ppm/10 sec.
Delayed Toxicity - Suspected carcinogen.

### Fire
**Fire Extinguishing Agents:** Use alcohol foam, carbon dioxide or dry chemical. Water may be ineffective because only dilution with 23 volumes of water renders ethylene oxide nonflammable. Water may be used to cool fire-exposed containers.
**Behaviour in Fire:** Flashback may occur along vapour trail. Can burn inside enclosures with no oxygen.
**Ignition Temperature:** 429°C.   **Burning Rate:** 3.5 mm/min.

### Reactivity
**With Water:** Reacts exothermically unless large dilution achieved.
**With Common Materials:** May polymerize violently with evolution of heat on contact with anhydrous chlorides of iron, tin and aluminum, oxides of iron and aluminum and alkali metal hydroxides. Can react with acids, bases, alcohols, ammonia, copper, magnesium perchlorate and potassium.
**Stability:** Stable, within the limits of the foregoing.

### Environment
**Water:** Prevent entry into water intakes and waterways. Aquatic toxicity rating = 10 to 100 ppm/96 h/TLm/freshwater; 90 mg/L/24 h/goldfish/$LD_{50}$/freshwater; BOD: 6%, 5 days.
**Land-Air:** No information.
**Food Chain Concentration Potential:** None.

## EMERGENCY MEASURES

### Special Hazards
FLAMMABLE; POISON. May polymerize violently.

### Immediate Responses
Keep non-involved people away from spill site. Issue warnings: "FLAMMABLE; POISON". Call Fire Department. Evacuate from downwind. Eliminate all sources of ignition. Avoid contact and inhalation. Work from upwind and use water spray to control vapour. Stop or reduce discharge, if this can be done without risk. If material is leaking from container, it will inevitably catch fire. Due to the material's ability to burn within vessel with no oxygen, depressurization should be attempted. The best method is to displace ethylene oxide through the leak with water. Contact manufacturer or supplier for assistance. Dike to prevent runoff from rainwater or water application. Notify environmental authorities.

### Protective Clothing and Equipment
Respiratory protection - self-contained breathing apparatus and full protective clothing. Gloves - rubber or plastic. Acid suit - jacket and pants, rubber or plastic. Boots - high, rubber (pants worn outside boots). Leather should not be used.

### Fire and Explosion
Use dry chemical, alcohol, foam or carbon dioxide to extinguish. Water may be ineffective, but may be used to cool fire-exposed containers and to protect men effecting shutoff. Flashback may occur along vapour trail and into vessels if pressure reduced.

### First Aid
Move victim out of spill area to fresh air. Call for medical assistance, but start first aid at once. Inhalation: give artificial respiration if breathing has stopped; give oxygen if breathing is laboured. Contact: remove contaminated clothing and immediately flush eyes and skin with plenty of warm water. Ingestion: give milk or water to conscious victim and induce vomiting. Keep victim warm and quiet. If medical assistance is not immediately available, transport victim to hospital, clinic or doctor.

## ENVIRONMENTAL PROTECTION MEASURES

### Response

**Water**
1. Contact manufacturer or supplier for advice.
2. Notify environmental authorities to discuss disposal and cleanup of contaminated materials.

**Land-Air**
1. Contact manufacturer or supplier for advice.
2. Dike to prevent runoff from rainwater or water application.
3. If liquid, remove material with nitrogen displacement and place in appropriate containers.
4. Recover undamaged containers.
5. Notify environmental authorities to discuss disposal and cleanup of contaminated materials.

### Disposal
1. Contact manufacturer or supplier for advice on disposal.
2. Contact environmental authorities for advice on disposal.

ETHYLENE OXIDE   $CH_2CH_2O$

# 2-ETHYLHEXANOL    $CH_3(CH_2)_3CH(C_2H_5)CH_2OH$

## IDENTIFICATION

UN No. 2282 - hexanols

### Common Synonyms
2-ETHYLHEXYL ALCOHOL
OCTYL ALCOHOL
2-ETHYL-1-HEXANOL

### Observable Characteristics
Colourless liquid. Unpleasant, musty odour.

### Manufacturers
BASF Canada Limited, Montreal, Quebec.

### Transportation and Storage Information
**Shipping State:** Liquid.
**Classification:** Not regulated.
**Inert Atmosphere:** No requirement.
**Venting:** Open (flame arrester).
**Pump Type:** Standard, flammable liquid types.
**Label(s):** None required.
**Storage Temperature:** Ambient.
**Hose Type:** Natural rubber, polyethylene, butyl.
**Grades or Purity:** Technical, 99 to 99.7%.
**Containers and Materials:** Cans, drums, tank cars, tank trucks; steel.

### Physical and Chemical Characteristics
**Physical State** (20°C, 1 atm): Liquid.
**Solubility (Water):** 0.1 g/100 mL (20°C).
**Molecular Weight:** 130.2
**Vapour Pressure:** 0.36 mm Hg (20°C); 1.0 mm Hg (33°C).
**Boiling Point:** 179 to 185°C.
**Floatability (Water):** Floats.
**Odour:** Unpleasant, musty (0.08 to 0.14 ppm, odour threshold).
**Flash Point:** 85°C (o.c.); 73°C (c.c.).
**Vapour Density:** 4.5
**Specific Gravity:** 0.83 (20°C).
**Colour:** Colourless.
**Explosive Limits:** 0.88 to 9.7%.
**Melting Point:** -76°C.

## HAZARD DATA

### Human Health
**Symptoms:** <u>Inhalation and Ingestion</u>: nausea, vomiting, headache, dizziness; <u>Contact</u>: skin and eyes - mildly irritating.
**Toxicology:** Moderately toxic by ingestion and skin contact.
**TLV®** - No information.    **LC$_{50}$** - No information.    **LD$_{50}$** - Oral: rat = 3.2 g/kg
**Short-term Inhalation Limits** - No information.    **Delayed Toxicity** - No information.

### Fire
**Fire Extinguishing Agents:** Use foam, carbon dioxide or dry chemical. Water may be ineffective.
**Behaviour in Fire:** No information.
**Ignition Temperature:** 231°C.    **Burning Rate:** 4 mm/min.

### Reactivity
**With Water:** No reaction.
**With Common Materials:** Can react with oxidizing materials.
**Stability:** Stable.

### Environment
**Water:** Prevent entry into water intakes and waterways. Fish toxicity: 19 ppm/24 h/brine shrimp/TLm/saltwater; BOD: 88%, 5 days (theoretical).
**Land-Air:** No information.
**Food Chain Concentration Potential** None.

## EMERGENCY MEASURES

### Special Hazards
COMBUSTIBLE.

### Immediate Responses
Keep non-involved people away from spill site. Call Fire Department. Avoid contact. Stop discharge if safe to do so. Notify manufacturer. Notify environmental authorities.

### Protective Clothing and Equipment
In fires, Respiratory protection – use self-contained breathing apparatus. Goggles – (mono), tight-fitting. Face shield may be worn, but must not replace goggles. Gloves – rubber or plastic. Coveralls or acid suit (jacket and pants).

### Fire and Explosion
Use dry chemical, alcohol foam or carbon dioxide to extinguish fire. Water may be ineffective on fire, but may be used to cool fire-exposed containers.

### First Aid
Move victim out of spill area to fresh air. Call for medical assistance, but start first aid at once. Contact: remove contaminated clothing and wash eyes and affected skin thoroughly with warm water. Ingestion: give milk or water to conscious victim to drink. Keep victim warm and quiet. If medical assistance is not immediately available, transport victim to hospital, clinic or doctor.

## ENVIRONMENTAL PROTECTION MEASURES

### Response

**Water**
1. Stop or reduce discharge if safe to do so.
2. Contact manufacturer or supplier for advice.
3. If possible, contain discharge by booming.
4. If floating, skim and remove.
5. Notify environmental authorities to discuss disposal and cleanup of contaminated materials.

**Land-Air**
1. Stop or reduce discharge if safe to do so.
2. Contact manufacturer or supplier for advice.
3. Contain spill by diking with earth or other barrier.
4. Remove material with pumps or vacuum equipment and place in appropriate containers.
5. Recover undamaged containers.
6. Absorb residual liquid on natural or synthetic sorbents.
7. Notify environmental authorities to discuss disposal and cleanup of contaminated materials.

### Disposal
1. Contact manufacturer or supplier for advice on disposal.
2. Contact environmental authorities for advice on disposal.
3. Incinerate (approval of environmental authorities required).

**2-ETHYLHEXANOL** $CH_3(CH_2)_3CH(C_2H_5)CH_2OH$

# FENITROTHION  $C_9H_{12}NO_5PS$

## IDENTIFICATION

UN No. 2783
Danger Group According to Percentage of Active
Substance: Group III, Solid 45 to 100%
Liquid 10 to 100%

### Common Synonyms
O,O-DIMETHYL 0-4-NITRO-M-TOLYL PHOSPHOROTHIOATE

### Common Trade Names
FOLITHION, SUMITHION, NOVATHION

(An insecticide used largely in Canada for spruce budworm control).

### Observable Characteristics
Yellowish to yellowish-brown liquid or solid. Slight odour.

### Manufacturers
Sumitomo Shoji Canada Ltd., Toronto, Ontario.
Chemagro Ltd., Mississauga, Ontario.

### Transportation and Storage Information
**Shipping State:** Solid or liquid (formulation).
**Classification:** None.
**Inert Atmosphere:** No requirement.
**Venting:** Open.
**Pump Type:** No information.

**Label(s):** Not regulated.
**Storage Temperature:** Ambient.
**Hose Type:** No information.

**Grades or Purity:** Various as described below.
**Containers and Materials:** Cans, drums; aluminum stainless steel, lined steel.

### Physical and Chemical Characteristics
**Physical State** (20°C, 1 atm): Liquid (technical).
**Solubility** (Water): Insoluble (technical) EC and WP are dispersible.
**Molecular Weight:** 277.2
**Vapor Pressure:** 0.000006 mm Hg (20°C).
**Boiling Point:** 140-145°C (0.1 mm Hg) decomposes.

**Floatability** (Water): Sinks, EC and WP dispersible in water.
**Odour:** Slight.
**Flash Point:** No information.
**Vapour Density:** No information.
**Specific Gravity:** 1.32-1.34 (technical).

**Colour:** Yellow to yellowish-brown.
**Explosive Limits:** SN may form explosive mixtures in air.
**Melting Point:** 0.3°C (technical).

## HAZARD DATA

### Human Health
**Symptoms:** Inhalation, Ingestion or Contact (skin): readily absorbed through skin; nausea, salivation, tearing, abdominal cramps, vomiting, sweating, slow pulse, muscular tremors.
**Toxicology:** Highly toxic by ingestion and skin contact.
**TLV®** - No information.
**Short-term Inhalation Limits** - No information.

**$LC_{50}$** - No information.
**Delayed Toxicity** - No information.

**$LD_{50}$** - Oral: rat = 0.25 g/kg
**$LD_{50}$** - Oral: cat = 0.14 g/kg

### Fire
**Fire Extinguishing Agents:** Foam, carbon dioxide or dry chemical.
**Behaviour in Fire:** Releases toxic fumes.
**Ignition Temperature:** No information.

**Burning Rate:** No information.

### Reactivity
**With Water:** No reaction.
**With Common Materials:** No information.
**Stability:** Stable.

### Environment
**Water:** Prevent entry into water intakes or waterways.
**Land-Air:** $LD_{50}$ - Oral: Wild bird = 0.025 g/kg; chicken = 0.28 g/kg; duck = 1.19 g/kg.
**Food Chain Concentration Potential:** Decomposes in tissue to other compounds (Desmethylsumithion and dimethylphosphorothioic acid).

## EMERGENCY MEASURES

### Special Hazards
POISON.

### Immediate Responses
Keep non-involved people away from spill site. Stop or reduce discharge if safe to do so. Notify manufacturer or supplier. Dike to contain material or water runoff. Notify environmental authorities.

### Protective Clothing and Equipment
In fires or confined spaces - Respiratory Protection - self-contained breathing apparatus and totally encapsulated suit. Otherwise, approved pesticide respirator and impervious outer clothing.

### Fire and Explosion
Use carbon dioxide, foam or dry chemical to extinguish. Releases toxic fumes in fires.

### First Aid
Move victim out of spill area to fresh air. Call for medical assistance, but start first aid at once. Inhalation: if breathing has stopped, give artificial respiration (not mouth-to-mouth method); if laboured, give oxygen. Contact: skin - remove contaminated clothing and flush affected areas with plenty of water; eyes - irrigate with plenty of water. Ingestion: give water to conscious victim to drink and induce vomiting; in the case of petroleum distillates, do not induce vomiting for fear of aspiration and chemical pneumonia. If medical assistance is not immediately available, transport victim to hospital, doctor, or clinic.

## ENVIRONMENTAL PROTECTION MEASURES

### Response

**Water**
1. Stop or reduce discharge if safe to do so.
2. Contact manufacturer or supplier for advice.

**Floats**
3. If possible contain discharge by booming.
4. If floating, skim and remove.

**Sinks or mixes**
3. If possible contain discharge by damming or water diversions.
4. Dredge or vacuum pump to remove contaminants, liquids and contaminated bottom sediments.

5. Notify environmental authorities to discuss disposal and cleanup of contaminated materials.

**Land-Air**
1. Stop or reduce discharge if safe to do so.
2. Contact manufacturer or supplier for advice.
3. Contain spill by diking with earth or other barrier.
4. If liquid, remove material with pumps or vacuum equipment and place in appropriate containers.
5. If solid, remove material by manual or mechanical means.
6. Recover undamaged containers.
7. Absorb residual liquid on natural or synthetic sorbents.
8. Remove contaminated soil for disposal.
9. Notify environmental authorities to discuss cleanup and disposal of contaminated materials.

### Disposal
1. Contact manufacturer or supplier for advice on disposal.
2. Contact environmental authorities for advice on disposal.

### Available Formulations
**Technical Grade:** Purity: 95%+.
Properties: insoluble liquid, sinks in water.

**Formulations:**

Type:
- EC - emulsifiable concentrate
- LI - liquid
- SN - solution
- WP - wettable powder

Purity:
- 80%
- 95+%
- typically 96% in organic solvents.
- 40%

Properties:
- dispersible in water
- insoluble in water.
- combustible.
- low combustibility.

FENITROTHION  $C_9H_{12}NO_5PS$

# FERRIC CHLORIDE  FeCl$_3$

## IDENTIFICATION

UN No. Solid 1173
Solution 2583

### Common Synonyms
IRON CHLORIDE
FERRIC TRICHLORIDE
FERRIC PERCHLORIDE
IRON TRICHLORIDE
IRON PERCHLORIDE
IRON (III) CHLORIDE

### Observable Characteristics
Brown to black solid or dark reddish-brown liquid (solution). Odourless or slight acidic odour.

### Manufacturers
No Canadian Manufacturers
Canadian Supplier:
Diversey Canada Limited, Mississauga, Ontario.

### Transportation and Storage Information
**Shipping State:** Liquid (39% solution); solid (96%).
**Classification:** Corrosive.
**Inert Atmosphere:** No requirement.
**Venting:** Open.
**Pump Type:** Gear, centrifugal; rubber-lined, plastic-lined.

**Label(s):** Black and white label - CORROSIVE SOLID; Class 8, Group II; SOLUTION; Class 5, Group III.
**Storage Temperature:** Ambient.
**Hose Type:** Natural rubber, polyethylene, polypropylene, PVC.

**Grades or Purity:** Solution (42° Bé; 39%); solid, anhydrous, 96%.
**Containers and Materials:** Solution - drums, tank cars, tank trucks. Anhydrous (solid) - drums.

### Physical and Chemical Characteristics
**Physical State** (20°C, 1 atm): Solid.
**Solubility** (Water): 74.4 g/100 mL (0°C); 535.7 g/100 mL (100°C).
**Molecular Weight:** 162.2
**Vapour Pressure:** No information.
**Boiling Point:** 315°C (decomposes).

**Floatability** (Water): Sinks and mixes.
**Odour:** Odourless or slight acidic odour.
**Flash Point:** Not flammable.
**Vapour Density:** No information.
**Specific Gravity:** 2.9 (anhydrous) (20°C); 1.4 (42° Bé solution).

**Colour:** Anhydrous - brown to black; Solution - dark reddish-brown.
**Explosive Limits:** Not flammable.
**Melting Point:** Approximately 300°C.

## HAZARD DATA

### Human Health
**Symptoms:** <u>Contact</u>: eyes - irritation; skin - irritation, dermatitis, staining and burning. <u>Ingestion</u>: may cause burns of the mucous membranes and severe irritation of the gastrointestinal tract. <u>Inhalation</u>: nose and throat irritation.
**Toxicology:** Moderately to highly toxic by ingestion. Highly toxic by contact.
TLV® 1 mg/m$^3$ (as Fe)  LC$_{50}$ - No information.  LD$_{50}$ - Oral: mouse = 1.28 g/kg
Short-term Inhalation Limits - 2 mg/m$^3$ (as Fe)  Delayed Toxicity - None known.
(15 min).

### Fire
**Fire Extinguishing Agents:** Not combustible. Most fire extinguishing agents, except water, may be used in fires involving ferric chloride.
**Behaviour in Fire:** When heated to decomposition in fire, it may release fumes of hydrogen chloride.
**Ignition Temperature:** Not combustible.  **Burning Rate:** Not combustible.

### Reactivity
**With Water:** Reacts exothermically producing corrosive HCl fumes.
**With Common Materials:** Ferric chloride is a strong oxidizing agent and solutions are very corrosive to iron, copper and most metals. Violent reaction with allyl chloride and sodium and potassium metals.
**Stability:** Stable.

### Environment
**Water:** Prevent entry into water intakes and waterways. Fish toxicity: 4 ppm (as Fe)/24 to 96 h/striped bass larvae/LC$_{50}$/freshwater; 15 ppm/96 h/<u>Daphnia magna</u> (water flea)/TLm; BOD: None.
**Land-Air:** No information.
**Food Chain Concentration Potential:** None.

## EMERGENCY MEASURES

### Special Hazards
CORROSIVE.

### Immediate Responses
Keep non-involved people away from spill site. Issue warning: "CORROSIVE". Avoid contact and inhalation. Stop or reduce discharge, if this can be done without risk. Contain spill by diking with earth or other available material. Contact manufacturer for advice. Notify environmental authorities.

### Protective Clothing and Equipment
<u>Goggles</u> - (mono), tight-fitting. If face shield is used, it must not replace goggles. <u>Gloves</u> - rubber. <u>Boots</u> - high, rubber (pants worn outside boots). Ferric chloride attacks leather boots and shoes rapidly. <u>Outerwear</u> - coveralls, aprons; rubber or vinyl. Respiratory protection required only in the case of fire, where hydrogen chloride may be present.

### Fire and Explosion
Not combustible. Most fire extinguishing agents, except water, may be used. When heated to decomposition in fire, fumes of hydrogen chloride may be released.

### First Aid
Move victim out of spill area to fresh air. Call for medical assistance, but start first aid at once. <u>Contact</u>: eyes - irrigate with plenty of water for at least 15 minutes; skin - remove contaminated clothing and wash affected areas with plenty of water. <u>Ingestion</u>: give conscious victim milk or water to drink. Keep warm and quiet. If medical assistance is not immediately available, transport victim to hospital, doctor or clinic.

## ENVIRONMENTAL PROTECTION MEASURES

### Response

**Water**
1. Stop or reduce discharge if safe to do so.
2. Contact manufacturer or supplier for advice.
3. If possible, contain discharge by damming or water diversion.
4. Dredge or vacuum pump to remove contaminants, liquids and contaminated bottom sediments.
5. Notify environmental authorities to discuss disposal and cleanup of contaminated materials.

**Land-Air**
1. Stop or reduce discharge if safe to do so.
2. Contact manufacturer or supplier for advice.
3. Dike to prevent runoff from rainwater or water application.
4. Remove material by manual or mechanical means.
5. Recover undamaged containers.
6. Notify environmental authorities to discuss disposal and cleanup of contaminated materials.

### Disposal
1. Contact manufacturer or supplier for advice on disposal.
2. Contact environmental authorities for advice on disposal.

FERRIC CHLORIDE   $FeCl_3$

# FLUORINE  $F_2$

## IDENTIFICATION  UN No. 1045

| Common Synonyms | Observable Characteristics | Manufacturers |
|---|---|---|
| None. | Pale yellow compressed gas. Sharp, choking, irritating odour. | No Canadian manufacturers. Canadian suppliers: Air Products, Brampton, Ontario. Originating from: Air Products and Chemicals Incorporated, Specialty Gas Department, Allentown, PA, USA |

### Transportation and Storage Information

**Shipping State:** Liquid (compressed gas).
**Classification:** Poison, oxidizer.
**Inert Atmosphere:** No requirement.
**Venting:** Safety relief.
**Pump Type:** No information.

**Label(s):** White label - POISON. Yellow label - OXIDIZER; Class 2.3, 5.1.
**Storage Temperature:** Ambient.
**Hose Type:** No information.

**Grades or Purity:** >98% $F_2$.
**Containers and Materials:** Cylinders, tank trailers; steel.

### Physical and Chemical Characteristics

**Physical State** (20°C, 1 atm): Gas.
**Solubility (Water):** Reacts to form HF.
**Molecular Weight:** 38.0
**Vapour Pressure:** 18 100 mm Hg (0°C); 31 300 mm Hg (20°C).
**Boiling Point:** -188°C.

**Floatability (Water):** Liquefied fluorine sinks; liquid and gas react to form HF.
**Odour:** Sharp, choking, irritating; (0.035 ppm, odour threshold).
**Flash Point:** Not flammable.
**Vapour Density:** 1.3

**Colour:** Pale yellow.
**Explosive Limits:** Not flammable.
**Melting Point:** -219.6°C.
**Specific Gravity:** (liquid) 1.5 (-188°C).

## HAZARD DATA

### Human Health

**Symptoms: Inhalation:** irritation and ulceration of mucous membranes, coughing fits, difficulty breathing, respiratory damage, coma, death. **Ingestion:** painful burning, ulceration of mucous membranes, intense thirst, difficulty breathing, coughing, nausea and vomiting, convulsions, coma, death. **Contact:** skin - burning, blisters, profound tissue damage; eyes - burning, irreparable corneal damage, loss of vision.
**Toxicology:** Highly toxic by inhalation, ingestion and contact.
TLV® (inhalation) 1 ppm; 2 mg/m³.   $LC_{50}$ - Inhalation: rat = 185 ppm/1 h   $LD_{50}$ - No information.
Short-term Inhalation Limits - 2 ppm; 4 mg/m³   $LC_{50}$ - Inhalation: mouse = 150 ppm/1 h
(15 min).   Delayed Toxicity - No information.

### Fire

**Fire Extinguishing Agents:** Once a fire has started with fluorine as the oxidizer, there is no effective way of stopping it other than shutting off the source of fluorine. Clear the area and allow the fire to burn until all of the fluorine is consumed. Do not attempt to extinguish the fire with water or chemicals, since these would act as additional fuel.
**Behaviour in Fire:** Fluorine acts exothermically with most materials.   **Burning Rate:** Not combustible.
**Ignition Temperature:** Not combustible.

### Reactivity

**With Water:** Reacts exothermically to produce HF and oxygen.
**With Common Materials:** Reacts violently with many materials; such as most organic matter, hydrogen-containing molecules, oxides of sulfur, nitrogen, phosphorous halides, alkaline metals and alkaline earths. Reacts violently with metals and their sulfides, phosphides, oxides, hydrides, acetylides and carbides. Reacts violently with other halogens and halogen acids, phosphorus, sulfur, hydrazine, coke, potassium nitrate, carbon tetrachloride, silicates, alkenes, alkylbenzenes, carbon disulfide, aluminum, manufactured gas and many polymers.
**Stability:** Stable only under normal circumstances, reacts with many materials.

### Environment

**Water:** Prevent entry into water intakes and waterways. Harmful to aquatic life in very low concentrations. **Fish toxicity:** 2.3 ppm/tns/trout/TLm/freshwater; **BOD:** No information.
**Land-Air:** No information.   **Food Chain Concentration Potential:** None.

## EMERGENCY MEASURES

**Special Hazards**

POISON. OXIDIZER. CORROSIVE.

**Immediate Responses**

Keep non-involved people away from spill site. Issue warning: "POISON and OXIDIZER". Call Fire Department (note caution re: use of water). Consider evacuation of downwind area. Contact manufacturer or supplier immediately for guidance and assistance. Control of discharge should be attempted only by experienced persons. Work from upwind. Notify environmental authorities.

**Protective Clothing and Equipment**

Respiratory protection - self-contained breathing apparatus and totally encapsulated suit. <u>Gloves</u> - leather. <u>Boots</u> - high rubber.

**Fire and Explosion**

Water should be used only to cool containers exposed to fire. Do NOT attempt to extinguish fire with water or chemicals.

**First Aid**

Move victim out of spill site to fresh air. Call for medical assistance, but start first aid at once. <u>Inhalation:</u> if breathing has stopped or is laboured, give artificial respiration (not mouth-to-mouth method). Supplemental oxygen should be given also. <u>Contact:</u> eyes - irrigate with plenty of water; skin - flush affected areas with plenty of water and remove contaminated clothing immediately. If medical assistance is not immediately available, transport victim to hospital, doctor or clinic.

## ENVIRONMENTAL PROTECTION MEASURES

**Response**

**Water**
1. Stop or reduce discharge if safe to do so.
2. Contact manufacturer or supplier for advice.
3. Notify environmental authorities to discuss disposal and cleanup of contaminated materials.

**Land-Air**
1. Stop or reduce discharge if safe to do so.
2. Contact manufacturer or supplier for advice.
3. Recover undamaged containers.
4. Notify environmental authorities to discuss disposal and cleanup of contaminated materials.

**Disposal**
1. Contact manufacturer or supplier for advice on disposal.
2. Contact environmental authorities for advice on disposal.

FLUORINE $F_2$

# FLUOROCHLOROMETHANES

## IDENTIFICATION

UN No. 1028 dichlorodifluoromethane
1018 chlorodifluoromethane
1022 chlorotrifluoromethane

### Common Synonyms
11; $CCl_3F$ - Trichlorofluoromethane (Freon 11)
12; $Cl_2F_2C$ - Dichlorodifluoromethane (Freon 12; Halon 12)
13; $ClF_3C$ - Chlorotrifluoromethane (Freon 13)
21; $Cl_2FCH$ - Dichlorofluoromethane (Freon 21)
22; $ClF_2CH$ - Chlorodifluoromethane (Freon 22)

### Observable Characteristics
Colourless, practically odourless gases.

### Manufacturers
Du Pont Canada, Maitland, Ont.
Allied Chemical, Amherstburg, Ont.

### Transportation and Storage Information
**Shipping State:** Liquefied (compressed gases).
**Classification:** Non-flammable gases.
**Inert Atmosphere:** No requirement.
**Venting:** Safety-relief.
**Label(s):** Green and white label - NON-FLAMMABLE GAS; Class 2.2.
**Storage Temperature:** Ambient.
**Hose Type:** No information.
**Pump Type:** No information.
**Grades or Purity:** Technical and refrigerant.
**Containers and Materials:** Cylinders; steel.

### Physical and Chemical Characteristics
**Physical State** (20!C, 1 atm): Gas.
**Solubility** (Water): (11) 0.11 g/100 mL (25°C); (12) 0.028 g/100 mL (25°C); (13) 0.009 g/100 mL (25°C); (21) 0.95 g/100 mL (20°C); (22) 0.30 g/100 mL (25°C).
**Molecular Weight:** (11) 137.4; (12) 120.9; (13) 104.5; (21) 102.9; (22) 86.5.
**Vapour Pressure:** (11) 687 mm Hg; (12) 4 250 mm Hg; (13) 2 400 mm Hg; (21) 1 216 mm Hg; (22) 6 800 mm Hg (all at 20°C).
**Boiling Point:** (11) 23.8°C; (12) -29.8°C; (13) -81.4°C; (21) 8.9°C; (22) -40.7°C.
**Floatability** (Water): Sink and boil.
**Odour:** Odourless.
**Flash Point:** Not flammable.
**Vapour Density:** (11) No information; (12) 4.2; (13) 3.6; (21) 3.6; (22) 3.1.
**Specific Gravity:** (11) 1.49 (0°C); (12) 1.33 (20°C) liquid; (13) 1.35 (-40°C) liquid; (21) 1.42 (0°C); (22) 1.4 (-40.7°C) liquid.
**Colour:** Colourless.
**Explosive Limits:** Not flammable.
**Melting Point:** (11) -111.1°C; (12) -158.0°C; (13) -181.0°C; (21) -135.0°C; (22) -157.4°C.

## HAZARD DATA

### Human Health
**Symptoms:** <u>Inhalation</u>: dizziness, loss of coordination, narcosis, asphyxia, nausea, vomiting. <u>Contact</u>: skin - defatting; liquid causes frostbite, irritation and dermatitis.
**Toxicology:** Low toxicity by inhalation and contact.
TLV® - (11) 1 000 ppm; 5 600 mg/m³; (12) 1 000 ppm; 4 950 mg/m³; (21) 10 ppm; 40 mg/m³; (22) 1 000 ppm; 3 500 mg/m³.
$LC_{50}$ - (11) Inhalation: mouse = 10 000 ppm/30 min
$LC_{50}$ - (12) Inhalation: rat = 80 000 ppm/30 min
Delayed Toxicity - No information.
$LD_{50}$ - (11) Intraperitoneal: mouse = 1.74 g/kg
Short-term Inhalation Limits - (12) 1 250 ppm; 6 200 mg/m³; (22) 1 250 ppm; 4 375 mg/m³ 15 min).

### Fire
**Fire Extinguishing Agents:** Not combustible. Most fire extinguishing agents may be used in fires involving fluorochloromethanes.
**Behaviour in Fire:** Toxic by-products are formed on heating to decomposition; highly toxic chlorides and fluorides.
**Ignition Temperature:** Not combustible.
**Burning Rate:** Not combustible.

### Reactivity
**With Water:** No reaction.
**With Common Materials:** (12) (13) react with aluminum.
**Stability:** Stable.

### Environment
**Water:** Prevent entry into water intakes and waterways. Aquatic toxicity rating (12) and (22): >1 000 ppm/96 h/TLm/freshwater; BOD: No information.
**Land-Air:** No information.
**Food Chain Concentration Potential:** None.

## EMERGENCY MEASURES

**Special Hazards**

**Immediate Responses**

Keep non-involved people away from spill site. If involved in fire, evacuate downwind. Notify manufacturer or supplier. Notify environmental authorities.

**Protective Clothing and Equipment**

In fires or confined spaces Respiratory protection – self-contained breathing apparatus and totally encapsulated suit. Otherwise, protective clothing as required.

**Fire and Explosion**

Not combustible. Most fire extinguishing agents may be used on fires involving fluorochloromethanes. Toxic by-products are formed on heating to decomposition, highly toxic chlorides and fluorides.

**First Aid**

Move victim out of spill site to fresh air. Call for medical assistance, but start first aid at once. Inhalation: give artificial respiration if breathing has stopped; give oxygen if breathing is laboured. Contact: skin – remove contaminated clothing, flush affected areas with plenty of water and treat as for frostbite; eyes – irrigate with plenty of water. If medical assistance is not immediately available, transport victim to doctor, clinic or hospital.

## ENVIRONMENTAL PROTECTION MEASURES

**Response**

**Water**
1. Stop or reduce discharge if safe to do so.
2. Contact manufacturer or supplier for advice.
3. Notify environmental authorities to discuss disposal and cleanup of contaminated materials.

**Land-Air**
1. Stop or reduce discharge if safe to do so.
2. Contact manufacturer or supplier for advice.
3. Remove material by manual or mechanical means.
4. Notify environmental authorities to discuss disposal and cleanup of contaminated materials.

**Disposal**
1. Contact manufacturer or supplier for advice on disposal.
2. Contact environmental authorities for advice on disposal.

FLUOROCHLOROMETHANES

# FLUOSILICIC ACID   $H_2SiF_6 \cdot xH_2O$

## IDENTIFICATION

UN No. 1778

### Common Synonyms

HYDROFLUOSILICIC ACID
HEXAFLUOROSILICIC ACID
SILICOFLUORIC ACID
HYDROSILICOFLUORIC ACID
SAND ACID

### Observable Characteristics

Colourless to white liquid. Sharp, acidic odour.

### Manufacturers

Canadian Industries Ltd.,
Courtright, Ontario.
Cominco Ltd., Trail, B.C.
International Minerals and Chemicals,
Corporation (Canada) Ltd.,
Port Maitland, Ontario.

### Transportation and Storage Information

**Shipping State:** Liquid.
**Classification:** Corrosive liquid.
**Inert Atmosphere:** No requirement.
**Venting:** Open.
**Pump Type:** Centrifugal; rubber-lined, without packing; PVC, propylene or hard rubber.

**Label(s):** Black and white label - CORROSIVE LIQUID; Class 8, Group II.
**Storage Temperature:** Ambient.
**Hose Type:** Natural rubber, Hysunite, Hypalon.

**Grades or Purity:** 22 to 30% solutions in water.
**Containers and Materials:** Drums, tank cars, tank trucks; steel, rubber or plastic-lined.

### Physical and Chemical Characteristics

**Physical State** (20°C, 1 atm): Liquid.
**Solubility (Water):** Soluble in all proportions.
**Molecular Weight:** 144 (for solute only).
**Vapour Pressure:** No information.
**Boiling Point:** 100 to 105°C (decomposes).

**Floatability (Water):** Sinks and mixes.
**Odour:** Sharp, acidic.
**Flash Point:** Not flammable.
**Vapour Density:** No information.
**Specific Gravity:** 1.2 to 1.3 (20°C).

**Colour:** Colourless.
**Explosive Limits:** Not flammable.
**Melting Point:** -31 to -20°C.

## HAZARD DATA

### Human Health

**Symptoms: Inhalation:** irritation of mucous membranes, watering of eyes, salivation, coughing fits, ulceration of mucous membranes (nose and throat), difficulty breathing, headache, nausea, coma. **Contact:** skin - burning, blistering, and profound damage to tissues; eyes - watering, burning, irreparable damage. **Ingestion:** irritation and painful burning of lips, mouth and throat, ulceration of mucous membranes, pain in swallowing, intense thirst, painful stomach cramps and distension of stomach, difficulty breathing, coughing, nausea and vomiting, convulsions, coma.
**Toxicology:** Highly toxic by inhalation, ingestion and contact.
TLV® (inhalation) 1 ppm; 2 mg/m³ (as F).   LC$_{50}$ - No information.   LD$_{50}$ - No information.
Short-term Inhalation Limits - 2 ppm; 4 mg/m³   Delayed Toxicity - No information.   LD$_{Lo}$ - Oral: guinea pig = 0.2 g/kg
(15 min) (as F).

### Fire

**Fire Extinguishing Agents:** Not combustible; most fire extinguishing agents may be used on fires involving fluosilicic acid, except water.
**Behaviour in Fire:** If decomposing (boiling), as in fire, corrosive and irritating HF fumes are liberated.
**Ignition Temperature:** Not combustible.   **Burning Rate:** Not combustible.

### Reactivity

**With Water:** Soluble, with evolution of heat and toxic HF fumes.
**With Common Materials:** Vigorously attacks most common metals, including steel, with evolution of hydrogen gas. May react with aluminum, alkalis, cyanides and magnesium.
**Stability:** Decomposes with heat (evolution of HF).

### Environment

**Water:** Prevent entry into water intakes and waterways. Harmful to aquatic life.
**Land-Air:** No information.
**Food Chain Concentration Potential:** None.

## EMERGENCY MEASURES

### Special Hazards
CORROSIVE.

### Immediate Responses
Keep non-involved people away from spill site. Issue warning: "CORROSIVE". Contact supplier or manufacturer for advice. Contain spill by diking to prevent runoff. Avoid contact and inhalation of fumes. Stop or reduce discharge, if this can be done without risk. Notify environmental authorities.

### Protective Clothing and Equipment
Respiratory protection - self-contained breathing apparatus and full protective clothing. Gloves - (gauntlet) rubber or neoprene. Boots - high, rubber (pants worn outside boots). Outerwear - acid-resistant aprons or suits.

### Fire and Explosion
Not combustible; most fire extinguishing agents may be used on fires involving fluosilicic acid, except water.

### First Aid
Move victim out of spill area to fresh air. Call for medical assistance, but start first aid at once. Inhalation: if breathing has stopped, give artificial respiration (not mouth-to-mouth method); if laboured, give oxygen. Contact: eyes - irrigate with plenty of water for at least 15 minutes; skin - flush with plenty of water and remove contaminated clothing. Ingestion: give conscious victim milk or water to drink. Do not induce vomiting. If medical assistance is not immediately available, transport victim to hospital, doctor or clinic.

## ENVIRONMENTAL PROTECTION MEASURES

### Response

**Water**
1. Stop or reduce discharge if safe to do so.
2. Contact manufacturer or supplier for advice.
3. If possible, contain discharge by damming or water diversion.
4. Dredge or vacuum pump to remove contaminants, liquids and contaminated bottom sediments.
5. Notify environmental authorities to discuss disposal and cleanup of contaminated materials.

**Land-Air**
1. Stop or reduce discharge if safe to do so.
2. Contact manufacturer or supplier for advice.
3. Contain spill by diking with earth or other barrier.
4. Remove material with pumps or vacuum equipment and place in appropriate containers.
5. Recover undamaged containers.
6. Neutralize soil and contaminated waters with sodium bicarbonate.
7. Notify environmental authorities to discuss disposal and cleanup of contaminated materials.

### Disposal
1. Contact manufacturer or supplier for advice on disposal.
2. Contact environmental authorities for advice on disposal.

FLUOSILICIC ACID   $H_2SiF_6 \cdot xH_2O$

# FORMALDEHYDE   HCHO (aqueous)

## IDENTIFICATION

UN No. 1198
        2209

### Common Synonyms
FORMALITH, FORMALIN
FORMALDEHYDE SOLUTION
FORMIC ALDEHYDE
METHANAL, METHYLENE OXIDE
OXOMETHANE

### Observable Characteristics
Clear, colourless liquid. Pungent, irritating odour.

### Manufacturers
Celanese Canada Ltd., Edmonton, Alberta.
Reichhold Chemicals Ltd., North Bay, Ont.; Thunder Bay, Ontario.
Bakelite Thermosets Ltd., Belleville, Ont.

### Transportation and Storage Information
**Shipping State:** Liquid (aqueous solution).
**Classification:** Not regulated.
**Inert Atmosphere:** No requirement.
**Venting:** Pressure-vacuum.
**Pump Type:** Centrifugal; stainless steel.

**Label(s):** None. Class 9.1, Group III.
**Storage Temperature:** Ambient.
**Hose Type:** Polyethylene, butyl, Hypalon, Viton, natural rubber.

**Grades or Purity:** Pure formaldehyde is not available commercially, because of its tendency to polymerize. It is sold as aqueous solutions containing 37 to 50% formaldehyde by weight and 0 to 15% methanol.
**Containers and Materials:** Drums, tank cars, tank trucks; stainless steel.

### Physical and Chemical Characteristics
**Physical State** (20°C, 1 atm): Liquid.
**Solubility** (Water): Soluble in all proportions.
**Molecular Weight:** 30.0 (solute only).
**Vapour Pressure:** 10 mm Hg (-88°C) (pure); 760 mm Hg (-19.5°C).
**Boiling Point:** -20°C (pure); 101°C (37% solution).

**Floatability** (Water): Sinks and mixes.
**Odour:** Pungent, irritating (0.8 to 1 ppm; odour threshold).
**Flash Point:** 37% (no methanol) 85°C (c.c.). (with 15% methanol) 50°C (c.c.); 50% (1.5% methanol) 68.5°C (c.c.).
**Vapour Density:** 1.1
**Specific Gravity:** 37% solution 1.1 (25°C); liquefied pure 1.1 (25°C).

**Colour:** Colourless.
**Explosive Limits:** 7 to 73%.
**Melting Point:** 37% solution, -15°C; -118°C (pure).

## HAZARD DATA

### Human Health
**Symptoms:** Inhalation: vapour very irritating to eyes and respiratory tract, sore throat, coughing, allergenic effects, bronchitis. Contact: eyes - liquid can cause severe burns skin - repeated contact causes a tanning effect and dermatitis. Ingestion: severe irritation of the gastrointestinal tract; nausea, coma, abdominal pain.
**Toxicology:** Highly toxic by all routes.
TLV® - (inhalation) 1 ppm; 1.5 mg/m³.
Short-term Inhalation Limits - 2 ppm; 3 mg/m³.

$LC_{50}$ - No information.
$LC_{Lo}$ - Inhalation: rat = 250 ppm/4 h
Delayed Toxicity - Suspected carcinogen.

$LD_{50}$ - Oral: rat = 0.8 g/kg

### Fire
**Fire Extinguishing Agents:** Use carbon dioxide, alcohol foam, dry chemical or water spray. Water may be used to cool fire-exposed containers and disperse vapours.
**Behaviour in Fire:** Formaldehyde vapours may be evolved.   **Burning Rate:** No information.
**Ignition Temperature:** 424°C (37% methanol free).

### Reactivity
**With Water:** No reaction.
**With Common Materials:** Reacts violently with nitrogen dioxide, (perchloric acid and aniline) and performic acid.
**Stability:** Stable.

### Environment
**Water:** Prevent entry into water intakes and waterways. Harmful to aquatic life in low concentrations. Fish toxicity: 25 ppm/96 h/channel catfish/TLm/freshwater; 15 mg/m³/24 h/striped bass larvae/$LC_{50}$/freshwater; 30 to 1 000 ppm/48 h/shrimp/$LC_{50}$/saltwater; BOD: 33 to 106%; 5 days.
**Land-Air:** No information.
**Food Chain Concentration Potential:** None.

## EMERGENCY MEASURES

### Special Hazards
POISON.

### Immediate Responses
Keep non-involved people away from spill site. Call Fire Department. Avoid contact and inhalation. Eliminate all ignition sources. Stay upwind and use water spray to control vapours. Stop or reduce discharge, if this can be done without risk. Contact supplier for guidance. Dike or dam spill to prevent runoff. Notify environmental authorities.

### Protective Clothing and Equipment
Respiratory protection - self-contained breathing apparatus. Acid suit - jacket and pants, rubber or plastic, or coveralls. Boots - high, rubber (pants worn outside boots). Gloves - rubber or plastic.

### Fire and Explosion
Use water spray, dry chemical, alcohol foam or carbon dioxide. Cool fire-exposed containers with water spray.

### First Aid
Move victim out of spill area to fresh air. Call for medical assistance, but start first aid at once. Inhalation: give artificial respiration if breathing has stopped (not mouth-to-mouth method); give oxygen if breathing is laboured. Contact: remove contaminated clothing. Wash eyes and skin with plenty of warm water for at least 15 minutes. Ingestion: give warm water to conscious victim to drink. Keep warm and quiet. If medical assistance is not immediately available, transport victim to hospital, clinic or doctor.

## ENVIRONMENTAL PROTECTION MEASURES

### Response

**Water**
1. Stop or reduce discharge if safe to do so.
2. Contact manufacturer or supplier for advice.
3. If possible, contain discharge by damming or water diversion.
4. Dredge or vacuum pump to remove contaminants, liquids and contaminated bottom sediments.
5. Notify environmental authorities to discuss disposal and cleanup of contaminated materials.

**Land-Air**
1. Stop or reduce discharge if safe to do so.
2. Contact manufacturer or supplier for advice.
3. Dike to prevent runoff from rainwater or water application.
4. Remove material with pumps or vacuum equipment and place in appropriate containers.
5. Adsorb residual liquid on natural or synthetic sorbents.
6. Notify environmental authorities to discuss disposal and cleanup of contaminated materials.

### Disposal
1. Contact manufacturer or supplier for advice on disposal.
2. Contact environmental authorities for advice on disposal.
3. Incinerate (approval of environmental authorities required).

FORMALDEHYDE   HCHO (aqueous)

# FORMIC ACID   HCOOH

## IDENTIFICATION

UN No. 1779

### Common Synonyms
FORMYLIC ACID
HYDROGEN CARBOXYLIC ACID
METHANOIC ACID

### Observable Characteristics
Colourless fuming liquid.
Pungent, penetrating odour.

### Manufacturers
No Canadian Manufacturers.
US Manufacturers: Celanese Chemical Co., New York, NY
Du Pont, Wilmington, DC
Union Carbide, New York, NY

### Transportation and Storage Information
**Shipping State:** Liquid.
**Classification:** Corrosive.
**Inert Atmosphere:** No requirement.
**Venting:** Pressure-vacuum.
**Pump Type:** No information.

**Label(s):** Black and white label - CORROSIVE.
**Storage Temperature:** Ambient.
**Hose Type:** No information.

**Grades or Purity:** Technical, 85 to 95%.
**Containers and Materials:** Carboys, drums, bulk by tank cars, tank trucks; stainless steel.

### Physical and Chemical Characteristics
**Physical State** ($20°C$, 1 atm): Liquid.
**Solubility (Water):** Miscible in all proportions.
**Molecular Weight:** 46.0
**Vapour Pressure:** 35 mm Hg ($20°C$); 54 mm Hg ($30°C$).
**Boiling Point:** $100.8°C$.

**Floatability (Water):** Sinks and mixes.
**Odour:** Pungent, penetrating (21 ppm, odour threshold).
**Flash Point:** 100%, $69°C$; 90%, $50°C$ (c.c.).
**Vapour Density:** 1.6
**Specific Gravity:** 1.22 ($20°C$).

**Colour:** Colourless.
**Explosive Limits:** 18 to 57% (90% solution only).
**Melting Point:** $8.4°C$.

## HAZARD DATA

### Human Health
**Symptoms:** Inhalation: irritation of respiratory tract, cough, bronchitis, difficulty breathing. Ingestion: nausea, vomiting, dizziness, unconsciousness. Contact: skin - irritation and burns; eyes - irritation and burns.
**Toxicology:** Moderately toxic by all routes.
**TLV®** - 5 ppm; 9 mg/m$^3$.
**Short-term Inhalation Limits** - No information.
**LC$_{50}$** - No information.
**Delayed Toxicity** - No information.

**LD$_{50}$** - Oral: rat = 1.21 g/kg
**LD$_{50}$** - Oral: mouse = 1.10 g/kg

### Fire
**Fire Extinguishing Agents:** Use carbon dioxide, dry chemical or alcohol foam. Water may be used to cool fire-exposed containers and disperse vapours.
**Behaviour in Fire:** No information.
**Ignition Temperature:** 100%, $539°C$; 90%, $434°C$.
**Burning Rate:** 0.5 mm/min.

### Reactivity
**With Water:** No reaction, soluble.
**With Common Materials:** Reacts violently with alkalis, furfuryl alcohol and hydrogen peroxide. Reacts with sulfuric acid to produce CO.
**Stability:** Stable.

### Environment
**Water:** Prevent entry into water intakes and waterways. Fish toxicity: 175 mg/L/24 h/bluegill/TLm/freshwater; 120 ppm/48 h/Daphnia magna/TLm/freshwater; 2 500 mg/L/tns/Gammarus pulex/LD$_{100}$/freshwater; BOD: 2 to 27%, 5 days.
**Land-Air:** No information.
**Food Chain Concentration Potential:** None.

## EMERGENCY MEASURES

**Special Hazards**

CORROSIVE. FLAMMABLE.

**Immediate Responses**

Keep non-involved people away from spill site. Issue warning: "CORROSIVE". Call Fire Department. Avoid contact and inhalation. Eliminate all sources of ignition. Call manufacturer or supplier for advice. Dike to contain material. Notify environmental authorities.

**Protective Clothing and Equipment**

Respiratory protection - self-contained breathing apparatus and totally encapsulated suit.

**Fire and Explosion**

Use carbon dioxide, dry chemical or alcohol foam to extinguish. Water may be used to cool fire-exposed containers and disperse vapours.

**First Aid**

Move victim out of spill site to fresh air. Call for medical assistance, but start first aid at once. **Inhalation:** if breathing has stopped, give artificial respiration; if laboured, give oxygen. **Contact:** skin - remove contaminated clothing and flush affected areas with plenty of water; eyes - irrigate with water. **Ingestion:** give water to conscious victim to drink; do not induce vomiting. If medical assistance is not immediately available, transport victim to hospital, doctor or clinic.

## ENVIRONMENTAL PROTECTION MEASURES

**Response**

**Water**
1. Stop or reduce discharge if safe to do so.
2. Contact manufacturer or supplier for advice.
3. If possible, contain discharge by damming or water diversion.
4. Dredge or vacuum pump to remove contaminants, liquids and contaminated bottom sediments.
5. Notify environmental authorities to discuss disposal and cleanup of contaminated materials.

**Land-Air**
1. Stop or reduce discharge if safe to do so.
2. Contact manufacturer or supplier for advice.
3. Contain spill by diking with earth or other barrier.
4. Remove material with pumps or vacuum equipment in appropriate containers.
5. Recover undamaged containers.
6. Adsorb residual liquid on natural or synthetic sorbents.
7. Notify environmental authorities to discuss disposal and cleanup of contaminated materials.

**Disposal**

1. Contact manufacturer or supplier for advice on disposal.
2. Contact environmental authorities for advice on disposal.

FORMIC ACID   HCOOH

# FURFURAL  OC$_3$H$_3$C CHO

## IDENTIFICATION  UN No. 1199

### Common Synonyms
FURAL
FURFURALDEHYDE
2-FURALDEHYDE
2-FURAN CARBONAL
ANT OIL
BRAN OIL
PYROMUCIC ALDEHYDE

### Observable Characteristics
Colourless to yellow to reddish-brown, liquid. Almond-like odour.

### Manufacturers
No Canadian manufacturers.
Canadian supplier:
The Quaker Oats Co. of Canada Ltd., Peterborough, Ont.

Originating from:
The Quaker Oats Company, Chicago, Ill.

### Transportation and Storage Information

**Shipping State:** Liquid.
**Classification:** Combustible liquid.
**Inert Atmosphere:** No requirement.
**Venting:** Pressure-vacuum.
**Pump Type:** Gear, centrifugal. Steel, stainless steel.

**Label(s):** Red and white label - COMBUSTIBLE LIQUID; Class 3.3, Group II.
**Storage Temperature:** Ambient.
**Hose Type:** Butyl, polyethylene, polypropylene.

**Grades or Purity:** Technical, refined.
**Containers and Materials:** Drums, tank cars, tank trucks; steel, stainless steel.

### Physical and Chemical Characteristics

**Physical State** (20°C, 1 atm)**:** Liquid.
**Solubility** (Water)**:** 8.3 g/100 mL (20°C); 19.9 g/100 mL (90°C).
**Molecular Weight:** 96.1
**Vapour Pressure:** 1 mm Hg (20°C); 3 mm Hg (30°C); 10 mm Hg (50°C).
**Boiling Point:** 161.7°C.

**Floatability** (Water)**:** Sinks.
**Odour:** Almond-like (0.25 ppm odour threshold).
**Flash Point:** 60°C (c.c.); 67°C (o.c.).
**Vapour Density:** 3.3
**Specific Gravity:** 1.16 (20°C).

**Colour:** Colourless to yellow (pure) will darken to reddish-brown on exposure to light.
**Explosive Limits:** 2.1 to 19.3%.
**Melting Point:** -36.5°C.

## HAZARD DATA

### Human Health
**Symptoms:** Inhalation: irritation of throat, respiratory system, headache, lung congestion. Contact: skin - absorbed readily, dermatitis and eczema; eyes - irritation. Ingestion: sore throat, abdominal pain, vomiting, diarrhea.
**Toxicology:** Moderately toxic by inhalation, ingestion and skin contact.
TLV® (skin) 2 ppm; 8 mg/m$^3$.
Short-term Inhalation Limits - 10 ppm; 40 mg/m$^3$
LC$_{50}$ - No information.
LC$_{Lo}$ - Inhalation: rat = 153 ppm/4 h
Delayed Toxicity - No information.
LD$_{50}$ - Oral: rat = 0.127 g/kg
(15 min).

### Fire
**Fire Extinguishing Agents:** Water spray, carbon dioxide, dry chemical, or alcohol foam. Water may be used to cool fire-exposed containers and disperse vapours.
**Behaviour in Fire:** No information.  **Burning Rate:** 2.6 mm/min.
**Ignition Temperature:** 316°C.

### Reactivity
**With Water:** No reaction.
**With Common Materials:** Reacts explosively with mineral acids or alkalis.
**Stability:** Stable.

### Environment
**Water:** Prevent entry into water intakes and waterways. Harmful to aquatic life in low concentrations. Fish toxicity: 24 ppm/96 h/bluegill/TLm/freshwater; 1.2 to 32 mg/L/24 h/bluegill/TLm/freshwater; 24 to 44 mg/L/24 to 96 h/mosquito fish/TLm/freshwater; BOD: 28 to 77%; 5 days.
**Land-Air:** No information.
**Food Chain Concentration Potential:** None.

## EMERGENCY MEASURES

### Special Hazards
COMBUSTIBLE. VIOLENT REACTION WITH ACIDS AND ALKALIS.

### Immediate Responses
Keep non-involved people away from spill site. Issue warning: "COMBUSTIBLE". Call Fire Department. Avoid contact and inhalation. Contain spill by diking. Stop or reduce discharge, if this can be done without risk. Notify manufacturer or supplier. Notify environmental authorities.

### Protective Clothing and Equipment
Respiratory protection - self-contained breathing apparatus and totally encapsulated suit. Gloves - rubber. Boots - high, rubber (pants worn outside boots).

### Fire and Explosion
Moderate fire and explosion hazard when exposed to heat and flame. Use water spray, carbon dioxide, dry chemical or alcohol foam to extinguish. Water may be used to cool fire-exposed containers and disperse vapours.

### First Aid
Move victim out of spill area to fresh air. Call for medical assistance, but start first aid at once. Inhalation: if breathing has stopped, give artificial respiration; if laboured, give oxygen. Contact: remove contaminated clothing immediately; flush eyes and affected skin with plenty of warm water. Ingestion: give water or milk to conscious victim to drink and induce vomiting. Keep warm and quiet. If medical assistance is not immediately available, transport victim to hospital, doctor or clinic.

## ENVIRONMENTAL PROTECTION MEASURES

### Response

**Water**
1. Stop or reduce discharge if safe to do so.
2. Contact manufacturer or supplier for advice.
3. If possible, contain discharge by damming or water diversion.
4. Dredge or vacuum pump to remove contaminants, liquids and contaminated bottom sediments.
5. Notify environmental authorities to discuss disposal and cleanup of contaminated materials.

**Land-Air**
1. Stop or reduce discharge if safe to do so.
2. Contact manufacturer or supplier for advice.
3. Dike to prevent runoff from rainwater or water application.
4. Remove material with pumps or vacuum equipment and place in appropriate containers.
5. Recover undamaged containers.
6. Absorb residual liquid on natural or synthetic sorbents.
7. Notify environmental authorities to discuss disposal and cleanup of contaminated materials.

### Disposal
1. Contact manufacturer or supplier for advice on disposal.
2. Contact environmental authorities for advice on disposal.
3. Incinerate (approval of environmental authorities required).

FURFURAL   $OC_3H_3$ CHO

# GASOLINE

## IDENTIFICATION

**Common Synonyms**
PETROL, AUTOMOTIVE FUEL
AV-GAS
- leaded contains tetraethyl lead
- lead-free may contain other compounds.

**Observable Characteristics**
Colourless (or dyed red, purple) liquid.
Typical gasoline odour.

**Manufacturers**
Universally available.

**UN No. 1203**

### Transportation and Storage Information

**Shipping State:** Liquid.
**Classification:** Flammable liquid.
**Inert Atmosphere:** No requirement.
**Venting:** Open (flame arrester) or pressure-vacuum.
**Pump Type:** Standard.

**Label(s):** Red and white label - FLAMMABLE LIQUID; Class 3.2, Group II.
**Storage Temperature:** Ambient.
**Hose Type:** Standard.

**Grades or Purity:** Various octane ratings or use ratings. Leaded or lead-free.
**Containers and Materials:** Cans, drums, tank cars, tank trucks; steel.

### Physical and Chemical Characteristics

**Physical State** (20°C, 1 atm): Liquid.
**Solubility** (Water): 1 to 100 ppm/100 mL water.
**Molecular Weight:** Mixture of materials.
**Vapour Pressure:** 300 to 600 mm Hg (20°C).
**Boiling Point:** 40 to 200°C.

**Floatability** (Water): Floats.
**Odour:** Gasoline (0.25 ppm, odour threshold).
**Flash Point:** -43°C (c.c.) (up to 60 octane); -38°C (c.c.) up to 100 octane; -46°C (c.c.) aviation.
**Vapour Density:** 3 to 4
**Specific Gravity:** 0.75 to 0.85 at 20°C.

**Colour:** Colourless to (dyed red or purple).
**Explosive Limits** 1.4 to 7.6%.
**Melting Point:** -90 to -75°C.

## HAZARD DATA

### Human Health

**Symptoms:** Inhalation: vapours cause rapid breathing, excitability, staggering, headache, fatigue, nausea and vomiting, dizziness, drowsiness, narcosis, convulsions, coma. Ingestion: gastrointestinal irritation, dizziness, fatigue, loss of consciousness, coma. Contact: skin - dryness, cracking, irritation; eyes - watering, stinging and inflammation.
**Toxicology:** Moderately toxic by inhalation.
TLV® - 300 ppm; 900 mg/m³.
Short-term Inhalation Limits - 500 ppm; 1 500 mg/m³ (15 min).

$LC_{50}$ - No information.
$LC_{Lo}$ - Inhalation: man = 900 ppm/1 h
Delayed Toxicity - No information.

$LD_{50}$ - No information.

### Fire

**Fire Extinguishing Agents:** Foam, carbon dioxide, dry chemical. Water may be ineffective and cause fire to spread, but may be used to cool fire-exposed containers.
**Behaviour in Fire:** Flashback may occur along vapour trail. **Burning Rate:** 4 mm/min.
**Ignition Temperature:** 280°C, up to 60 octane; 440°C, 100 to 130 octane (aviation grade); 471°C, 115 to 145 octane (aviation grade).

### Reactivity

**With Water:** No reaction.
**With Common Materials:** Can react vigorously with oxidizing materials.
**Stability:** Stable.

### Environment

**Water:** Prevent entry into water intakes and waterways. Harmful to aquatic life. Fish toxicity: 90 ppm/24 h/juvenile American shad/TLm/freshwater; 91 mg/L/24 h/juvenile American shad/TLm/saltwater; 5 to 40 ppm/96 h/rainbow trout/TLm/freshwater; BOD: 8%, 5 days.
**Land-Air:** No information.
**Food Chain Concentration Potential:** None.

## EMERGENCY MEASURES

**Special Hazards**
FLAMMABLE.

**Immediate Responses**
Keep non-involved people away from spill site. Issue warning: "FLAMMABLE". Call Fire Department. Eliminate all sources of ignition. Notify manufacturer. Dike to prevent runoff. Shut off leak, if safe to do so. Notify environmental authorities.

**Protective Clothing and Equipment**
Protective clothing as required.

**Fire and Explosion**
Use foam, dry chemical or carbon dioxide to extinguish. Water may be ineffective and cause fire to spread, but may be used to cool fire-exposed containers. Flashback may occur along vapour trail.

**First Aid**
Move victim out of spill area to fresh air. Call for medical assistance, but start first aid at once. <u>Inhalation:</u> if breathing has stopped give artificial respiration; if laboured, give oxygen. <u>Ingestion:</u> give water to conscious victim to drink; do not induce vomiting. <u>Contact:</u> skin – remove contaminated clothing and wash affected areas with plenty of warm water; eyes – irrigate with plenty of water. If medical assistance is not immediately available, transport victim to hospital, doctor or clinic.

## ENVIRONMENTAL PROTECTION MEASURES

**Response**

**Water**
1. Stop or reduce discharge if safe to do so.
2. Contact manufacturer or supplier for advice.
3. If possible, contain discharge by booming.
4. If floating, skim and remove.
5. Notify environmental authorities to discuss disposal and cleanup of contaminated materials.

**Land-Air**
1. Stop or reduce discharge if safe to do so.
2. Contact manufacturer or supplier for advice.
3. Contain spill by diking with earth or other barrier.
4. Remove material with pumps or vacuum equipment and place in appropriate containers.
5. Adsorb residual liquid on natural or synthetic sorbents.
6. Remove contaminated soil for disposal.
7. Notify environmental authorities to discuss disposal and cleanup of contaminated materials.

**Disposal**
1. Contact manufacturer or supplier for advice on disposal.
2. Contact environmental authorities for advice on disposal.
3. Incinerate (approval of environmental authorities required).

GASOLINE

# GLYCERINE  CH2OHCHOHCH2OH

## IDENTIFICATION

### Common Synonyms
GLYCEROL
GLYCYL ALCOHOL
1,2,3-PROPANETRIOL
1,2,3-TRIHYDROXYPROPANE

### Observable Characteristics
Colourless oily liquid. Odourless.

### Manufacturers
Lever Detergents Ltd., Toronto, Ont.
Proctor & Gamble Co. of Canada Ltd., Hamilton, Ont.
Colgate-Palmolive Limited, Toronto, Ont.
Emery Industries (Canada) Ltd., Toronto, Ont.
Dow Chemical Canada - from Dow Chemical, Freeport, TX.

### Transportation and Storage Information
**Shipping State:** Liquid.
**Classification:** None.
**Inert Atmosphere:** No requirement.
**Venting:** Open (flame arrester) or pressure-vacuum.
**Pump Type:** Centrifugal, gear; steel or stainless steel.

**Label(s):** Not required. Voluntary label - COMBUSTIBLE LIQUID.
**Storage Temperature:** Ambient.
**Hose Type:** Polyethylene, Viton, butyl, Hypalon, Teflon, natural rubber.

**Grades or Purity:** CP, 99.5%; USP, 96%.
**Containers and Materials:** Drums, tank cars, tank trucks; steel or stainless steel.

### Physical and Chemical Characteristics
**Physical State** (20°C, 1 atm): Liquid.
**Solubility** (Water): Soluble in all proportions.
**Molecular Weight:** 92.1
**Vapour Pressure:** 0.0025 mm Hg (50°C).
**Boiling Point:** 290°C.

**Floatability** (Water): Sinks and mixes.
**Odour:** Odourless.
**Flash Point:** 177°C (o.c.); 160°C (c.c.).
**Vapour Density:** 3.2
**Specific Gravity:** 1.26 (20°C).

**Colour:** Colourless.
**Explosive Limits:** Not explosive in air.
**Melting Point:** 17.9°C.

## HAZARD DATA

### Human Health
**Symptoms:** Inhalation; irritation of respiratory tract. Ingestion: restlessness, nausea, vomiting, diarrhea, fever. Contact: eyes-irritation.
**Toxicology:** Moderately toxic by inhalation, ingestion and contact.
**TLV®** - (inhalation) 10 mg/m$^3$ (mist).   $LD_{50}$ - Oral: rat = 12.6 g/kg
**Short-term Inhalation Limits** - No information.   $LC_{50}$ - No information.
**Delayed Toxicity** - No information.

### Fire
**Fire Extinguishing Agents:** Use alcohol foam, dry chemical or carbon dioxide. Water or foam may cause frothing.
**Behaviour in Fire:** No information.   **Burning Rate:** 0.9 mm/min.
**Ignition Temperature:** 370°C.

### Reactivity
**With Water:** No reaction; soluble.
**With Common Materials:** Can react with powerful oxidizers. Reacts violently with acetic anhydride, calcium hypochlorite, chlorine, chromic anhydride, chromium oxide, potassium permanganate, silver perchlorate and sodium peroxide.
**Stability:** Stable.

### Environment
**Water:** Prevent entry into water intakes and waterways. Fish toxicity: >5 000 mg/L/24 h/goldfish/$LD_{50}$/freshwater; Aquatic toxicity rating = >1 000 ppm/96 h/TLm/freshwater; BOD: 61 to 87%, 5 days.
**Land-Air:** No information.
**Food Chain Concentration Potential:** None.

## EMERGENCY MEASURES

**Special Hazards**

COMBUSTIBLE.

**Immediate Responses**

Keep non-involved people away from spill site. Issue warning: "COMBUSTIBLE". Call Fire Department. Stop or reduce discharge, if safe to do so. Contain spill by diking. Notify manufacturer. Notify environmental authorities.

**Protective Clothing and Equipment**

Goggles - (mono), tight-fitting, and rubber gloves. Protective outerwear as required.

**Fire and Explosion**

Use alcohol foam, dry chemical or carbon dioxide. Water or foam may cause frothing.

**First Aid**

Move victim out of spill site to fresh air. Call for medical assistance, but start first aid at once. Contact: remove contaminated clothing; wash eyes and skin thoroughly with plenty of water. Ingestion: give water to conscious victim to drink. If medical assistance is not immediately available, transport victim to hospital, doctor or clinic.

## ENVIRONMENTAL PROTECTION MEASURES

**Response**

**Water**
1. Stop or reduce discharge if safe to do so.
2. Contact manufacturer or supplier for adice.
3. If possible, contain discharge by damming or water diversion.
4. Dredge or vacuum pump to remove contaminants, liquids and contaminated bottom sediments.
5. Notify environmental authorities to discuss disposal and cleanup of contaminated materials.

**Land-Air**
1. Stop or reduce discharge if safe to do so.
2. Contact manufacturer or supplier for advice.
3. Contain spill by diking with earth or other barrier.
4. When liquid, remove material with pumps or vacuum equipment and place in appropriate containers.
5. When solid, remove material by manual or mechanical means.
6. Recover undamaged containers.
7. Adsorb residual liquid on natural or synthetic sorbents.
8. Notify environmental authorities to discuss disposal and cleanup of contaminated materials.

**Disposal**
1. Contact manufacturer or supplier for advice on disposal.
2. Contact environmental authorities for advice on disposal.

GLYCERINE   $CH_2OHCHOHCH_2OH$

# n-HEXANE  $C_6H_{14}$

## IDENTIFICATION  UN No. 1208

| Common Synonyms | Observable Characteristics | Manufacturers | |
|---|---|---|---|
| HEXANE | Clear, colourless liquid. Mild, gasoline-like odour. | ESSO Chemicals, Sarnia, Ontario.<br>Texaco Canada, Mississauga, Ontario.<br>Canadian suppliers:<br>Hercules Canada, Varennes, Québec.<br>Polysar, Sarnia, Ontario.<br>Stanchem, Montreal, Québec. | Originating from:<br>Phillips Petroleum, USA<br>Union Oil, USA. |

### Transportation and Storage Information

**Shipping State:** Liquid.
**Classification:** Flammable liquid.
**Inert Atmosphere:** No requirement.
**Venting:** Open (flame arrester) or pressure-vacuum.
**Pump Type:** Centrifugal gear; explosion-proof.

**Label(s):** Red and white label - FLAMMABLE LIQUID; Class 3.1, Group II.
**Storage Temperature:** Ambient.
**Hose Type:** Reinforced antistatic rubber, or neoprene.

**Grades or Purity:** 85, 95 and 99%.
**Containers and Materials:** Cans, drums, tank cars, tank trucks; steel.

### Physical and Chemical Characteristics

**Physical State** (20°C, 1 atm): Liquid.
**Solubility** (Water): 0.00095 g/100 mL (20°C).
**Molecular Weight:** 86.2
**Vapour Pressure:** 120 mm Hg (20°C); 190 mm Hg (30°C).
**Boiling Point:** 68.7 °C.

**Floatability** (Water): Floats.
**Odour:** Mild, gasoline-like.
**Flash Point:** -22°C (c.c.).
**Vapour Density:** 2.8
**Specific Gravity:** 0.66 (20°C) (liquid).

**Colour:** Colourless.
**Explosive Limits:** 1.1 to 7.5%.
**Melting Point:** -94 to -96°C.

## HAZARD DATA

### Human Health

**Symptoms:** Inhalation: irritation of respiratory tract, dizziness, numbness of extremities, intoxication, unconsciousness. Ingestion: irritation of gastrointestinal tract, nausea, vomiting, headache, depression. Contact: skin - defatting and dermatitis; eyes - irritation.
**Toxicology:** Low toxicity by all routes.
TLV® - 50 ppm; 180 mg/m³.    $LC_{50}$ - No information.    $LD_{50}$ - No information.
Short-term Inhalation Limits - No information.    $LC_{Lo}$ - Inhalation: mouse = 120 000 mg/m³.    $LD_{Lo}$ - Intraperitoneal: rat = 9.1 g/kg
Delayed Toxicity - No information.

### Fire

**Fire Extinguishing Agents:** Use dry chemical and carbon dioxide. Water may be ineffective but may be used to cool fire-exposed containers.
**Behaviour in Fire:** Flashback may occur along vapour trail.
**Ignition Temperature:** 223°C.    **Burning Rate:** 7.3 mm/min.

### Reactivity

**With Water:** No reaction.
**With Common Materials:** Can react vigorously with oxidizing materials.
**Stability:** Stable.

### Environment

**Water:** Prevent entry into water intakes and waterways. Aquatic toxicity rating = >1 000 ppm/96 h/TLm/freshwater; BOD: 221%; 5 days.
**Land-Air:** No information.
**Food Chain Concentration Potential:** None.

## EMERGENCY MEASURES

**Special Hazards**
FLAMMABLE.

**Immediate Responses**
Keep non-involved people away from spill site. Issue warning: "FLAMMABLE". Call Fire Department. Eliminate all sources of ignition. Stop or reduce discharge, if safe to do so. Dike to contain the spill. Notify manufacturer. Notify environmental authorities.

**Protective Clothing and Equipment**
Protective outerwear as required.

**Fire and Explosion**
Use carbon dioxide or dry chemical to extinguish. Water may be ineffective, but may be used to cool fire-exposed containers.

**First Aid**
Move victim out of spill area to fresh air. Call for medical assistance, but start first aid at once. <u>Inhalation:</u> if breathing has stopped, give artificial respiration; if laboured, give oxygen. <u>Ingestion:</u> give water to conscious victim to drink, do not induce vomiting. <u>Contact:</u> skin - remove contaminated clothing and wash affected areas with plenty of water; eyes - irrigate with plenty of water. If medical assistance is not immediately available, transport victim to doctor, hospital or clinic.

## ENVIRONMENTAL PROTECTION MEASURES

**Response**

**Water**
1. Stop or reduce discharge if safe to do so.
2. Contact manufacturer or supplier for advice.
3. If possible, contain discharge by damming or water diversion.
4. If possible, contain discharge by booming.
5. If floating, skim and remove.
6. Notify environmental authorities to discuss disposal and cleanup of contaminated materials.

**Land-Air**
1. Stop or reduce discharge if safe to do so.
2. Contact manufacturer or supplier for advice.
3. Contain spill by diking with earth or other barrier.
4. Remove material with pumps or vacuum equipment and place in appropriate containers.
5. Recover undamaged containers.
6. Notify environmental authorities to discuss disposal and cleanup of contaminated materials.

**Disposal**
1. Contact manufacturer or supplier for advice on disposal.
2. Contact environmental authorities for advice on disposal.
3. Incinerate (approval of environmental authorities required).

n-HEXANE   $C_6H_{14}$

# HYDRAZINE $N_2H_4$

## IDENTIFICATION

**Common Synonyms**
DIAMIDE
DIAMINE
HYDRAZINE, BASE
HYDRAZINE, AQUEOUS SOLUTION
HYDRAZINE, HYDRATE

**Observable Characteristics**
Anhydrous - colourless fuming liquid.
Hydrates - colourless to yellow liquids.
Ammoniacal odour.

**UN No.** 2029 anhydrous
2030 24% solution

**Manufacturers**

### Transportation and Storage Information

**Shipping State:** Liquid.
**Classification:** Corrosive, poison.
**Inert Atmosphere:** Padded.
**Venting:** Pressure-vacuum.
**Pump Type:** No information.

**Label(s):** Black and white label - POISON.
**Storage Temperature:** Ambient.
**Hose Type:** No information.

**Grades or Purity:** Anhydrous 100%; hydrates 35 to 64% water solutions.
**Containers and Materials:** Carboys, drums; steel or aluminum.

### Physical and Chemical Characteristics

**Physical State** (20°C, 1 atm): Liquid.
**Solubility (Water):** Soluble in all proportions.
**Molecular Weight:** 32.1
**Vapour Pressure:** 16 mm Hg (20°C); 25 mm Hg (30°C).
**Boiling Point:** Anhydrous 113.5°C; 24% solution 102°C.

**Floatability (Water):** Floats and mixes (saltwater); sinks and mixes (freshwater).
**Odour:** Ammoniacal (3 to 4 ppm), odour threshold.
**Flash Point:** 52°C (o.c.); 37.8°C (c.c.)
**Vapour Density:** 1.04 to 1.1
**Specific Gravity:** 1.004 (anhydrous) (20°C).

**Colour:** Colourless to yellow.
**Explosive Limits:** 2.9 to 98%.
**Melting Points:** 1.5°C (anhydrous); -14°C (24% solution).

## HAZARD DATA

### Human Health

**Symptoms:** Inhalation: sore throat, coughing, laboured breathing, dizziness. Contact: skin - redness, pain, absorbed; eyes - temporary blindness, redness and pain. Ingestion: headache, nausea, abdominal pain, diarrhea, dizziness; symptoms similar to inhalation.
**Toxicology:** Highly toxic by all routes.
TLV® (skin) 0.1 ppm; 0.1 mg/m³.
Short-term Inhalation Limits - No information.
LC$_{50}$ - Inhalation: rat = 570 ppm/4 h
LC$_{50}$ - Inhalation: mouse = 252 ppm/4 h
Delayed Toxicity - Suspected carcinogen.
LD$_{50}$ - Intraperitoneal: rat = 0.76 g/kg
LD$_{Lo}$ - Oral: mouse = 0.41 g/kg

### Fire

**Fire Extinguishing Agents:** Use foam, carbon dioxide or dry chemical. Flooding with water may be necessary to prevent reignition. Water may be used to cool fire-exposed containers and knock down vapours.
**Behaviour in Fire:** Releases toxic NO$_X$ fumes. Can ignite spontaneously in air on contact with porous materials such as earth, asbestos, wood or cloth.
**Ignition Temperature:** Vary with contact of different materials: iron-rust 23°C, black iron 132°C, stainless steel 156°C; glass 270°C.
**Burning Rate:** 1 mm/min (est.)

### Reactivity

**With Water:** No reaction; soluble.
**With Common Materials:** Reacts violently with (alkali metals and ammonia), chlorine, chromates, cupric oxide and salts, fluorine, hydrogen peroxide, metallic oxides, nickel, nitric acid, nitrous oxide, oxygen, potassium dichromate and sodium dichromate.
**Stability:** May spontaneously ignite under various situations.

### Environment

**Water:** Prevent entry into water intakes and waterways. Fish toxicity: 146 ppm/0.5 h/rainbow trout/LD$_{100}$/freshwater; 136 ppm/22 h/carp/LD$_{100}$/freshwater.
**Land-Air:** No information.
**Food Chain Concentration Potential:** No information.

## EMERGENCY MEASURES

### Special Hazards
CORROSIVE. POISON. Flammable. May spontaneously ignite under certain conditions.

### Immediate Responses
Keep non-involved people away from spill site. Issue warnings: "CORROSIVE, POISON". Call Fire Department. Avoid contact and inhalation. Evacuate from downwind. Notify manufacturer or supplier. Dike to contain material. Notify environmental authorities.

### Protective Clothing and Equipment
Respiratory protection - self-contained breathing apparatus and totally encapsulated suit.

### Fire and Explosion
Use foam, carbon dioxide, or dry chemical to extinguish. Flooding with water may be necessary to prevent reignition. Water may be used to cool fire-exposed containers and knock down vapours. Releases toxic $NO_x$ fumes in fires.

### First Aid
Move victim out of spill site to fresh air. Call for medical assistance, but start first aid at once. Inhalation: if breathing has stopped, give artificial respiration, (not mouth-to-mouth method) if laboured, give oxygen. Contact: skin - remove contaminated clothing and flush affected areas with plenty of water; eyes - irrigate with water. Ingestion: give conscious victim water to drink; do not induce vomiting. If medical assistance is not immediately available, transport victim to doctor, hospital or clinic.

## ENVIRONMENTAL PROTECTION MEASURES

### Response

**Water**
1. Stop or reduce discharge if safe to do so.
2. Contact manufacturer or supplier for advice.
3. If possible, contain discharge by damming or water diversion.
4. Dredge or vacuum pump to remove contaminants, liquids and contaminated bottom sediments.
5. Notify environmental authorities to discuss disposal and cleanup of contaminated materials.

**Land-Air**
1. Stop or reduce discharge if safe to do so.
2. Contact manufacturer or supplier for advice.
3. Contain spill by diking with earth or other barrier.
4. With great caution, remove material with pumps or vacuum equipment and place in appropriate containers.
5. Recover undamaged containers.
6. Notify environmental authorities to discuss disposal and cleanup of contaminated materials.

### Disposal
1. Contact manufacturer or supplier for advice on disposal.
2. Contact environmental authorities for advice on disposal.

HYDRAZINE   $N_2H_4$

# HYDROCHLORIC ACID   HCl (aqueous)

## IDENTIFICATION

UN No. 1789

### Common Synonyms
MURIATIC ACID
HYDROGEN CHLORIDE (aqueous)
CHLOROHYDRIC ACID

### Observable Characteristics
Water-white to pale yellow liquid. Sharp, irritating odour. Fumes in humid air.

### Manufacturers
Canadian Industries Limited, Cornwall, Ont., Becancour, Que.
Dow Chemical Canada Inc.,
Sarnia, Ont., Fort Saskatchewan, Alta.
Du Pont Canada Limited, Maitland, Ont.
FMC Chemicals, Squamish, B.C.
Canadian Occidental, Vancouver, B.C.
St. Anne Chemicals, Macawick, N.B.

### Transportation and Storage Information
**Shipping State:** Liquid (aqueous solution).
**Classification:** Corrosive liquid.
**Inert Atmosphere:** No requirement.
**Venting:** Open.
**Pump Type:** Gear, centrifugal, diaphragm. Rubber or plastic-lined.

**Label(s):** White and black label - CORROSIVE LIQUID; Class 8, Group II.
**Storage Temperature:** Ambient.
**Hose Type:** Natural rubber, polyethylene, polypropylene, PVC, etc.

**Grades or Purity:** Commercial strengths; 18°Bé (27.9% HCl); 20°Bé (31.5% HCl); 22°Bé (35.2% HCl); 23°Bé (37.1% HCl).
**Containers and Materials:** Drums, tank cars, tank trucks; steel, rubber-lined.

### Physical and Chemical Characteristics
**Physical State** (20°C, 1 atm): Liquid.
**Solubility (Water):** Soluble in all proportions.
**Molecular Weight:** 36.5 (HCl gas).
**Vapour Pressure:** 25 mm Hg (25°C) (20°Bé)
**Boiling Point:** 83°C (20°Bé); 62°C (22°Bé); 51°C (23°Bé); 98°C (18°Bé).

**Floatability (Water):** Sinks and mixes.
**Odour:** Sharp, irritating (1 to 10 ppm, odour threshold).
**Flash Point:** Not flammable.
**Vapour Density:** 1.3
**Specific Gravity:** 1.14 (18°Bé); 1.16 (20°Bé); 1.18 (22°Bé); 1.19 (23°Bé) (15.5°C).

**Colour:** Colourless to pale yellow.
**Explosive Limits:** Not flammable.
**Melting Point:** -42°C (18°Bé); -53°C (20°Bé); -66°C (22°Bé); -74°C (23°Bé).

## HAZARD DATA

### Human Health
**Symptoms: Inhalation:** Vapours can produce severe irritation of the upper respiratory tract, coughing, burning of throat and choking. **Contact: eyes** - severe irritation of eyes and lids; burning; visual impairment may result; **skin** - can cause serious burns. **Ingestion:** very corrosive, can cause serious internal injury. May be fatal.
**Toxicology:** Moderately toxic by all routes.
TLV® - (inhalation) 5 ppm; 7 mg/m³.  $LC_{50}$ - Inhalation: rat = 3 124 ppm/30 min   $LD_{50}$ - Oral: rabbit = 0.9 g/kg
Short-term Inhalation Limits - No information.   Delayed Toxicity - None known.

### Fire
**Fire Extinguishing Agents:** Not combustible. Most fire extinguishing agents may be used on fires involving hydrochloric acid. Water spray may be used sparingly to knock down vapours.
**Behaviour in Fire:** When heated, toxic and corrosive hydrogen chloride gas is released.
**Ignition Temperature:** Not combustible.   **Burning Rate:** Not combustible.

### Reactivity
**With Water:** Soluble, with evolution of heat.
**With Common Materials:** Reacts violently with acetic anhydride, ammonium hydroxide, calcium phosphide, chlorosulfonic acid, ethylenediamine, oleum, perchloric acid, propylene oxide, sodium hydroxide, sulfuric acid and vinyl acetate. Reacts with metals forming flammable $H_2$ gas.
**Stability:** Stable.

### Environment
**Water:** Prevent entry into water intakes and waterways. Harmful to aquatic life. Fish toxicity: 282 ppm/96 h/mosquito fish/TLm/freshwater; 100 to 300 mg/L/48 h/shrimp/$LC_{50}$/saltwater; BOD: None.
**Land-Air:** No information.
**Food Chain Concentration Potential:** None.

## EMERGENCY MEASURES

**Special Hazards**
CORROSIVE.

**Immediate Responses**
Keep non-involved people away from spill site. Issue warning: "CORROSIVE". Contact manufacturer for guidance and assistance. Avoid contact and inhalation. Stop or reduce discharge if this can be done without risk. Contain spill by diking to prevent runoff. Notify environmental authorities.

**Protective Clothing and Equipment**
Respiratory protection - self-contained breathing apparatus and acid suit (rubber). Gloves - rubber. Boots - high, rubber (pants worn outside boots).

**Fire and Explosion**
Not combustible. Most fire extinguishing agents may be used on fires involving hydrochloric acid. Water may be used sparingly to knock down vapours.

**First Aid**
Move victim out of spill area to fresh air. Call for medical assistance, but start first aid at once. Contact: immediately irrigate eyes and flush skin with plenty of water for at least 30 minutes while removing contaminated clothing. Inhalation: if breathing has stopped, give artificial respiration; if laboured, give oxygen. Ingestion: if victim is conscious, give as much water or milk as possible to dilute acid. Do not induce vomiting. If medical assistance is not immediately available, transport victim to hospital, doctor or clinic.

## ENVIRONMENTAL PROTECTION MEASURES

**Response**

**Water**
1. Stop or reduce discharge if safe to do so.
2. Contact manufacturer or supplier for advice.
3. If possible, contain discharge by damming or water diversion.
4. Notify environmental authorities to discuss disposal and cleanup of contaminated materials.

**Land-Air**
1. Stop or reduce discharge if safe to do so.
2. Contact manufacturer or supplier for advice.
3. Contain spill by diking with earth or other barrier.
4. Remove material with pumps or vacuum equipment and place in appropriate containers.
5. Adsorb residual liquid on natural or synthetic sorbents.
6. Contaminated soil can be treated with lime or soda ash (calcium hydroxide) to render it neutral.
7. Notify environmental authorities to discuss disposal and cleanup of contaminated materials.

**Disposal**
1. Contact manufacturer or supplier for advice on disposal.
2. Contact environmental authorities for advice on disposal.

HYDROCHLORIC ACID   HCl (aqueous)

# HYDROFLUORIC ACID   HF aqueous

## IDENTIFICATION

UN No. 1790

| Common Synonyms | Observable Characteristics | Manufacturers |
|---|---|---|
| HYDROGEN FLUORIDE (aqueous)<br>FLUORHYDRIC ACID | Clear, colourless liquid. Strong, irritating odour. | Allied Chemical Ltd., Valleyfield, Quebec, Amherstburg, Ontario.<br>Alcan Smelters and Chemicals, Jonquière, Que. |

### Transportation and Storage Information

**Shipping State:** Liquid (aqueous solution).
**Classification:** Corrosive, poison.
**Inert Atmosphere:** No requirement.
**Venting:** Pressure-vacuum.
**Pump Type:** Centrifugal; stainless steel; all propylene.

**Label(s):** White and black label - CORROSIVE. White label - POISON; Class 8, 2.3.
**Storage Temperature:** Ambient.
**Hose Type:** Reinforced Halon TFE or Hypalon.

**Grades or Purity:** Technical, 48%; CP, 70%.
**Containers and Materials:** Bottles, drums, tank cars, tank trucks; polylined.

### Physical and Chemical Characteristics

**Physical state** (20°C, 1 atm): Liquid.
**Solubility (Water):** Soluble in all proportions.
**Molecular Weight:** 20.0 (for HF gas).
**Vapour Pressure:** 25 mm Hg (20°C) 48%; 100 mm Hg (15°C), 125 mm Hg (20°C) 70%.
**Boiling Point:** 67°C (70% solution) 108°C (48% solution); 112°C (38% solution).

**Floatability (Water):** Sinks and mixes.
**Odour:** Strong, irritating (5 ppm, odour threshold).
**Flash Point:** Not flammable.
**Vapour Density:** 0.7
**Specific Gravity:** 1.26 (70%); 1.19 (48%) (25°C).

**Colour:** Colourless.
**Explosive Limits:** Not flammable.
**Melting Point:** -37°C (48%); -70°C (70%).

## HAZARD DATA

### Human Health

**Symptoms: Contact:** severe burns to skin and eyes. **Inhalation:** may severely burn respiratory tract, rapid lung inflammation and congestion. **Ingestion:** severe tissue burns. Can be fatal by all routes.
**Toxicology:** Highly toxic by all routes.
**TLV®** – (inhalation) 3 ppm; 2.5 mg/m$^3$ (as F).
Short-term Inhalation Limits – 6 ppm; 5 mg/m$^3$ (15 min).

$LC_{50}$ - Inhalation: rat = 1 276 ppm/1 h
Delayed Toxicity - No information.

$LD_{50}$ - No information.
$LDL_o$ - Intraperitoneal: rat = 0.025 g/kg

### Fire

**Fire Extinguishing Agents:** Not combustible. Water may be used on fires involving hydrofluoric acid.
**Behaviour in Fire:** Not combustible. Toxic and corrosive HF vapours emitted.
**Ignition Temperature:** Not combustible.
**Burning Rate:** Not combustible.

### Reactivity

**With Water:** Soluble with evolution of heat.
**With Common Materials:** Reacts violently with calcium oxide, chlorosulfonic acid with acetic anhydride, ammonium hydroxide, ethylenediamine, fluorine, oleum, propylene oxide, sodium, sodium hydroxide, sulfuric acid and vinyl acetate.
**Stability:** Stable.

### Environment

**Water:** Prevent entry into water intakes and waterways. **Fish toxicity:** 60 ppm/tns/fish/lethal/freshwater; >300 mg/L as NaF/48 h/brine shrimp/$LC_{50}$/saltwater; BOD: None.
**Land-Air:** No information.
**Food Chain Concentration Potential:** None.

## EMERGENCY MEASURES

### Special Hazards
CORROSIVE. POISON. TOXIC.

### Immediate Responses
Keep non-involved people away from spill area. Issue warning: "CORROSIVE, POISON". Contact supplier for advice. Avoid contact and inhalation. Stop or reduce discharge, if this can be done without risk. Stay upwind and if water spray is used to control vapours; dike runoff. Notify environmental authorities.

### Protective Clothing and Equipment
Respiratory protection - self-contained breathing apparatus and totally encapsulated suit or acid suit. Boots - high, neoprene (pants worn outside boots). Gloves - neoprene.

### Fire and Explosion
Not combustible. Water may be used on fires involving hydrofluoric acid.

### First Aid
Move victim out of spill area to fresh air. Call for medical assistance, but start first aid at once. Inhalation: give artificial respiration if breathing has stopped (not mouth-to-mouth method); give oxygen if breathing is laboured. Contact: remove contaminated clothing; wash eyes and skin with plenty of warm water for at least 15 minutes. Ingestion: give milk or water to conscious victim to drink. DO NOT INDUCE VOMITING. If vomiting occurs spontaneously, give more water to further dilute the chemical. Keep victim warm and quiet. If medical assistance is not immediately available, transport victim to hospital, clinic or doctor.

## ENVIRONMENTAL PROTECTION MEASURES

### Response

**Water**
1. Stop or reduce discharge if safe to do so.
2. Contact manufacturer or supplier for advice.
3. If possible, contain discharge by damming or water diversion.
4. Notify environmental authorities to discuss disposal and cleanup of contaminated materials.

**Land-Air**
1. Stop or reduce discharge if safe to do so.
2. Contact manufacturer or supplier for advice.
3. Contain spill by diking with earth or other barrier.
4. Remove material with pumps or vacuum equipment and place in appropriate containers.
5. Absorb residual liquid on natural or synthetic sorbents.
6. Contaminated soil can be treated with lime (calcium hydroxide) to render it neutral.
7. Notify environmental authorities to discuss disposal and cleanup of contaminated materials.

### Disposal
1. Contact manufacturer or supplier for advice on disposal.
2. Contact environmental authorities for advice on disposal.

HYDROFLUORIC ACID   HF aqueous

# HYDROGEN  H₂

## IDENTIFICATION

UN No. 1049 (compressed gas)

| Common Synonyms | Observable Characteristics | Manufacturers |
|---|---|---|
| | Colourless gas. Odourless. | Canadian manufacturers: Syncrude Canada, Fort McMurray, Alberta. Canadian Fertilizers, Medicine Hat, Alberta. Alberta Gas Chemicals, Medicine Hat, Alberta. Cominco, Carselands, Alberta. |

| Transportation and Storage Information | | |
|---|---|---|
| **Shipping State:** Gas (compressed) and liquid (compressed gas). **Classification:** Flammable gas. **Inert Atmosphere:** No requirement. **Venting:** Safety relief. | **Label(s):** Red label - FLAMMABLE GAS; Class 2.1. **Storage Temperature:** Ambient. **Hose Type:** Gas - braided-high pressure. | **Grades or Purity:** Technical; pure from 99.8% to ultra pure. **Containers and Materials:** As gas in cylinders, tube trailers; steel. As liquid in cargo or portable tanks. |

| Physical and Chemical Characteristics | | |
|---|---|---|
| **Physical State** (20°C, 1 atm): Gas. **Solubility** (Water): 0.00015 g/100 mL (20°C). **Molecular Weight:** 2.0 **Vapour Pressure:** 8 590 mm Hg (-241°C). **Boiling Point:** -252.8°C. | **Floatability** (Water): Floats and boils. **Odour:** Odourless. **Flash Point:** <-50°C. **Vapour Density:** 0.07 (gas) (25°C). **Specific Gravity:** (liquid) 0.07 (-250°C). | **Colour:** Colourless. **Explosive Limits:** 4.0 to 75%. **Melting Point:** -259.2°C. |

## HAZARD DATA

### Human Health
**Symptoms:** Inhalation: acts as an asphyxiant by displacing air; resulting in cyanosis, diminished mental alertness, impairment and collapse. Contact: skin - frostbite.
**Toxicology:** Low toxicity by all routes.                   LC₅₀ - No information.                   LD₅₀ - No information.
TLV® - Asphyxiant.
Short-term Inhalation Limits - No information.               Delayed Toxicity - No information.

### Fire
**Fire Extinguishing Agents:** Shut off leak before attempting to extinguish fire. Use carbon dioxide, dry chemical, water or halons to extinguish. Water may be used to cool fire-exposed containers.
**Behaviour in Fire:** Containers may rupture violently.
**Ignition Temperature:** 500°C.     **Burning Rate:** 9.9 mm/min.

### Reactivity
**With Water:** No reaction.
**With Common Materials:** Can react vigorously with oxidizing materials. Reacts violently with bromine, chlorine, chlorine trifluoride, fluorine, lithium, nitrogen trifluoride, 1-pentol, and finely divided platinum in air.
**Stability:** Stable.

### Environment
**Water:** No information.
**Land-Air:** No information.
**Food Chain Concentration Potential:** No information.

## EMERGENCY MEASURES

**Special Hazards**
FLAMMABLE.

**Immediate Responses**
Keep non-involved people away and upwind from spill site. Issue warning: "FLAMMABLE". CALL FIRE DEPARTMENT. Eliminate all sources of ignition. Notify manufacturer. Avoid contact with liquid. Stop or reduce discharge, if safe to do so. Notify environmental authorities.

**Protective Clothing and Equipment**
Outer protective clothing as required.

**Fire and Explosion**
Shut off leak before attempting to extinguish fire. Use carbon dioxide, dry chemical, water or halons to extinguish. Water may be used to cool fire-exposed containers. Containers may rupture violently.

**First Aid**
Move victim out of spill site to fresh air. Call for medical assistance, but start first aid at once. <u>Inhalation</u>: if breathing has stopped, give artificial respiration; if laboured, give oxygen. <u>Contact</u>: skin and eyes – remove contaminated clothing, flush affected areas with plenty of water and treat as for frostbite. If medical assistance is not immediately available, transport victim to doctor, clinic or hospital.

## ENVIRONMENTAL PROTECTION MEASURES

**Response**

**Water**
1. Stop or reduce discharge if safe to do so.
2. Contact manufacturer or supplier for advice.
3. Notify environmental authorities to discuss disposal and cleanup of contaminated materials.

**Land-Air**
1. Stop or reduce discharge if safe to do so.
2. Contact manufacturer or supplier for advice.
3. Notify environmental authorities to discuss disposal and cleanup of contaminated materials.

**Disposal**
1. Contact manufacturer or supplier for advice on disposal.
2. Contact environmental authorities for advice on disposal.

HYDROGEN  $H_2$

# HYDROGEN CHLORIDE   HCl (anhydrous)

## IDENTIFICATION

**UN No. 1050**

### Common Synonyms
HYDROCHLORIC ACID (anhydrous)

### Observable Characteristics
Clear, colourless gas. Pungent, suffocating odour. Fumes in contact with humid air.

### Manufacturers
Dow Chemical Canada Inc., Sarnia, Ontario.
Canadian Industries Ltd.,
Becancour, Quebec.
Du Pont Canada, Maitland, Ontario.

### Transportation and Storage Information
**Shipping State:** Liquid (compressed gas).
**Classification:** Poisonous gas.
**Inert Atmosphere:** No requirement.
**Venting:** Safety relief.

**Label(s):** White Label - POISONOUS GAS; Class 2.3.
**Storage Temperature:** Ambient.
**Hose Type:** Monel, stainless steel. Flexible braided, pressure.

**Grades or Purity:** Technical, 97.5 to 99%.
**Containers and Materials:** Cylinders, tank cars; steel, stainless steel.

### Physical and Chemical Characteristics
**Physical State** (20°C, 1 atm): Gas.
**Solubility (Water):** Soluble 82 g/100 mL (0°C); 56.1 g/100 mL (60°C).
**Molecular Weight:** 36.5
**Vapour Pressure:** 1 500 mm Hg (−76.6°C); 30 400 mm Hg (17°C).
**Boiling Point:** −85.0°C

**Floatability (Water):** Liquid, sinks and mixes.
**Odour:** Highly irritating, pungent and suffocating. (1 to 10 ppm, odour threshold).
**Flash Point:** Not flammable.
**Vapour Density:** 1.3
**Specific Gravity:** (liquid) 1.19 (−85°C).

**Colour:** Colourless.
**Explosive Limits:** Not flammable.
**Melting Point:** −114.2°C.

## HAZARD DATA

### Human Health
**Symptoms: Inhalation:** vapour produces severe irritation of the upper respiratory tract, coughing, burning of throat, choking and possibly death. **Contact: eyes** - causes severe burns and damage; skin - burns. **Ingestion:** severe irritation, burning of throat and gastrointestinal tract. May be fatal.
**Toxicology:** Toxic by all routes.
TLV* - (inhalation) 5 ppm; 7 mg/m$^3$.   $LC_{50}$ - Inhalation: rat = 4 701 ppm/30 min.   $LD_{50}$ - Oral: rabbit = 0.9 g/kg (solution).
Short-term Inhalation Limits - No information.   Delayed Toxicity - None known.

### Fire
**Fire Extinguishing Agents:** Not combustible. Water may be used on fires involving hydrogen chloride. Water may be used to cool fire-exposed containers and knock down vapours.
**Behaviour in Fire:** Not combustible.
**Ignition Temperature:** Not combustible.   **Burning Rate:** Not combustible.

### Reactivity
**With Water:** Soluble; forms hydrochloric acid.
**With Common Materials:** Reacts violently with calcium carbide, cesium carbide, lithium silicide, magnesium boride, mercuric sulfate and sodium.
**Stability:** Stable.   Corrosive to metals when wet.

### Environment
**Water:** Prevent entry into water intakes and waterways. Harmful to aquatic life in low concentrations. Fish toxicity: 282 ppm/96 h/mosquito fish/TLm/ freshwater; 100 to 330 ppm/48 h/shrimp/$LC_{50}$/saltwater; BOD: None.
**Land-Air:** No information.
**Food Chain Concentration Potential:** None.

## EMERGENCY MEASURES

### Special Hazards
CORROSIVE.

### Immediate Responses
Keep non-involved people away from spill site. Issue warning: "CORROSIVE". Call Fire Department. Evacuate from downwind. Contact manufacturer. Avoid all contact and inhalation. Notify environmental authorities.

### Protective Clothing and Equipment
Respiratory protection - self-contained breathing apparatus and totally encapsulated suit. Boots - high, rubber (pants worn outside boots). Gloves - rubber.

### Fire and Explosion
Not combustible. Water may be used on fires involving hydrogen chloride. Water may also be used to cool fire-exposed containers and knock down vapours.

### First Aid
Move victim out of spill area to fresh air. Call for medical assistance, but start first aid at once. Contact: immediately flush eyes and skin with plenty of warm water for 30 minutes or more and remove contaminated clothing. Inhalation: if not breathing, give artificial respiration (not mouth-to-mouth method); if breathing is laboured, give oxygen. Ingestion: give milk or water to conscious victim. Do not induce vomiting. Keep warm and quiet. If medical assistance is not immediately available, transport victim to hospital, doctor or clinic.

## ENVIRONMENTAL PROTECTION MEASURES

### Response

**Water**
1. Stop or reduce discharge if safe to do so.
2. Contact manufacturer or supplier for advice.
3. If possible, contain discharge by damming or water diversion.
4. Dredge or vacuum pump to remove contaminants, liquids and contaminated bottom sediments.
5. Notify environmental authorities to discuss disposal and cleanup of contaminated materials.

**Land-Air**
1. Stop or reduce discharge if safe to do so.
2. Contact manufacturer or supplier for advice.
3. Dike to prevent runoff from rainwater or water application.
4. Recover undamaged containers.
5. Neutralize contaminated soil with lime (calcium hydroxide).
6. Notify environmental authorities to discuss disposal and cleanup of contaminated materials.

### Disposal
1. Contact manufacturer or supplier for advice on disposal.
2. Contact environmental authorities for advice on disposal.

**HYDROGEN CHLORIDE    HCl (anhydrous)**

# HYDROGEN FLUORIDE   HF (anhydrous)

## IDENTIFICATION

UN No. 1052

### Common Synonyms
HYDROFLUORIC ACID (anhydrous)

### Observable Characteristics
Colourless liquid or gas. Pungent, suffocating odour. Fumes in humid air.

### Manufacturers
Allied Chemical Ltd., Amherstburg, Ontario, Valleyfield, Quebec. Alcan Smelters and Chemicals, Jonquière, Quebec.

### Transportation and Storage Information
**Shipping State:** Liquid (compressed gas).
**Classification:** Corrosive liquid, poison.
**Inert Atmosphere:** No requirement.
**Venting:** Safety relief.

**Label(s):** White and black label - CORROSIVE. White label - POISON; Class 8, 2.3.
**Storage Temperature:** Ambient.
**Hose Type:** Halon TFE plastic hose, Hypalon.

**Grades or Purity:** Technical, 99.0%.
**Containers and Materials:** Cylinders, tank cars; steel, nickel-steel alloys.

### Physical and Chemical Characteristics
**Physical State** (20°C, 1 atm): Gas.
**Solubility (Water):** Soluble in all proportions.
**Molecular Weight:** 20.0
**Vapour Pressure:** 775 mm Hg (20°C); 1 200 mm Hg (33°C).
**Boiling Point:** 19.5°C

**Floatability (Water):** Soluble.
**Odour:** Irritating, pungent (0.1 ppm, odour threshold)
**Flash Point:** Not flammable.
**Vapour Density:** 1.86 (25°C).
**Specific Gravity:** 0.96 (liquid) (20°C).

**Colour:** Colourless.
**Explosive Limits:** Not flammable.
**Melting Point:** -83.4°C.

## HAZARD DATA

### Human Health
**Symptoms:** Inhalation: irritation of mucous membranes, pain in throat, difficulty breathing, headache, fatigue, shock, coma. Contact: skin - readily absorbed, extreme burns, blisters and damage to tissue; eyes - burning, irreparable damage. Ingestion: irritation and burning of lips, mouth and throat, intense thirst, coughing, difficulty breathing, nausea and vomiting, shock, convulsions, coma. Can be fatal by all routes.
**Toxicology:** Highly toxic by inhalation, ingestion and skin absorption.
TLV• - (inhalation) 3 ppm; 2.5 mg/m$^3$ (as F).     $LC_{50}$ - Inhalation: rat = 1 276 ppm/1 h      $LD_{50}$ - No information.
Short-term Inhalation Limits - 6 ppm; 5 mg/m$^3$    Delayed Toxicity - No information.             $LD_{Lo}$ - Intraperitoneal: rat = 0.025 g/kg
(15 min) (as F).

### Fire
**Fire Extinguishing Agents:** Not combustible; however, water may be used on fires involving HF. Water may be used to cool fire-exposed containers and knock down vapours.
**Behaviour in Fire:** Not combustible.           **Burning Rate:** Not combustible.
**Ignition Temperature:** Not combustible.

### Reactivity
**With Water:** Soluble; forming hydrofluoric acid.
**With Common Materials:** Reacts violently with arsenic trioxide and phosphorus pentoxide.
**Stability:** Stable.

### Environment
**Water:** Prevent entry into water intakes and waterways. Harmful to aquatic life in very low concentrations. Fish toxicity: 60 ppm/tns/fish/lethal/freshwater; >300 ppm/48 h/shrimp/$LC_{50}$/saltwater.
**Land-Air:** No information.
**Food Chain Concentration Potential:** No information.

## EMERGENCY MEASURES

**Special Hazards**

CORROSIVE. POISON.

**Immediate Responses**

Keep non-involved people away from spill area. Issue warning: "CORROSIVE, POISON". Call Fire Department. Contact manufacturer or supplier for advice. Avoid contact and inhalation. Stop or reduce discharge, if this can be done without risk. Stay upwind and if water spray is used to control vapours, dike runoff. Notify environmental authorities.

**Protective Clothing and Equipment**

Respiratory protection - self-contained breathing apparatus and totally encapsulated acid suit are mandatory. Boots - high, neoprene (pants worn outside boots). Gloves - neoprene.

**Fire and Explosion**

Not combustible. Water may be used to fight fires involving HF, and to cool fire-exposed containers and knock down vapours.

**First Aid**

Move victim out of spill area to fresh air. Call for medical assistance, but start first aid at once. Inhalation: if breathing has stopped, give artificial respiration (not mouth-to-mouth method); if laboured, give oxygen. Contact: remove contaminated clothing; wash eyes and skin with plenty of warm water for at least 15 minutes. Ingestion: give milk or water to conscious victim. Do not induce vomiting. If vomiting occurs, give more water to further dilute the chemical. Keep patient warm and quiet. If medical assistance is not immediately available, transport victim to hospital, clinic or doctor.

## ENVIRONMENTAL PROTECTION MEASURES

**Response**

**Water**
1. Stop or reduce discharge if safe to do so.
2. Contact manufacturer or supplier for advice.
3. If possible, contain discharge by damming or water diversion.
4. Notify environmental authorities to discuss disposal and cleanup of contaminated materials.

**Land-Air**
1. Stop or reduce discharge if safe to do so.
2. Contact manufacturer or supplier for advice.
3. Contain spill by diking with earth or other barrier.
4. Remove material with pumps or vacuum equipment and place in appropriate containers.
5. Absorb residual liquid on natural or synthetic sorbents.
6. Contaminated soil can be treated with lime (calcium hydroxide) to render it neutral.
7. Notify environmental authorities to discuss disposal and cleanup of contaminated materials.

**Disposal**
1. Contact manufacturer or supplier for advice on disposal.
2. Contact environmental authorities for advice on disposal.

**HYDROGEN FLUORIDE    HF (anhydrous)**

# HYDROGEN PEROXIDE   H₂O₂ (aqueous)

## IDENTIFICATION

UN No. 2015, >60%
2014, 8 to 60%

### Common Synonyms
PEROXIDE
HYDROGEN DIOXIDE
HYDROPEROXIDE

### Observable Characteristics
Clear, colourless liquid. Slightly sharp odour.

### Manufacturers
No Canadian manufacturers.
Canadian suppliers:
B.C. Forest Products, Crofton, B.C.
Canadian Industries Ltd., Cornwall, Ont.
Du Pont Canada, Maitland, Ontario.

Originating from:
FMC Corp., USA

E.I. Du Pont de Nemours, USA

### Transportation and Storage Information

**Shipping State:** Liquid (aqueous solution).
**Classification:** Oxidizer and corrosive.
**Inert Atmosphere:** No requirements.
**Venting:** Safety relief or pressure-vacuum.
**Pump Type:** Aluminum, or all plastic; stainless steel.

**Label(s):** Yellow label - OXIDIZER; Class 5.1, Group I. Black and white label - CORROSIVE; Class 8, Group I.
**Storage Temperature:** Ambient.
**Hose Type:** Polyethylene, PVC (Tygon), 304 or 316 braided stainless steel.

**Grades or Purity:** Many concentrations 27.5, 30, 50, 70, 98% as aqueous solutions.
**Containers and Materials:** Drums (aluminum, polyethylene); tank cars, tank trucks; aluminum.

### Physical and Chemical Characteristics

**Physical State** (20°C, 1 atm): Liquid (aqueous solution).
**Solubility (Water):** Soluble in all proportions.
**Molecular Weight:** 34.0 (for solute only).
**Vapour Pressure:** 27.5%, 25 mm Hg; 35%, 23 mm Hg; 50%, 18 mm Hg; 70%, 11 mm Hg (30°C).
**Boiling Point:** 27.5%, 105°C; 35%, 108°C; 50%, 114°C; 70%, 125°C; 150.2°C (pure).

**Floatability (Water):** Mixes.
**Odour:** Slightly sharp.
**Flash Point:** Not flammable.
**Vapour Density:** 1.2
**Specific Gravity:** 1.13 (35%); 1.20 (50%); 1.29 (70%) all at (20°C); 1.45 (pure).

**Colour:** Colourless.
**Explosive Limits:** Not flammable.
**Melting Point:** (35%) -32.8°C; (50%) -50°C; (70%) -39°C; -0.41°C (pure).

## HAZARD DATA

### Human Health

**Symptoms: Inhalation:** sore throat, coughing, laboured breathing, headache, nausea, weakness. **Ingestion:** pain in swallowing, intense thirst, nausea and vomiting, difficulty breathing, convulsions, coma. **Contact:** skin - burning, painful blistering; shock; eyes - stinging, burning, opaqueness of cornea.
**Toxicology:** Highly toxic by all routes.
TLV® - (inhalation) 1 ppm; 1.5 mg/m³. LC$_{50}$ - Inhalation: rat = 2 000 mg/m³/4 h (90%). LD$_{50}$ - Skin: rat = 4.06 g/kg (90%).
Short-term Inhalation Limits - 2 ppm; 3 mg/m³  Delayed Toxicity - None known.
(15 min).

### Fire

**Fire Extinguishing Agents:** Not combustible. In fires involving H₂O₂ use water; H₂O₂ may cause other firefighting agents to be ineffective. Use water to cool fire-exposed containers.
**Behaviour in Fire:** Can cause ignition of combustible material on contact. May decompose with explosive violence on contact with iron, copper, chromium, brass, bronze, lead, silver, manganese and their salts.
**Ignition Temperature:** Not combustible.  **Burning Rate:** Not combustible.

### Reactivity

**With Water:** No reaction. Soluble.
**With Common Materials:** Reacts violently with acetic acid, acetic anhydride, acetone, antimony and arsenic trisulfide, brass, bronze, t-butylalcohol, cellulose, charcoal, chlorosulfonic acid, chromium, copper, cupric sulfide, ethyl alcohol, ferrous sulfide, formic acid, organic matter, hydrazine, iron, lead, lead sulfide, magnesium, manganese, manganese dioxide, mercuric oxide, nitric acid, platinum, potassium, potassium permanganate, silver, sodium, and sodium iodate.
**Stability:** Stable, within the limits of the foregoing. Higher concentrations are less stable and more reactive.

### Environment

**Water:** Prevent entry into water intakes and waterways. Fish toxicity:  40 ppm/tns/(fingerling trout)/toxic/saltwater; BOD: None.
**Land-Air:** No information.
**Food Chain Concentration Potential:** None.

## EMERGENCY MEASURES

**Special Hazards**

OXIDIZER. CORROSIVE. Reactive.

**Immediate Responses**

Keep non-involved people away from spill site. Issue warning: "OXIDIZER, CORROSIVE". Call Fire Department. Contact manufacturer or supplier for advice and possible assistance. Avoid contact and inhalation. Stop or reduce discharge, if this can be done without risk. Contain spill by diking with earth. Notify environmental authorities.

**Protective Clothing and Equipment**

Respiratory protection - self-contained breathing apparatus. Boots - neoprene, or rubber (pants worn outside boots). Gloves - rubber, neoprene or PVC. Outerwear - rubber, neoprene or PVC. Flammable clothing, such as cotton, rayon, leather or wool, should not be worn when in contact with high-strength hydrogen peroxide.

**Fire and Explosion**

Not combustible. In fires involving $H_2O_2$, use water; $H_2O_2$ may cause other firefighting agents to be ineffective. Use water to cool fire-exposed containers.

**First Aid**

Move victim out of spill area to fresh air. Call for medical assistance, but start first aid at once. Contact: eyes - irrigate with plenty of water; skin - remove contaminated clothing and flush affected areas with plenty of water. Ingestion: give water to conscious victim to drink. Inhalation: remove at once from further exposure. If medical assistance is not immediately available, transport victim to hospital, doctor or clinic.

## ENVIRONMENTAL PROTECTION MEASURES

**Response**

**Water**
1. Stop or reduce discharge if safe to do so.
2. Contact manufacturer or supplier for advice.
3. If possible, contain discharge by damming or water diversion.
4. Notify environmental authorities to discuss disposal and cleanup of contaminated materials.

**Land-Air**
1. Stop or reduce discharge if safe to do so.
2. Contact manufacturer or supplier for advice.
3. Contain spill by diking with earth.
4. If concentrated, dilute with at least an equal volume of water.
5. Remove material with pumps or vacuum equipment and place in appropriate containers.
6. Recover undamaged containers.
7. Notify environmental authorities to discuss disposal and cleanup of contaminated materials.

**Disposal**
1. Contact manufacturer or supplier for advice on disposal.
2. Contact environmental authorities for advice on disposal.

HYDROGEN PEROXIDE   $H_2O_2$ (aqueous)

# HYDROGEN SULFIDE   $H_2S$

## IDENTIFICATION

| Common Synonyms | Observable Characteristics | UN No. 1053 Manufacturers |
|---|---|---|
| SULPHURETTED HYDROGEN<br>HYDROGEN SULPHIDE<br>SULFUR HYDRIDE | Colourless gas. Offensive rotten-egg odour. | Thio-Pet Chemicals Limited,<br>Fort Saskatchewan, Alta.<br>Laurentide Chemicals & Sulphur Ltd.,<br>Montreal East, Que. |

### Transportation and Storage Information

**Shipping State:** Liquid (compressed gas).
**Classification:** Flammable, poisonous gas.
**Inert Atmosphere:** No requirement.
**Venting:** Safety relief.

**Label(s):** Red label - FLAMMABLE GAS; Class 3.
White label - POISON; Class 2.3.
**Storage Temperature:** Ambient.

**Grades or Purity:** Technical, 98.5%; $H_2S$ present in many concentrations from gas wells.
**Containers and Materials:** Cylinders, tank cars; steel.

### Physical and Chemical Characteristics

**Physical state** (20°C, 1 atm): Gas.
**Solubility** (Water): 0.5 g/100 mL (20°C).
**Molecular Weight:** 34.1
**Vapour Pressure:** 7 600 mm Hg (-0.4°C); 15 200 mm Hg (25.5°C).
**Boiling Point:** -60.2°C.

**Floatability** (Water): Liquid floats, boils and dissolves.
**Odour:** Rotten-eggs (0.005 ppm, odour threshold).
**Flash Point:** <-50°C.
**Vapour Density:** 1.2 (0°C).
**Specific Gravity:** 0.95 (-60°C).

**Colour:** Colourless.
**Explosive Limits:** 4.0 to 44%.
**Melting Point:** -83 to -86°C.

## HAZARD DATA

### Human Health

**Symptoms:** Insidious poison, since sense of smell may be fatigued and fail to give warning of high concentrations. Inhalation: irritation of nose, throat and eyes, sneezing, headache, dizziness, nausea and vomiting, pale complexion, cold sweat, diarrhea, muscular weakness, drowsiness, unconsciousness and death. Contact: skin - irritation, painful inflammation; eyes - irritation, watering, inflammation.
**Toxicology:** Highly toxic by inhalation and contact.
TLV* - (inhalation) 10 ppm; 14 mg/m³.     $LC_{50}$ - Inhalation: mouse = 673 ppm/1 h     $LD_{50}$ - No information.
Short-term Inhalation Limits - 15 ppm; 21 mg/m³     $LC_{50}$ - Inhalation: rat = 444 ppm.
(15 min).     Delayed Toxicity - No information.

### Fire

**Fire Extinguishing Agents:** Stop flow of gas. Use water to keep fire-exposed containers cool.
**Behaviour in Fire:** Flashback may occur along vapour trail. Emits fumes of $SO_x$ when heated to decomposition.
**Ignition Temperature:** 260°C.     **Burning Rate:** (Liquid) 2.3 mm/min.

### Reactivity

**With Water:** No reaction; soluble.
**With Common Materials:** Can react vigorously with oxidizing materials. Reacts violently with chlorine monoxide, chromic anhydride, copper, fluorine, acetaldehyde, lead dioxide, nitric acid, nitrogen trichloride, nitrogen trifluoride.
**Stability:** Stable.

### Environment

**Water:** Prevent entry into water intakes and waterways. Harmful to aquatic life in low concentrations. Fish toxicity: 1.38 ppm/48 h/fathead minnow/TLm/freshwater; 10 mg/L/96 h/goldfish/$LC_{100}$/freshwater; 0.86 ppm/24 h/trout/$LC_{100}$/freshwater; BOD: Not available.
**Land-Air:** No information.
**Food Chain Concentration Potential:** None.

## EMERGENCY MEASURES

**Special Hazards**
FLAMMABLE. POISON. Odour sensation lost with high concentrations or long period of exposure.

**Immediate Responses**
Keep non-involved people away from spill site. Issue warning: "FLAMMABLE, POISON". CALL FIRE DEPARTMENT. Call manufacturer for guidance and assistance. Eliminate all ignition sources. If necessary, evacuate people downwind. Work from upwind. Use water spray to control vapours and to cool containers. Notify environmental authorities.

**Protective Clothing and Equipment**
Respiratory protection - self contained breathing apparatus and totally encapsulated suit. Gloves - rubber. Boots - rubber.

**Fire and Explosion**
Stop flow of gas. Use water to cool fire-exposed containers. Flash back may occur along vapour trail. Emits toxic $SO_x$ fumes when heated to decomposition.

**First Aid**
Move victim out of spill site to fresh air. Call medical assistance, but start first aid at once. Inhalation: if breathing has stopped, give artificial respiration (not mouth-to-mouth method); if laboured, give oxygen. Keep patient warm. Contact: skin - remove contaminated clothing and flush affected areas with plenty of water; eyes - irrigate with plenty of water. If medical assistance is not immediately available, transport victim to hospital, doctor or clinic.

## ENVIRONMENTAL PROTECTION MEASURES

**Response**

*Water-Land-Air*
Because of the properties of hydrogen sulfide, its shipment in refrigerated, pressurized tank cars with specialized valve and relief devices, it is inadvisable for an inexperienced person to try to control an emergency situation.
1. Contact the manufacturer, who will suggest appropriate action or may be able to arrange for on-the-site assistance. Do not attempt to control a leak or deal with an emergency situation without this guidance.
2. Response possibilities include: allowing dispersal into the atmosphere.
3. Vapours can be controlled with fog or water spray.
4. Notify environmental authorities to discuss disposal and cleanup of contaminated materials.

**Disposal**
1. Contact manufacturer or supplier for advice on disposal.
2. Contact environmental authorities for advice on disposal.

HYDROGEN SULFIDE  $H_2S$

# ISOPROPYL ALCOHOL  CH$_3$CHOHCH$_3$

## IDENTIFICATION

**Common Synonyms**
ISOPROPANOL
DIMETHYLCARBINOL
2-PROPANOL
sec-PROPYL ALCOHOL
RUBBING ALCOHOL

**Observable Characteristics**
Clear, colourless liquid. Alcohol odour.

**UN No. 1219**

**Manufacturers**
Gulf Oil Canada Ltd.,
Montreal, Quebec.
Shell Canada Ltd.,
Corunna, Ontario.

### Transportation and Storage Information

**Shipping State:** Liquid.
**Classification:** Flammable liquid.
**Inert Atmosphere:** No requirement.
**Venting:** Open (flame arrester) or pressure-vacuum.
**Pump Type:** Gear, centrifugal, flammable liquid types.

**Label(s):** Red label - FLAMMABLE LIQUID; Class 3.2, Group II.
**Storage Temperature:** Ambient.
**Hose Type:** Polyethylene, polypropylene, butyl, Hypalon, natural rubber.

**Grades or Purity:** Commercial, technical, 91%, 95%, 99%.
**Containers and Materials:** Drums, tank cars, tank trucks; steel.

### Physical and Chemical Characteristics

**Physical State (20°C, 1 atm):** Liquid.
**Solubility (Water):** Soluble in all proportions.
**Molecular Weight:** 60.1
**Vapour Pressure:** 32 mm Hg (20°C); 44 mm Hg (25°C); 57 mm Hg (30°C).
**Boiling Point:** 82.4°C.

**Floatability (Water):** Floats and mixes.
**Odour:** Alcohol-like (40 ppm, odour threshold).
**Flash Point:** 12°C (c.c.); 14°C (c.c. 88% solution).
**Vapour Density:** 2.1
**Specific Gravity:** 0.79 (20°C).

**Colour:** Colourless.
**Explosive Limits:** 2.3% to 12.7%.
**Melting Point:** -86 to -89°C.

## HAZARD DATA

### Human Health

**Symptoms:** Contact: eyes - severely irritating; skin - defatting and cracking. Inhalation: causes depression of the central nervous system, nausea, vomiting, depressed respiration, abdominal pain. Ingestion: dizziness, headache, nausea, vomiting.
**Toxicology:** Low toxicity by contact and moderate toxicity by ingestion.
TLV* (inhalation, skin) 400 ppm; 980 mg/m$^3$.   LC$_{50}$ - No information.   LD$_{50}$ - Oral: rat = 5.84 g/kg
Short-term Inhalation Limits - 500 ppm,   Delayed Toxicity - No information.
1 225 mg/m$^3$ (15 min).

### Fire

**Fire Extinguishing Agents:** For fires use dry chemical, carbon dioxide or alcohol foam. Water may be ineffective but may be used to cool fire-exposed containers.
**Behaviour in Fire:** Flashback may occur along vapour trail.
**Ignition Temperature:** 399°C.   **Burning Rate:** 2.3 mm/min.

### Reactivity

**With Water:** No reaction; soluble.
**With Common Materials:** Can react vigorously with oxidizing materials. Can react violently with oleum, phosgene and potassium-t-butoxide.
**Stability:** Stable.

### Environment

**Water:** Prevent entry into water intakes and waterways. Harmful to aquatic life. Fish toxicity: 900 to 1 100 ppm/24 h/chub/critical/freshwater; 1 400 mg/L/48 h/brine shrimp/LC$_{50}$/saltwater; Aquatic toxicity rating = 100 to 1 000 ppm/96 h/TLm/freshwater; BOD: 129 to 159%; 5 days.
**Land-Air:** No information.
**Food Chain Concentration Potential:** None.

## EMERGENCY MEASURES

**Special Hazards**

FLAMMABLE.

**Immediate Responses**

Keep non-involved people away from spill site. Issue warning: "FLAMMABLE". CALL FIRE DEPARTMENT. Eliminate all ignition sources. Notify manufacturer or supplier. Dike to contain spill and prevent runoff. Stop or reduce discharge, if this can be done without risk. Notify environmental authorities.

**Protective Clothing and Equipment**

Respiratory protection - self-contained breathing apparatus (in fire). Gloves - rubber or plastic. Outer clothing - suitable for situation, coveralls, etc. Boots - rubber.

**Fire and Explosion**

Use dry chemical, carbon dioxide or alcohol foam. Water may be ineffective on fire, may be used to cool fire-exposed containers. Flash back may occur along vapour trail.

**First Aid**

Move victim out of spill area to fresh air. Call for medical assistance, but start first aid at once. Contact: remove contaminated clothing; wash eyes and skin with plenty of water. Inhalation: give artificial respiration if breathing has stopped; oxygen, if breathing is laboured. Ingestion: give water to conscious victim and induce vomiting. If medical assistance is not immediately available, transport victim to hospital, doctor or clinic.

## ENVIRONMENTAL PROTECTION MEASURES

**Response**

**Water**
1. Stop or reduce discharge if safe to do so.
2. Contact manufacturer or supplier for advice.
3. If possible, contain discharge by damming or water diversion.
4. Dredge vacuum pump to remove contaminants liquids and contaminated bottom sediments.
5. Notify environmental authorities to discuss disposal and cleanup of contaminated materials.

**Land-Air**
1. Stop or reduce discharge if safe to do so.
2. Contact manufacturer or supplier for advice.
3. Contain spill by diking with earth or other barrier.
4. Dike to prevent runoff from rainwater or water application.
5. Remove material with pumps or vacuum equipment and place in appropriate containers.
6. Recover undamaged containers.
7. Absorb residual liquid on natural or synthetic sorbents.
8. Notify environmental authorities to discuss disposal and cleanup of contaminated materials.

**Disposal**
1. Contact manufacturer or supplier for advice on disposal.
2. Contact environmental authorities for advice on disposal.

ISOPROPYL ALCOHOL    $CH_3CHOHCH_3$

# KEROSENE

## IDENTIFICATION

| Common Synonyms | Observable Characteristics | Manufacturers |
|---|---|---|
| STOVE (RANGE) OIL<br>JET FUEL: JP-1<br>ILLUMINATING OIL<br>KEROSINE<br>RANGE OIL<br>FUEL OIL NO. 1<br>COAL OIL | Colourless or pale yellow liquid. Gasoline-like odour. | Universally available. |

UN No. 1223

### Transportation and Shipping Information

**Shipping State:** Liquid.
**Classification:** Flammable liquid.
**Inert Atmosphere:** No requirement.
**Venting:** Open (flame arrester).
**Pump Type:** Gear or centrifugal.

**Label(s):** Red label - FLAMMABLE LIQUID; Class 3.3, Group II.
**Storage Temperature:** Ambient.
**Hose Type:** Neoprene, polyethylene, Viton.

**Grades or Purity:** Mixture of light hydrocarbon distillates.
**Containers and Materials:** Drums, tank cars, tank trucks; steel.

### Physical and Chemical Characteristics

**Physical State** (20°C, 1 atm): Liquid.
**Solubility** (Water): 0.0002 to 0.0004 g/100 mL (20°C).
**Molecular Weight:** Variable.
**Vapour Pressure:** ~5 mm Hg (20°C).
**Boiling Point:** 150 to 300°C.

**Floatability** (Water): Floats.
**Odour:** Gasoline-like (odour threshold about 1 ppm).
**Flash Point:** 43 to 72°C.
**Vapour Density:** 4.5
**Specific Gravity:** 0.8 to 0.85 (20°C).

**Colour:** Colourless or pale yellow.
**Explosive Limits:** 0.7 to 5%.
**Melting Point:** -43 to -49°C.

## HAZARD DATA

### Human Health

**Symptoms:** Inhalation: nausea, vomiting, coughing, headache. Ingestion: nausea, vomiting, weakness, dizziness, slow and shallow respiration, convulsions, unconsciousness. Contact: skin - irritating and defatting; eyes - irritation.
**Toxicology:** Very low toxicity by all routes.
**TLV®** - No information.
**Short-term Inhalation Limits** - No information.
**$LC_{50}$** - No information.
**Delayed Toxicity** - No information.
**$LD_{50}$** - Oral: guinea pig = 20 g/kg

### Fire

**Fire Extinguishing Agents:** Use dry chemical, foam, carbon dioxide. Water may be ineffective but may be used to cool fire-exposed containers.
**Behaviour in Fire:** No information.
**Ignition Temperature:** 210°C.
**Burning Rate:** 4 mm/min.

### Reactivity

**With Water:** No reaction.
**With Common Materials:** Can react with strong oxidizing agents.
**Stability:** Stable.

### Environment

**Water:** Prevent entry into water intakes and waterways. Fish toxicity: 2 990 ppm/24 h/bluegill/freshwater; 100-1 000 ppm/96 h/rainbow trout/$LC_{50}$/freshwater;
**BOD:** 53%, 5 days.
**Land-Air:** No information.
**Food Chain Concentration Potential:** None.

## EMERGENCY MEASURES

### Special Hazards
FLAMMABLE.

### Immediate Responses
Keep non-involved people away from spill site. Issue warning: "FLAMMABLE". CALL FIRE DEPARTMENT. Eliminate all sources of ignition. Notify supplier or manufacturer. Stop or reduce discharge, if this can be done without risk. Dike spill area to contain runoff. Notify environmental authorities.

### Protective Clothing and Equipment
Respiratory protection - in fires or closed spaces, use self-contained breathing apparatus; otherwise, Goggles or face shield. Coveralls. Rubber gloves. Rubber boots (high) - wear pants outside boots.

### Fire and Explosion
Use dry chemical, foam or carbon dioxide to extinguish. Water may be used to cool fire-exposed containers.

### First Aid
Move victim out of spill area to fresh air. Call for medical assistance, but start first aid at once. Inhalation: give artificial respiration if breathing has stopped, give oxygen if breathing is laboured. Contact: eyes - irrigate with plenty of water for at least 15 minutes; skin - remove contaminated clothing and flush affected areas with plenty of water. Ingestion: give conscious victim water to drink, do not induce vomiting. If medical assistance is not immediately available, transport victim to hospital, doctor or clinic.

## ENVIRONMENTAL PROTECTION MEASURES

### Response

**Water**
1. Stop or reduce discharge if safe to do so.
2. Contact manufacturer or supplier for advice.
3. If possible, contain discharge by booming.
4. If floating, skim and remove.
5. Notify environmental authorities to discuss disposal and cleanup of contaminated materials.

**Land-Air**
1. Stop or reduce discharge if safe to do so.
2. Contact manufacturer or supplier for advice.
3. Contain spill by diking with earth or other barrier.
4. Remove material with pumps or vacuum equipment and place in appropriate containers.
5. Recover undamaged containers.
6. Absorb residual liquid on natural or synthetic sorbents.
7. Notify environmental authorities to discuss disposal and cleanup of contaminated materials.

### Disposal
1. Contact manufacturer or supplier for advice on disposal.
2. Contact environmental authorities for advice on disposal.
3. Incinerate (approval of environmental authorities required).

**KEROSENE**

# LATEX

## IDENTIFICATION

| Common Synonyms | Observable Characteristics | Manufacturers |
|---|---|---|
| ARCONITRILE BUTADIENE RUBBER<br>PLASTIC LATEX<br>LATEX, LIQUID SYNTHETIC<br>SYNTHETIC RUBBER LATEX<br>NATURAL RUBBER LATEX | White liquid. Variable and characteristic odour. | Universally available. |

### Transportation and Shipping Information

**Shipping State:** Liquid.
**Classification:** None.
**Inert Atmosphere:** No requirement.
**Venting:** Open.
**Pump Type:** No information.

**Label(s):** Not regulated.
**Storage Temperature:** Ambient.
**Hose Type:** No information.

**Grades or Purity:** Regular and concentrated.
**Containers and Materials:** Drums, tank cars, tank trucks; steel.

### Physical and Chemical Characteristics

**Physical State** (20°C, 1 atm): Liquid.
**Solubility** (Water): Insoluble; but may mechanically mix.
**Molecular Weight:** Variable.
**Vapour Pressure:** No information.
**Boiling Point:** >200°C.

**Floatability** (Water): Sinks in freshwater; floats in saltwater (may mechanically mix).
**Odour:** Variable and characteristic.
**Flash Point:** Not flammable.
**Vapour Density:** No information.
**Specific Gravity:** 1.06 (25°C).

**Colour:** White.
**Explosive Limits:** Not flammable.
**Melting Point:** >150°C.

## HAZARD DATA

### Human Health

**Symptoms:** <u>Inhalation</u>: irritation of mucous membranes. <u>Ingestion</u>: dangerous to ingest, as coagulation may occur internally. <u>Contact</u>: skin - irritation, eyes - irritation.
**Toxicology:** Low toxicity by all routes.
**TLV®:** No information.
**Short-term Inhalation Limits** - No information.

$LC_{50}$ - No information.
**Delayed Toxicity** - No information.

$LD_{50}$ - No information.

### Fire

**Fire Extinguishing Agents:** Not flammable. Most fire extinguishing agents may be used on fires involving latex.
**Behaviour in Fire:** Splatters when burning.
**Ignition Temperature:** No information.

**Burning Rate:** No information.

### Reactivity

**With Water:** No reaction.
**With Common Materials:** No information.
**Stability:** Stable.

### Environment

**Water:** Prevent entry into water intakes and waterways.
**Land-Air:** No information.
**Food Chain Concentration Potential:** No information.

## EMERGENCY MEASURES

### Special Hazards

### Immediate Responses
Keep non-involved people away from spill site. Call Fire Department. Stop or reduce discharge if safe to do so. Notify manufacturer. Notify environmental authorities.

### Protective Clothing and Equipment
In fires Respiratory protection - self-contained breathing apparatus; otherwise protective clothing as required.

### Fire and Explosion
Not flammable. Most fire extinguishing agents may be used on fires involving latex. Splatters when burning.

### First Aid
Move victim out of spill site to fresh air. Call for medical assistance, but start first aid at once. Inhalation: if breathing has stopped give artificial respiration; if laboured, give oxygen. Contact: skin - remove contaminated clothing and wash affected areas with plenty of water; eyes - irrigate with water. Ingestion: give water to conscious victim to drink; get prompt medical assistance. If medical assistance is not immediately available, transport victim to doctor, clinic or hospital.

## ENVIRONMENTAL PROTECTION MEASURES

### Response

**Water**

Floating
1. Stop or reduce discharge if safe to do so.
2. Contact manufacturer or supplier for advice.
3. If possible, contain discharge by booming.
4. If floating, skim and remove.
5. Notify environmental authorities to discuss disposal and cleanup of contaminated materials.

Sinking
1. Stop or reduce discharge if safe to do.
2. Contact manufacturer or supplier for advice.
3. If possible, contain discharge by damming or water diversion.
4. Dredge or vacuum pump to remove contaminants, liquids and contaminated bottom sediments.
5. Notify environmental authorities to discuss disposal and cleanup of contaminated materials.

**Land-Air**
1. Stop or reduce discharge if safe to do so.
2. Contact manufacturer or supplier for advice.
3. Contain spill by diking with earth or other barrier.
4. Remove material by manual or mechanical means.
5. Recover undamaged containers.
6. Notify environmental authorities to discuss disposal and cleanup of contaminated materials.

### Disposal
1. Contact manufacturer or supplier for advice on disposal.
2. Contact environmental authorities for advice on disposal.
3. Incinerate (approval of environmental authorities required).
4. May be dumped in a municipal landfill (approval of environmental authorities required).

LATEX

# LEAD ACETATE  Pb(CH$_3$COO)$_2$·3H$_2$O

## IDENTIFICATION

### Common Synonyms
LEAD ACETATE TRIHYDRATE
SUGAR OF LEAD
PLUMBOUS ACETATE

### Observable Characteristics
White to brown or grey powder or flakes. Odourless.

### Manufacturers
Anachemia Chemicals Ltd., Montreal, Que., Toronto, Ont.

UN No. 1616

### Transportation and Storage Information
**Shipping State:** Solid.
**Classification:** Poison.
**Inert Atmosphere:** No requirement.
**Venting:** Open.
**Label(s):** White label - POISON; Class 6.1, Group III.
**Storage Temperature:** Ambient.
**Grades or Purity:** Technical, 97%.
**Containers and Materials:** Multiwall paper bags, drums.

### Physical and Chemical Characteristics
**Physical State** (20°C, 1 atm): Solid.
**Solubility** (Water): 45.6 g/100 mL (15°C); 200 g/100 mL (100°C).
**Molecular Weight:** 379.3
**Vapour Pressure:** No information.
**Boiling Point:** Decomposes at 200°C.
**Floatability** (Water): Sinks and mixes.
**Odour:** Odourless.
**Flash Point:** Not flammable.
**Vapour Density:** No information.
**Specific Gravity:** 2.55 (20°C).
**Colour:** White to brown or grey.
**Explosive Limits:** Not flammable.
**Melting Point:** Loses water at 75°C.

## HAZARD DATA

### Human Health
**Symptoms:** <u>Inhalation:</u> of dust, mist or fumes, irritation of nose and eyes, headache, stomach cramps and fatigue. <u>Ingestion:</u> metallic taste in mouth, constriction of throat, stomach pains, nausea, vomiting, diarrhea, convulsions, coma. <u>Contact:</u> skin - irritation, inflammation; eyes - inflammation.
**Toxicology:** High toxic by ingestion.
TLV® (inhalation) 0.15 mg/m$^3$ (as Pb).
Short-term Inhalation Limits - 0.45 mg/m$^3$ (15 min) (as Pb).
LC$_{50}$ - No information.
Delayed Toxicity - Cumulative poison. Suspected carcinogen.
LD$_{50}$ - Intraperitoneal: guinea pig = 0.22 g/kg
TD$_{Lo}$ - Oral: rat = 8.52 g/kg

### Fire
**Fire Extinguishing Agents:** Most fire extinguishing agents may be used. Use water sparingly.
**Behaviour in Fire:** In fires, toxic PbO fumes are released.
**Ignition Temperature:** >280°C.
**Burning Rate:** No information.

### Reactivity
**With Water:** No reaction; soluble.
**With Common Materials:** Reacts violently with KBrO$_3$ (potassium bromate).
**Stability:** Stable.

### Environment
**Water:** Prevent entry into water intakes and waterways. Fish toxicity: 7.5 mg/L (as Pb)/4 days/minnow/TLm/soft water.
**Land-Air:** No information.
**Food Chain Concentration Potential:** Fish and terrestrial animals are capable of concentrating lead.

## EMERGENCY MEASURES

**Special Hazards**

POISON.

**Immediate Responses**

Keep non-involved people away from spill site. Issue warning: "POISON". Avoid contact and inhalation. Dike to prevent runoff from rainwater or water application. Lightly wet down dry spillage to prevent wind drift of dust. Notify manufacturer or supplier. Notify environmental authorities.

**Protective Clothing and Equipment**

In fire or enclosed spaces, Respiratory protection - self-contained breathing apparatus. Otherwise, dust respirators (with suitable filters) or metal fume respirators for normal situations. Rubber gloves and boots will prevent contact. Frequent changes of clothing and footwear should be provided for contaminated clothing, and footwear should be washed immediately after contact.

**Fire and Explosion**

Most fire extinguishing agents may be used. Use water sparingly. In fires, toxic PbO fumes are released.

**First Aid**

Move victim out of spill site to fresh air. Call for medical assistance, but start first aid at once. Inhalation: if breathing has stopped give artificial respiration (not mouth-to-mouth method); if laboured, give oxygen. Contact: skin - remove contaminated clothing, and flush affected areas with plenty of water; eyes - irrigate with plenty of water. Ingestion: give water to conscious victim to drink, and induce vomiting. If medical assistance is not immediately available, transport victim to doctor, clinic or hospital.

## ENVIRONMENTAL PROTECTION MEASURES

**Response**

**Water**
1. Stop or reduce discharge if safe to do so.
2. Contact manufacturer or supplier for advice.
3. If possible, contain discharge by damming or water diversion.
4. Dredge or vacuum pump to remove contained contaminants and contaminated bottom sediments.
5. Notify environmental authorities to discuss disposal and cleanup of contaminated materials.

**Land-Air**
1. Stop or reduce discharge if safe to do so.
2. Contact manufacturer or supplier for advice.
3. Dike to prevent runoff from rainwater or water application.
4. Remove material by manual or mechanical means.
5. Broken and empty bags or containers should be handled carefully to avoid scattering of dust.
6. Remove contaminated soil for disposal.
7. Notify environmental authorities to discuss disposal and cleanup of contaminated materials.

**Disposal**
1. Contact manufacturer or supplier for advice on disposal.
2. Contact environmental authorities for advice on disposal.

LEAD ACETATE  Pb$(CH_3COO)_2 \cdot 3H_2O$

# LEAD CHROMATE  $PbCrO_4$

## IDENTIFICATION

### Common Synonyms
CHROME YELLOW
LEAD CHROMATE (IV)
PARIS YELLOW

### Observable Characteristics
Yellow crystals. Odourless.

### Manufacturers
Hercules Canada, St. Jean, Quebec.
Reed, Ajax, Ontario.

### Transportation and Storage Information
**Shipping State:** Solid.
**Classification:** None.
**Inert Atmosphere:** No requirement.
**Venting:** Closed.
**Label(s):** Not regulated.
**Storage Temperature:** Ambient.
**Grades or Purity:** Technical.
**Containers and Materials:** Kegs, drums; steel.

### Physical and Chemical Characteristics
**Physical State (20°C, 1 atm):** Solid.
**Solubility (Water):** 0.000 007 to 0.000 017 g/100 mL (20°C).
**Molecular Weight:** 323.2
**Vapour Pressure:** No information.
**Boiling Point:** Decomposes at 844°C.
**Flotability (Water):** Sinks.
**Odour:** Odourless.
**Flash Point:** Not flammable.
**Vapour Density:** No information.
**Specific Gravity:** 6.1
**Colour:** Yellow.
**Explosive Limits:** Not flammable.
**Melting Point:** 844°C; decomposes.

## HAZARD DATA

### Human Health
**Symptoms:** Inhalation: irritation of nose and eyes, headache, stomach cramps, fatigue. Ingestion: metallic taste in mouth, stomach pains, nausea and vomiting, diarrhea, convulsions, coma; Contact: skin - irritation and inflammation; eyes - watering, irritation and inflammation.
**Toxicology:** Highly toxic by ingestion.
**TLV® (inhalation)** 0.05 mg/m³ (as Cr).    $LC_{50}$ - No information.    $LD_{50}$ - Intraperitoneal: guinea pig = 0.40 g/kg
Short-term Inhalation Limits - No information.    Delayed Toxicity - Suspected carcinogen.

### Fire
**Fire Extinguishing Agents:** Not combustible; most fire extinguishing agents may be used on fires involving lead chromate.
**Behaviour in Fire:** Decomposes at 844°C, liberating oxygen and toxic lead compounds.
**Ignition Temperature:** Not combustible.    **Burning Rate:** Not combustible.

### Reactivity
**With Water:** No reaction.
**With Common Materials:** May react with ferric ferrocyanide.
**Stability:** Stable.

### Environment
**Water:** Prevent entry into water intakes or waterways.
**Land-Air:** No information.
**Food Chain Concentration Potential:** May be concentrated (as lead) in tissue.

## EMERGENCY MEASURES

**Special Hazards**

POISON.

**Immediate Responses**

Keep non-involved people away from spill site. Avoid contact and inhalation of fumes. Stop or reduce discharge if this can be done without risk. Notify supplier and environmental authorities.

**Protective Clothing and Equipment**

Respiratory protection - in confined spaces or fires - self-contained breathing apparatus and totally encapsulated suit, otherwise protective clothing as required. Gloves: rubber or plastic.

**Fire and Explosion**

Not combustible, most fire extinguishing agents may be used on fires involving lead chromate. At high temperatures lead chromate decomposes yielding oxygen and toxic lead compounds.

**First Aid**

Move victim out of spill area to fresh air. Call for medical assistance, but start first aid at once. Inhalation: give artificial respiration if breathing has stopped; if laboured, give oxygen. Contact: skin and eyes - remove contaminated clothing and flush affected areas with plenty of water. Ingestion: give conscious victim plenty of water to drink and induce vomiting. If medical assistance is not immediately available, transport victim to hospital, doctor or clinic.

## ENVIRONMENTAL PROTECTION MEASURES

**Response**

**Water**
1. Stop or reduce discharge if safe to do so.
2. Contact manufacturer or supplier for advice.
3. If possible, contain discharge by damming or water diversion.
4. Dredge or vacuum pump to remove contaminants and contaminated bottom sediments.
5. Notify environmental authorities to discuss disposal and cleanup of contaminated materials.

**Land-Air**
1. Stop or reduce discharge if safe to do so.
2. Contact manufacturer or supplier for advice.
3. Contain spill by diking with earth or other material.
4. Remove material by manual or mechanical means.
5. Recover undamaged containers.
6. Notify environmental authorities to discuss disposal and cleanup of contaminated materials.

**Disposal**
1. Contact manufacturer or supplier for advice on disposal.
2. Contact environmental authorities for advice on disposal.

LEAD CHROMATE  $PbCrO_4$

# LEAD NITRATE   Pb(NO$_3$)$_2$

## IDENTIFICATION

UN No. 1469

| Common Synonyms | Observable Characteristics | Manufacturers |
|---|---|---|
| NITRIC ACID, LEAD II SALT | White crystals. Odourless. | No Canadian manufacturers. |

### Transportation and Storage Information

**Shipping State:** Solid.
**Classification:** Oxidizing material. Poison.
**Inert Atmosphere:** No requirement.
**Venting:** Open.

**Label(s):** Yellow label - OXIDIZER; Class 5.1, Group II; White label - POISON; Class 6.1, Group II.
**Storage Temperature:** Ambient.

**Grades or Purity:** Technical, 98+%.
**Containers and Materials:** Multiwall paper bags and drums.

### Physical and Chemical Characteristics

**Physical State** (20°C, 1 atm.): Solid.
**Solubility** (Water): 37.7 g/100 mL (0°C); 127 g/100 mL (100°C).
**Molecular Weight:** 331.2
**Vapour Pressure:** No information.
**Boiling Point:** Decomposes at 470°C.

**Floatability** (Water): Sinks and mixes.
**Odour:** Odourless.
**Flash Point:** Not flammable.
**Vapour Density:** No information.
**Specific Gravity:** 4.53 (30°C).

**Colour:** White.
**Explosive Limits:** Not flammable.
**Melting Point:** Decomposes at 470°C.

## HAZARD DATA

### Human Health

**Symptoms:** Inhalation: of dust, mist or fumes, irritation of nose and eyes, headache, stomach cramps and fatigue. Ingestion: metallic taste in mouth, constriction of throat, stomach pains, nausea, vomiting, diarrhea, convulsions, coma. Contact: skin - irritation, inflammation; eyes - inflammation.
**Toxicology:** Highly toxic by ingestion.
TLV® (inhalation) 0.15 mg/m$^3$ (as Pb).
Short-term Inhalation Limits - 0.45 mg/m$^3$ (15 min) (as Pb).

LC$_{50}$ - No information.
Delayed Toxicity - Cumulative poison.

LD$_{50}$ - No information.
LD$_{Lo}$ - Oral: guinea pig = 0.5 g/kg

### Fire

**Fire Extinguishing Agents:** Not combustible. Most fire extinguishing agents may be used. Use water sparingly.
**Behaviour in Fire:** Not combustible. When heated to decomposition, can emit toxic nitrogen oxide and lead oxide fumes.
**Ignition Temperature:** Not flammable.
**Burning Rate:** Not flammable.

### Reactivity

**With Water:** No reaction; soluble.
**With Common Materials:** Strong oxidizer, can ignite organic materials. Reacts violently with ammonium thiocyanate, carbon and lead hypophosphite.
**Stability:** Stable.

### Environment

**Water:** Prevent entry into water intakes and waterways. Fish toxicity: 240 ppm/48 h/mosquito fish/TLm/freshwater.
**Land-Air:** No information.
**Food Chain Concentration Potential:** Fish and terrestrial animals are capable of concentrating lead.

## EMERGENCY MEASURES

**Special Hazards**

OXIDIZER. POISON.

**Immediate Responses**

Keep non-involved people away from spill site. Issue warning: "OXIDIZER. POISON". Avoid contact and inhalation. Call Fire Department. Dike to prevent runoff from rainwater or water application. Lightly wet down dry spillage to prevent wind drift or dust. Notify manufacturer or supplier. Notify environmental authorities.

**Protective Clothing and Equipment**

In fire or enclosed spaces, Respiratory protection - self-contained breathing apparatus. Otherwise, dust respirators (with suitable filters) or metal fume respirators for normal situations. Gloves and boots - rubber. Frequent changes of clothing and footwear should be provided. Contaminated clothing and footwear should be washed immediately after contact.

**Fire and Explosion**

Not combustible. Most fire extinguishing agents may be used. Use water sparingly.

**First Aid**

Move victim out of spill site to fresh air. Call for medical assistance, but start first aid at once. Inhalation: if breathing has stopped give artificial respiration (not mouth-to-mouth method); if laboured, give oxygen. Contact: skin - remove contaminated clothing, and flush affected areas with plenty of water; eyes - irrigate with plenty of water. Ingestion: give water to conscious victim to drink, and induce vomiting. If medical assistance is not immediately available, transport victim to doctor, clinic or hospital.

## ENVIRONMENTAL PROTECTION MEASURES

**Response**

**Water**
1. Stop or reduce discharge if safe to do so.
2. Contact manufacturer or supplier for advice.
3. If possible, contain discharge by damming or water diversion.
4. Dredge or vacuum pump to remove contaminants, liquids and contaminated bottom sediments.
5. Notify environmental authorities to discuss disposal and cleanup of contaminated materials.

**Land-Air**
1. Stop or reduce discharge if safe to do so.
2. Contact manufacturer or supplier for advice.
3. Dike to prevent runoff from rainwater or water application.
4. Remove material by manual or mechanical means.
5. Broken and empty bags or containers should be handled carefully to avoid scattering of dust.
6. Remove contaminated soil for disposal.
7. Notify environmental authorities to discuss disposal and cleanup of contaminated materials.

**Disposal**
1. Contact manufacturer or supplier for advice on disposal.
2. Contact environmental authorities for advice on disposal.

LEAD NITRATE   $Pb(NO_3)_2$

# LEAD OXIDES
(black) $Pb_2O$
(red) $Pb_3O_4$
(yellow) $PbO$

## IDENTIFICATION

### Common Synonyms

(Yellow) PbO
Lead Oxide, Yellow
Plumbous Oxide
Lead Monoxide

(Red) $Pb_3O_4$
Red Lead
Minium
Plumboplumbic Oxide
Lead Oxide, Red
Lead Tetroxide

(Black) $Pb_2O$
Lead Oxide, Black
Litharge, Leaded

### Observable Characteristics
Yellow, red or black powder. Odourless.

### Manufacturers
Canada Metal Co. Ltd., Toronto, Ont., Winnipeg, Man.
Carter White Lead Co. of Canada Ltd., Montreal, Que.
Metalex Products, Vancouver, B.C.

### Transportation and Storage Information
**Shipping State:** Solid.
**Classification:** Not regulated.
**Inert Atmosphere:** No requirement.
**Venting:** Open.
**Label(s):** Not regulated. Voluntary white label - POISON.
**Storage Temperature:** Ambient.
**Grades or Purity:** Yellow (litharge) - assay 98% and battery grades, 50 to 95% PbO. Red (red lead) - 85,95,97,98% PbO. Black - usually a mixture of Pb and PbO.
**Containers and Materials:** Multiwall paper bags, cans, drums.

### Physical and Chemical Characteristics
**Physical State** (20°C, 1 atm): Solid.
**Solubility** (Water): Yellow 0.0017 g/100 mL (20°C); black insoluble; red 0.0026 g/100 mL (25°C).
**Molecular Weight:** Yellow, 223.2; red, 685.6; black, 430.4
**Vapour Pressure:** No information.
**Boiling Point:** Yellow, >886°C; red decomposes 500°C; black, decomposes upon heating.
**Floatability** (Water): Sinks.
**Odour:** Odourless.
**Flash Point:** Not flammable.
**Vapour Density:** No information.
**Specific Gravity:** Yellow, 9.2 to 9.7; red, 8.8 to 9.2; black, 8.3 (20°C).
**Colour:** Yellow, red, black.
**Explosive Limits:** Not flammable.
**Melting Point:** Yellow, 888°C; red decomposes 500 to 530°C; black decomposes with heat.

## HAZARD DATA

### Human Health
**Symptoms:** Inhalation: of dust, mist or fumes; irritation of nose and eyes, headache, stomach cramps and fatigue. Ingestion: metallic taste in mouth, constriction of throat, stomach pains, vomiting, diarrhea, convulsions and coma. Contact: skin - irritation, inflammation; eyes - inflammation.
**Toxicology:** Highly toxic by ingestion.
TLV* - (inhalation) 0.15 mg/m³ (as Pb).
Short-term Inhalation Limits - 0.45 mg/m³ (15 min).
$LC_{50}$ - No information.
Delayed Toxicity - Cumulative poison.
$LD_{50}$ - Intraperitoneal: guinea pig = 0.22 g/kg
$LD_{Lo}$ - Intraperitoneal: rat = 0.43 g/kg

### Fire
**Fire Extinguishing Agents:** Not combustible. Most fire extinguishing may be used on fires involving lead oxides.
**Behaviour in Fire:** Toxic Pb and PbO and fumes may be released in fires or when heated to decomposition.
**Ignition Temperature:** Not combustible.
**Burning Rate:** Not combustible.

### Reactivity
**With Water:** No reaction.
**With Common Materials:** In general, Pb oxides are oxidizing agents. Reacts violently with aluminum, hydrogen trisulfide, sodium, sulfur trioxide, and titanium.
**Stability:** Stable.

### Environment
**Water:** Prevent entry into water intakes and waterways. Fish toxicity: (Yellow) >56 000 ppm/96 h/mosquito fish/TLm/turbid water.
**Land-Air:** No information.
**Food Chain Concentration Potential:** Fish and terrestrial animals are capable of concentrating lead.

## EMERGENCY MEASURES

### Special Hazards
POISON. OXIDIZER.

### Immediate Responses
Keep non-involved people away from spill site. Issue warning: "POISON". Notify manufacturer or supplier. Avoid contact or inhalation. Dike to prevent runoff from rainwater or water application. Lightly wet down dry spillage to prevent wind drift of dust. Notify environmental authorities.

### Protective Clothing and Equipment
In fires or enclosed spaces, Respiratory protection - self-contained breathing apparatus, otherwise dust respirators (with suitable filters) for protection against dust under normal conditions. Rubber gloves and boots will prevent contact. Frequent changes of clothing should be provided. Contaminated clothing and footwear to be washed immediately after contact.

### Fire and Explosion
Not combustible, most fire extinguishing agents may be used.

### First Aid
Move victim out of spill site to fresh air. Call for medical assistance, but start first aid at once. Inhalation: if breathing has stopped give artificial respiration (not mouth-to-mouth method); if laboured, give oxygen. Contact: skin - remove contaminated clothing, and flush affected areas with plenty of water; eyes - irrigate with plenty of water. Ingestion: give water to conscious victim to drink, and induce vomiting. If medical assistance is not immediately available, transport victim to doctor, clinic or hospital.

## ENVIRONMENTAL PROTECTION MEASURES

### Response

**Water**
1. Stop or reduce discharge if safe to do so.
2. Contact manufacturer or supplier for advice.
3. If possible, contain discharge by damming or water diversion.
4. Dredge or vacuum pump to remove contaminants, liquids and contaminated bottom sediments.
5. Notify environmental authorities to discuss disposal and cleanup of contaminated materials.

**Land-Air**
1. Stop or reduce discharge if safe to do so.
2. Contact manufacturer or supplier for advice.
3. Dike to prevent runoff from rainwater or water application.
4. Remove material by manual or mechanical means.
5. Broken paper bags should be handled carefully to avoid scattering of dust.
6. Remove contaminated soil for disposal.
7. Notify environmental authorities to discuss disposal and cleanup of contaminated materials.

### Disposal
1. Contact manufacturer or supplier for advice on disposal.
2. Contact environmental authorities for advice on disposal.

LEAD OXIDES    (black) $Pb_2O$
(red) $Pb_3O_4$
(yellow) $PbO$

# MAGNESIUM HYDROXIDE   Mg(OH)$_2$

## IDENTIFICATION

| Common Synonyms | Observable Characteristics | Manufacturers |
|---|---|---|
| MAGNESIUM HYDRATE<br>MAGNESIUM MAGNA<br>MILK OF MAGNESIA | White powder. Odourless. | No Canadian manufacturers.<br>Canadian suppliers:<br>Dow Chemical Canada Inc.,<br>Sarnia, Ontario.<br>Chorney Chemical, Toronto, Ontario.<br>Frank E. Dempsey, Toronto, Ontario. | Originating from:<br>Dow Chemical, Luddington, Mich.<br>Barcroft, USA.<br>Reheis Chemical, USA. |

### Transportation and Storage Information

**Shipping State:** Solid.
**Classification:** None.
**Inert Atmosphere:** No requirement.
**Venting:** Open.

**Label(s):** Not regulated.
**Storage Temperature:** Ambient.

**Grades or Purity:** Technical.
**Containers and Materials:** Multiwall paper bags, drums.

### Physical and Chemical Characteristics

**Physical State** (20°C, 1 atm): Solid.
**Solubility** (Water): 0.18 g/100 mL (25°C); 0.04 g/100 mL (100°C).
**Molecular Weight:** 58.3
**Vapour Pressure:** No information.
**Boiling Point:** Loses water at 350°C.

**Floatability** (Water): Sinks.
**Odour:** Odourless.
**Flash Point:** Not flammable.
**Vapour Density:** No information.
**Specific Gravity:** 2.36 (20°C).

**Colour:** White.
**Explosive Limits:** Not flammable.
**Melting Point:** Loses water at 350°C.

## HAZARD DATA

### Human Health

**Symptoms: Contact:** skin and eyes - irritation and redness. **Ingestion:** abdominal pain and diarrhea.
**Toxicology:** Low toxicity by all routes.
**TLV®** - No information.    **LC$_{50}$** - No information.    **LD$_{50}$** - No information.
**Short-term Inhalation Limits** - No information.    **Delayed Toxicity** - No information.

### Fire

**Fire Extinguishing Agents:** Not combustible. Most fire extinguishing agents may be used on fires involving magnesium hydroxide.
**Behaviour in Fire:** Upon heating, decomposes to magnesium oxide (MgO) and water vapour.
**Ignition Temperature:** Not combustible.    **Burning Rate:** Not combustible.

### Reactivity

**With Water:** No reaction.
**With Common Materials:** Reacts violently with maleic anhydride and phosporus. Reacts with acids.
**Stability:** Stable.

### Environment

**Water:** Prevent entry into water intakes and waterways.
**Land-Air:** No information.
**Food Chain Concentration Potential:** No information.

## EMERGENCY MEASURES

**Special Hazards**

**Immediate Responses**

Keep non-involved people away from spill site. Notify manufacturer. Dike to prevent runoff. Notify environmental authorities.

**Protective Clothing and Equipment**

In fires or confined spaces, Respiratory protection - use self-contained breathing apparatus. Otherwise, protective clothing and equipment as required.

**Fire and Explosion**

Not combustible. Most fire extinguishing agents may be used.

**First Aid**

Move victim out of spill site to fresh air. Call for medical assistance, but start first aid at once. Contact: skin - remove contaminated clothing and wash affected areas with plenty of water; eyes - irrigate with plenty of water. Ingestion: give plenty of water to conscious victim to drink. If medical assistance is not immediately available, transport victim to hospital, doctor or clinic.

## ENVIRONMENTAL PROTECTION MEASURES

**Response**

**Water**
1. Stop or reduce discharge if safe to do so.
2. Contact manufacturer or supplier for advice.
3. If possible, contain discharge by damming or water diversion.
4. Dredge or vacuum pump to remove contaminants, liquids and contaminated bottom sediments.
5. Notify environmental authorities to discuss disposal and cleanup of contaminated materials.

**Land-Air**
1. Stop or reduce discharge if safe to do so.
2. Contact manufacturer or supplier for advice.
3. Contain spill by diking with earth or other barrier.
4. Remove material by manual or mechanical means.
5. Recover undamaged containers.
6. Notify environmental authorities to discuss disposal and cleanup of contaminated materials.

**Disposal**
1. Contact manufacturer or supplier for advice on disposal.
2. Contact environmental authorities for advice on disposal.

MAGNESIUM HYDROXIDE    $Mg(OH)_2$

# MALATHION   $(CH_3O)_2P(S)SCH(COOC_2H_5)CH_2COOC_2H_5$

## IDENTIFICATION

UN No. 2783
Danger Group According to Percentage of Active Substance
Group III    liquid 30 to 100%

### Common Synonyms
S-1,2-BIS (ETHOXYCARBONYL)ETHYL
O,O-DIMETHYL PHOSPHORODITHIOATE
**Common Trade Names**
CALMATHION, CELTHION, CYTHION, MALATIOZOL
(A commonly-used general purpose insecticide.)

### Observable Characteristics
Yellow to dark brown liquid or white to brown powder. Skunk-like odour.

### Manufacturers
Canadian Industries Ltd., (C.I.L.), Montreal, Quebec.
Chipman Chemicals Ltd., Stoney Creek, Ont.
Cyanamid Canada, Willowdale, Ont.

### Transportation and Storage Information
**Shipping State:** Liquid or solid (wettable powder).
**Classification:** None.
**Inert Atmosphere:** No requirement.
**Venting:** Closed.
**Pump Type:** No information.

**Label(s):** Not regulated.
**Storage Temperature:** Ambient.
**Hose Type:** No information.

**Grades or Purity:** Various purities as detailed below.
**Containers and Materials:** Glass bottles, cans, drums; steel.

### Physical and Chemical Characteristics
**Physical State** (20°C, 1 atm): Liquid (technical).
**Solubility** (Water): 0.0145 g/100 mL (technical).
**Molecular Weight:** 330 (technical).
**Vapour Pressure:** 0.00004 mm Hg (30°C).
**Boiling Point:** 156°C (at 0.7 mm Hg pressure).

**Floatability** (Water): Technical sinks, SN floats.
**Odour:** Skunk-like.
**Flash Point:** 54°C (c.c.) SN; 163°C (c.c.) technical; 22 to 32°C (c.c.) EC; PP is also flammable.
**Vapour Density:** No information.
**Specific Gravity:** 1.23, 25°C (technical).

**Colour:** Liquid, yellow to dark brown; powder, white to brown.
**Explosive Limits:** Technical not flammable; SN, PP and EC are flammable.
**Melting Point:** 2.9°C (technical).

## HAZARD DATA

### Human Health
**Symptoms:** Inhalation: headache, tightness in chest, wheezing, throat constriction. Contact: eyes - irritation; skin - irritation; may be absorbed. Ingestion: abdominal cramps, nausea, vomiting, diarrhea.
**Toxicology:** Moderately by ingestion and contact.
TLV - (skin) 10 mg/m$^3$          LC$_{50}$ - No information.          LD$_{50}$ - Oral: rat = 0.885 g/kg
Short-term Inhalation Limits - No information.   Delayed Toxicity - No information.       LD$_{50}$ - Oral: man = 0.857 g/kg

### Fire
**Fire Extinguishing Agents:** Most fire extinguishing agents may be used on fires involving malathion. Water spray should be used to cool fire-exposed containers.
**Behaviour in Fire:** In fires or when heated to decomposition releases toxic SO$_x$, PO$_x$ and other fumes.
**Ignition Temperature:** Variable.    **Burning Rate:** No information.

### Reactivity
**With Water:** No reaction; only WC insoluble.
**With Common Materials:** May react with oxidizing agents.
**Stability:** Stable.

### Environment
**Water:** Prevent entry into water intakes and waterways. Fish toxicity: 16 ppm/96 h/fathead minnow/LC$_{50}$/freshwater; 0.2 ppm/96 h/rainbow trout/LC$_{50}$/freshwater; 0.12 ppm/24 h/bluegill/LC$_{50}$/freshwater; 0.33 to 1 ppm/48 h/brown shrimp/LC$_{50}$/saltwater; 0.032/24 h/grass shrimp/LC$_{50}$/saltwater.
**Land-Air:** LD$_{50}$/exposed for 5 days/Oral: duck/>5 000 ppm; LD$_{50}$/exposed for 5 days/Oral: pheasant/2 500 to 4 500 ppm.
**Food Chain Concentration Potential:** No information.

## EMERGENCY MEASURES

### Special Hazards
POISON.

### Immediate Responses
Keep non-involved people away from spill site. Stop or reduce discharge if safe to do so. Notify manufacturer or supplier. Dike to contain material or water runoff. Notify environmental authorities.

### Protective Clothing and Equipment
In fires or confined spaces - Respiratory Protection - self-contained breathing apparatus and totally encapsulated suit. Otherwise, approved pesticide respirator and impervious outer clothing.

### Fire and Explosion
Use carbon dioxide, foam or dry chemical to extinguish. Releases toxic fumes in fires.

### First Aid
Move victim out of spill site to fresh air. Call for medical assistance, but start first aid at once. Inhalation: if breathing has stopped, give artificial respiration (not mouth-to-mouth method); if laboured, give oxygen. Contact: skin - remove contaminated clothing and flush affected areas with plenty of water; eyes - irrigate with plenty of water. Ingestion: give water to conscious victim to drink and induce vomiting; in the case of petroleum distillates, do not induce vomiting for fear of aspiration and chemical pneumonia. If medical assistance is not immediately available, transport victim to hospital, doctor, or clinic.

## ENVIRONMENTAL PROTECTION MEASURES

### Response

**Water**
1. Stop or reduce discharge if safe to do so.
2. Contact manufacturer or supplier for advice.

Floats / Sinks or mixes
3. If possible contain discharge by booming.
3. If possible contain discharge by damming or water diversions.
4. If floating, skim and remove.
4. Dredge or vacuum pump to remove contaminants, liquids and contaminated bottom sediments.
5. Notify environmental authorities to discuss disposal and cleanup of contaminated materials.

**Land-Air**
1. Stop or reduce discharge if safe to do so.
2. Contact manufacturer or supplier for advice.
3. Contain spill by diking with earth or other barrier.
4. If liquid, remove material with pumps or vacuum equipment and place in appropriate containers.
5. If solid, remove material by manual or mechanical means.
6. Recover undamaged containers.
7. Absorb residual liquid on natural or synthetic sorbents.
8. Remove contaminated soil for disposal.
9. Notify environmental authorities to discuss cleanup and disposal of contaminated materials.

### Disposal
1. Contact manufacturer or supplier for advice on disposal.
2. Contact environmental authorities for advice on disposal.

### Available Formulations

**Technical Grade:** Purity: typically 95% malathion.
Properties: combustible.

**Formulations:**

| Type: | Purity: | Properties: |
|---|---|---|
| DU - dust or wettable powder | - typically 4% malathion, remainder inerts (e.g. clay, talc). | - dispersible in water, low combustibility |
| PP - pressurized product | - typically 3% malathion | - combustible and explosive |
| EC - emulsifiable concentrate | - typically 20% malathion | - combustible |
| SN - solution | - typically 3% malathion, large percentage of petroleum distillates | - combustible, may be flammable; floats on water |

MALATHION    $(CH_3O)_2P(S)SCH(COOC_2H_5)CH_2COOC_2H_5$

# MALEIC ANHYDRIDE (CHCO)₂O

## IDENTIFICATION

UN No. 2215

### Common Synonyms
TOXILIC ANHYDRIDE
cis-BUTENEDIOIC ANHYDRIDE
2,5-FURANDIONE
MALEIC ACID, ANHYDRIDE

### Observable Characteristics
Colourless to white, crystals or solid. Choking acrid odour.

### Manufacturers
Monsanto Canada, Montreal, Quebec.
BASF Canada, Cornwall, Ontario.

### Transportation and Storage Information
**Shipping State:** Solid.
**Classification:** Corrosive.
**Inert Atmosphere:** No requirement.
**Venting:** Open.

**Label(s):** Black and white label: CORROSIVE; Class 8, Group III.
**Storage Temperature:** Ambient.

**Grades or Purity:** Commercial; 99.5%.
**Containers and Materials:** Drums; fibre and steel.

### Physical and Chemical Characteristics
**Physical State** (20°C, 1 atm): Solid.
**Solubility** (Water): Reacts ~16 g/100 mL (30°C).
**Molecular Weight:** 98.1
**Vapour Pressure:** 0.00005 mm Hg (20°C); 0.0002 mm Hg (30°C).
**Boiling Point:** 200 to 202°C.

**Floatability** (Water): Reacts, sinks or floats depending on material.
**Odour:** Choking, acrid (0.3 to 0.5 ppm).
**Flash Point:** 102°C (c.c.)
**Vapour Density:** 3.4
**Specific Gravity:** 1.3 (60°C); 0.93 (20°C).

**Colour:** Colourless to white.
**Explosive Limits:** 1.4 to 7.1%.
**Melting Point:** 53°C.

## HAZARD DATA

### Human Health
**Symptoms:** Inhalation: irritation of throat and respiratory tract, bronchitis; Ingestion: abdominal pain, vomiting, nausea. Contact: skin - serious burns; eyes - severe burning.
**Toxicology:** Highly toxic upon contact with skin and eyes.
TLV® 0.25 ppm, 1 mg/m³.    LC₅₀ - No information.    LD₅₀ - Oral:rat = 0.48 g/kg
Short-term Inhalation Limits - No information.    Delayed Toxicity - No information.

### Fire
**Fire Extinguishing Agents:** Use alcohol foam or carbon dioxide. Water or foam may cause frothing.
**Behaviour in Fire:** No information.    **Burning Rate:** 1.4 mm/min.
**Ignition Temperature:** 477°C.

### Reactivity
**With Water:** Reacts, forming maleic acid.
**With Common Materials:** Can react with oxiding materials. Reacts violently with alkaline metals, amines, calcium hydroxide, lithium, potassium hydroxide, pyridine, sodium and sodium hydroxide.
**Stability:** Stable.

### Environment
**Water:** Prevent entry into water intakes and waterways. Fish toxicity: 150 ppm/24 h/sunfish/TLm/freshwater; 240 mg/L/24 to 96 h/mosquito fish/TLm/freshwater; BOD: 40 to 60%, 5 days.
**Land-Air:** No information.
**Food Chain Concentration Potential:** No information.

## EMERGENCY MEASURES

**Special Hazards**

CORROSIVE. Combustible.

**Immediate Responses**

Keep non-involved people away from spill site. Issue warning "CORROSIVE". Call Fire Department. Notify manufacturer or supplier. Notify environmental authorities.

**Protective Clothing and Equipment**

In fires or confined spaces, Respiratory protection - self-contained breathing apparatus and totally encapsulated suit. Otherwise, protective clothing as required.

**Fire and Explosion**

Use alcohol foam or carbon dioxide to extinguish. Water or foam may cause frothing.

**First Aid**

Move victim out of spill site to fresh air. Call for medical assistance, but start first aid at once. Inhalation: give artificial respiration if breathing has stopped (not mouth-to-mouth method); oxygen if breathing is laboured. Contact: skin - remove contaminated clothing and flush affected areas with plenty of water; eyes - irrigate with plenty of water. Ingestion: give conscious victim water to drink. If medical assistance is not immediately available, transport victim to hospital, doctor or clinic.

## ENVIRONMENTAL PROTECTION MEASURES

**Response**

**Water**
1. Stop or reduce discharge if safe to do so.
2. Contact manufacturer or supplier for advice.
3. If possible, contain discharge by damming or water diversion.
4. Dredge or vacuum pump to remove contaminants, liquids and contaminated bottom sediments.
5. Notify environmental authorities to discuss disposal and cleanup of contaminated materials.

**Land-Air**
1. Stop or reduce discharge if safe to do so.
2. Contact manufacturer or supplier for advice.
3. Contain spill by diking with earth or other barrier.
4. Remove material by manual or mechanical means.
5. Recover undamaged containers.
6. Notify environmental authorities to discuss disposal and cleanup of contaminated materials.

**Disposal**
1. Contact manufacturer or supplier for advice on disposal.
2. Contact environmental authorities for advice on disposal.
3. Incinerate (approval of environmental authorities required).

---

BILL MICHAEL, TECHNICIAN

British Columbia
**Ministry of Environment, Lands and Parks**
Pollution Prevention
BC Environment
Southern Interior Region

Mailing Address:
201 - 3547 Skaha Lake Road
Penticton BC V2A 7K2

Telephone: (250) 490-8227
Facsimile: (250) 492-1314
e-mail: bill.michael@gems9.gov.bc.ca

---

MALEIC ANHYDRIDE  $(CHCO)_2O$

# MCPA   $CH_3ClC_6H_3OCH_2COOH$

## IDENTIFICATION

**UN No.** 2765
**Danger Group According to Percentage of Active Substance:** Group III, Liquid 35 to 100%

### Common Synonyms
(4-CHLORO-O-TOLYLOXY) ACETIC ACID

### Common Trade Names
MCPA AMINE
(A herbicide used for the control of broadleaf weeds.)

### Observable Characteristics
Light brown solid or colourless liquids.

### Manufacturers
Ciba-Geigy Canada Ltd., Etobicoke, Ont.
Chipman Inc. Stoney Creek, Ontario.
Interprovincial Co-op., Saskatoon, Sask.
Uniroyal Chemical, Elmira, Ontario.

### Transportation and Storage Information
**Shipping State:** Solid or liquid (formulation).
**Classification:** None.
**Inert Atmosphere:** No requirement.
**Venting:** Open.
**Pump Type:** No information.

**Label(s):** Not regulated.
**Storage Temperature:** Ambient.
**Hose Type:** No information.

**Grades or Purity:** Various as shown below.
**Containers and Materials:** Glass bottles, cans, drums; steel.

### Physical and Chemical Characteristics
**Physical State** (20°C, 1 atm): Solid (technical).
**Solubility (Water):** 0.083 g/100 mL (solid) (25°C). SN, EC soluble in water.
**Molecular Weight:** 200.6
**Vapour Pressure:** No information.
**Boiling Point:** No information.

**Floatability (Water):** Sinks; SN and EC mix.
**Odour:** No information.
**Flash Point:** Not flammable.
**Vapour Density:** No information.
**Specific Gravity:** 1.56 (25°C) technical.

**Colour:** Colourless (liquids); light brown (solid).
**Explosive Limits:** Not flammable.
**Melting Point:** 99 to 107°C (technical).

## HAZARD DATA

### Human Health
**Symptoms: Ingestion and Skin Contact:** readily absorbed by skin; burning pain, abdominal pain, flushing of skin, vomiting, muscular tremors, abnormal temperature, lethargy and muscular weakness.
**Toxicology:** Highly toxic by skin contact; moderately toxic by ingestion.
**TLV®** No information.   $LD_{50}$ - Oral: rat = 0.70 g/kg
**Short-term Inhalation Limits** - No information.   $LC_{50}$ - No information.   $LD_{50}$ - Oral: mouse = 0.55 g/kg
Delayed Toxicity - No information.

### Fire
**Fire Extinguishing Agents:** Use foam, carbon dioxide, or dry chemical.
**Behaviour in Fire:** Releases toxic fumes.
**Ignition Temperature:** No information.   **Burning Rate:** No information.

### Reactivity
**With Water:** No reaction.
**With Common Materials:** No information.
**Stability:** Stable.

### Environment
**Water:** Prevent entry into water intakes and waterways. Aquatic toxicity rating = 10 to 100 ppm/96 h/TLm/freshwater. Fish toxicities: bluegill/$LC_{50}$/freshwater; 75 mg/L/45 h/longnose killifish/$LC_{50}$/saltwater.
**Land-Air:** No information.
**Food Chain Concentration Potential:** No information.

## EMERGENCY MEASURES

**Special Hazards**
POISON.

**Immediate Responses**
Keep non-involved people away from spill site. Stop or reduce discharge if safe to do so. Notify environmental authorities. Dike to contain material or water runoff. Notify manufacturer or supplier.

**Protective Clothing and Equipment**
In fires or confined spaces - Respiratory Protection - self-contained breathing apparatus and totally encapsulated suit. Otherwise, approved pesticide respirator and impervious outer clothing.

**Fire and Explosion**
Use carbon dioxide, foam or dry chemical to extinguish. Releases toxic fumes in fires.

**First Aid**
Move victim out of spill area to fresh air. Call for medical assistance, but start first aid at once. Inhalation: if breathing has stopped, give artificial respiration (not mouth-to-mouth method); if laboured, give oxygen. Contact: skin - remove contaminated clothing and flush affected areas with plenty of water; eyes - irrigate with plenty of water. Ingestion: give water to conscious victim to drink and induce vomiting; in the case of petroleum distillates, do not induce vomiting for fear of aspiration and chemical pneumonia. if medical assistance is not immediately available, transport victim to hospital, doctor, or clinic.

## ENVIRONMENTAL PROTECTION MEASURES

**Response**

**Water**
1. Stop or reduce discharge if safe to do so.
2. Contact manufacturer or supplier for advice.

**Floats**
3. If possible contain discharge by booming.
4. If floating, skim and remove.

**Sinks or mixes**
3. If possible contain discharge by damming or water diversions.
4. Dredge or vacuum pump to remove contaminants, liquids and contaminated bottom sediments.

5. Notify environmental authorities to discuss disposal and cleanup of contaminated materials.

**Land-Air**
1. Stop or reduce discharge if safe to do so.
2. Contact manufacturer or supplier for advice.
3. Contain spill by diking with earth or other barrier.
4. If liquid, remove material with pumps or vacuum equipment and place in appropriate containers.
5. If solid, remove material by manual or mechanical means.
6. Recover undamaged containers.
7. Absorb residual liquid on natural or synthetic sorbents.
8. Remove contaminated soil for disposal.
9. Notify environmental authorities to discuss cleanup and disposal of contaminated materials.

**Disposal**
1. Contact manufacturer or supplier for advice on disposal.
2. Contact environmental authorities for advice on disposal.

**Available Formulations**

**Technical Grade:** Purity: 85 to 95%.
Properties: insoluble in water.

**Formulations:**

| Type: | Purity: | Properties: |
|---|---|---|
| SN - solution (diethanolamine salt) | - typically 50% | - soluble in water, combustible |
| SN - solution (sodium salt) | - typically 40% | - soluble in water |
| EC - emulsifiable concentrate (ester) | - typically 50% | - soluble in water, combustible |

Other Possible Ingredients Found in Formulations: Dicamba, mecoprop, linuron, bromacil.

MCPA    $CH_3ClC_6H_3OCH_2COOH$

# MERCURY  Hg

## IDENTIFICATION

| Common Synonyms | Observable Characteristics | Manufacturers |
|---|---|---|
| QUICKSILVER<br>HYDRARGYRUM | Silvery liquid. Odourless. | Johnson Matthey and Mallory Ltd., Brampton, Ontario. |

| Transportation and Storage Information | | |
|---|---|---|
| **Shipping State:** Liquid.<br>**Classification:** None.<br>**Inert Atmosphere:** No requirement.<br>**Venting:** Open.<br>**Pump Type:** No information. | **Label(s):** Not regulated.<br>**Storage Temperature:** Ambient.<br>**Hose Type:** No information. | **Grades or Purity:** Pure.<br>**Containers and Materials:** Bottles, flasks, drums. |

| Physical and Chemical Characteristics | | |
|---|---|---|
| **Physical State** (20°C, 1 atm)**:** Liquid.<br>**Solubility** (Water)**:** 0.025 g/100 mL (30°C).<br>**Molecular Weight:** 200.6<br>**Vapour Pressure:** 0.0012 mm Hg (20°C); 1 mm Hg (126°C).<br>**Boiling Point:** 356.6°C. | **Floatability** (Water)**:** Sinks.<br>**Odour:** Odourless.<br>**Flash Point:** Not flammable.<br>**Vapour Density:** 7.0<br>**Specific Gravity:** 13.6 at 20°C. | **Colour:** Silvery.<br>**Explosive Limits:** Not flammable.<br>**Melting Point:** -38.9°C. |

## HAZARD DATA

### Human Health

**Symptoms:** Inhalation: metallic taste, rapid and difficult breathing, coughing; Ingestion: metallic taste, intense thirst, pain in swallowing, abdominal pain, nausea and vomiting, diarrhea, trembling extremities. Contact: skin - irritation, inflammation, blisters; eyes - irritation, watering, edema of eyelids.
**Toxicology:** Highly toxic by all routes.
TLV® (skin - Hg vapour) 0.05 mg/m$^3$.   LC$_{50}$ - No information.   LD$_{50}$ - No information.
Short-term Inhalation Limits - 0.15 mg/m$^3$ (15 min)   LC$_{Lo}$ - Inhalation: rabbit = 29 mg/m$^3$/30 h   TD$_{Lo}$ - Intraperitoneal: rat = 0.4 g/kg
as Hg vapour.   Delayed Toxicity - Known long-term effects to central nervous system.

### Fire

**Fire Extinguishing Agents:** Not combustible. Most fire extinguishing agents may be used on fires involving mercury.
**Behaviour in Fire:** Vapourizes readily emitting toxic Hg fumes.
**Ignition Temperature:** Not combustible.   **Burning Rate:** Not combustible.

### Reactivity

**With Water:** No reaction.
**With Common Materials:** Reacts violently with acetylene, ammonia, chlorine, chlorine dioxide, methyl azide and sodium carbide.
**Stability:** Stable.

### Environment

**Water:** Prevent entry into water intakes and waterways. Harmful to aquatic life. Fish toxicity: 0.075 mg/m$^3$/48 h/prawn/LC$_{50}$/saltwater; 5.7 mg/m$^3$/48 h/shrimp/LC$_{50}$/saltwater; EPA criterion = 0.064 µg/L/24 h/freshwater; EPA criterion = 0.025 µg/L/24 h/saltwater; BOD: None.
**Land-Air:** No information.
**Food Chain Concentration Potential:** Most biota are known to concentrate mercury.

## EMERGENCY MEASURES

**Special Hazards**

POISON.

**Immediate Responses**

Keep non-involved people away from spill site. Notify manufacturer or supplier. Dike to contain spill. Notify environmental authorities.

**Protective Clothing and Equipment**

In fires or confined spaces, Respiratory protection - self-contained breathing apparatus and totally encapsulated suit.

**Fire and Explosion**

Not combustible. Most fire extinguishing agents may be used on fires involving mercury. Vapourizes readily emitting toxic Hg fumes.

**First Aid**

Move victim out of spill site to fresh air. Call for medical assistance, but start first aid at once. Inhalation: give artificial respiration if necessary. Contact: skin - remove contaminated clothing and wash affected areas with plenty of water; eyes - irrigate with plenty of water. Ingestion: give water or milk to conscious victim to drink, and induce vomiting. If medical assistance is not immediately available, transport victim to doctor, clinic or hospital.

## ENVIRONMENTAL PROTECTION MEASURES

**Response**

**Water**
1. Stop or reduce discharge if safe to do so.
2. Contact manufacturer or supplier for advice.
3. If possible, contain discharge by damming or water diversion.
4. Dredge or vacuum pump to remove contaminants, liquids and contaminated bottom sediments.
5. Notify environmental authorities to discuss disposal and cleanup of contaminated materials.

**Land-Air**
1. Stop or reduce discharge if safe to do so.
2. Contact manufacturer or supplier for advice.
3. Contain spill by diking with earth or other barrier.
4. Remove material by manual or mechanical means.
5. Absorb residual liquid on natural or synthetic sorbents.
6. Remove contaminated soil for disposal.
7. Notify environmental authorities to discuss disposal and cleanup of contaminated materials.

**Disposal**
1. Contact manufacturer or supplier for advice on disposal.
2. Contact environmental authorities for advice on disposal.

MERCURY    Hg

# METHANE  CH$_4$

## IDENTIFICATION

**UN No.** 1971 - gas
1972 - liquid

### Common Synonyms

MARSH GAS
SEWAGE GAS
NATURAL GAS*
METHYL HYDRIDE

*Natural gas is mostly methane, but contains other gases.

### Observable Characteristics

Colourless or liquid. Odourless when pure (commercial forms may have odour compounds added).

### Manufacturers

Canadian supplier:
Canadian Liquid Air Co. Ltd.,
Montreal, Que.

### Transportation and Storage Information

**Shipping State:** Liquid (compressed gas).
**Classification:** Flammable gas.
**Inert Atmosphere:** No requirement.
**Venting:** Safety relief.
**Pump Type:** No information.

**Label(s):** Red label - FLAMMABLE GAS; Class 2.1.
**Storage Temperature:** Ambient.
**Hose Type:** No information.

**Grades or Purity:** Technical, 95%, Btu grade.
**Containers and Materials:** Cylinders, tank cars, tank trucks; steel.

### Physical and Chemical Characteristics

**Physical State** (20°C, 1 atm): Gas.
**Solubility** (Water): 0.0024 g/100 mL (20°C).
**Molecular Weight:** 16.0
**Vapour Pressure:** Gas.
**Boiling Point:** -161.5°C.

**Floatability** (Water): Floats and boils (liquid).
**Odour:** Odourless.
**Flash Point:** -188°C.
**Vapour Density:** 0.42 to 0.56
**Specific Gravity:** (Liquid) 0.42 (-164°C).

**Colour:** Colourless.
**Explosive Limits:** 5.0 to 15.0%.
**Melting Point:** -182 to -184°C.

## HAZARD DATA

### Human Health

**Symptoms:** Inhalation: asphyxiation, dizziness, difficult breathing, nausea and vomiting, exhaustion, loss of consciousness, convulsions. Contact: skin - feeling of intense cold, frostbite; eyes - stinging pain, watering of eyes, inflammation.
**Toxicology:** Asphyxiant.
**TLV®**= simple asphyxiant.        **LC$_{50}$ - Man:** tolerance limit = 1 000 ppm        **LD$_{50}$** - No information.
**Short-term Inhalation Limits** - No information.   (18 mg/m$^3$).
                                    Delayed Toxicity - None known.

### Fire

**Fire Extinguishing Agents:** Shut off flow of gas if safe to do so. In case of fire use carbon dioxide or dry chemical to extinguish.
**Behaviour in Fire:** Flash back may occur along vapour trail.
**Ignition Temperature:** 537°C.    **Burning Rate:** 12.5 mm/min.

### Reactivity

**With Water:** No reaction.
**With Common Materials:** Reacts with strong oxidizers. Reacts violently with bromine pentafluoride, chlorine, chlorine dioxide, liquid oxygen, nitrogen trifluoride, oxygen difluoride.
**Stability:** Stable.

### Environment

**Water:** Prevent entry into water intakes and waterways. Not seriously harmful to aquatic life. BOD: 304% (35 days).
**Land-Air:** No information.
**Food Chain Concentration Potential:** None.

## EMERGENCY MEASURES

**Special Hazards**

FLAMMABLE.

**Immediate Responses**

Keep non-involved people away from spill site. Issue warning: "FLAMMABLE". CALL FIRE DEPARTMENT. Contact supplier for advice. Eliminate all sources of ignition. Stop or reduce discharge if this can be done without risk. Notify environmental authorities.

**Protective Clothing and Equipment**

Respiratory protection - in fires, use self-contained breathing apparatus. Otherwise: Gloves - rubber or plastic. Coveralls or acid suit - (jacket and pants).

**Fire and Explosion**

Shut off flow of gas if safe to do so. Use carbon dioxide or dry chemical to extinguish. Flash back may occur along vapour trail.

**First Aid**

Move victim out of spill site to fresh air. Call for medical assistance, but start first aid at once. Inhalation: give artificial respiration if breathing has stopped; oxygen if breathing is laboured. Contact: (liquified methane) skin and eyes - treat as for frostbite; remove contaminated clothing and wash eyes and skin thoroughly with plenty of water. If medical assistance is not immediately available, transport victim to hospital, clinic or doctor.

## ENVIRONMENTAL PROTECTION MEASURES

**Response**

**Water-Land-Air**
1. Stop or reduce discharge if safe to do so.
2. Contact manufacturer or supplier for advice.
3. Notify environmental authorities to discuss disposal and cleanup of contaminated materials.

**Disposal**
1. Contact manufacturer or supplier for advice on disposal.
2. Contact environmental authorities for advice on disposal.
3. Incinerate (approval of environmental authorities required).

METHANE  $CH_4$

# METHANOL  CH₃OH

## IDENTIFICATION

UN No. 1230

### Common Synonyms

METHYL ALCOHOL
WOOD ALCOHOL
WOOD NAPHTHA
WOOD SPIRIT
CARBINOL
METHYL HYDROXIDE

### Observable Characteristics

Clear colourless, liquid. Alcohol-like odour.

### Manufacturers

Alberta Gas Chemical, Medicine Hat, Alta.
Celanese Canada, Cornwall, Ontario.

### Transportation and Storage Information

**Shipping State:** Liquid.
**Classification:** Flammable, poisonous liquid.
**Inert Atmosphere:** No requirement.
**Venting:** Open (flame arrester) or pressure-vacuum.
**Pump Type:** No information.

**Label(s):** Red and white label - FLAMMABLE; Class 3.2, Group II. Black and white label - POISON; Class 6.1, Group II.
**Storage Temperature:** Ambient.
**Hose Type:** No information.

**Grades or Purity:** Pure, 99.9%; crude.
**Containers and Materials:** Cans, drums, tank cars, tank trucks; steel.

### Physical and Chemical Characteristics

**Physical State** (20°C, 1 atm): Liquid.
**Solubility (Water):** Soluble in all proportions.
**Molecular Weight:** 32.0
**Vapour Pressure:** 92 mm Hg (20°C); 160 mm Hg (30°C).
**Boiling Point:** 64.5°C.

**Floatability (Water):** Floats and mixes.
**Odour:** Alcohol-like.
**Flash Point:** 11°C (c.c.).
**Vapour Density:** 1.1
**Specific Gravity:** 0.79 at 20°C.

**Colour:** Colourless.
**Explosive Limits:** 6.0% to 36.0%.
**Melting Point:** -93 to -98°C.

## HAZARD DATA

### Human Health

**Symptoms:** Inhalation: asphyxia, headache, fatigue, nausea and vomiting, exhaustion, loss of consciousness, convulsions. Contact: skin - readily absorbed producing symptoms similar to inhalation; eyes - irritation, damage, blurred vision. Ingestion: headache, fatigue, nausea and vomiting, blindness and death.
**Toxicology:** Low toxicity by contact; moderate by inhalation.
TLV® - (skin) 200 ppm; 260 mg/m³.
Short-term Inhalation Limits - 250 ppm; 310 mg/m³ (15 min).
LC₅₀ - No information.
TCLo - Inhalation: human = 86 000 mg/m³
Delayed Toxicity - None known.
LD₅₀ - Oral: rat = 13 g/kg

### Fire

**Fire Extinguishing Agents:** Use alcohol foam. Water may be ineffective.
**Behaviour in Fire:** Flashback may occur along vapour trail.
**Ignition Temperature:** 385°C.
**Burning Rate:** 1.7 mm/min.

### Reactivity

**With Water:** No reaction; soluble.
**With Common Materials:** Can react vigorously with oxidizing agents. Reacts violently with chromic anhydride, lead perchlorite, perchloric acid, phosphorus trioxide, and (hydroxides and chloroform).
**Stability:** Stable.

### Environment

**Water:** Prevent entry into water intakes and waterways. Fish toxicity: 13 680 ppm/96 h/rainbow trout fingerling/LC₅₀/freshwater; 1 700 ppm/96 h/brown shrimp/LC₅₀/saltwater; 17 000 ppm/24 h/creek chub/LC₁₀₀/freshwater; BOD: 48 to 124%; 5 days.
**Land-Air:** No information.
**Food Chain Concentration Potential:** None.

## EMERGENCY MEASURES

**Special Hazards**

FLAMMABLE. POISON.

**Immediate Responses**

Keep non-involved people away from spill site. Issue warnings: "FLAMMABLE; POISON". CALL FIRE DEPARTMENT. Eliminate all sources of ignition. Notify manufacturer. Stop leak, if safe to do so. Dike to contain material. Notify environmental authorities.

**Protective Clothing and Equipment**

Protective outerwear as required.

**Fire and Explosion**

Use alcohol foam to extinguish; water may be ineffective. Flash back may occur along vapour trail.

**First Aid**

Move victim out of spill site to fresh air. Call for medical assistance, but start first aid at once. <u>Inhalation</u>: if breathing has stopped give artificial respiration (not mouth-to-mouth method); if laboured, give oxygen. <u>Contact</u>: skin - remove contaminated clothing, and flush affected aras with plenty of water; eyes - irrigate with plenty of water. <u>Ingestion</u>: give plenty of water to conscious victim to drink and induce vomiting. If medical assistance is not immediately available, transport victim to doctor, hospital or clinic.

## ENVIRONMENTAL PROTECTION MEASURES

**Response**

**Water**
1. Stop or reduce discharge if safe to do so.
2. Contact manufacturer or supplier for advice.
3. If possible, contain discharge by damming or water diversion.
4. Notify environmental authorities to discuss disposal and cleanup of contaminated materials.

**Land-Air**
1. Stop or reduce discharge if safe to do so.
2. Contact manufacturer or supplier for advice.
3. Contain spill by diking with earth or other barrier.
4. Remove material with pumps or vacuum equipment and place in appropriate containers.
5. Recover undamaged containers.
6. Absorb residual liquid on natural or synthetic sorbents.
7. Notify environmental authorities to discuss disposal and cleanup of contaminated materials.

**Disposal**

1. Contact manufacturer or supplier for advice on disposal.
2. Contact environmental authorities for advice on disposal.
3. Incinerate (approval of environmental authorities required).

METHANOL    $CH_3OH$

# METHYL ACRYLATE  $CH_2{:}CHCOOCH_3$

## IDENTIFICATION

### Common Synonyms
ACRYLIC ACID, METHYL ESTER
METHYL 2-PROPENOATE
METHYL PROPENATE
PROPENOIC ACID, METHYL ESTER

### Observable Characteristics
Colourless liquid. Sweet, sharp odour.

### Manufacturers
UN No. 1919

No Canadian manufacturers.
Celanese Chem. Co., New York, NY.
Rohm and Haas Co., Philadelphia, PA.

### Transportation and Storage Information
**Shipping State:** Liquid.
**Classification:** Flammable liquid.
**Inert Atmosphere:** Air must be present.
**Venting:** Open (flame arrester).
**Pump Type:** No information.

**Label(s):** Red and white label - FLAMMABLE LIQUID; Class 3.2, Group II.
**Storage Temperature:** Ambient (if inhibited, <4°C if no inhibiter).
**Hose Type:** No information.

**Grades or Purity:** Technical 98.5% (minimum purity); 99.9% (with inhibitor; hydroquinone and its methyl ether, in presence of air).
**Containers and Materials:** Drums, tank cars; steel.

### Physical and Chemical Characteristics
**Physical State** (20°C, 1 atm): Liquid.
**Solubility** (Water): 5.2 g/100 mL (20°C).
**Molecular Weight:** 86.1
**Vapour Pressure:** 70 mm Hg (20°C); 110 mm Hg (30°C).
**Boiling Point:** 80 to 81°C.

**Floatability** (Water): Floats.
**Odour:** Sweet, sharp.
**Flash Point:** -3°C (o.c.)
**Vapour Density:** 3.0
**Specific Gravity:** 0.96 (20°C).

**Colour:** Colourless.
**Explosive Limits:** 2.8 to 25%.
**Melting Point:** -76.5°C.

## HAZARD DATA

### Human Health
**Symptoms:** <u>Inhalation</u>: irritation of mucous membranes, headache, nausea, cyanosis, dizziness, fatigue, diarrhea. <u>Ingestion</u>: irritation of lips, mouth and throat, pain in swallowing, stomach and abdominal pain, nausea and vomiting, diarrhea, shock, convulsions. <u>Contact</u>: skin - readily absorbed, itching and irritation, inflammation, blisters; eyes - irritation, watering of eyes, inflammation, chemical burns and lesions.
**Toxicology:** Highly toxic by inhalation and ingestion; moderately toxic by contact.
TLV® (skin) 10 ppm; 35 mg/m³.    $LC_{50}$ - No information.    $LD_{50}$ - Oral; rat = 0.3 g/kg
Short-term Inhalation Limits - No information.    $LC_{Lo}$ - Inhalation: rat = 1 000 ppm/4 h
Delayed Toxicity - No information.

### Fire
**Fire Extinguishing Agents:** Foam, dry chemical or carbon dioxide. Water may be ineffective but should be used to keep fire-exposed containers cool and disperse vapours.
**Behaviour in Fire:** At temperatures above 21°C polymerization may take place, and if in a closed container, may cause a violent rupture. Flash back may occur along vapour trail.
**Ignition Temperature:** 468°C.    **Burning Rate:** No information.

### Reactivity
**With Water:** No reaction; slightly soluble.
**With Common Materials:** Can react vigorously with oxidizing agents.
**Stability:** Polymerizes at high temperatures or under certain conditions.

### Environment
**Water:** Prevent entry into water intakes and waterways. Harmful to aquatic life. Aquatic toxicity rating = 100 to 1 000 ppm/96 h/TLm/freshwater.
**Land-Air:** No information.
**Food Chain Concentration Potential:** None.

## EMERGENCY MEASURES

**Special Hazards**

FLAMMABLE. Polymerizes under certain conditions.

**Immediate Responses**

Keep non-involved people away from spill site. Issue warning: "FLAMMABLE". CALL FIRE DEPARTMENT. Eliminate all sources of ignition. Notify manufacturer or supplier. Dike to prevent runoff of material. Notify environmental authorities.

**Protective Clothing and Equipment**

Respiratory protection - self-contained breathing apparatus and totally encapsulated suit.

**Fire and Explosion**

Use foam, dry chemical or carbon dioxide to extinguish. Water may be ineffective, but should be used to cool fire-exposed containers and disperse vapours. At temperatures above 21°C, polymerization may take place and, if in a closed container, violent rupture may occur. Flash back may occur along vapour trail.

**First Aid**

Move victim out of spill site to fresh air. Call for medical assistance, but start first aid at once. Inhalation: if breathing has stopped give artificial respiration; if laboured, give oxygen. Contact: skin - remove contaminated clothing, and flush affected aras with plenty of water; eyes - irrigate with plenty of water. Ingestion: give plenty of water to conscious victim to drink and induce vomiting. If medical assistance is not immediately available, transport victim to doctor, hospital or clinic.

## ENVIRONMENTAL PROTECTION MEASURES

**Response**

**Water**

Dissolved
1. Stop or reduce discharge if safe to do so.
2. Contact manufacturer or supplier for advice.
3. If possible, contain discharge by damming or water diversion.
4. Dredge or vacuum pump to remove contaminants, liquids and contaminated bottom sediments.
5. Notify environmental authorities to discuss disposal and cleanup of contaminated materials.

Undissolved
1. Stop or reduce discharge if safe to do so.
2. Contact manufacturer or supplier for advice.
3. If possible, contain spill by booming.
4. If floating, skim and remove.
5. Notify environmental authorities to discuss disposal and cleanup of contaminated materials.

**Land-Air**
1. Stop or reduce discharge if safe to do so.
2. Contact manufacturer or supplier for advice.
3. Contain spill by diking with earth or other barrier.
4. Remove material with pumps or vacuum equipment and place in appropriate containers.
5. Recover undamaged containers.
6. Absorb residual liquid on natural or synthetic sorbents.
7. Notify environmental authorities to discuss disposal and cleanup of contaminated materials.

**Disposal**

1. Contact manufacturer or supplier for advice on disposal.
2. Contact environmental authorities for advice on disposal.
3. Incinerate (approval of environmental authorities required).

METHYL ACRYLATE   $CH_2:CHCOOCH_3$

# METHYLAMINES

## IDENTIFICATION

| Common Synonyms | Observable Characteristics | UN No. 1061 Anhydrous |
| --- | --- | --- |
| MONOMETHYLAMINE ($CH_3NH_2$) Aminomethane, Carbinamine<br>DIMETHYLAMINE ($CH_3)_2NH$<br>TRIMETHYLAMINE ($CH_3)_3N$<br>Methanamine, N,N-Dimethylamine | Colourless gases or liquids. Ammonia-like odours. | 1235 Aqueous solution<br><br>**Manufacturers**<br>Chinook Chemicals, Sombra, Ontario. |

### Transportation and Storage Information

**Shipping State:** Liquid (compressed gas or aqueous solution).
**Classification:** Flammable liquid or gas.
**Inert Atmosphere:** No requirement.
**Venting:** Safety relief.
**Label(s):** Red label - FLAMMABLE LIQUID OR GAS.
**Storage Temperature:** Ambient.
**Grades or Purity:** Anhydrous, 96.5%+; aqueous solutions 25, 30, 35, 40%.
**Containers and Materials:** Anhydrous - cylinders. Solutions - drums, tank cars, tank trucks; steel, stainless steel.

### Physical and Chemical Characteristics

**Physical State** (20°C, 1 atm): Gas.
**Solubility (Water):** Soluble in all proportions.
**Molecular Weight:** 31.1 (mono); 45.1 (di); 59.1 (tri).
**Vapour Pressure:** (mono) 2 356 mm Hg; (di) 1 292 mm Hg; (tri) 1 444 mm Hg (all at 20°C).
**Boiling Point:** (mono) -6.5°C; (di) 7.4°C; (tri) 2.9°C.
**Specific Gravity:** (mono) 0.77 (-70°C); (di) 0.68 (0°C); (tri) 0.66 (-5°C) (anhydrous forms). Solutions: 25% (mono) 0.93 (15°C); 40%, 0.89 (15°C). 25% (di) 0.92 (15°C); 40%, 0.88 (15°C). 25% (tri) 0.92 (15°C); 40% 0.89 (15°C).
**Floatability (Water):** Liquid or solutions float and mix.
**Odour:** Ammonia-like, fishy (0.02 to 3.3 ppm, odour threshold).
**Flash Point:** gases <0°C. 30% (mono) solution 0.28°C; 40%, -13°C. 25% (di) solution 5.6°C; 40%, -15.6°C; 25% (tri) solution 5°C (all o.c.).
**Vapour Density:** (mono) 1.07; (di) 1.55; (tri) 2.04.
**Colour:** Colourless.
**Explosive Limits:** 4.9 to 20.7% (mono); 2.8 to 14.4% (di); 2.0 to 11.6% (tri).
**Melting Point:** (anhydrous forms)(mono) -92.5°C; (di) -92 to -96°C; (tri) -177 to -124°C. 25% mono solution -31°C; 40%, -37°C. 25% di solution -17°C; 40%, -37°C. 25% tri solution +6°C; 40%, +4°C.

## HAZARD DATA

### Human Health

**Symptoms:** Inhalation: irritation of nose and eyes, sneezing, coughing. Ingestion: burning sensation, pain in swallowing, nausea and vomiting, stomach cramps, rapid breathing, diarrhea. Contact: skin - painful burning, ulceration and shock; eyes - painful irritation, intense watering, burns and irreparable damage.
**Toxicology:** Moderately toxic by all routes.
TLV® (mono) 10 ppm; 12 mg/m³, (di) 10 ppm; 18 mg/m³.
Short-term Inhalation Limits - No information.
LC50 - No information.
Delayed Toxicity - No information.
LD50 - Oral: rat = 0.698 g/kg (di)
LDLo - Subcutaneous: rat = 0.20 g/kg (mono)

### Fire

**Fire Extinguishing Agents:** Stop flow of gas. Use water spray, carbon dioxide, dry chemical and alcohol foam on water solution. Use water to cool fire-exposed containers.
**Behaviour in Fire:** Flash back may occur along vapour trail.
**Ignition Temperature:** 430°C (mono); 400°C (di); 109°C (tri).
**Burning Rate:** No information.

### Reactivity

**With Water:** No reaction; soluble.
**With Common Materials:** Can react vigorously with oxidizing materials. Aqueous solutions are corrosive to many metals.
**Stability:** Stable.

### Environment

**Water:** Prevent entry into water intakes and waterways. Harmful to aquatic life. Fish toxicity: (mono) 10 to 30 ppm, (di) 30 to 50 ppm/24 h/creek chub/TLm/ freshwater; Aquatic toxicity rating = 10 to 100 ppm/96 h/TLm/freshwater; BOD: 130% (di), 5 days.
**Land-Air:** No information.
**Food Chain Concentration Potential:** None.

## EMERGENCY MEASURES

**Special Hazards**

FLAMMABLE. Aqueous solutions are corrosive.

**Immediate Responses**

Keep non-involved people away from spill site. Issue warning: "FLAMMABLE". CALL FIRE DEPARTMENT. Eliminate all sources of ignition. Notify manufacturer or supplier. Dike to prevent runoff from water application. Notify environmental authorities.

**Protective Clothing and Equipment**

Respiratory protection - self-contained breathing apparatus and totally encapsulated suit.

**Fire and Explosion**

Stop flow of gas. Use water spray, carbon dioxide, dry chemical or alcohol foam on water solutions. Use water to cool fire-exposed containers. Flash back may occur along vapour trail.

**First Aid**

Move victim out of spill site to fresh air. Call for medical assistance, but start first aid at once. Inhalation: if breathing has stopped, give artificial respiration (not mouth-to-mouth method); if laboured, give oxygen. Contact: skin - remove contaminated clothing and flush affected areas with plenty of water; eyes - irrigate with plenty of water. Ingestion: give conscious victim plenty of water to drink; do not induce vomiting. If medical assistance is not immediately available, transport victim to hospital, doctor or clinic.

## ENVIRONMENTAL PROTECTION MEASURES

**Response**

**Water**
1. Stop or reduce discharge if safe to do so.
2. Contact manufacturer or supplier for advice.
3. If possible, contain discharge by damming or water diversion.
4. Dredge or vacuum pump to remove contaminants, liquids and contaminated bottom sediments.
5. Notify environmental authorities to discuss disposal and cleanup of contaminated materials.

**Land-Air**
1. Stop or reduce discharge if safe to do so.
2. Contact manufacturer or supplier for advice.
3. Dike to prevent runoff from rainwater or water application.
4. Remove material with pumps or vacuum equipment and place in appropriate containers.
5. Recover undamaged containers.
6. Absorb residual liquid on natural or synthetic sorbents.
7. Notify environmental authorities to discuss disposal and cleanup of contaminated materials.

**Disposal**
1. Contact manufacturer or supplier for advice on disposal.
2. Contact environmental authorities for advice on disposal.

METHYLAMINES

# METHYL CHLORIDE   CH₃Cl

## IDENTIFICATION   UN No. 1063

| Common Synonyms | Observable Characteristics | Manufacturers |
|---|---|---|
| CHLOROMETHANE<br>MONOCHLOROMETHANE | Colourless gas or liquid. Sweet, ethereal odour. | No Canadian manufacturer.<br>Canadian suppliers:<br>Domtar, Longford Mills, Ontario.<br>Dow Chemical Canada, Inc, Sarnia, Ont.<br><br>Ethyl Canada, Sarnia, Ontario. | Originating from:<br>Conoco Chem., USA<br>Dow Chemical, Plaquemine, LA,<br>Freeport, TX,<br>Ethyl, USA |

### Transportation and Storage Information

**Shipping State:** Liquefied compressed gas.
**Classification:** Poisonous gas and flammable liquid.
**Inert Atmosphere:** No requirement.
**Venting:** Safety relief.
**Pump Type:** Positive displacement; carbon steel, stainless steel.

**Label(s):** White label - POISONOUS GAS; Class 2.3. Red label - FLAMMABLE LIQUID; Class 3.1.
**Storage Temperature:** Ambient.
**Hose Type:** Teflon, Viton, bronze, stainless steel.

**Grades or Purity:** Refrigeration grade; technical grade. 99.5+%.
**Containers and Materials:** Cylinders, tank cars, tank trucks; steel, stainless steel.

### Physical and Chemical Characteristics

**Physical State** (20°C, 1 atm): Gas.
**Solubility:** Slight; 0.72 g/100 mL (0°C); 0.43 g/100 mL (25°C).
**Molecular Weight:** 50.5
**Vapour Pressure:** 3800 mm Hg (20°C); 5090 mm Hg (30°C).
**Boiling Point:** -24.2°C.

**Floatability (Water):** Floats and boils.
**Odour:** Faintly sweet, ether-like; (21 mg/m³, odour threshold).
**Flash Point:** Below -46°C
**Vapour Density:** 1.8
**Specific Gravity:** (Liquid) 0.92 (20°C); 0.99 (-25°C).

**Colour:** Colourless.
**Explosive Limits:** 10.7 to 17.4%.
**Melting Point:** -97.7°C.

## HAZARD DATA

### Human Health

**Symptoms: Inhalation:** asphyxia; headache, fatigue, mental confusion, nausea and vomiting, giddiness, exhaustion, loss of consciousness, convulsions. **Contact:** skin - feeling of cold, pain, frostbite; eyes - stinging, watering, inflammation.
**Toxicology:** Repeated exposures which do not at once cause serious symptoms may be followed after a few days by more severe effects. TLV® - (inhalation) 50 ppm, 105 mg/m³. LC₅₀ - Inhalation: rat = 152 000 mg/m³ for 30 min. LD₅₀ - No information. Short-term Inhalation Limits - 100 ppm; 205 mg/m³ (15 min). Delayed Toxicity - Sublethal or lethal effects may occur several days after exposure. Symptoms may not appear until several hours after exposure.

### Fire

**Fire Extinguishing Agents:** Do not attempt to put out fire until leak has been turned off. In fires, use dry chemical, carbon dioxide or foam. Water may be used to knock down vapours or cool fire-exposed containers.
**Behaviour in Fire:** Flashback may occur along vapour trail. Decomposition from heat or fire can produce toxic chloride fumes.
**Ignition Temperature:** 632°C.   **Burning Rate:** 2.2 mm/min.

### Reactivity

**With Water:** No reaction.
**With Common Materials:** Can react violently with powdered aluminum, magnesium, sodium and other alkali metals.
**Stability:** Stable.

### Environment

**Water:** Prevent entry into water intakes or waterways. Aquatic toxicity rating = >1 000 ppm/96 h/TLm/freshwater; BOD: None.
**Land-Air:** No information.
**Food Chain Concentration Potential:** None.

## EMERGENCY MEASURES

### Special Hazards
FLAMMABLE.

### Immediate Responses
Keep non-involved people away from spill site. Issue warning: "FLAMMABLE". Call Fire Department. Eliminate all sources of ignition. Call manufacturer for advice. Avoid contact and inhalation. In fire, work from upwind. If water spray used, dike area to contain toxic runoff. Stop or reduce discharge, if this can be done without risk. Notify environmental authorities.

### Protective Clothing and Equipment
Respiratory protection - self-contained breathing apparatus. Acid suit - jacket and pants, rubber or plastic. Boots - high, rubber; pants worn outside boots. Gloves - rubber or plastic.

### Fire and Explosion
Do not attempt to put out fire until leak has been shut off. For fires, use dry chemical, carbon dioxide or foam. Use water spray to knock down vapour and cool fire-exposed containers. Flashback may occur along vapour trail. Decomposition from heat or fire can produce toxic chloride fumes.

### First Aid
Move victim out of spill area to fresh air. Call for medical assistance, but start first aid at once. Inhalation: give artificial respiration if breathing has stopped; give oxygen if breathing is laboured. Contact: remove contaminated clothing; wash eyes and affected skin with plenty of warm water. Do not rub areas which appear to have been frostbitten. If medical assistance is not immediately available, transport victim to hospital, doctor or clinic. Victim should be hospitalized for observation because of slowly developing symptoms.

## ENVIRONMENTAL PROTECTION MEASURES

### Response

**Water**
1. Stop or reduce discharge if safe to do so.
2. Contact manufacturer or supplier for advice.
3. If possible, contain discharge by damming or water diversion.
4. Dredge or vacuum pump to remove contaminants, liquids and contaminated bottom sediments.
5. Notify environmental authorities to discuss disposal and cleanup of contaminated materials.

**Land-Air**
1. Stop or reduce discharge if safe to do so.
2. Contact manufacturer or supplier for advice.
3. Dike to prevent runoff from rainwater or water application.
4. Recover undamaged containers.
5. Notify environmental authorities to discuss disposal and cleanup of contaminated materials.

### Disposal
1. Contact manufacturer or supplier for advice on disposal.
2. Contact environmental authorities for advice on disposal.

METHYL CHLORIDE $CH_3Cl$

# METHYL ETHYL KETONE   $CH_3COC_2H_5$

## IDENTIFICATION

**Common Synonyms**

MEK
2-BUTANONE
ETHYL METHYL KETONE

**Observable Characteristics**

Colourless liquid. Sweet, fragrant.

**UN No. 1193**

**Manufacturers**

Shell Canada Limited, Montreal, Que.
Gulf Oil Canada Limited, Montreal, Que.
Esso Chemicals, Sarnia, Ont.

### Transportation and Storage Information

**Shipping State:** Liquid.
**Classification:** Flammable liquid.
**Inert Atmosphere:** No requirement.
**Venting:** Open (flame arrester) or pressure-vacuum.
**Pump Type:** Gear, centrifugal; explosion-proof motors.

**Label(s):** Red label - FLAMMABLE LIQUID; Class 3.2, Group II.
**Storage Temperature:** Ambient.
**Hose Type:** Polyethylene, butyl, polypropylene.

**Grades or Purity:** Technical, 99.5+%.
**Containers and Materials:** Drums, tank cars, tank trucks; steel.

### Physical and Chemical Characteristics

**Physical State** (20°C, 1 atm): Liquid.
**Solubility** (Water): 35.3 g/100 mL (10°C); 26.3 (20°C); 19.0 g/100 mL (90°C).
**Molecular Weight:** 72.1
**Vapour Pressure:** 77.5 mm Hg (20°C).
**Boiling Point:** 79.6°C.

**Floatability** (Water): Floats and mixes.
**Odour:** Sweet, fragrant (2.0 to 6.0 ppm, odour threshold).
**Flash Point:** -5.6°C (o.c.); -9.0°C (c.c.).
**Vapour Density:** 2.4
**Specific Gravity:** 0.81 (20°C).

**Colour:** Colourless.
**Explosive Limits:** 1.7 to 11.4%.
**Melting Point:** -85 to 87°C.

## HAZARD DATA

### Human Health

**Symptoms:** Inhalation: coughing, shortness of breath, headache, dizziness, irritation of respiratory tract; Ingestion: irritation of digestive tract, nausea and vomiting, headache and dizziness. Contact: skin - defatting and dermatitis; eyes - irritation and burning.
**Toxicology:** Highly toxic by ingestion, moderately toxic by contact.
TLV® - (inhalation) 200 ppm; 590 mg/m³.
Short-term Inhalation Limits - 300 ppm; 885 mg/m³ (15 min).
$LC_{50}$ - No information.
Delayed Toxicity - No information.
$TC_{Lo}$ - Inhalation: human = 100 ppm/5 min.
$LD_{50}$ - Oral: rat = 3.4 g/kg

### Fire

**Fire Extinguishing Agents:** Alcohol foam, dry chemical, carbon dioxide. Water may be ineffective, but may be used to cool fire-exposed containers.
**Behaviour in Fire:** Flashback may occur along vapour trail.
**Ignition Temperature:** 516°C.
**Burning Rate:** 4.1 mm/min.

### Reactivity

**With Water:** No reaction, soluble.
**With Common Materials:** Can react with oxidizing materials. Reacts violently with chlorosulfonic acid, oleum and potassium-t-butoxide.
**Stability:** Stable.

### Environment

**Water:** Prevent entry into water intakes and waterways. Fish toxicity: 5 640 mg/L/48 h/bluegill/TLm/freshwater; 500 mg/L/24 h/goldfish/$LC_{50}$/freshwater; 5 600 mg/L/48 h/mosquito fish/TLm/freshwater; Aquatic toxicity rating = >1 000 ppm/96 h/TLm/freshwater; BOD: 151 to 224%, 5 days.
**Land-Air:** No information.
**Food Chain Concentration Potential:** None.

## EMERGENCY MEASURES

### Special Hazards
FLAMMABLE.

### Immediate Responses
Keep non-involved people away from spill site. Issue warning: "FLAMMABLE". CALL FIRE DEPARTMENT. Eliminate all ignition sources. Call manufacturer or supplier for advice. Contain spill by diking. Prevent runoff into sewers or watercourses. Avoid inhalation or contact. Stop or reduce discharge, if safe to do so. Notify environmental authorities.

### Protective Clothing and Equipment
Respiratory protection - self-contained breathing apparatus, and full protective clothing. Gloves - rubber. Boots - high, rubber (pants worn outside boots).

### Fire and Explosion
Use dry chemical, alcohol foam or carbon dioxide to extinguish. Water may be ineffective, but may be used to cool fire-exposed containers. Flash back may occur along vapour trail.

### First Aid
Move victim out of spill area, to fresh air. Call for medical assistance, but start first aid at once. Inhalation: if breathing has stopped, give artificial respiration; if laboured, give oxygen. Contact: eyes - irrigate with plenty of water; skin - wash affected areas with plenty of water and remove contaminated clothing. Ingestion: give plenty of water to conscious victim to drink and induce vomiting. If medical assistance is not immediately available, transport victim to hospital, doctor or clinic.

## ENVIRONMENTAL PROTECTION MEASURES

### Response

**Water**
1. Stop or reduce discharge if safe to do so.
2. Contact manufacturer or supplier for advice.
3. If possible, contain discharge by booming.
4. If floating, skim and remove.
5. Notify environmental authorities to discuss disposal and cleanup of contaminated materials.

**Land-Air**
1. Stop or reduce discharge if safe to do so.
2. Contact manufacturer or supplier for advice.
3. Contain spill by diking with earth or other barrier.
4. Remove material with pumps or vacuum equipment and place in appropriate containers.
5. Recover undamaged containers.
6. Absorb residual liquid on natural or synthetic sorbents.
7. Notify environmental authorities to discuss disposal and cleanup of contaminated materials.

### Disposal
1. Contact manufacturer or supplier for advice on disposal.
2. Contact environmental authorities for advice on disposal.
3. Incinerate (approval of environmental authorities required).

METHYL ETHYL KETONE  $CH_3COC_2H_5$

# METHYLCYCLOPENTADIENYL MANGANESE TRICARBONYL   $C_9H_7O_3Mn$

## IDENTIFICATION

### Common Synonyms
METHYLCYCLOPENTADIENYLMANGANESE-
TRICARBONYL
ETHYL MMT
ANTIKNOCK
MMT

### Observable Characteristics
Orange liquids. Sensitive to light. Faint, pleasant, herbaceous odour.

### Manufacturers
Ethyl Corporation of Canada Limited, Corunna, Ontario; Toronto, Ontario.
U.S. manufacturer:
Ethyl Corporation, Houston, TX

### Transportation and Storage Information
**Shipping State:** Liquid.
**Classification:** Not regulated. Voluntary poison label.
**Inert Atmosphere:** No requirement.
**Venting:** Pressure-vacuum.
**Pump Type:** Standard types (e.g. centrifugal). For LP grades, flammable solvent equipment required.
**Label(s):** Voluntary white label - POISON.
**Storage Temperature:** Ambient.
**Hose Type:** Standard neoprene hydrocarbon hose. Also flexible stainless steel, steel, Teflon.

**Grades or Purity:** Ethyl MMT (neat), Ethyl MMT/LP 62, Ethyl MMT/LP 46.
**Containers and Materials:** Cans, drums; steel.

### Physical and Chemical Characteristics
**Physical State (20°C, 1 atm):** Liquid.
**Solubility (Water):** 0.007 g/100 mL (25°C).
**Molecular Weight:** 218.1
**Vapour Pressure:** 0.051 mm Hg (20°C); 9.3 mm Hg (100°C).
**Boiling Point:** 233°C (for neat MMT).
**Floatability (Water):** Sink.
**Odour:** Faint, pleasant, herbaceous.
**Flash Point:** Ethyl MMT, 96°C (c.c.); Ethyl MMT/LP 62, 90°C (c.c.); Ethyl MMT/LP 46, 65°C (c.c.)
**Vapour Density:** No information.
**Specific Gravity:** Ethyl MMT, 1.38 at 20°C; Ethyl MMT/LP 62-1.11; Ethyl MMT/LP 46-1.02.
**Colour:** Dark orange.
**Explosive Limits:** Not determined for neat MMT; solvents in other grades may be explosive.
**Melting Point:** -2.2 to 1.5°C.

## HAZARD DATA

### Human Health
**Symptoms:** Inhalation: difficulty breathing, headache, metallic taste, nausea. Ingestion: symptoms similar to inhalation. Contact: skin - readily absorbed; eyes - irritation.
**Toxicology:** Highly toxic by inhalation and ingestion; moderately toxic by contact.
TLV® - (skin) 0.2 mg/m³ (as Mn).   LC$_{50}$ - (Inhalation) rat = 247 mg/m³/1 h   LD$_{50}$ - Oral: rat = 0.058 g/kg
Short-term Inhalation Limits - 0.6 mg/m³ (15 min)   Delayed Toxicity - No information.
(as Mn).

### Fire
**Fire Extinguishing Agents:** Dry chemical, carbon dioxide, foam, water spray.
**Behaviour in Fire:** Decomposes slowly at 200°C; fairly rapidly at 300°C yielding toxic gases.
**Ignition Temperature:** No information.   **Burning Rate:** No information.

### Reactivity
**With Water:** No reaction.
**With Common Materials:** No information.
**Stability:** Stable.

### Environment
**Water:** Prevent entry into water intakes and waterways.   **BOD:** No information.
**Land-Air:** No information.
**Food Chain Concentration Potential:** No information.

## EMERGENCY MEASURES

### Special Hazards
POISON. FLAMMABLE.

### Immediate Responses
Keep non-involved people away from spill site. Issue warnings: "POISON; FLAMMABLE". Call fire department. Eliminate all ignition sources. Call manufacturer for guidance and assistance. Stop or reduce discharge, if this can be done without risk. Avoid contact and inhalation. Contain spill area by diking. Notify environmental authorities.

### Protective Clothing and Equipment
Respiratory protection - use self-contained breathing apparatus and totally encapsulated suit. Gloves - neoprene or vinyl. Boots - rubber (pants worn outside boots).

### Fire and Explosion
Use dry chemical, carbon dioxide, foam or water spray to extinguish.

### First Aid
Move victim out of spill area to fresh air. Call for medical assistance, but start first aid at once. Inhalation: if breathing has stopped give artificial respiration (not mouth-to-mouth method); if laboured, give oxygen. Contact: eyes irrigate with plenty of water; skin - flush with plenty water and remove contaminated clothing. Ingestion: give conscious victim plenty of water to drink and induce vomiting. Keep victim warm and quiet. If medical assistance is not immediately available, transport victim to hospital, doctor or clinic.

## ENVIRONMENTAL PROTECTION MEASURES

### Response

**Water**
1. Stop or reduce discharge if safe to do so.
2. Contact manufacturer or supplier for advice.
3. If possible, contain discharge by damming or water diversion.
4. Dredge or vacuum pump to remove contaminants, liquids and contaminated bottom sediments.
5. Notify environmental authorities to discuss disposal and cleanup of contaminated materials.

**Land-Air**
1. Stop or reduce discharge if safe to do so.
2. Contact manufacturer or supplier for advice.
3. Contain spill by diking with earth or other barrier.
4. Remove material with pumps or vacuum equipment and place in appropriate containers.
5. Absorb residual liquid on natural or synthetic sorbents.
6. Remove contaminated soil for disposal.
7. Notify environmental authorities to discuss disposal and cleanup of contaminated materials.

### Disposal
1. Contact manufacturer or supplier for advice on disposal.
2. Contact environmental authorities for advice on disposal.

METHYLCYCLOPENTADIENYL MANGANESE TRICARBONYL   $C_9H_7O_3Mn$

# METHYLENE CHLORIDE   $CH_2Cl_2$

## IDENTIFICATION   UN No. 1593

| Common Synonyms | Observable Characteristics | Manufacturers |
|---|---|---|
| DICHLOROMETHANE<br>METHYLENEDICHLORIDE<br>METHANE DICHLORIDE | Colourless liquid. Ethereal odour. | No Canadian manufacturers.<br>Canadian supplier:<br>Dow Chemical Canada Inc., Sarnia, Ontario.<br>Stanchem, Montreal, Quebec.<br><br>Originating from:<br>PPG Industries, USA |

### Transportation and Storage Information

**Shipping State:** Liquid.
**Classification:** Poison.
**Inert Atmosphere:** No requirement.
**Venting:** Pressure-vacuum.
**Pump Type:** Centrifugal or positive displacement; steel or stainless steel.

**Label(s):** Black and white label – POISON; Class 6.1, Group III.
**Storage Temperature:** Ambient.
**Hose Type:** Stainless steel; Teflon, bronze. Viton or cross-linked polyethylene may be used with caution.

**Grades or Purity:** Technical.
**Containers and Materials:** Drums, tank cars, tank trucks; steel.

### Physical and Chemical Characteristics

**Physical State** (20°C, 1 atm): Liquid.
**Solubility:** 2.0 g/100 mL (20°C); 1.67 g/100 mL (25°C).
**Molecular Weight:** 84.9
**Vapour Pressure:** 349 mm Hg (20°C); 500 mm Hg (30°C).
**Boiling Point:** 39 to 41°C.

**Floatability (Water):** Sinks.
**Odour:** Ethereal (25 to 220 ppm, odour threshold).
**Flash Point:** Not flammable.
**Vapour Density:** 2.9
**Specific Gravity:** 1.33 (20°C).

**Colour:** Colourless.
**Explosive Limits:** 12 to 22%.
**Melting Point:** -96.7°C.

## HAZARD DATA

### Human Health

**Symptoms:** Inhalation: irritation of respiratory tract, headache, dizziness, nausea, unconsciousness; Ingestion: abdominal pain, readily absorbed producing symptoms similar to inhalation. Contact: skin - readily absorbed with symptoms similar to inhalation; eyes - irritation, slight burns.
**Toxicology:** Moderately toxic by all routes.
TLV® - (inhalation) 100 ppm; 350 mg/m$^3$.   $LC_{50}$ - Inhalation: rat = 88 000 mg/m$^3$/0.5 h   $LD_{50}$ - Oral: rat = 1.67 g/kg
Short-term Inhalation Limits - 500 ppm;   Delayed Toxicity - No information.
1 400 mg/m$^3$ (15 min).

### Fire

**Fire Extinguishing Agents:** Most fire extinguishing agents may be used on fires involving methylene chloride, including water spray or fog and foam. Water may be used to cool fire-exposed containers.
**Behaviour in Fire:** When heated to decomposition, emits toxic fumes of hydrogen and chloride.
**Ignition Temperature:** 556°C.   **Burning Rate:** No information.

### Reactivity

**With Water:** No reaction.
**With Common Materials:** Reacts violently with lithium, sodium/potassium alloy and potassium-t-butoxide. May react with aluminum, magnesium and their alloys.
**Stability:** Stable.

### Environment

**Water:** Prevent entry into water intakes and waterways. Aquatic toxicity rating = 100 to 1 000 ppm/96 h/TLm/freshwater; BOD: None.
**Land-Air:** No information.
**Food Chain Concentration Potential:** None.

## EMERGENCY MEASURES

**Special Hazards**

POISON.

**Immediate Responses**

Keep non-involved people away from spill site. Issue warning: "POISON". Call Fire Department. Call manufacturer for assistance. Avoid contact and inhalation. Dike to prevent runoff. Notify environmental authorities.

**Protective Clothing and Equipment**

In fires or enclosed spaces, Respiratory protection - self-contained breathing apparatus and totally encapsulated suit. Otherwise, suitable respirator and protective outerwear as required.

**Fire and Explosion**

Most fire extinguishing agents may be used on fires involving methylene chloride. Water may be used to cool fire-exposed containers. When heated to decomposition, emits toxic fumes of hydrogen and chloride.

**First Aid**

Move victim out of spill area to fresh air. Call for medical assistance, but start first aid at once. Inhalation: if breathing has stopped, give artificial respiration (not mouth-to-mouth method); if laboured, give oxygen. Contact: skin - remove contaminated clothing and flush affected areas with plenty of water; eyes - irrigate with plenty of water for at least 15 minutes. Ingestion: give plenty of water to drink. Do not induce vomiting. If medical assistance is not immediately available, transport victim to hospital, doctor or clinic.

## ENVIRONMENTAL PROTECTION MEASURES

**Response**

Water
1. Stop or reduce discharge if safe to do so.
2. Contact manufacturer or supplier for advice.
3. If possible, contain discharge by damming or water diversion.
4. Dredge or vacuum pump to remove contaminants, liquids and contaminated bottom sediments.
5. Notify environmental authorities to discuss disposal and cleanup of contaminated materials.

Land-Air
1. Stop or reduce discharge is safe to do so.
2. Contact manufacturer or supplier for advice.
3. Contain spill by diking with earth or other barrier.
4. Remove material with pumps or vacuum equipment and place in appropriate containers.
5. Recover undamaged containers.
6. Adsorb residual liquid on natural or synthetic sorbents.
7. Remove contaminated soil for disposal.
8. Notify environmental authorities to discuss disposal and cleanup of contaminated materials.

**Disposal**

1. Contact manufacturer or supplier for advice on disposal.
2. Contact environmental authorities for advice on disposal.

METHYLENE CHLORIDE $CH_2Cl_2$

# METHYL ISOBUTYL KETONE ($(CH_3)_2CHCH_2COCH_3$)

## IDENTIFICATION

UN No. 1245

### Common Synonyms
HEXONE
4-METHYL-2-PENTANONE
ISOBUTYL METHYL KETONE
MIBK
2-METHYL-4-PENTANONE

### Observable Characteristics
Colourless liquid. Sharp, pleasant odour.

### Manufacturers
Shell Canada, Montreal, Quebec.
Gulf Canada, Montreal, Quebec.

### Transportation and Storage Information
**Shipping State:** Liquid.
**Classification:** Flammable liquid.
**Inert Atmosphere:** No requirement.
**Venting:** Open (flame arrester) or pressure-vacuum.
**Pump Type:** Gear, centrifugal; explosion-proof motors.
**Label(s):** Red and white label - FLAMMABLE LIQUID; Class 3.2, Group II.
**Storage Temperature:** Ambient.
**Hose Type:** Polyethylene, butyl, polypropylene.
**Grades or Purity:** Technical, 98.5%.
**Containers and Materials:** Cans, drums, tank cars; steel.

### Physical and Chemical Characteristics
**Physical State** (20°C, 1 atm): Liquid.
**Solubility** (Water): 1.8 g/100 mL (20°C).
**Molecular Weight:** 100.2
**Vapour Pressure:** 6 mm Hg (20°C); 10 mm Hg (30°C).
**Boiling Point:** 116 to 119°C.
**Floatability** (Water): Floats.
**Odour:** Sharp, pleasant (0.1 to 7.8 ppm).
**Flash Point:** 18°C (c.c.).
**Vapour Density:** 3.5
**Specific Gravity:** 0.80 (20°C).
**Colour:** Colourless.
**Explosive Limits:** 1.2 to 8.0%.
**Melting Point:** -80 to -85°C.

## HAZARD DATA

### Human Health
**Symptoms:** Inhalation: irritation of respiratory tract, headache, nausea, dizziness, unconsciousness, CNS depressant. Ingestion: irritation of gastrointestinal tract, nausea and vomiting. Contact: skin - defatting and irritation; eyes - irritation and burning sensation.
**Toxicology:** Moderately toxic by all routes.
TLV® (inhalation) 50 ppm; 205 mg/m³.
Short-term Inhalation Limits - 75 ppm; 300 mg/m³ (15 min).
$LC_{50}$ - No information.
Delayed Toxicity - No information.
$LC_{Lo}$ - Inhalation: rat = 4 000 ppm/15 min.
$LD_{50}$ - Oral: rat = 2.08 g/kg

### Fire
**Fire Extinguishing Agents:** Alcohol foam, dry chemical, carbon dioxide. Water may be ineffective, but may be used to cool fire-exposed containers.
**Behaviour in Fire:** Flash back may occur along vapour trail.
**Ignition Temperature:** 448°C.
**Burning Rate:** No information.

### Reactivity
**With Water:** No reaction.
**With Common Materials:** Can react vigorously with reducing or oxidizing materials. Reacts violently with potassium-t-butoxide.
**Stability:** Stable.

### Environment
**Water:** Prevent entry into water intakes and waterways. Aquatic toxicity rating = >1 000 ppm/96 h/TLm/freshwater; Fish toxicity: 460 mg/L/24 h/goldfish/ $LC_{50}$/freshwater; BOD: 12 to 60%, 5 days.
**Land-Air:** No information.
**Food Chain Concentration Potential:** None.

## EMERGENCY MEASURES

**Special Hazards**
FLAMMABLE.

**Immediate Responses**
Keep non-involved people away from spill site. Issue warning "FLAMMABLE". CALL FIRE DEPARTMENT. Call manufacturer for assistance. Dike to prevent runoff. Call environmental authorities.

**Protective Clothing and Equipment**
In fires Respiratory protection – self-contained breathing apparatus. Otherwise, suitable respirator and outer protective clothing as required.

**Fire and Explosion**
Use alcohol foam, dry chemical or carbon dioxide to extinguish. Water may be ineffective, but may be used to cool fire-exposed containers. Flash back may occur along vapour trail.

**First Aid**
Move victim out of spill site. Call for medical assitance but start first aid at once. Inhalation: if breathing has stopped, give artificial respiration; if laboured, give oxygen. Contact: skin – remove contaminated clothing and flush affected areas with plenty of water; eyes – irrigate with plenty of water. Ingestion: give conscious victim plenty of water to drink. If medical assistance is not immediately available, transport victim to doctor, hospital or clinic.

## ENVIRONMENTAL PROTECTION MEASURES

**Response**

**Water**
1. Stop or reduce discharge if safe to do so.
2. Contact manufacturer or supplier for advice.
3. If possible, contain discharge by booming.
4. If floating, skim and remove.
5. Notify environmental authorities to discuss disposal and cleanup of contaminated materials.

**Land-Air**
1. Stop or reduce discharge if safe to do so.
2. Contact manufacturer or supplier for advice.
3. Contain spill by diking with earth or other barrier.
4. Remove material with pumps or vacuum equipment and place in appropriate containers.
5. Recover undamaged containers.
6. Absorb residual liquid on natural or synthetic sorbents.
7. Notify environmental authorities to discuss disposal and cleanup of contaminated materials.

**Disposal**
1. Contact manufacturer or supplier for advice on disposal.
2. Contact environmental authorities for advice on disposal.
3. Incinerate (approval of environmental authorities required).

METHYL ISOBUTYL KETONE   $(CH_3)_2CHCH_2COCH_3$

# METHYL MERCAPTAN $CH_3SH$

## IDENTIFICATION

UN No. 1064

### Common Synonyms

METHANETHIOL
MERCAPTOMETHANE
(Methyl mercaptan is frequently added to natural gas to provide odour).

### Observable Characteristics

Colourless gas. Strong unpleasant (rotten cabbage) odour.

### Manufacturers

Canadian supplier:
Canadian Liquid Air Co. Ltd.,
Montreal, Que.

### Transportation and Storage Information

**Shipping State:** Liquid (compressed gas).
**Classification:** Flammable gas.
**Inert Atmosphere:** No requirement.
**Venting:** Safety relief.

**Label(s):** Red label - FLAMMABLE GAS; Class 3. White label - POISON; Class 2.3.
**Storage Temperature:** Ambient.

**Grades or Purity:** 98.0% purity; 99.5+% reagent.
**Containers and Materials:** Cylinders, tank cars; steel.

### Physical and Chemical Characteristics

**Physical State** (20°C, 1 atm): Gas.
**Solubility** (Water): 2.4 g/100 mL (15°C).
**Molecular Weight:** 48.1
**Vapour Pressure:** 760 mm Hg (6.8°C); 1 520 mm Hg (26°C); 3 800 mm Hg (60°C).
**Boiling Point:** 6 to 7.6 °C.

**Floatability** (Water): Floats and boils.
**Odour:** Strongly unpleasant (rotten cabbage), (0.001 to 0.002 ppm, odour threshold).
**Flash Point:** -17.8°C.
**Vapour Density:** 1.7
**Specific Gravity:** 0.87 (21°C) (liquid).

**Colour:** Colourless.
**Explosive Limits:** 3.9 to 21.8%.
**Melting Point:** -123°C.

## HAZARD DATA

### Human Health

**Symptoms:** Inhalation: irritating to respiratory tract; may cause dizziness, suffocation, headache and vomiting; CNS depressant, muscular weakness, tremors, unconsciousness and respiratory paralysis. Contact: skin and eyes - irritation and frostbite.
**Toxicology:** Moderately toxic by inhalation.
TLV® (inhalation) 0.5 ppm; 1 mg/m³.
Short-term Inhalation Limits - No information.

$LC_{50}$ - Inhalation: rat = 675 ppm
Delayed Toxicity - No information.

$LD_{50}$ - Subcutaneous: mouse = 0.0024 g/kg

### Fire

**Fire Extinguishing Agents:** Alcohol foam, carbon dioxide, dry chemical. Water may be used to cool fire-exposed containers and knock down vapours.
**Behaviour in Fire:** Emits highly toxic $SO_X$ fumes in high temperatures. Flash back may occur along vapour trail.
**Ignition Temperature:** No information.
**Burning Rate:** 3.8 mm/min.

### Reactivity

**With Water:** Reacts with water to produce toxic $SO_X$ fumes.
**With Common Materials:** Can react vigorously with oxidizing materials. Reacts with acids to produce toxic and flammable $SO_X$ vapours.
**Stability:** Stable.

### Environment

**Water:** Prevent entry into water intakes and waterways; toxic to aquatic life. Fish toxicity: 1.0 ppm/120 h/<u>Daphnia</u> sp./TLm/freshwater; 0.55 to 0.9 mg/L/ Salmonides/TLm/freshwater.
**Land-Air:** No information.
**Food Chain Concentration Potential:** No information.

## EMERGENCY MEASURES

**Special Hazards**

FLAMMABLE; POISON.

**Immediate Responses**

Keep non-involved people away from spill site. Issue warnings: "FLAMMABLE; POISON". CALL FIRE DEPARTMENT. Eliminate all sources of ignition. Contact supplier and request advice or assistance. Avoid contact and inhalation. Stop or reduce discharge, if safe to do so. Notify environmental authorities.

**Protective Clothing and Equipment**

Respiratory protection - self-contained breathing apparatus, and totally encapsulated suit.

**Fire and Explosion**

Use alcohol foam, dry chemical or carbon dioxide to extinguish fire. Use water to cool fire-exposed containers. Emits highly toxic $SO_x$ fumes in high temperatures. Flash back may occur along vapour trail.

**First Aid**

Move victim out of spill area to fresh air. Call for medical assistance, but start first aid at once. Inhalation: give artificial respiration if breathing has stopped (not mouth-to-mouth method); and oxygen if breathing is laboured. Contact: remove contaminated clothing and wash eyes and skin thoroughly with plenty of warm water. Keep victim warm and quiet. If medical assistance is not immediately available, transport victim to hospital, clinic or doctor.

## ENVIRONMENTAL PROTECTION MEASURES

**Response**

**Water**
1. Stop or reduce discharge if safe to do so.
2. Contact manufacturer or supplier for advice.
3. Notify environmental authorities to discuss disposal and cleanup of contaminated materials.

**Land-Air**
1. Stop or reduce discharge if safe to do so.
2. Contact manufacturer or supplier for advice.
3. Dike to prevent runoff from rainwater or water application.
4. Recover undamaged containers.
5. Notify environmental authorities to discuss disposal and cleanup of contaminated materials.

**Disposal**
1. Contact manufacturer or supplier for advice on disposal.
2. Contact environmental authorities for advice on disposal.

METHYL MERCAPTAN   $CH_3SH$

# METHYL METHACRYLATE  $CH_2:C(CH_3)COOCH_3$

## IDENTIFICATION

UN No. 1247 (monomer, inhibited)

### Common Synonyms
METHACRYLIC ACID, METHYL ESTER
METHYL-α-METHACRYLATE
METHYL 2-METHYL-2-PROPENOATE
MME
METHACRYLATE MONOMER

### Observable Characteristics
Colourless liquid. Sharp, sweet odour.

### Manufacturers
No Canadian manufacturers.
Canadian suppliers:
Chemacryl Plastics, Niagara Falls, Ont.
DuPont, Canada, Toronto, Ontario.
Rohm & Haas Canada, Toronto, Ontario.

Originating from:
Cy-Ro Industries, USA
El duPont de Nemours, USA
Rohm & Haas, USA

### Transportation and Storage Information
**Shipping State:** Liquid.
**Classification:** Flammable liquid.
**Inert Atmosphere:** No requirement.
**Venting:** Pressure-vacuum.
**Pump Type:** No information.

**Label(s):** Red and white label - FLAMMABLE LIQUID; Class 3.2, Group II.
**Storage Temperature:** Ambient.
**Hose Type:** No information.

**Grades or Purity:** Technical (inhibited: hydroquinone 22 to 65 ppm; hydroquinone methyl ester 22 to 120 ppm; dimethyl-t-butylphenol 45 to 65 ppm).
**Containers and Materials:** Drums, tank cars, trucks; steel.

### Physical and Chemical Characteristics
**Physical State** (20°C, 1 atm): Liquid.
**Solubility** (Water): 1.5 g/100 mL (25°C).
**Molecular Weight:** 100.1
**Vapour Pressure:** 28 mm Hg (20°C); 40 mm Hg (26°C); 49 mm Hg (30°C).
**Boiling Point:** 100 to 101°C.

**Floatability** (Water): Floats.
**Odour:** Sharp, sweet, (0.05 to 0.21 ppm).
**Flash Point:** 10°C (o.c.).
**Vapour Density:** 3.5
**Specific Gravity:** 0.94 (20°C).

**Colour:** Colourless.
**Explosive Limits:** 1.7 to 8.2%.
**Melting Point:** -48 to -50°C.

## HAZARD DATA

### Human Health
**Symptoms:** Inhalation: headache, drowsiness, nausea, irritability, narcosis. Contact: skin - irritation and dermatitis; eyes - irritation.
**Toxicology:** Moderately toxic by all routes.  LC$_{50}$ - Inhalation: rat = 3 750 ppm   LD$_{50}$ - Oral: guinea pig = 6.3 g/kg
TLV® 100 ppm; 410 mg/m³.
Short-term Inhalation Limits - 125 ppm; 510 mg/m³ Delayed Toxicity - No information.
(15 min).

### Fire
**Fire Extinguishing Agents:** Foam, carbon dioxide, dry chemical. Water may be ineffective but may be used to cool fire-exposed containers and knock down vapours.
**Behaviour in Fire:** Flash back may occur along vapour trail. At high temperatures (as in fire) polymerization may occur and cause container rupture.
**Ignition Temperature:** 421°C.   **Burning Rate:** 2.5 mm/min.

### Reactivity
**With Water:** No reaction.
**With Common Materials:** Can react with oxidizing materials. Reacts violently with benzoyl peroxide.
**Stability:** Stable.

### Environment
**Water:** Prevent entry into water intakes and waterways. Harmful to aquatic life. Fish toxicity: 232 to 368 mg/L/24 to 96 h/bluegill/TLm/freshwater; Aquatic toxicity rating = 100 to 1 000 ppm/96 h/TLm/freshwater; BOD: 14%, 5 days.
**Land-Air:** No information.
**Food Chain Concentration Potential:** No information.

## EMERGENCY MEASURES

### Special Hazards
FLAMMABLE.

### Immediate Responses
Keep non-involved people away from spill site. Issue warning: "FLAMMABLE". CALL FIRE DEPARTMENT. Eliminate all sources of ignition. Contact manufacturer for assistance. Dike to prevent runoff. Notify environmental authorities.

### Protective Clothing and Equipment
In fires, Respiratory protection – self-contained breathing apparatus; and protective outerwear as required.

### Fire and Explosion
Foam, carbon dioxide or dry chemical. Water may be ineffective but may be used to cool fire-exposed containers and knock down vapours. Flash back may occur along vapour trail. At high temperatures (as in fire) polymerization may occur and cause container rupture.

### First Aid
Move victim out of spill site to fresh air. Call for medical assistance, but start first aid at once. Inhalation: if breathing has stopped, give artificial respiration; if laboured, give oxygen. Contact: skin – remove contaminated clothing and flush affected area with plenty of water; eyes – irrigate with plenty of water. Ingestion: give conscious victim plenty of water to drink and induce vomiting. If medical assistance is not immediately available, transport victim to hospital, doctor or clinic.

## ENVIRONMENTAL PROTECTION MEASURES

### Response

**Water**
1. Stop or reduce discharge if safe to do so.
2. Contact manufacturer or supplier for advice.
3. If possible, contain discharge by booming.
4. If floating, skim and remove.
5. Notify environmental authorities to discuss disposal and cleanup of contaminated materials.

**Land-Air**
1. Stop or reduce discharge if safe to do so.
2. Contact manufacturer or supplier for advice.
3. Contain spill by diking with earth or other barrier.
4. Remove material with pumps or vacuum equipment and place in appropriate containers.
5. Recover undamaged containers.
6. Absorb residual liquid on natural or synthetic sorbents.
7. Notify environmental authorities to discuss disposal and cleanup of contaminated materials.

### Disposal
1. Contact manufacturer or supplier for advice on disposal.
2. Contact environmental authorities for advice on disposal.
3. Incinerate (approval of environmental authorities required).

METHYL METHACRYLATE    $CH_2=C(CH_3)COOCH_3$

# MORPHOLINE   $OCH_2CH_2NHCH_2CH_2$

## IDENTIFICATION   UN No. 2054

### Common Synonyms
DIETHYLENIMIDE OXIDE
TETRAHYDRO-2H-1,4-OXAZINE
TETRAHYDRO-p-ISOXAZINE
TETRAHYDRO-1,4-OXAZINE
TETRAHYDRO-1,4-ISOXAZINE

### Observable Characteristics
Colourless, oily liquid with fishy ammonia odour.

### Manufacturers
Canadian supplier:
Canadian Industries Ltd., Toronto, Ont.
McArthur Chemical, Montreal, Ont.
Texaco Chemical Canada, Toronto, Ont.
BASF Canada, Montreal, Quebec.
Kingsley and Keith, Montreal, Quebec.

Originating from:
Texaco, United Kingdom.

BASF, W. Germany.
Chemische Werke
Huls, W. Germany.

**Grades or Purity:** Technical, 98%.
**Containers and Materials:** Drums, tank cars; steel.

### Transportation and Storage Information
**Shipping State:** Liquid.
**Classification:** Flammable liquid.
**Inert Atmosphere:** No requirement.
**Venting:** Open.
**Pump Type:** No information.

**Label(s):** Red and white label – FLAMMABLE LIQUID; Class 3.3, Group II.
**Storage Temperature:** Ambient.
**Hose Type:** No information.

### Physical and Chemical Characteristics
**Physical State (20°C, 1 atm):** Liquid.
**Solubility (Water):** Soluble in all proportions.
**Molecular Weight:** 87.1
**Vapour Pressure:** 4.3 mm Hg (10°C); 8.0 mm Hg (20°C); 13.4 mm Hg (30°C).
**Boiling Point:** 127 to 129°C.

**Floatability (Water):** Mixes.
**Odour:** Fishy (0.01 to 0.14 ppm, odour threshold).
**Flash Point:** 38°C (o.c.).
**Vapour Density:** 3.0
**Specific Gravity:** 1.00 at 20°C (liquid).

**Colour:** Colourless.
**Explosive Limits:** 2.0 to 11.2%.
**Melting Point:** -3 to -5°C.

## HAZARD DATA

### Human Health
**Symptoms: Contact:** liquid causes skin and eye burns, readily absorbed by skin. **Inhalation:** vapours may cause nausea, headache and irritation to respiratory tract. **Ingestion:** abdominal pain, diarrhea, vomiting.
**Toxicology:** Highly toxic by skin absorption and moderately toxic by ingestion.
TLV® (skin) 20 ppm; 70 mg/m³.
Short-term Inhalation Limits – 30 ppm; 105 m g/m³ (15 min).

$LC_{50}$ – Inhalation: mouse = 1 320 mg/m³.
Delayed Toxicity – May produce liver and kidney damage.

$LD_{50}$ – Oral: rat = 1.05 g/kg

### Fire
**Fire Extinguishing Agents:** Water spray, alcohol foam, dry chemical or carbon dioxide. Use water to cool fire-exposed containers and disperse vapours.
**Behaviour in Fire:** Flash back may occur along vapour trail. In fires, toxic $NO_x$ fumes are produced.
**Ignition Temperature:** 310°C.  **Burning Rate:** 1.9 mm/min.

### Reactivity
**With Water:** No reaction, soluble.
**With Common Materials:** Can react with oxidizing materials.
**Stability:** Stable.

### Environment
**Water:** Prevent entry into water intakes and waterways. Aquatic toxicity rating = 100 to 1 000 ppm/96 h/TLm/freshwater; 300 ppm/96 h/bluegill/$TL_{50}$/freshwater; BOD: 2%, 5 days.
**Land-Air:** No information.
**Food Chain Concentration Potential:** None.

## EMERGENCY MEASURES

**Special Hazards**
FLAMMABLE.

**Immediate Responses**
Keep non-involved people away from spill site. Issue warning: "FLAMMABLE". CALL FIRE DEPARTMENT. Eliminate all sources of ignition. Avoid contact and inhalation. Call manufacturer or supplier. Dike to prevent runoff. Call environmental authorities.

**Protective Clothing and Equipment**
Respiratory protection - self-contained breathing apparatus and totally encapsulated suit.

**Fire and Explosion**
Use water spray, alcohol foam, dry chemical or carbon dioxide to extinguish fire. Water may be used to cool fire-exposed containers and knock down vapours. Flash back may occur along vapour trail. In fires, toxic $NO_x$ fumes are produced.

**First Aid**
Move victim out of spill area. Call for medical assistance, but start first aid at once. Inhalation: if breathing has stopped, give artificial respiration; if laboured, give oxygen. Contact: skin - remove contaminated clothing and flush affected area with plenty of water; eyes - irrigate with plenty of water. Ingestion: give plenty of water to conscious victim to drink. If medical assistance is not immediately available, transport victim to hospital, doctor or clinic.

## ENVIRONMENTAL PROTECTION MEASURES

**Response**

**Water**
1. Stop or reduce discharge if safe to do so.
2. Contact manufacturer or supplier for advice.
3. If possible, contain discharge by damming or water diversion.
4. Dredge or vacuum pump to remove contaminants, liquids and contaminated bottom sediments.
5. Notify environmental authorities to discuss disposal and cleanup of contaminated materials.

**Land-Air**
1. Stop or reduce discharge if safe to do so.
2. Contact manufacturer or supplier for advice.
3. Dike to prevent runoff from rainwater or water application.
4. Remove material with pumps or vacuum equipment and place in appropriate containers.
5. Recover undamaged containers.
6. Absorb residual liquid on natural or synthetic sorbents.
7. Notify environmental authorities to discuss disposal and cleanup of contaminated materials.

**Disposal**
1. Contact manufacturer or supplier for advice on disposal.
2. Contact environmental authorities for advice on disposal.

MORPHOLINE   $OCH_2CH_2NHCH_2CH_2$

# NAPHTHA SOLVENT

## IDENTIFICATION

| | | |
|---|---|---|
| | **Observable Characteristics**<br>Colourless liquid. Gasoline-like odour. | UN No. 1255 - petroleum<br>1256 - solvent<br>1300 - V, M&P<br>**Manufacturers**<br>Esso Chemical Canada, Div. of Imperial Oil Ltd., Sarnia, Ont.<br>Gulf Oil Canada Ltd., Montreal, Que.<br>Shell Canada Limited, Montreal, Que.; Sarnia, Ont.; North Burnaby, B.C. |

**Common Synonyms**

PETROLEUM SOLVENT — MINERAL SPIRITS
PETROLEUM ETHER — IOSOL
PETROLEUM SPIRITS — LIGHT LIGROIN
DRY CLEANERS NAPHTHA — CLEANING SOLVENTS
LIGHT NAPHTHA — SAFETY SOLVENT
BENZIN — WHITE SPIRITS

**Naphtha:** A generic term applied to refined, partly refined, or unrefined, petroleum products and liquid products of natural gas which distill between the temperatures of 35 and 205°C.

### Transportation and Storage Information

**Shipping State:** Liquid.
**Classification:** Flammable liquid.
**Inert Atmosphere:** No requirement.
**Venting:** Open (flame arrester) or pressure-vacuum.
**Pump Type:** Gear, centrifugal, etc.

**Label(s):** Red label - FLAMMABLE LIQUID; Class 3.2.
**Storage Temperature:** Ambient.
**Hose Type:** Buna-N, polyethylene, Viton and synthetic rubber.

**Grades or Purity:** By flashpoint, petroleum naphtha f.p. >38°C; naphtha solvent f.p. ~43°C; naphtha vm&p -7 to -13°C.
**Containers and Materials:** Drums, tank cars, tank trucks; steel.

### Physical and Chemical Characteristics

**Physical State** (20°C, 1 atm): Liquid.
**Solubility** (Water): Insoluble.
**Molecular Weight:** Variable, mixtures of hydrocarbons.
**Vapour Pressure:** Variable, 0 to 67 mm Hg (38°C).
**Boiling Point:** 30 to 202°C (variable).

**Floatability** (Water): Floats.
**Odour:** Gasoline-like (~5 ppm, odour threshold).
**Flash Point:** Variable, -20 to 50°C.
**Vapour Density:** Variable 2.5 to 4.8.
**Specific Gravity:** 0.75-0.87 (20°C) (variable).

**Colour:** Colourless.
**Explosive Limits:** 0.8 to 7.0%.
**Melting Point:** <-30°C.

## HAZARD DATA

### Human Health

**Symptoms:** Inhalation and Ingestion: nausea, vomiting, coughing, irritation of respiratory tract, weakness, dizziness, unconsciousness. Contact: eyes - slightly irritating, skin - dermatitis.
**Toxicology:** Moderately toxic by inhalation.
**TLV®** (inhalation) No information.
**Short-term Inhalation Limits** - No information.

LC$_{50}$ - No information.
LC$_{Lo}$ - Inhalation: rat = 1 600 ppm
Delayed Toxicity - No information.

LD$_{50}$ - Oral: rat = 0.5 to 5.0 g/kg
LD$_{Lo}$ - Intraperitoneal: mammal = 2.5 g/kg

### Fire

**Fire Extinguishing Agents:** Use foam, carbon dioxide or dry chemical. Water may be ineffective but may be used to cool fire-exposed containers and to protect personnel.
**Behaviour in Fire:** Flash back may occur along vapour trail.
**Ignition Temperature:** 229 to 293°C.
**Burning Rate:** 4 mm/min.

### Reactivity

**With Water:** No reaction.
**With Common Materials:** Can react with oxidizing materials. Reacts violently with chromates.
**Stability:** Stable.

### Environment

**Water:** Prevent entry into water intakes and waterways. Toxic to fish at approximately 10 ppm. BOD: Not available.
**Land-Air:** No information.

## EMERGENCY MEASURES

**Special Hazards**
FLAMMABLE.

**Immediate Responses**
Keep non-involved people away from spill site. Issue warning: "FLAMMABLE". CALL FIRE DEPARTMENT. Eliminate all sources of ignition. Call supplier or manufacturer. Call environmental authorities.

**Protective Clothing and Equipment**
Respiratory protection - in fires - self-contained breathing apparatus, otherwise; Coveralls. Gloves - polyethylene or suitable plastic. Boots - high (not natural rubber). Goggles - or face shield.

**Fire and Explosion**
Use foam, dry chemical or carbon dioxide to extinguish. Water may be ineffective, but may be used to cool fire-exposed containers and protect personnel. Flash back may occur along vapour trail.

**First Aid**
Move victim out of spill site to fresh air. Call for medical assistance, but start first aid at once. Inhalation: if breathing has stopped, give artificial respiration; if laboured, give oxygen. Contact: skin - remove contaminated clothing, and flush affected areas with plenty of water; eyes - irrigate with plenty of water. Ingestion: do not induce vomiting in conscious victim. Keep victim warm and quiet. If medical assistance is not immediately available, transport victim to doctor, hospital or clinic.

## ENVIRONMENTAL PROTECTION MEASURES

**Response**

**Water**
1. Stop or reduce discharge if safe to do so.
2. Contact manufacturer or supplier for advice.
3. If possible, contain discharge by booming.
4. If floating, skim and remove.
5. Notify environmental authorities to discuss disposal and cleanup of contaminated materials.

**Land-Air**
1. Stop or reduce discharge if safe to do so.
2. Contact manufacturer or supplier for advice.
3. Contain spill by diking with earth or other barrier.
4. Remove material with pumps or vacuum equipment and place in appropriate containers.
5. Recover undamaged containers.
6. Absorb residual liquid on natural or synthetic sorbents.
7. Notify environmental authorities to discuss disposal and cleanup of contaminated materials.

**Disposal**
1. Contact manufacturer or supplier for advice on disposal.
2. Contact environmental authorities for advice on disposal.
3. Incinerate (approval of environmental authorities required).

NAPHTHA SOLVENT

# NAPHTHALENE  $C_{10}H_8$

## IDENTIFICATION

**Common Synonyms**
NAPHTHALIN
TAR CAMPHOR
MOTH FLAKES
MOTH BALLS

**Observable Characteristics**
Colourless powder or flakes. Aromatic "mothball" odour.

**UN No. 1334** refined or crude

**Manufacturers**
Record Chemical Co. Ltd., Montreal, Que.
Kent Laboratories Ltd., Vancouver, B.C.

### Transportation and Storage Information

**Shipping State:** Solid.
**Classification:** Flammable solid.
**Inert Atmosphere:** No requirement.
**Venting:** Open.

**Label(s):** Red and white label - FLAMMABLE SOLID; Class 4.1, Group III.
**Storage Temperature:** Ambient.

**Grades or Purity:** Pure: mp = 80°C; crude: mp = 74 to 80°C.
**Containers and Materials:** Cans, drums; steel.

### Physical and Chemical Characteristics

**Physical State** (20°C, 1 atm): Solid.
**Solubility (Water):** 0.0034 g/100 mL (25°C).
**Molecular Weight:** 128.2
**Vapour Pressure:** 1 mm Hg (53°C).
**Boiling Point:** 218°C (sublimes).

**Floatability (Water):** Sinks. Naphthalene in some forms includes sufficient air pockets to allow floatation.
**Odour:** Aromatic mothball odour (0.03 ppm, odour threshold).
**Flash Point:** 88°C (o.c.); 79°C (c.c.).
**Vapour Density:** 4.4
**Specific Gravity:** 1.15 (20°C).

**Colour:** Colourless.
**Explosive Limits:** 0.9 to 5.9%.
**Melting Point:** 80.2°C, volatilizes at room temperature.

## HAZARD DATA

### Human Health

**Symptoms:** Inhalation: vapours may cause coughing, headache, mental confusion, nausea and vomiting. Ingestion: nausea, vomiting, diarrhea, general fatigue and symptoms similar to inhalation. Contact: skin - irritation and dermatitis; eyes - irritation and watering.

**Toxicology:**
TLV® (inhalation) 10 ppm; 50 mg/m³.
Short-term Inhalation Limits - 15 ppm; 75 mg/m³ (15 min).

$LC_{50}$ - Not available.
**Delayed Toxicity** - No information.

$LD_{50}$ - Oral: rat = 1.78 g/kg

### Fire

**Fire Extinguishing Agents:** Water spray, carbon dioxide, dry chemical or foam. Water spray may cause extensive foaming.
**Behaviour in Fire:** No information.
**Ignition Temperature:** 526°C.

**Burning Rate:** 4.3 mm/min.

### Reactivity

**With Water:** No reaction.
**With Common Materials:** May react with oxidizing materials. Reacts violently with chromic anhydride.
**Stability:** Stable.

### Environment

**Water:** Prevent entry into water intakes and waterways. Fish toxicity: 150 ppm/96 h/sunfish/TLm/freshwater; Aquatic toxicity rating = 1 to 10 ppm/96 h/TLm/freshwater; 150 to 220 mg/L/48 h/mosquito fish/TLm/freshwater; 1.8 ppm/72 h/fingerling salmon/critical/saltwater; BOD: 0%, 5 days; 192%, 25 days.
**Land-Air:** No information.
**Food Chain Concentration Potential:** None.

## EMERGENCY MEASURES

**Special Hazards**
COMBUSTIBLE.

**Immediate Responses**
Keep non-involved people away from spill site. Issue warning: "FLAMMABLE". CALL FIRE DEPARTMENT. Avoid contact and inhalation. Call manufacturer. Stop or reduce discharge if safe to do so. Notify environmental authorities.

**Protective Clothing and Equipment**
Respiratory protection - in fires or enclosed spaces, self-contained breathing apparatus; otherwise: Goggles -(mono), tight fitting. If face shield is used, it must not replace goggles. Gloves - rubber. Boots - rubber (pants worn outside boots). Coveralls or acid suit - rubber, if in fire or major spill.

**Fire and Explosion**
Use water spray, dry chemical, alcohol foam or carbon dioxide to extinguish. Water spray may cause extensive foaming.

**First Aid**
Move victim out of spill area to fresh air. Call for medical assistance, but start first aid at once. Inhalation: if breathing has stopped, give artificial respiration; if laboured, give oxygen. Contact: skin and eyes – remove contaminated clothing and flush affected areas with plenty of water. Ingestion: give water to conscious victim. If medical assistance is not immediately available, transport victim to hospital, doctor or clinic.

## ENVIRONMENTAL PROTECTION MEASURES

**Response**

**Water**
1. Stop or reduce discharge if safe to do so.
2. Contact manufacturer or supplier for advice.
3. If possible, contain discharge by damming or water diversion.
4. Dredge or vacuum pump to remove contaminants, liquids and contaminated bottom sediments.
5. Notify environmental authorities to discuss disposal and cleanup of contaminated materials.

**Land-Air**
1. Stop or reduce discharge if safe to do so.
2. Contact manufacturer or supplier for advice.
3. Dike to prevent runoff from rainwater or water application.
4. Remove material by manual or mechanical means.
5. Recover undamaged containers.
6. Notify environmental authorities to discuss disposal and cleanup of contaminated materials.

**Disposal**
1. Contact manufacturer or supplier for advice on disposal.
2. Contact environmental authorities for advice on disposal.

NAPHTHALENE $C_{10}H_8$

# NATURAL GAS

## IDENTIFICATION

UN No. 1971 compressed
1972 refrigerated liquid

**Common Synonyms**
LNG
GAS
Typical composition 85% methane, 10% ethane, 3% propane, 2% butane, pentanes and others. Unprocessed varieties may contain up to 50% $H_2S$, which is very toxic.

**Observable Characteristics**
Colourless, odourless gas if pure. Commercial varieties have methyl mercaptan added as an odour.

**Manufacturers**
Universally available.

**Transportation and Storage Information**
**Shipping State:** Gas or liquid (compressed gas or refrigerated).
**Classification:** Flammable gas.
**Inert Atmosphere:** No requirement.
**Venting:** Safety-relief.

**Label(s):** Red label - FLAMMABLE GAS.
**Storage Temperature:** Ambient or at -162°C.

**Grades or Purity:** Variable.
**Containers and Materials:** Cylinders, tank trucks, tank cars, tank vessels; steel.

**Physical and Chemical Characteristics**
**Physical State** (20°C, 1 atm): Gas.
**Solubility** (Water): 0.006 g/mL (20°C).
**Molecular Weight:** Variable.
**Vapour Pressure:** 2 900 mm Hg (-140°C); 16 600 mm Hg (-100°C).
**Boiling Point:** -162 to -130°C.

**Floatability** (Water): Floats and boils.
**Odour:** Odourless unless methyl mercaptan has been added.
**Flash Point:** <-50°C.
**Vapour Density:** 0.7 to 1.40
**Specific Gravity:** (liquid) 0.4 to 0.5 (at -162 to -130°C).

**Colour:** Colourless.
**Explosive Limits:** 3.8 to 17%.
**Melting Point:** -182 to -150°C.

## HAZARD DATA

**Human Health**
**Symptoms:** Inhalation: asphyxiant, headache, laboured breathing, unconsciousness. Contact: skin - frostbite; eyes - frostbite.
**Toxicology:** Low toxicity by all routes.
**TLV®** (inhalation) No information.    $LC_{50}$ - No information.    $LD_{50}$ - No information.
**Short-term Inhalation Limits** - No information.    **Delayed Toxicity** - No information.

**Fire**
**Fire Extinguishing Agents:** Stop flow of gas. Most fire extinguishing agents may be used. Water should be used to cool fire-exposed containers.
**Behaviour in Fire:** Containers may rupture violently in fires.
**Ignition Temperature:** 482 to 632°C.    **Burning Rate:** No information.

**Reactivity**
**With Water:** No reaction.
**With Common Materials:** No information.
**Stability:** Stable.

**Environment**
**Water:** Prevent entry into water intakes and waterways. No information on aquatic toxicity.
**Land-Air:** No information.
**Food Chain Concentration Potential:** None.

## EMERGENCY MEASURES

**Special Hazards**
FLAMMABLE.

**Immediate Responses**
Keep non-involved people away from spill site. Issue warning: "FLAMMABLE". CALL FIRE DEPARTMENT. Notify manufacturer or supplier. Notify environmental authorities.

**Protective Clothing and Equipment**
In fires Respiratory protection - self-contained breathing apparatus; otherwise, protective outer clothing as required.

**Fire and Explosion**
Stop flow of gas. Most fire extinguishing agents may be used. Water should be used to cool fire-exposed containers. Containers may rupture violently in fires.

**First Aid**
Move victim out of spill site to fresh air. Call for medical assistance, but start first aid at once. Inhalation: if breathing has stopped, give artificial respiration; if laboured, give oxygen. Contact: skin - remove contaminated clothing, flush affected areas with plenty of water and treat as for frostbite; eyes - irrigate with plenty of water. If medical assistance is not immediately available, transport victim to hospital, doctor or clinic.

## ENVIRONMENTAL PROTECTION MEASURES

**Response**

**Water-Land-Air**
1. Stop or reduce discharge if safe to do so.
2. Contact manufacturer or supplier for advice.
3. Notify environmental authorities to discuss disposal and cleanup of contaminated materials.

**Disposal**
1. Contact manufacturer or supplier for advice on disposal.
2. Contact environmental authorities for advice on disposal.

NATURAL GAS

# NITRIC ACID  HNO₃

## IDENTIFICATION

UN No. 2032 fuming
2031 >40% acid
1760 <40% acid

### Common Synonyms

RED FUMING NITRIC ACID (RFNA)
8 to 17% NO₂
WHITE FUMING NITRIC ACID (WFNA)
0.1 to 0.4% NO₂

### Observable Characteristics

Clear, colourless to slightly yellow liquid (aqueous solution). Strong, pungent odour. Fumes in humid air.

### Manufacturers

Canadian Industries Ltd.,
Carseland, Alta., Courtwright, Ont.,
McMasterville, Que.
Cyanamid of Canada Ltd, Niagara Falls, Ont.
Esso Chemical Canada, Redwater, Alta.
Nitrochem, Maitland, Ontario.

### Transportation and Storage Information

**Shipping State:** Liquid (aqueous solution).
**Classification:** Corrosive, oxidizer, poison.
**Inert Atmosphere:** No requirement.
**Venting:** Open or pressure-vacuum.
**Pump Type:** Gear, centrifugal; stainless steel.

**Label(s):** <70%, Black and white label - CORROSIVE; Class 8, Group I; fuming. Yellow label - OXIDIZER; Class 5.1, Group I. White label - POISON; Class 6.1, Group I.
**Storage Temperature:** Ambient.
**Hose Type:** Flexible stainless steel, Teflon-lined, or comparable type.

**Grades or Purity:** 36°Bé, 52.3% HNO₃; 40°Bé, 67.38% min. HNO₃; 48.5°Bé, 95.1% min. HNO₃. Red fuming (strengths >85.7); White fuming (strengths >97.5).
**Containers and Materials:** Tank cars, tank trucks, drums; stainless steel, aluminum.

### Physical and Chemical Characteristics

**Physical State** (20°C, 1 atm): Liquid (aqueous solution).
**Solubility (Water):** Soluble in all proportions.
**Molecular Weight:** 63.0 (solute).
**Vapour Pressure:** 7 mm Hg (20°C) (70% solution); 42 mm Hg (20°C) (100% solution).

**Floatability (Water):** Sinks and mixes.
**Odour:** Strong, pungent.
**Flash Point:** Not flammable.
**Vapour Density:** 2.2
**Boiling Point:** 119°C (40°Bé); 121°C (42°Bé). Higher grades decompose at 86°C.

**Colour:** Colourless to light yellow.
**Explosive Limits:** Not flammable.
**Melting Point:** 36°Bé, -19.5°C; 40°Bé, -24.5°C; 42°Bé, -33.0°C; 48.5°Bé, -52.0°C.
**Specific Gravity:** 1.33 (36°Bé); 1.38 (40°Bé); 1.41 (42°Bé); 1.50 (48.5°Bé).

## HAZARD DATA

### Human Health

**Symptoms:** Inhalation: irritation of mucous membranes, difficulty breathing, coughing fits, nausea, and muscular weakness. Ingestion: irritation and burning, intense thirst, abdominal cramps, nausea and vomiting, difficulty breathing, convulsions and coma. Contact: skin - yellow discolouration, burning sensation, inflammation, blisters; eyes - burning, watering, ulceration, loss of sight.
**Toxicology:** Highly toxic to skin eyes and mucous membranes.
TLV - (inhalation) 2 ppm; 5 mg/m³.
Short-term Inhalation Limits - 4 ppm; 10 mg/m³ (15 min).
$LC_{50}$ - Inhalation: rat = 65 ppm/4 h red fuming (NO₂); Inhalation: rat = 242 ppm/ 30 min white fuming (NO₂)
$LD_{50}$ - Oral: rat = 0.05 to 0.5 g/kg
$LD_{Lo}$ - Oral: human = 0.43 g/kg
Delayed Toxicity - No information.

### Fire

**Fire Extinguishing Agents:** Not combustible. Use water spray in fires involving nitric acid. Water spray may be used to knock down vapours.
**Behaviour in Fire:** In fires toxic NO$_x$ fumes are evolved.
**Ignition Temperature:** Not combustible.
**Burning Rate:** Not combustible.

### Reactivity

**With Water:** No reaction; very soluble. Mixes exothermically and may release toxic NO$_x$ fumes.
**With Common Materials:** Strong oxidizer; may react with many organics to cause fires or release toxic NO$_x$ fumes. Reacts violently with acetic acid, acetic anhydride, (acetone & acetic acid), acetylene, acrolein, acrylonitrile, allyl alcohol, allyl chloride, ammonia, ammonium hydroxide, aniline, arsine, calcium hypophosphite, carbon, chlorosulfonic acid, cresol, cumene, cyanides, cyclic ketones, cyclohexanone, epichlorohydrin, ethyl alcohol, ferrous oxide, fluorine, furfuryl alcohol, hydrazine, hydrogen iodide, hydrogen peroxide, hydrogen sulfide, lithium, magnesium, manganese, nitrobenzene, oleum, phosphine, phosphorus, phosphorus trichloride, phthalic acid, phthalic anhydride, potassium hypophosphite, propylene oxide, pyridine, sodium, sodium hydroxide, sulfamic acid, (sulfuric acid & glycerides), titanium, vinyl acetate and zinc.
**Stability:** Reactive and oxidizing.

### Environment

**Water:** Prevent entry into water intakes and waterways. Fish toxicity: 72 ppm/96 h/mosquito fish/TLm/freshwater; Aquatic toxicity rating = 10 to 100 ppm/ 96 h/TLm; 750 ppm/48 h/goldfish/$LC_{100}$/freshwater; BOD: None.
Food Chain Concentration Potential: None.

## EMERGENCY MEASURES

**Special Hazards**

CORROSIVE. OXIDIZER. POISON. Reactive.

**Immediate Responses**

Keep non-involved people away from spill site. Issue warnings: "CORROSIVE; OXIDIZER; POISON". Call Fire Department. Avoid contact and inhalation. Contact manufacturer for guidance and assistance. Evacuate from downwind if oxides of nitrogen present. Stop or reduce discharge, if this can be done without risk. Contain spill by diking to prevent runoff. Notify environmental authorities.

**Protective Clothing and Equipment**

Respiratory protection - self-contained breathing apparatus and protective clothing. Gloves - rubber. Boots - high, rubber (pants worn outside boots).

**Fire and Explosion**

Not combustible. Use water spray on fires involving nitric acid. Water spray may be used to knock down vapours. In fires, toxic $NO_x$ fumes are evolved.

**First Aid**

Move victim from spill site to fresh air. Call for medical assistance, but start first aid at once. Inhalation: if breathing has stopped, give artificial respiration (not mouth-to-mouth method); if laboured, give oxygen. Contact: eyes - immediately irrigate with plenty of water; skin - remove contaminated clothing and flush with plenty of water. Ingestion: do not induce vomiting. Keep victim warm and quiet. If medical assistance is not immediately available, transport victim to hospital, doctor or clinic.

## ENVIRONMENTAL PROTECTION MEASURES

**Response**

**Water**
1. Stop or reduce discharge if safe to do so.
2. Contact manufacturer or supplier for advice.
3. Notify environmental authorities to discuss disposal and cleanup of contaminated materials.

**Land-Air**
1. Stop or reduce discharge if safe to do so.
2. Contact manufacturer or supplier for advice.
3. Contain spill by diking with earth or other barrier.
4. Remove material with pumps or vacuum equipment and place in appropriate containers.
5. Recover undamaged containers.
6. Soils contaminated by nitric acid might be neutralized with lime (environmental authorities' approval required).
7. Notify environmental authorities to discuss disposal and cleanup of contaminated materials.

**Disposal**

1. Contact manufacturer or supplier for advice on disposal of material.
2. Contact environmental authorities for advice on disposal.

NITRIC ACID   $HNO_3$

# NITRILOTRIACETIC ACID  $N(CH_2COOH)_3$

## IDENTIFICATION

### Common Synonyms
AMINOTRIACETIC ACID
TRIGLYCINE
NTA
TRIGLYCOLLAMIC ACID
N,N - bis (CARBOXYMETHYLGLYSINE)

### Observable Characteristics
White solid. Odourless.

### Manufacturers
Clough Chemical Co. Ltd.,
St. Jean, Quebec.

### Transportation and Storage Information
**Shipping State:** Solid.
**Classification:** Not regulated.
**Inert Atmosphere:** No requirement.
**Venting:** Open.
**Label(s):** Not regulated.
**Storage Temperature:** Ambient.
**Grades or Purity:** Commercial 99.5 +%.
**Containers and Materials:** Cans, drums; steel.

### Physical and Chemical Characteristics
**Physical State** (20°C, 1 atm): Solid.
**Solubility** (Water): 0.13 g/100mL (22.5°C)
**Molecular Weight:** 191.2
**Vapour Pressure:** No information.
**Boiling Point:** Decomposes 240°C.
**Floatability** (Water): Sinks.
**Odour:** Odourless.
**Flash Point:** Not flammable.
**Vapour Density:** No information.
**Specific Gravity:** >1 at 20°C (solid).
**Colour:** White.
**Explosive Limits:** Not flammable.
**Melting Point:** Decomposes 240°C.

## HAZARD DATA

### Human Health
**Symptoms: Contact:** skin and eyes - irritation - little information available.
**Toxicology:** No information.
**TLV®:** No information.
**Short-term Inhalation Limits** - No information.
**LC$_{50}$** - No information.
**Delayed Toxicity** - Suspected carcinogen.
**LD$_{50}$** - Oral: rat = 1.47 g/kg

### Fire
**Fire Extinguishing Agents:** Most fire extinguishing agents may be used on fires involving nitrilotriacetic acid.
**Behaviour in Fire:** Releases toxic fumes when heated.
**Ignition Temperature:** No information.
**Burning Rate:** No information.

### Reactivity
**With Water:** No reaction.
**With Common Materials:** No information.
**Stability:** Stable.

### Environment
**Water:** Prevent entry into water intakes and waterways. **Fish toxicity:** 340 ppm/24 h/guppy/lethal concentration/freshwater; BOD: 0%, 5 days.
**Land-Air:** No information.
**Food Chain Concentration Potential:** None.

## EMERGENCY MEASURES

### Special Hazards

### Immediate Responses
Keep non-involved people away from spill site. Call Fire Department. Avoid contact and inhalation. Call manufacturer. Call environmental authorities.

### Protective Clothing and Equipment:
In fires or confined spaces <u>Respiratory protection</u> - self-contained breathing apparatus; otherwise, protective clothing as required.

### Fire and Explosion
Most fire extinguishing agents may be used on fires involving nitrilotriacetic acid. Releases toxic fumes when heated.

### First Aid
Move victim out of spill site to fresh air. Call for medical assistance, but start first aid at once. <u>Contact</u>: skin - remove contaminated clothing and flush affected areas with plenty of water; eyes - irrigate with plenty of water. <u>Ingestion</u>: give conscious victim water to drink. If medical assistance is not immediately available, transport victim to doctor, hospital or clinic.

## ENVIRONMENTAL PROTECTION MEASURES

### Response

**Water**
1. Stop or reduce discharge if safe to do so.
2. Contact manufacturer or supplier for advice.
3. If possible contain discharge by damming or water diversion.
4. Dredge or vacuum pump to remove contaminants, liquids and contaminated bottom sediments.
5. Notify environmental authorities to discuss disposal and cleanup of contaminated materials.

**Land-Air**
1. Stop or reduce discharge if safe to do so.
2. Contact manufacturer or supplier for advice.
3. If possible, contain discharge by damming or water diversion.
4. Remove materials by manual or mechanical means.
5. Recover undamaged containers.
6. Notify environmental authorities to discuss disposal and cleanup of contaminated materials.

### Disposal
1. Contact manufacturer or supplier for advice on disposal.
2. Contact environmental authorities for advice on disposal.

NITRILOTRIACETIC ACID   $N(CH_2COOH)_3$

# NITROGLYCERINE   $CH_2NO_3CHNO_3CH_2NO_3$

## IDENTIFICATION

| | | |
|---|---|---|
| **Common Synonyms** | **Observable Characteristics** | UN No. 0143 (desensitized) |
| NG | Colourless to yellow liquid. | 1204 (<1% nitroglycerin in alcohol) |
| 1,2,3-PROPANETRIOL TRINITRATE | | |
| GLYCERYL TRINITRATE | | **Manufacturers** |
| BLASTING OIL | | Canadian Industries Ltd., |
| GLYCEROL, NITRIC ACID TRIESTER | | (C-I-L) Montreal, Quebec. |
| NITROGLYCEROL, BLASTING GELATIN | | DuPont of Canada Ltd., Montreal, Quebec. |
| TRINITROGLYCEROL, NITROL | | |

### Transportation and Storage Information

**Shipping State:** Liquid; Solid (liquid absorbed on inert materials).
**Classification:** Explosive, Poison.
**Inert Atmosphere:** No requirement.
**Venting:** Open.

**Label(s):** Orange and black label – EXPLOSIVE; Class 1.1, Group D. White label – POISON; Class 6.1 (for desensitized). Red and white label – FLAMMABLE LIQUID; Class 3.2, Group II (for <1% nitroglycerin in alcohol).

**Grades or Purity:** Inerted < 5% nitroglycerin; <1% solution nitroglycerin in alcohol.
**Containers and Materials:** Special containers.

### Physical and Chemical Characteristics

**Physical State** (20°C, 1 atm): Liquid.
**Solubility** (Water): 0.18 g/100 mL (20°C).
**Molecular Weight:** 227.1
**Vapour Pressure:** 0.00025 mm Hg (20°C); 0.0073 mm Hg (50°C); 0.098 mm Hg (80°C).
**Boiling Point:** 260°C (explodes).

**Floatability** (Water): Sinks.
**Odour:** No information.
**Flash Point:** Explosive.
**Vapour Density:** 7.8
**Specific Gravity:** 1.60 (20°C).

**Colour:** Colourless to yellow.
**Explosive Limits:** Explosive.
**Melting Point:** 3 to 13°C.

## HAZARD DATA

### Human Health

**Symptoms:** Inhalation: headache, flushing of skin, vomiting, dizziness, collapse, cyanosis, convulsions, coma and respiratory paralysis. Ingestion: symptoms similar to inhalation. Contact: skin – readily absorbed yielding symptoms similar to inhalation; eyes – burning and irritation.
**Toxicology:** Highly toxic by all routes.
TLV® (skin) 0.05 ppm; 0.5 mg/m$^3$.    LC$_{50}$ – No information.    LD$_{50}$ – No information.
Short-term Inhalation Limits – 0.1 ppm; 1 mg/m$^3$    Delayed Toxicity – No information.    LD$_{Lo}$ – Subcutaneous: rabbit = 0.4 g/kg
(15 min).

### Fire

**Fire Extinguishing Agents:** Material typically would explode in any situation involving fire.
**Behaviour in Fire:** Explodes.
**Ignition Temperature:** 260°C explodes.    **Burning Rate:** Detonation velocity 7 600 m/s

### Reactivity

**With Water:** No reaction.
**With Common Materials:** No information
**Stability:** Unstable, when shocked or exposed to heat or flame, may explode.

### Environment

**Water:** Prevent entry into water intakes and waterways. Fish toxicity: 26 mg/L/tns/Daphnia magna/LD$_{Lo}$/freshwater; BOD: No information.
**Land-Air:** No information.
**Food Chain Concentration Potential:** No information.

## EMERGENCY MEASURES

**Special Hazards**
EXPLOSIVE. POISON. FLAMMABLE. Can be detonated by shock, heat or fire.

**Immediate Responses**
Keep non-involved people away from spill site. Issue warning: "EXPLOSIVE; POISON; FLAMMABLE". Evacuate area. Notify manufacturer and supplier. Notify environmental authorities.

**Protective Clothing and Equipment:**
Respiratory protection - self-contained breathing apparatus and totally encapsulated suit.

**Fire and Explosion**
Material typically would explode in any situation involving fire.

**First Aid**
Move victim out of spill site to fresh air. Call for medical assistance, but start first aid at once. <u>Inhalation:</u> if breathing has stopped, give artificial respiration (not mouth-to-mouth method); if laboured, give oxygen. <u>Contact:</u> skin - remove contaminated clothing and flush affected areas with plenty of water; eyes - irrigate with plenty of water. <u>Ingestion:</u> give conscious victim water to drink and induce vomiting. If medical assistance is not immediately available, transport victim to doctor, hospital or clinic.

## ENVIRONMENTAL PROTECTION MEASURES

**Response**

**Water-Land-Air**
1. Stop or reduce discharge if safe to do so.
2. Contact manufacturer or supplier for advice.
3. Notify environmental authorities to discuss disposal and cleanup of contaminated materials.

**Disposal**
1. Contact manufacturer or supplier for advice on disposal.
2. Contact environmental authorities for advice on disposal.

NITROGLYCERINE   $CH_2NO_3CHNO_3CH_2NO_3$

# NONYL PHENOL

## IDENTIFICATION

| Common Synonyms | Observable Characteristics | Manufacturers |
|---|---|---|
| NONYLPHENOL<br>2,6-DIMETHYL-4-HEPTYLPHENOL<br>o-NONYL PHENOL<br>p-NONYL PHENOL | Pale yellow liquid.<br>Medicinal odour. | Domtar (CDC Division),<br>Longford Mills, Ontario.<br>Hart Chemical,<br>Guelph, Ontario. |

**Transportation and Storage Information**

**Shipping State:** Liquid.
**Classification:** None.
**Inert Atmosphere:** No requirement.
**Venting:** Open.
**Pump Type:** No information.

**Label(s):** Not regulated.
**Storage Temperature:** Ambient.
**Hose Type:** No information.

**Grades or Purity:** 90% para-isomer plus 4% ortho-isomer and 5% 2,4-dinonylphenol.
**Containers and Materials:** Drums, tank cars; steel, stainless steel.

**Physical and Chemical Characteristics**

**Physical State (20°C, 1 atm):** Liquid.
**Solubility (Water):** Insoluble.
**Molecular Weight:** 220.3
**Vapour Pressure:** No information.
**Boiling Point:** 290 to 304°C.

**Floatability (Water):** Floats.
**Odour:** Medicinal.
**Flash Point:** 141°C (c.c.).
**Vapour Density:** 7.6
**Specific Gravity:** 0.95 (20°C).

**Colour:** Pale yellow.
**Explosive Limits:** No information.
**Melting Point:** -10 to 2°C.

## HAZARD DATA

**Human Health**

**Symptoms:** Inhalation: sore throat, coughing, laboured breathing, unconsciousness. Contact: skin - severely irritating, redness, pain, burns; eyes - redness, pain; Ingestion: sore throat, abdominal pain, nausea, diarrhea.
**Toxicology:** Moderately toxic by ingestion and contact. $LD_{50}$ - Oral: rat = 1.62 g/kg
TLV - No information.   $LC_{50}$ - No information.
Short-term Inhalation Limits - No information.   Delayed Toxicity - No information.

**Fire**

**Fire Extinguishing Agents:** Use alcohol foam, carbon dioxide or dry chemical. Water or foam may cause frothing.
**Behaviour in Fire:** No information.   **Burning Rate:** No information.
**Ignition Temperature:** No information.

**Reactivity**

**With Water:** No reaction.
**With Common Materials:** Can react with oxidizing materials.
**Stability:** Stable.

**Environment**

**Water:** Prevent entry into water intakes and waterways. Aquatic toxicity rating = 10 to 100 ppm/96h/TLm/freshwater.
**Land-Air:** No information.
**Food Chain Concentration Potential:** None.

## EMERGENCY MEASURES

**Special Hazards**

COMBUSTIBLE. CORROSIVE.

**Immediate Responses**

Keep non-involved people away from spill site. Call Fire Department. Avoid contact and inhalation. Notify manufacturer. Dike to contain material. Notify environmental authorities.

**Protective Clothing and Equipment**

In fires or confined spaces - Respiratory protection - self-contained breathing apparatus and totally encapsulated suit. Otherwise, protective clothing as required.

**Fire and Explosion**

Use alcohol foam, carbon dioxide or dry chemical to extinguish. Water or foam may cause frothing.

**First Aid**

Move victim out of spill site to fresh air. Call for medical assistance, but start first aid at once. Inhalation: if breathing has stopped, give artificial respiration; if laboured, give oxygen. Contact: skin - remove contaminated clothing and flush affected areas with plenty of water; eyes - irrigate with plenty of water. Ingestion: give water to conscious victim to drink. If medical assistance is not immediately available, transport victim to hospital, doctor or clinic.

## ENVIRONMENTAL PROTECTION MEASURES

**Response**

**Water**
1. Stop or reduce discharge if safe to do so.
2. Contact manufacturer or supplier for advice.
3. If possible, contain discharge by booming.
4. If floating, skim and remove.
5. Notify environmental authorities to discuss disposal and cleanup of contaminated materials.

**Land-Air**
1. Stop or reduce discharge if safe to do so.
2. Contact manufacturer or supplier for advice.
3. Contain spill by diking with earth or other barrier.
4. Remove material with pumps or vacuum equipment and place in appropriate containers.
5. Recover undamaged containers.
6. Absorb residual liquid on natural or synthetic sorbents.
7. Notify environmental authorities to discuss disposal and cleanup of contaminated materials.

**Disposal**
1. Contact manufacturer or supplier for advice on disposal.
2. Contact environmental authorities for advice on disposal.

NONYL PHENOL

# OILS, CRUDE

## IDENTIFICATION

| Common Synonyms | Observable Characteristics | UN No. 1267 Manufacturers |
|---|---|---|
| PETROLEUM CRUDE | Dark brown to black oily liquids. Acrid, tarry, offensive odour. | Universally available. |

### Transportation and Storage Information

**Shipping State:** Liquid.
**Classification:** Not regulated.
**Inert Atmosphere:** No requirement.
**Venting:** Open (flame arrester).
**Pump Type:** Centrifugal, gear, etc.

**Label(s):** None.
**Storage Temperature:** Ambient.
**Hose Type:** Neoprene, butyl, Viton.

**Grades or Purity:** Wide variety, depending on oil field where produced.
**Containers and Materials:** Tankers, tank cars, tank trucks; steel.

### Physical and Chemical Characteristics

**Physical State** (20°C, 1 atm): Liquid.
**Solubility** (Water): 0.001 mg/100 mL.
**Molecular Weight:** Various (mixture of hydrocarbons).
**Vapour Pressure:** >3 mm Hg (20°C) (typically).
**Boiling Point:** 150 to 300°C.

**Floatability** (Water): Floats.
**Odour:** Acrid, tarry, offensive (if high in $H_2S$).
**Flash Point:** (-10 to 20°C).
**Vapour Density:** Variable (typically 3 to 5).
**Specific Gravity:** 0.8-0.9 (15°C).

**Colour:** Dark brown to black.
**Explosive Limits:** Variable 1 to 7%.
**Melting Point:** Various (-60 to -20°C).

## HAZARD DATA

### Human Health

**Symptoms: Inhalation:** dizziness, headache. **Ingestion:** nausea and vomiting. **Contact: skin** – irritation; **eyes** – irritation.
**Toxicology:** Generally low toxicity if little $H_2S$ is present.
TLV® (as a mineral oil mist) 5 mg/m$^3$     $LD_{50}$ - No information.     $LD_{50}$ - No information.
Short-term Inhalation Limits – 10 mg/m$^3$ (15 min)     Delayed Toxicity - No information.
(as mineral oil mist).

### Fire

**Fire Extinguishing Agents:** Use dry chemical, foam or carbon dioxide. Water may be ineffective, but should be used to cool fire-exposed containers.
**Behaviour in Fire:** Can give off toxic fumes.     **Burning Rate:** 4 mm/min.
**Ignition Temperature:** Variable; typically >400°C.

### Reactivity

**With Water:** No reaction; insoluble.
**With Common Materials:** May react with oxidizing materials.
**Stability:** Stable.

### Environment

**Water:** Prevent entry into water intakes and waterways. Fish toxicities: 1 to 30 mg/L/96 h/rainbow trout/$LC_{50}$/freshwater; 330 mg/L/48 h/brown shrimp/$LC_{50}$/saltwater.
**Land-Air:** Fouling to landscape.
**Food Chain Concentration Potential:** None.

## EMERGENCY MEASURES

**Special Hazards**
COMBUSTIBLE.

**Immediate Responses**
Keep non-involved people away from spill site. Call Fire Department. Stop or reduce discharge, if this can be done without risk. Contain spill, if possible, to prevent or reduce runoff and pollution. Notify supplier and appropriate environmental authorities.

**Protective Clothing and Equipment**
Goggles or face shield. Coveralls. Gloves - rubber. Boots - rubber.

**Fire and Explosion**
Use dry chemical, foam or carbon dioxide to extinguish. Water may be ineffective, but should be used to cool containers exposed to fire or heat.

**First Aid**
Move victim out of spill area to fresh air. Call for medical assistance, but start first aid at once. Contact: skin - remove oil-soaked clothing. Wipe off as much oil as possible. Wash skin thoroughly with warm water (and soap, if available) for at least 15 minutes; eyes - irrigate with water; Ingestion: give water to conscious victim to drink. Transport victim to hospital, doctor or clinic if necessary.

## ENVIRONMENTAL PROTECTION MEASURES

**Response**

**Water**
1. Stop or reduce discharge if safe to do so.
2. Contact manufacturer or supplier for advice.
3. Collect by skimming, pumping off or sorbing on sorbent.
4. Notify environmental authorities to discuss disposal and cleanup of contaminated materials.

**Land-Air**
1. Stop or reduce discharge if safe to do so.
2. Contact manufacturer or supplier for advice.
3. Contain spill by diking with earth or other barrier.
4. Prevent product from entering sewers.
5. Collect, pump off or use sorbents for residuals.
6. Collect recovered product in available containers.
7. Notify environmental authorities to discuss cleanup and disposal of contaminated materials.

**Disposal**
1. Contact manufacturer or supplier for advice on disposal.
2. Contact environmental authorities for advice on disposal.
3. Incinerate (approval of environmental authorities required).
4. Oil and slightly contaminated oil may be recycled in a refinery.

OILS, CRUDE

# OILS, FUEL (aviation)

## IDENTIFICATION

**Common Synonyms**
AVIATION GASOLINE
80, 100, 115, 130 (similar to gasoline, see gasoline).
JP-4, JP-5, JP-6. Type A, Type A-1, Type B, Type C (turbofuel A-1, A, B, C), (Jet A-1, A, B, C)

**Observable Characteristics**
Yellow - brown liquids.
Gasoline-like odours.

**UN No. 1863**

**Manufacturers**
Universally available.

**Transportation and Storage Information**
**Shipping State:** Liquid.
**Classification:** Flammable liquid.
**Inert Atmosphere:** No requirement.
**Venting:** Open (flame arrester).
**Pump Type:** Gear, centrifugal; steel, stainless steel.

**Label(s):** Red and black label – FLAMMABLE LIQUID.
**Storage Temperature:** Ambient.
**Hose Type:** Neoprene, Viton, polyethylene.

**Grades or Purity:** Type A-1, type A, type B, type C; JP-4, JP-5, JP-6.
**Containers and Materials:** Drums, tank cars, tank trucks; steel.

**Physical and Chemical Characteristics**
**Physical State (20°C, 1 atm):** Liquid.
**Solubility (Water):** 0.003 mg/L (20°C).
**Molecular Weight:** Variable (mixture of hydrocarbons).
**Vapour Pressure:** typically <93 mm Hg (20°C).
**Boiling Point:** (A, A-1) 163 to 259°C. (B) 72 to 235°C; (4, 5, 6) 121 to 260°C.

**Floatability (Water):** Float.
**Odour:** Gasoline-like.
**Flash Point:** (A-1, A) 43 to 66°C; (B, 4) -23 to -1°; (5) 35 to 63°C; (6) 38°C (o.c.).
**Vapour Density:** typically 3 to 5
**Specific Gravity:** (A-1) 0.806; (A) 0.816; (B) 0.764

**Colour:** Yellow-brown
**Explosive Limits:** Variable, JP-4, (1.3 to 8.0%; JP-6 (0.6 to 3.7%).
**Melting Point:** -60 to -40°C.

## HAZARD DATA

**Human Health**
**Symptoms:** Inhalation: dizziness, headache. Ingestion: nausea and vomiting. Contact: skin – irritation; eyes – irritation.
**Toxicology:** Low toxicity.
TLV® 300 ppm; 900 mg/m$^3$ (gasoline).
Short-term Inhalation Limits – 500 ppm; 1 500 mg/m$^3$ (gasoline).
$LC_{50}$ – No information.
Delayed Toxicity – No information.
$LD_{50}$ – No information.

**Fire**
**Fire Extinguishing Agents:** Foam, carbon dioxide or dry chemical. Water may be ineffective, but should be used to cool fire-exposed containers.
**Behaviour in Fire:** Flash back may occur along vapour trail.
**Ignition Temperature:** JP-4; 240°C; JP-5; 246°C; **Burning Rate:** 4 mm/min.
JP-6; 230°C.

**Reactivity**
**With Water:** No reaction.
**With Common Materials:** May react with oxidizing agents.
**Stability:** Stable.

**Environment**
**Water:** Prevent entry into water intakes and waterways. Fish toxicity: 2 ppm/96 h/bluegill/$LC_{50}$/freshwater; 100 ppm/96 h/grass shrimp/$LC_{50}$/saltwater.
**Land-Air:** No information.
**Food Chain Concentration Potential:** No information.

## EMERGENCY MEASURES

**Special Hazards**
FLAMMABLE.

**Immediate Responses**

Keep non-involved people away from spill site. Issue warning: "FLAMMABLE". Call fire department. Eliminate all sources of ignition. Notify supplier. Stop or reduce discharge if safe to do so. Dike to prevent runoff. Notify environmental authorities.

**Protective Clothing and Equipment:**

Gloves and Boots: neoprene, butyl rubber. Appropriate goggles or face shield.

**Fire and Explosion**

Use foam, carbon dioxide or dry chemical to extinguish. Water may be ineffective, but should be used to cool fire-exposed containers. Flashback may occur along vapour trail.

**First Aid**

Move victim out of spill site to fresh air. Call for medical assistance, but start first aid at once. Inhalation: if breathing has stopped, give artificial respiration; if laboured, give oxygen. Contact: skin - remove contaminated clothing and flush affected areas with soap and water; eyes - irrigate with water. Ingestion: give conscious victim water to drink; do not induce vomiting. If medical assistance is not immediately available, transport victim to doctor, hospital or clinic.

## ENVIRONMENTAL PROTECTION MEASURES

**Response**

**Water**
1. Stop or reduce discharge if safe to do so.
2. Contact manufacturer or supplier for advice.
3. If possible, contain spill by booming.
4. If floating, skim and remove.
5. Notify environmental authorities to discuss disposal and cleanup of contaminated materials.

**Land-Air**
1. Stop or reduce discharge if safe to do so.
2. Contact manufacturer or supplier for advice.
3. Contain spill by diking with earth or other barrier.
4. Remove material with pumps or vacuum equipment and place in appropriate containers.
5. Recover undamaged containers.
6. Adsorb residual liquid on natural or synthetic sorbents.
7. Notify environmental authorities to discuss cleanup and disposal of contaminated materials.

**Disposal**
1. Contact supplier for advice on disposal.
2. Contact environmental authorities for advice on disposal.
3. Incinerate (approval of environmental authorities required).
4. Oil and slightly contaminated oil may be recycled in a refinery.

OILS, FUEL (aviation)

# OILS, FUEL (distillates 1, 2 and 2D)

## IDENTIFICATION

**Common Synonyms**

KEROSENE (FUEL OIL NO. 1)
FUEL OIL NO. 1, 2, 2-D
DIESEL OIL LIGHT
HOME HEATING OIL    Fuel Oil No. 2 and 2-D
DIESEL OIL MEDIUM

**Observable Characteristics**

Oily liquids. Light brown to brown colour.
Characteristic diesel fuel-like odour.

UN No. 1233 kerosene (fuel oil no. 1)

**Manufacturers**

Universally available.

**Transportation and Storage Information**

**Shipping State:** Liquid.
**Classification:** Flammable liquid (kerosene).
**Inert Atmosphere:** No requirement.
**Venting:** Open (flame arrester).
**Pump Type:** Gear or centrifugal. Steel or stainless steel.

**Label(s):** Red label - FLAMMABLE LIQUID (kerosene).
**Storage Temperature:** Ambient.
**Hose Type:** Neoprene, Viton, polyethylene.

**Grades or Purity:** Kerosene, diesel fuel 1-D, diesel fuel 2; diesel fuel 2-D.
**Containers and Materials:** Drums, tank cars, tank trucks, tankers; steel.

**Physical and Chemical Characteristics**

**Physical state** (20°C, 1 atm)**:** Liquid.
**Solubility** (Water)**:** Insoluble (about 30 ppm).
**Molecular Weight:** Variable.
**Vapour Pressure:** <1 mm Hg (20°C).
**Boiling Point:** 150 to 350°C.

**Floatability** (Water)**:** Float.
**Odour:** Characteristic diesel-like (about 0.1 ppm, odour threshold).
**Flash Point:** (Fuel 1) 43 to 72°C; (Fuel 2) 52 to 96°C.
**Vapour Density:** about 3 to 5
**Specific Gravity:** 0.81 to 0.90 (20°C).

**Colour:** Light brown to brown.
**Explosive Limits:** 0.7 to 5% (fuel 1).
**Melting Point:** -18 to -46°C.

## HAZARD DATA

**Human Health**

**Symptoms:** Inhalation: dizziness, headache. Ingestion: nausea and vomiting. Contact: skin - irritation; eyes - irritation. pneumonitis. Dermatitis may result from prolonged and repeated skin exposure.
**Toxicology:** Low toxicity by all routes.
TLV® (inhalation) 5 mg/m$^3$ (mineral oil particulate mist).
Short-term Inhalation Limits - 10 mg/m$^3$ for 15 min (oil particulate mineral mist).

LC$_{50}$ - No information.
Delayed Toxicity - No information.

LD$_{50}$ - Oral: rat = 28 g/kg
LD$_{50}$ - Oral: rabbit = 0.2 g/kg

**Fire**

**Fire Extinguishing Agents:** Foam, carbon dioxide or dry chemical. Water may be ineffective but should be used to cool fire-exposed containers.
**Behaviour in Fire:** Flash back may occur along vapour trail.
**Ignition Temperature:** (Fuel 1) 210°C;    **Burning Rate:** 4 mm/min.
(Fuel 2) 257°C.

**Reactivity**

**With Water:** No reaction.
**With Common Materials:** Reacts with oxidizing agents.
**Stability:** Stable.

**Environment**

**Water:** Prevent entry into water intakes or waterways. Fish toxicity:  10 ppm/96 h/rainbow trout/lethal/LC$_{50}$/freshwater; 95 to 135 ppm/96 h/bluegill/LC$_{50}$/freshwater (fuel oil no. 2); 2 ppm/96 h/grass shrimp/LC$_{50}$/saltwater; BOD: Not available.
**Land-Air:** No information.
**Food Chain Concentration Potential:** None.

## EMERGENCY MEASURES

**Special Hazards**
FLAMMABLE.

**Immediate Responses**
Keep non-involved people away from spill site. Issue warning: "FLAMMABLE". CALL FIRE DEPARTMENT. Eliminate all sources of ignition. Notify supplier. Stop or reduce discharge if this can be done without risk. Dike to contain spill and prevent runoff. Notify environmental authorities.

**Protective Clothing and Equipment**
Gloves and Boots - neoprene, butyl rubber. Appropriate goggles or face shield.

**Fire and Explosion**
Use foam, carbon dioxide or dry chemical to extinguish. Water may be ineffective, but should be used to cool fire-exposed containers. Flash back may occur along vapour trail.

**First Aid**
Move victim out of spill site to fresh air. Call for medical assistance, but start first aid at once. <u>Inhalation</u>: give artificial respiration if necessary. <u>Contact</u>: skin - remove contaminated clothing follow by washing affected areas with soap and water; eyes - irrigate eyes with plenty of water. <u>Ingestion</u>: do not induce vomiting. If medical assistance is not immediately available, transport victim to hospital, doctor or clinic.

## ENVIRONMENTAL PROTECTION MEASURES

**Response**

**Water**
1. Stop or reduce discharge if safe to do so.
2. Contact manufacturer or supplier for advice.
3. If possible, contain discharge by booming.
4. If floating, skim material and remove.
5. Notify environmental authorities to discuss disposal and cleanup of contaminated materials.

**Land-Air**
1. Stop or reduce discharge if safe to do so.
2. Contact manufacturer or supplier for advice.
3. Contain spill by diking with earth or other barrier.
4. Remove material with pumps or vacuum equipment and place in appropriate containers.
5. Recover undamaged containers.
6. Absorb residual liquid on natural or synthetic sorbents.
7. Notify environmental authorities to discuss disposal and cleanup of contaminated materials.

**Disposal**
1. Contact supplier for advice on disposal.
2. Contact environmental authorities for advice on disposal.
3. Incinerate (approval of environmental authorities required).
4. Oil and slightly contaminated oil may be recycled in a refinery.

OILS, FUEL (distillates 1, 2 and 2D)

# OILS, FUEL (Residual Oils or Fuels 4, 5 and 6 Bunker)

## IDENTIFICATION

| Common Synonyms | Observable Characteristics | Manufacturers |
|---|---|---|
| RESIDUAL FUEL OIL NO. 4 (Bunker A)<br>RESIDUAL FUEL OIL NO. 5 (Bunker B)<br>RESIDUAL FUEL OIL NO. 6 (Bunker C)<br>BUNKER FUEL OIL | Heavy, oily, high-viscosity liquids. Brown to black. Characteristic diesel fuel-like odour. | Universally available. |

**Transportation and Storage Information**

**Shipping State:** Liquid.
**Classification:** None.
**Inert Atmosphere:** No requirement.
**Venting:** Open (flame arrester).
**Pump Type:** Gear or centrifugal.

**Label(s):** Not regulated.
**Storage Temperature:** Ambient or may be heated.
**Hose Type:** Neoprene, Viton, polyethylene.

**Grades or Purity:** Fuel oil (residual) No. 4, fuel oil (residual) No. 5, fuel oil (residual) No. 6
**Containers and Materials:** Drums, tank trucks tank cars, tankers; steel.

**Physical and Chemical Characteristics**

**Physical State** (20°C, 1 atm): Liquid.
**Solubility (Water):** Insoluble, <10 ppm.
**Molecular Weight:** Variable.
**Vapour Pressure:** No information; <1 mm Hg, 20°C approximately).
**Boiling Point:** 185 to 500°C.

**Floatability (Water):** Floats.
**Odour:** Characteristic diesel fuel-like (about 0.1 ppm, odour threshold).
**Flash Point:** (No. 4) 61 to 116°C; (No. 5 light) 69 to 169°C; (No. 5 heavy) 70 to 121°C; (No. 6) 66 to 132°C.
**Vapour Density:** Variable; 3 to 5
**Specific Gravity:** 0.90 to 1.05 (15.5°C).

**Colour:** Brown to black.
**Explosive Limits:** 1 to 5% (approximate).
**Melting Point:** -29 to 10°C.

## HAZARD DATA

**Human Health**

**Symptoms:** Inhalation: dizziness, headache. Ingestion: nausea and vomiting. Contact: skin - irritation; eyes - irritation.
**Toxicology:** Low toxicity by all routes.
TLV® (inhalation) 5 mg/m³ (mineral oil mist). LC$_{50}$ - No information. LD$_{50}$ - No information.
Short-term Inhalation Limits - 10 mg/m³ for 15 min (mineral oil particulate mist). Delayed Toxicity - No information.

**Fire**

**Fire Extinguishing Agents:** Dry chemical, foam or carbon dioxide. Water may be ineffective but may be used to cool fire-exposed containers.
**Behaviour in Fire:** In fires, toxic fumes may be emitted.
**Ignition Temperature:** No.4 263°C; No.6 407°C.  **Burning Rate:** 4 mm/min (approx.).

**Reactivity**

**With Water:** No reaction.
**With Common Materials:** May react with strong oxidizing agents.
**Stability:** Stable.

**Environment**

**Water:** Prevent entry into water intakes and waterways. Fish toxicity: 40 to 100 ppm/96 h/rainbow trout/LC$_{50}$/freshwater. Fuel oil no. 6, 26 ppm/96 h/grass shrimp/CL$_{50}$/saltwater; 21 ppm/96 h/mosquito fish/LC$_{50}$/freshwater; BOD: Not available.
**Land-Air:** No information.
**Food Chain Concentration Potential:** None.

## EMERGENCY MEASURES

**Special Hazards**

FLAMMABLE. COMBUSTIBLE.

**Immediate Responses**

Keep non-involved people away from spill site. Issue warning: "FLAMMABLE". CALL FIRE DEPARTMENT. Eliminate all sources of ignition. Notify supplier. Stop or reduce discharge if this can be done without risk. Dike spill area to contain runoff. Notify environmental authorities.

**Protective Clothing and Equipment**

Gloves and Boots - neoprene, butyl rubber. Appropriate goggles or face shield.

**Fire and Explosion**

Use foam, carbon dioxide or dry chemical to extinguish. Water may be ineffective, but should be used to cool fire-exposed containers. In fires, toxic fumes may be emitted.

**First Aid**

Move victim out of spill site to fresh air. Call for medical assistance, but start first aid at once. <u>Inhalation</u>: remove from exposure and give artificial respiration if necessary. <u>Contact</u>: skin - remove contaminated clothing and wash affected areas with soap and water; eyes - irrigate with plenty of water. <u>Ingestion</u>: give water to conscious victim to drink. If medical assistance is not immediately available, transport victim to hospital, doctor or clinic.

## ENVIRONMENTAL PROTECTION MEASURES

**Response**

**Water**
1. Stop or reduce discharge if safe to do so.
2. Contact manufacturer or supplier for advice.
3. If possible, contain discharge by booming.
4. If floating, skim and remove.
5. Notify environmental authorities to discuss disposal and cleanup of contaminated materials.

**Land-Air**
1. Stop or reduce discharge if safe to do so.
2. Contact manufacturer or supplier for advice.
3. Contain spill by diking with earth or other barrier.
4. Remove material with pumps or vacuum equipment and place in appropriate containers.
5. Recover undamaged containers.
6. Absorb residual liquid on natural or synthetic sorbents.
7. Notify environmental authorities to discuss disposal and cleanup of contaminated materials.

**Disposal**
1. Contact supplier for advice on disposal.
2. Contact environmental authorities for advice on disposal.
3. Incinerate (environmental authorities' approval required).
4. Oil and slightly contaminated oil may be recycled in a refinery.

OILS, FUEL (Residual Oils or Fuels 4, 5 and 6 Bunker)

# OLEUM  $H_2SO_4 + SO_3$

## IDENTIFICATION

UN No. 1831

### Common Synonyms

FUMING SULPHURIC ACID
SULPHURIC ACID, FUMING
DISULPHURIC ACID
(see sulfuric acid for dilute solutions).

### Observable Characteristics

Colourless to brown; heavy, oily liquid. Sharp, choking odour. Fumes on contact with moist air.

### Manufacturers

Canadian Industries Ltd., Montreal, Quebec.
Du Pont Canada Ltd., Montreal, Quebec.
Cominco Ltd., Vancouver, B.C.
Imperial Oil Ltd., Toronto, Ontario.
International Minerals and Chemicals Corp. Ltd., Toronto, Ontario.

### Transportation and Storage Information

**Shipping State:** Liquid.
**Classification:** Corrosive, poisonous liquid.
**Inert Atmosphere:** No requirement.
**Venting:** Open.
**Pump Type:** Centrifugal. Alloy 20, Durimet, Worthite.

**Label(s):** Black and white label - CORROSIVE; Class 8, Group I. White label - POISON; Class 6.1, Group I.
**Storage Temperature:** Ambient.
**Hose Type:** Up to 30%-polypropylene, Teflon-lined. Above 30%, flexible stainless steel or rigid piping with swivel joints.

**Grades or Purity:** 20 to 65% free $SO_3$ (104.5 to 114.6% equivalent $H_2SO_4$); technical, commercial.
**Containers and Materials:** Bottles, drums, tank cars, tank trucks; steel, stainless steel.

### Physical and Chemical Characteristics

**Physical State (20°C, 1 atm):** Liquid.
**Solubility (Water):** Soluble in water with vigorous reaction.
**Molecular Weight:** Variable.
**Vapour Pressure:** No information.
**Boiling Point:** Decomposes.

**Floatability (Water):** Reacts vigorously, releasing heat.
**Odour:** Sharp, choking (1 mg/m$^3$ odour threshold).
**Flash Point:** Not flammable.
**Vapour Density:** Approx. 2.8 (20°C).
**Specific Gravity:** 20% 1.88; 30% 1.92; 40% 1.95; 65% 1.98 (38°C).

**Colour:** Colourless to dark brown, depending on purity.
**Explosive Limits:** Not flammable.
**Melting Point:** -9°C, 20% free $SO_3$; 15.5°C, 30% free $SO_3$; 33.0°C, 40% free $SO_3$; and 3.6°C, 65% free $SO_3$.

## HAZARD DATA

### Human Health

**Symptoms:** Inhalation: sore throat, coughing, laboured breathing. Contact: eyes or skin - produces severe burns. Ingestion: sore throat, abdominal pain, nausea and vomiting.
**Toxicology:** Highly toxic by all routes.
**TLV\*** (inhalation) 1 mg/m$^3$ (as sulfuric acid).
**Short-term Inhalation Limits** - No information.

$LC_{50}$ - Inhalation: rat = 347 ppm/1 h
**Delayed Toxicity** - No information.

$LD_{50}$ - No information.

### Fire

**Fire Extinguishing Agents:** Not combustible. Use dry chemical on adjacent fires. Avoid use of water.
**Behaviour in Fire:** Not combustible. Toxic $SO_x$ fumes are released in fires.
**Ignition Temperature:** Not combustible.
**Burning Rate:** Not combustible.

### Reactivity

**With Water:** Reacts vigorously and exothermically producing toxic $SO_x$ fumes.
**With Common Materials:** May cause ignition by contact with combustible materials. Can react violently with reducing agents, acetic acid, acetic anhydride, acetonitrile, acrolein, acrylic acid, acrylonitrile, allyl alcohol, allyl chloride, ammonium hydroxide, aniline, n-butyraldehyde, cresol, cumene, diethylene glycol, epichlorohydrin, ethyl acetate, ethylene glycol, hydrochloric acid, hydrofluoric acid, isoprene, propylene oxide, pyridine, sodium hydroxide, stryene, vinyl acetate.
**Stability:** Stable (within the limits of the foregoing).

### Environment

**Water:** Prevent entry into water intakes and waterways. Harmful to aquatic life in very low concentrations. Aquatic toxicity = 10 to 100 ppm/96 h/TLm/freshwater; 43 ppm/48 h/prawn/$LC_{50}$/saltwater; 24 mg/L/24 h/bluegill/lethal/freshwater; BOD: None.
**Land-Air:** No information.
**Food Chain Concentration Potential:** None.

## EMERGENCY MEASURES

**Special Hazards**

CORROSIVE. POISON. Reacts with water. Reacts with many materials.

**Immediate Responses**

Keep non-involved people away from spill site. Issue warnings: "CORROSIVE; POISON". Call Fire Department. Evacuate from downwind. Avoid contact and inhalation. Contact manufacturer. Contain spill area by diking. Stop discharge if possible. Notify environmental authorities.

**Protective Clothing and Equipment**

Respiratory protection - self-contained breathing apparatus and totally encapsulated suit (acid suit, rubber). Gloves - gauntlet type, rubber. Boots - high, rubber (pants worn outside boots).

**Fire and Explosion**

Not combustible. Use dry chemical on adjacent fires. Avoid use of water. Toxic $SO_x$ fumes are released in fires.

**First Aid**

Move victim out of spill area to fresh air. Call for medical assistance, but start first aid at once. Inhalation: if breathing has stopped give artificial respiration (not mouth-to-mouth method); if laboured, give oxygen. Contact: eyes - irrigate immediately with plenty of water. If pain persists, continue irrigation; skin - remove contaminated clothing and flush affected areas with plenty of water. Ingestion: give conscious victim plenty of water to drink. Do not induce vomiting. If medical assistance is not immediately available, transport victim to hospital, doctor or clinic.

## ENVIRONMENTAL PROTECTION MEASURES

**Response**

**Water**
1. Stop or reduce discharge if safe to do so.
2. Contact manufacturer or supplier for advice.
3. If possible, contain discharge by damming or water diversion.
4. Notify environmental authorities to discuss disposal and cleanup of contaminated materials.

**Land-Air**
1. Stop or reduce discharge if safe to do so.
2. Contact manufacturer or supplier for advice.
3. Contain spill by diking with earth or other barrier.
4. Remove material with pumps or vacuum equipment and place in appropriate containers.
5. Remove material by manual or mechanical means.
6. Remove contaminated soil for disposal or neutralize with lime.
7. Notify environmental authorities to discuss disposal and cleanup of contaminated materials.

**Disposal**

1. Contact manufacturer or supplier for advice on disposal.
2. Contact environmental authorities for advice on disposal.

OLEUM  $H_2SO_4 + SO_3$

# OXYGEN $O_2$

## IDENTIFICATION

UN No. 1072 (compressed gas)

### Common Synonyms
LOX
LIQUID OXYGEN

### Observable Characteristics
Colourless gas or liquid. Odourless.

### Manufacturers
Canadian Liquid Air Ltd., Hamilton, Ontario.
Union Carbide Canada Ltd., Sault Ste.-Marie, Ontario.
Inco, Copper Cliff, Ontario.
Dofasco, Hamilton, Ontario.

### Transportation and Storage Information
**Shipping State:** Gas (compressed); liquid (compressed or liquefied gas).
**Classification:** Nonflammable compressed gas; oxidizer.
**Inert Atmosphere:** No requirement.
**Venting:** Safety relief.

**Label(s):** Green label - NON-FLAMMABLE GAS; Class 2.2. Yellow Label - OXIDIZER; Class 5.1.
**Storage Temperature:** Ambient.
**Hose Type:** Braided, high pressure, degreased for oxygen service.

**Grades or Purity:** 99.5 + %.
**Containers and Materials:** Cylinders, tank cars; steel (gas) cylinders, tank truck; steel (liquid).

### Physical and Chemical Characteristics
**Physical State** (20°C, 1 atm): Gas.
**Solubility** (Water): 0.004 g/100mL (25°C).
**Molecular Weight:** 32.0
**Vapour Pressure:** 1 200 mm Hg (-178°C); 31 600 mm Hg (-123°C);
**Boiling Point:** -183°C.

**Floatability** (Water): Floats and boils (liquid).
**Odour:** Odourless.
**Flash Point:** Not flammable.
**Vapour Density:** 1.1 (25°C).
**Specific Gravity:** Liquid, 1.14 (-183°C).

**Colour:** Colourless.
**Explosive Limits:** Not flammable.
**Melting Point:** -218.8°C.

## HAZARD DATA

### Human Health
**Symptoms:** Contact: with liquid may produce frostbite or burns to skin and eyes. Inhalation: generally no toxic effect, overdoses may cause irritation of mucous membranes due to dryness.
**Toxicology:** Relatively nontoxic.
TLV° - No information.
Short-term Inhalation Limits - No information.

$LC_{50}$ - No information.
$TC_{Lo}$ - Inhalation: human = 100%/14 h
Delayed Toxicity - No information.

$LD_{50}$ - Not pertinent.

### Fire
**Fire Extinguishing Agents:** Stop flow of gas before attempting to extinguish fire. Most fire extinguishing agents may be used on fires involving oxygen. Water may be used to cool fire-exposed containers.
**Behaviour in Fire:** Not flammable or combustible but vigorously supports combustion of a wide range of materials.
**Ignition Temperature:** Not flammable or combustible.
**Burning Rate:** Not flammable or combustible.

### Reactivity
**With Water:** No reaction.
**With Common Materials:** May react explosively with combustible materials. Reacts violently with aluminum compounds, ethers, hydrogen, trichloroethane, trichloroethylene, benzene, carbon monoxide, chlorinated hydrocarbons and many fuels; carbon tetrachloride, hydrazine, lithium hydride, magnesium, methane, methylene chloride, oil, paraformaldehyde and titanium.
**Stability:** Stable.

### Environment
**Water:** Little hazard.
**Land-Air:** Little hazard.
**Food Chain Concentration Potential:** None.

## EMERGENCY MEASURES

**Special Hazards**
OXIDIZER.

**Immediate Responses**
Keep non-involved people away from spill site. Issue warning: "OXIDIZER". Call Fire Department. Eliminate all sources of ignition. Call manufacturer or supplier. Avoid contact with liquid. Stop or reduce discharge if safe to do so. Notify environmental authorities.

**Protective Clothing and Equipment**
Coveralls, gloves (asbestos or non-flammable) and goggles (mono, tight fitting). All exposed clothing must be ventilated thoroughly after exposure (30 min) to eliminate oxygen.

**Fire and Explosion**
Stop flow of gas before attempting to extinguish fire. Most fire extinguishing agents may be used on fires involving oxygen. Water may be used to cool fire-exposed containers.

**First Aid**
Move victim out of spill area. Call for medical assistance, but start first aid at once. Inhalation: move victim to fresh air; apply artificial respiration if necessary. Contact: skin - remove contaminated clothing and flush affected areas with plenty of water; treat as for frostbite (liquid); eyes - irrigate with plenty of water. If medical assistance is not immediately available, transport victim to hospital, doctor or clinic.

## ENVIRONMENTAL PROTECTION MEASURES

**Response**

**Water-Land-Air**
1. Stop or reduce discharge if safe to do so.
2. Contact manufacturer or supplier for advice.
3. Allow to disperse.

**Disposal**
1. Contact manufacturer or supplier for advice on disposal.
2. Material can be safely dispersed into the environment.

OXYGEN  $O_2$

# PENTACHLOROPHENOL  $C_6Cl_5OH$

## IDENTIFICATION

UN No. NA 2020

### Common Trade Names and Synonyms

DOWICIDE 7
SANTOPHEN 20
PCP
PENCHLOROL
PENTA
PERMATOX
(Also see sodium pentachlorophenate).

### Observable Characteristics

White to light-brown crystalline solid or powder. Weak odour. Strong pungent odour when heated.

### Manufacturers

Uniroyal Chemicals Ltd., Edmonton, Alberta.

### Transportation and Storage Information

**Shipping State:** Solid.
**Classification:** Poison.
**Inert Atmosphere:** No requirement.
**Venting:** Open.

**Label(s):** White label - POISON.
**Storage Temperature:** Ambient.

**Grades or Purity:** 86 to 100%.
**Containers and Materials:** Bags, fibre drums, bulk by rail or truck.

### Physical and Chemical Characteristics

**Physical State** (20°C, 1 atm): Solid.
**Solubility (Water):** Slightly soluble. 0.0005 g/100 mL (0°C); 0.0014 g/100 mL (20°C); 0.0035 g/100 mL (50°C).
**Molecular Weight:** 266.4
**Vapour Pressure:** 0.00011 mm Hg (20°C); 40 mm Hg (211.2°C).
**Boiling Point:** Decomposes at 310°C.

**Floatability (Water):** Sinks.
**Odour:** Weak. Strong pungent odour when heated.
**Flash Point:** Not flammable.
**Vapour Density:** 9.2
**Specific Gravity:** 1.98 (22°C).

**Colour:** White to light brown.
**Explosive Limits:** Not flammable.
**Melting Point:** 188 to 191°C.

## HAZARD DATA

### Human Health

**Symptoms:** Inhalation: coughing, nausea, vomiting, abdominal cramps, sweating, fever, rapid pulse, convulsions, loss of consciousness. Contact: skin - may be absorbed, irritation, inflammation; eyes - irritation and watering.
**Toxicology:** Highly toxic by all routes.
TLV®- (skin) 0.5 mg/m³.
Short-term Inhalation/Skin Limits - 1.5 mg/m³ (15 min).

$LC_{50}$ - No information.
Delayed Toxicity - May cause liver and kidney damage.

$LD_{50}$ - Oral: rat = 0.050 g/kg

### Fire

**Fire Extinguishing Agents:** Use water spray, dry chemical, foam or carbon dioxide. Use water to cool fire-exposed containers.
**Behaviour in Fire:** When heated to decomposition, toxic fumes of chlorides are evolved.
**Ignition Temperature:** Not combustible.
**Burning Rate:** Not combustible.

### Reactivity

**With Water:** No reaction.
**With Common Materials:** No information.
**Stability:** Stable.

### Environment

**Water:** Prevent entry into water intakes and waterways. Fish toxicity: 5 ppm/3 h/trout/lethal/freshwater; 0.24 mg/L/24 h/fathead minnow/$LC_{50}$/freshwater; 0.267 mg/L/24 h/goldfish/$LC_{50}$/freshwater; EPA criteria: 6.2 mg/L/24 h (conc. should not exceed 14 μg/L fresh $H_2O$ at any time); BOD: Not available.
**Land-Air:** Toxic to plant life; 4 500 ppm/$LC_{50}$/mallard.
**Food Chain Concentration Potential:** No information.

## EMERGENCY MEASURES

**Special Hazards**

POISON.

**Immediate Responses**

Keep non-involved people away from spill site. Issue warning: "POISON". Avoid contact and inhalation. Dike to prevent runoff from rainwater or water application. Notify supplier. Notify environmental authorities.

**Protective Clothing and Equipment**

Respiratory protection - in fires or enclosed spaces use self-contained breathing apparatus and totally encapsulated suit. Gloves - gauntlet type, rubber, unlined. Boots - rubber, high (pants worn outside boots).

**Fire and Explosion**

Use water spray, dry chemical, foam or carbon dioxide to extinguish. Use water to cool fire-exposed containers. When heated to decomposition, toxic fumes of chlorides are evolved.

**First Aid**

Move victim out of spill area to fresh air. Call for medical assistance, but start first aid at once. Inhalation: if breathing has stopped, give artificial respiration (not mouth-to-mouth method); if laboured, give oxygen. Contact: skin and eyes - remove contaminated clothing and wash eyes and affected skin with plenty of water. Ingestion: give water to conscious victim to drink and induce vomiting. Keep victim warm and quiet. If medical assistance is not immediately available, transport victim to hospital, doctor or clinic.

## ENVIRONMENTAL PROTECTION MEASURES

**Response**

Water
1. Stop or reduce discharge if safe to do so.
2. Contact manufacturer or supplier for advice.
3. If possible, contain discharge by damming or water diversion.
4. Dredge or vacuum pump to remove liquids, contaminants and contaminated bottom sediments.
5. Notify environmental authorities to discuss disposal and cleanup of contaminated materials.

Land-Air
1. Stop or reduce discharge if safe to do so.
2. Contact manufacturer or supplier for advice.
3. Contain spill by diking with earth or other barrier.
4. Remove material by manual or mechanical means.
5. Recover undamaged containers.
6. Remove contaminated soil for disposal.
7. Notify environmental authorities to discuss disposal and cleanup of contaminated materials.

**Disposal**

1. Contact manufacturer or supplier for advice on disposal.
2. Contact environmental authorities for advice on disposal.

PENTACHLOROPHENOL   $C_6Cl_5OH$

# PENTAERYTHRITOL   C(CH$_2$OH)$_4$

## IDENTIFICATION

### Common Trade Names and Synonyms
2,2-bis (HYDROXYMETHYL)-1,3 - PROPANEDIOL
TETRAMETHYLOLMETHANE
PENTAERYTHRITE
TETRAHYDROXYMETHYLMETHANE
PE
mono PE, PENTEK
TETRAKIS (HYDROXYMETHYL) METHANE

### Observable Characteristics
White solid. Odourless.

### Manufacturers
Celanese Canada, Edmonton, Alberta.

### Transportation and Storage Information
**Shipping State:** Solid.
**Classification:** None.
**Inert Atmosphere:** No requirement.
**Venting:** Open.
**Label(s):** Not regulated.
**Storage Temperature:** Ambient.
**Grades or Purity:** Technical: 86 to 90%; involving 10 to 40% di-pentaerythritol. Pure: 98+ %.
**Containers and Materials:** Bags (multiwall paper); drums, carlots; steel.

### Physical and Chemical Characteristics
**Physical State** (20°C, 1 atm): Solid.
**Solubility** (Water): 5.56 g/100 mL (15°C).
**Molecular Weight:** 136.2
**Vapour Pressure:** No information.
**Boiling Point:** Sublimes (276°C).
**Floatability** (Water): Sinks and slowly mixes.
**Odour:** Odourless.
**Flash Point:** Not flammable.
**Vapour Density:** 4.7
**Specific Gravity:** 1.40 (25°C).
**Colour:** White.
**Explosive Limits:** Not flammable.
**Melting Point:** Pure, 260 to 269°C; test grade, 220-230°C (sublimes from 130°C at reduced pressure).

## HAZARD DATA

### Human Health
**Symptoms:** No information.
**Toxicology:** Low order of toxicity.
**TLV®-** (nuisance particulate) 10 mg/m$^3$.
**Short-term Inhalation** - (nuisance particulate) 20 mg/m$^3$.
**LC$_{50}$** - No information.
**Delayed Toxicity** - No information.
**LD$_{50}$ - Oral:** mouse = 25.5 g/kg
**LD$_{50}$ - Oral:** guinea pig = 11.3 g/kg

### Fire
**Fire Extinguishing Agents:** Most fire extinguishing agents may be used on fires involving pentaerythritol.
**Behaviour in Fire:** No information.
**Ignition Temperature:** 45°C (dust cloud).
**Burning Rate:** No information.

### Reactivity
**With Water:** No reaction; slightly soluble.
**With Common Materials:** May react with oxidizing materials. Reacts violently with acetic anhydride and nitric acid.
**Stability:** Stable.

### Environment
**Water:** prevent entry into water intakes and waterways. No information on fish toxicity available. BOD: 95 to 96%, 5 days.
**Land-Air:** No information.
**Food Chain Concentration Potential:** None.

## EMERGENCY MEASURES

**Special Hazards**
COMBUSTIBLE.

**Immediate Responses**
Keep non-involved people away from spill site. Call Fire Department. Notify supplier or manufacturer. Dike to prevent runoff. Notify environmental authorities.

**Protective Clothing and Equipment**
Protective outerwear as required.

**Fire and Explosion**
Most fire extinguishing agents may be used on fires involving pentaerythritol.

**First Aid**
Move victim out of spill area to fresh air. Call for medical assistance, but start first aid at once. <u>Inhalation</u>: move to fresh air. <u>Contact</u>: skin- remove contaminated clothing and flush affected areas with plenty of water; eyes - irrigate with plenty of water <u>Ingestion</u>: give conscious victim water to drink and induce vomiting. Transport victim to hospital, doctor or clinic if necessary.

## ENVIRONMENTAL PROTECTION MEASURES

**Response**

**Water**
1. Stop or reduce discharge if safe to do so.
2. Contact manufacturer or supplier for advice.
3. If possible, contain discharge by damming or water diversion.
4. Dredge or vacuum pump to remove contaminants, liquids and contaminated bottom sediments.
5. Notify environmental authorities to discuss disposal and cleanup of contaminated materials.

**Land-Air**
1. Stop or reduce discharge if safe to do so.
2. Contact manufacturer or supplier for advice.
3. Dike to prevent runoff from rainwater or water application.
4. Remove material by manual or mechanical means.
5. Recover undamaged containers.
6. Remove contaminated soil for disposal.
7. Notify environmental authorities to discuss disposal and cleanup of contaminated materials.

**Disposal**
1. Contact manufacturer or supplier for advice on disposal.
2. Contact environmental authorities for advice on disposal.
3. Incinerate (approval of environmental authorities required).

PENTAERYTHRITOL    $C(CH_2OH)_4$

# PERCHLOROETHYLENE $Cl_2C:CCl_2$

## IDENTIFICATION

UN No. 1897

### Common Trade Names and Synonyms
TETRACHLOROETHYLENE
CARBON BICHLORIDE
ETHYLENE TETRACHLORIDE
"per"
DRY CLEANING FLUID

### Observable Characteristics
Colourless liquid. Sweet, ethereal odour.

### Manufacturers
Dow Chemical Canada Inc., Sarnia, Ontario.
Canadian Industries Ltd. (C-I-L), Shawinigan, Quebec.

### Transportation and Storage Information
**Shipping State:** Liquid.
**Classification:** Poison.
**Inert Atmosphere:** No requirement.
**Venting:** Pressure-vacuum.
**Pump Type:** Centrifugal or positive displacement; steel or stainless steel.

**Label(s):** Black and white label - POISON; Class 6.1, Group III.
**Storage Temperature:** Ambient.
**Hose Type:** Seamless stainless steel, Teflon, bronze. Viton and cross-linked polyethylene may be used with caution.

**Grades or Purity:** Dry cleaning and industrial, 95+%.
**Containers and Materials:** Cans, drums, tank cars, tank trucks; steel.

### Physical and Chemical Characteristics
**Physical State** (20°C, 1 atm): Liquid.
**Solubility** (Water): 0.015 g/100 mL (25°C).
**Molecular Weight:** 165.8
**Vapour Pressure:** 14 mm Hg (20°C); 24 mm Hg (30°C), 45 mm Hg (40°C).
**Boiling Point:** 121°C.

**Floatability** (Water): Sinks.
**Odour:** Sweet etheral (5 to 50 ppm, odour threshold).
**Flash Point:** Not flammable.
**Vapour Density:** 5.8
**Specific Gravity:** 1.63 (20°C).

**Colour:** Colourless.
**Explosive Limits:** Not flammable.
**Melting Point:** -22.4°C.

## HAZARD DATA

### Human Health
**Symptoms: Inhalation:** irritation of nose, eyes and throat, headache, intoxication, narcosis, loss of consciousness and coma; **Ingestion:** gastrointestinal irritation, nausea, vomiting, diarrhea, drowsiness, narcosis. **Contact:** skin - absorbed, dry skin, inflammation, blisters; eyes - irritation, watering, inflammation.
**Toxicology:** Moderately toxic by all routes.
TLV● (skin) 50 ppm; 335 mg/m³.
Short-term Inhalation Limits - 200 ppm; 1 340 mg/m³.

$LC_{50}$ - 5 040 ppm (8 h) rat.
$TC_{Lo}$ - Inhalation: human = 10 000 mg/m³.
Delayed Toxicity - May cause liver and kidney damage.

$LD_{50}$ - Oral: rat = 2.60 g/kg

### Fire
**Fire Extinguishing Agents:** Not combustible. Most fire extinguishing agents may be used. Water may be used to cool fire-exposed containers.
**Behaviour in Fire:** When heated to decomposition, emits toxic fumes of chloride.
**Ignition Temperature:** Not combustible.
**Burning Rate:** Not combustible.

### Reactivity
**With Water:** No reaction.
**With Common Materials:** Reacts violently with barium, beryllium and lithium.
**Stability:** Stable.

### Environment
**Water:** Prevent entry into water intakes and waterways. Aquatic toxicity rating = 10 to 100 ppm/96 h/TLm/freshwater; BOD: 5 to 6%, 5 days.
**Land-Air:** No information.
**Food Chain Concentration Potential:** Known to be concentrated in tissue.

## EMERGENCY MEASURES

**Special Hazards**

POISON.

**Immediate Responses**

Keep non-involved people away from spill site. Issue warning: "POISON". Avoid contact and inhalation. Notify manufacturer or supplier. Shut off leak if safe to do so. Dike to contain material. Notify environmental authorities.

**Protective Clothing and Equipment**

In fires or enclosed spaces, Respiratory protection – self-contained breathing apparatus and totally encapsulated suit. Otherwise, outer protective clothing as required.

**Fire and Explosion**

Not combustible. Most fire extinguishing agents may be used on fires involving perchloroethylene. Water may be used to cool fire-exposed containers. When heated to decomposition, emits toxic fumes of chlorides.

**First Aid**

Move victim out of spill site to fresh air. Call for medical assistance, but start first aid at once. Inhalation: if breathing has stopped, give artificial respiration (not mouth-to-mouth method); if laboured, give oxygen. Contact: skin – remove contaminated clothing and flush affected areas with plenty of water; eyes – irrigate with plenty of water. Ingestion: give conscious victim plenty of water to drink; do not induce vomiting. If medical assistance is not immediately available, transport victim to hospital, doctor or clinic.

## ENVIRONMENTAL PROTECTION MEASURES

**Response**

**Water**
1. Stop or reduce discharge if safe to do so.
2. Contact manufacturer or supplier for advice.
3. If possible, contain discharge by damming or water diversion.
4. Dredge or vacuum pump to remove contaminated bottom sediments.
5. Notify environmental authorities to discuss disposal and cleanup of contaminated materials.

**Land-Air**
1. Stop or reduce discharge if safe to do so.
2. Contact manufacturer or supplier for advice.
3. Contain spill by diking with earth or other barrier.
4. Remove material with pumps or vacuum equipment and place in appropriate containers.
5. Recover undamaged containers.
6. Adsorb residual liquid on natural or synthetic sorbents.
7. Notify environmental authorities to discuss disposal and cleanup of contaminated materials.

**Disposal**

1. Contact manufacturer or supplier for advice on disposal.
2. Contact environmental authorities for advice on disposal.

PERCHLOROETHYLENE   $Cl_2C:CCl_2$

# PHENOL $C_6H_5OH$

## IDENTIFICATION

UN No. Solid 1671

### Common Synonyms
CARBOLIC ACID
OXYBENZENE
HYDROXYBENZENE
PHENIC ACID
PHENYLIC ACID
PHENYL HYDROXIDE

### Observable Characteristics
White to pink solid. Liquid at temperatures above 41°C. Sharp, medicinal, sweet, tarry odour.

### Manufacturers
Gulf Oil Canada Ltd.,
Montreal, Quebec.
Dow Chemical Canada Inc.,
Delta, B.C.

### Transportation and Storage Information
**Shipping State:** Solid or liquid (molten).
**Classification:** Poison.
**Inert Atmosphere:** No requirement.
**Venting:** Pressure-vacuum.

**Label(s):** White label - POISON; Class 6.1, Group II.
**Storage Temperature:** Ambient; solid; 54°C, molten.

**Grades or Purity:** Technical: 82 to 92%, containing cresol.
**Containers and Materials:** Drums, tank cars, tank trucks; stainless steel, lined carbon steel, nickel.

### Physical and Chemical Characteristics
**Physical State (20°C, 1 atm):** Solid.
**Solubility (Water):** In all proportions at 66°C; 8.2 g/100 mL (15°C).
**Molecular Weight:** 94.1
**Vapour Pressure:** 0.2 mm Hg (20°C); 0.35 mm Hg (25°C); 1 mm Hg (40°C).
**Boiling Point:** 182°C.

**Floatability (Water):** Floats and mixes.
**Odour:** Sharp, medicinal, sweet, tarry. (0.005 to 0.5 ppm odour threshold).
**Flash Point:** 85°C (o.c.), 79°C (c.c.).
**Vapour Density:** 3.2
**Specific Gravity:** 1.07 (20°C).

**Colour:** White to pink.
**Explosive Limits:** 1.8 LEL.
**Melting Point:** 41°C.

## HAZARD DATA

### Human Health
**Symptoms:** Inhalation: irritation of the nose, mouth, throat, difficulty breathing, headache, nausea, weakness, serious injury. **Contact:** skin - readily absorbed, severe burning, inflammation, painful blisters, unconsciousness, death; eyes - stinging, watering, severe burning, loss of vision. **Ingestion:** irritation and burning in throat, pain in swallowing, intense thirst, abdominal cramps, difficulty breathing, convulsions, coma, death.
**Toxicology:** Highly toxic by all routes.
TLV® (skin) 5 ppm; 19 mg/m³.
Short-term Inhalation Limits - (skin) 10 ppm; 38 mg/m³ for 15 min.

$LC_{50}$ - No information.
Delayed Toxicity - Causes liver and kidney damage.

$LD_{50}$ - Oral: rat = 0.414 g/kg
$LD_{Lo}$ - Oral: human = 0.14 g/kg

### Fire
**Fire Extinguishing Agents:** Use water spray, carbon dioxide, dry chemical or foam.
**Behaviour in Fire:** Emits toxic (phenol) vapours when heated.
**Ignition Temperature:** 715°C.
**Burning Rate:** 3.5 mm/min.

### Reactivity
**With Water:** No reaction; soluble.
**With Common Materials:** Can react with oxidizing materials. Reacts violently with (aluminum chloride and nitrobenzene), butadiene and calcium hypochlorite. When hot, reacts with aluminum, magnesium, lead and zinc.
**Stability:** Stable.

### Environment
**Water:** Harmful to aquatic life in very low concentrations. Prevent entry into water intakes and waterways. Aquatic toxicity rating: 10 to 100 ppm/96 h/TLm/freshwater; 23.5 ppm/14 h/shrimp/$LC_{50}$/saltwater; 9 mg/L/1 h/perch/$LC_{50}$/freshwater; 46 mg/L/24 h/goldfish/$LC_{50}$/freshwater; BOD: 140 to 180%, 5 days.
**Land-Air:** No information.
**Food Chain Concentration Potential:** None.

## EMERGENCY MEASURES

### Special Hazards
POISON.

### Immediate Responses
Keep non-involved people away from spill site. Issue warning: "POISON". Call Fire Department. Notify supplier. Avoid contact and inhalation. Stop or reduce discharge, if this can be done without risk. Dike spill to contain water runoff. Notify environmental authorities.

### Protective Clothing and Equipment
Respiratory protection - self-contained breathing apparatus and totally encapsulated suit. Boots - rubber. Gloves - rubber.

### Fire and Explosion
Use water spray, carbon dioxide, dry chemical or foam to extinguish. Emits toxic phenol vapours when heated.

### First Aid
Move victim out of spill site to fresh air. Call for medical assistance, but start first aid at once. Inhalation: if breathing has stopped, give artificial respiration (not mouth-to-mouth method); if laboured, give oxygen. Contact: skin and eyes - flush immediately with plenty of water for at least 30 minutes while removing contaminated clothing. Ingestion: give conscious victim water to drink; Induce vomiting immediately. If medical assistance is not immediately available, transport victim to hospital, doctor or clinic.

## ENVIRONMENTAL PROTECTION MEASURES

### Response

**Water**

Sinking
1. Stop or reduce discharge if safe to do so.
2. Contact manufacturer or supplier for advice.
3. If possible, contain discharge by damming or water diversion.
4. Dredge or vacuum pump to remove contaminants, liquids and contaminated bottom sediments.
5. Notify environmental authorities to discuss disposal and cleanup of contaminated materials.

Floating
1. Stop or reduce discharge if safe to do so.
2. Contact manufacturer or supplier for advice.
3. If possible, contain discharge by booming.
4. Skim and remove.
5. Notify environmental authorities to discuss disposal and cleanup of contaminated materials.

**Land-Air**
1. Stop or reduce discharge if safe to do so and using protective equipment.
2. Contact manufacturer or supplier for advice.
3. Contain spill by diking with earth or other barrier.
4. Remove material by manual or mechanical means.
5. Recover undamaged containers.
6. Adsorb residual liquid on natural or synthetic sorbents.
7. Remove contaminated soil for disposal.
8. Notify environmental authorities to discuss disposal and cleanup of contaminated materials.

### Disposal
1. Contact manufacturer or supplier for advice on disposal.
2. Contact environmental authorities for advice on disposal.
3. Incinerate (approval of environmental authorities required).

PHENOL  $C_6H_5OH$

# PHOSGENE  $COCl_2$

## IDENTIFICATION

UN No. 1076

### Common Synonyms

CARBONYL CHLORIDE
CARBON OXYCHLORIDE
CHLOROFORMYL CHLORIDE
CARBONIC DICHLORIDE

Note: PHOSGENE gas is a common combustion product resulting from the burning of many chlorinated compounds.

### Observable Characteristics

Colourless to light yellow gas. Odour sweet/sharp, like newly mown grass in dilute concentrations. Strong and stifling in heavy concentrations.

### Manufacturers

No Canadian manufacturer.
U.S. manufacturers:
Allied Chemical, Morristown, N.J.
Dow Chemical, Midland, MI
Upjohn, Kalamahzoo, MI

### Transportation and Storage Information

**Shipping State:** Liquid (compressed gas).
**Classification:** Poison and corrosive.
**Inert Atmosphere:** No requirement.
**Venting:** Safety relief.

**Label(s):** White label - POISON; Class 2.3. White and black label - CORROSIVE; Class 8.
**Storage Temperature:** Ambient.

**Grades or Purity:** Commercial, 99.0%.
**Containers and Materials:** Cylinders; steel.

### Physical and Chemical Characteristics

**Physical State** (20°C, 1 atm): Gas.
**Solubility** (Water): (moisture) Decomposes to produce $CO_2$ and HCl.
**Molecular Weight:** 98.9
**Vapour Pressure:** 568 mm Hg (0°C); 1 215 mm Hg (20°C); 1 418 mm Hg (25°C); 1 672 mm Hg (30°C).
**Boiling Point:** 7.6°C.

**Floatability** (Water): Liquid sinks and decomposes to produce $CO_2$ and HCl.
**Odour:** Sweet/sharp, like newly mown grass (dilute) (0.1 to 1.0 ppm odour threshold). Stiffling (heavy conc.). Sense of smell may be fatigued.
**Flash Point:** Not flammable.
**Vapour Density:** 3.5
**Specific Gravity:** 1.37 (liquid) (20°C).

**Colour:** Colourless to light yellow.
**Explosive Limits:** Not flammable.
**Melting Point:** -128°C.

## HAZARD DATA

### Human Health

**Symptoms:** Sense of smell may be fatigued, and fail to give warning of concentration. Not immediately irritating, even when fatal concentrations are inhaled. Inhalation: Dryness or burning in throat, numbness, chest pain, coughing, difficult breathing, cyanosis, headache, nausea, vomiting, pulmonary edema, shock and possible death. Contact: skin - burning and inflammation; eyes - burning.
**Toxicology:** Highly toxic by inhalation and irritation of eyes and mucous membranes.
TLV®- (inhalation) 0.10 ppm; 0.4 mg/m³.
$LC_{50}$ - Inhalation: human = 3 200 mg/m³.
$LC_{50}$ - Inhalation: rat = 50 ppm/30 min.
$LD_{50}$ - No information.
Delayed Toxicity - Symptoms may not appear immediately after exposure.
**Short-term Inhalation Limits** - No information.

### Fire

**Fire Extinguishing Agents:** Not combustible. Use water spray to cool fire-exposed containers and knock down vapours.
**Behaviour in Fire:** Not combustible. Toxic, corrosive (CO, $CO_2$, $Cl_2$, HCl, $COCl_2$) fumes are produced when heated to decomposition.
**Ignition Temperature:** Not combustible.
**Burning Rate:** Not combustible.

### Reactivity

**With Water:** Decomposes, to form hydrochloric acid and carbon dioxide.
**With Common Materials:** Reacts violently with aluminum, isopropyl alcohol, potassium and sodium.
**Stability:** Stable.

### Environment

**Water:** Prevent entry into water intakes and waterways. No information on fish toxicities. BOD: None.
**Land-Air:** No information.
**Food Chain Concentration Potential:** None.

## EMERGENCY MEASURES

**Special Hazards**

POISON. CORROSIVE.

**Immediate Responses**

Keep non-involved people away from spill site. Issue warnings: "POISON; CORROSIVE". Avoid contact. Evacuate area. Call Fire Department. Call manufacturer. Stop discharge if safe to do so. Stay upwind and use water spray to control vapour. Notify environmental authorities.

**Protective Clothing and Equipment**

Respiratory protection - self-contained breathing apparatus and totally encapsulated suit.

**Fire and Explosion**

Not combustible. Use water to cool fire-exposed containers and knock down vapours. Toxic, corrosive ($CO$, $CO_2$, $Cl_2$, $HCl$, $COCl_2$) fumes are produced when heated to decomposition.

**First Aid**

Move victim out of spill area to fresh air. Call for medical assistance, but start first aid at once. <u>Inhalation:</u> give artificial respiration (not mouth-to-mouth method) if breathing has stopped, and oxygen if breathing is laboured. Keep victim warm and maintain absolute rest. <u>Contact:</u> skin - remove contaminated clothing and flush affected areas with plenty of water; eyes - irrigate with plenty of water. If medical assistance is not immediately available, transport victim to hospital, doctor or clinic.

## ENVIRONMENTAL PROTECTION MEASURES

**Response**

**Water-Land-Air**

1. Contact manufacturer or supplier for advice.
2. Stop or reduce discharge if safe to do so (should only be attempted by an experienced person).
3. Use water spray to control vapours.
4. Notify environmental authorities.

**Disposal**

1. Contact manufacturer or supplier for advice on disposal.
2. Contact environmental authorities for advice on disposal.

PHOSGENE   $COCl_2$

# PHOSPHORIC ACID   $H_3PO_4$                                                UN No. 1805

## IDENTIFICATION

| Common Synonyms | Observable Characteristics | Manufacturers |
|---|---|---|
| ORTHOPHOSPHORIC ACID | Colourless liquid. Odourless. | International Minerals and Chem. Corp., Port Maitland, Ont.<br>Cominco Ltd., Trail and Kimberley, B.C.<br>Esso Chemical Canada, Redwater, Alta.<br>Belledune Fertilizer, Belledune, N.B. |

### Transportation and Storage Information

**Shipping State:** Liquid (aqueous solution) or >100 % $H_3PO_4$
**Classification:** Corrosive.
**Inert Atmosphere:** No requirement.
**Venting:** Open.
**Pump Type:** Positive displacement, gear, centrifugal; stainless steel.

**Label(s):** Black and white label – CORROSIVE; Class 8, Group III.
**Storage Temperature:** Ambient. If >80% may be heated
**Hose Type:** Polyethylene, butyl, Hypalon.

**Grades or Purity:** Aqueous solutions from 35 to 85%; 75 and 85% most common; 100% crystalline solid.
**Containers and Materials:** Carboys, drums, tank cars, tank trucks; stainless 316; rubber-lined tanks.

### Physical and Chemical Characteristics

**Physical State (20°C, 1 atm):** Liquid (aqueous solution), solid (100%).
**Solubility: (Water):** Completely soluble. 548 g/100 mL (solute) (20°C); aqueous solutions soluble in all proportions.
**Molecular Weight:** 98 (solute).
**Vapour Pressure:** 6.4 mm Hg (21°C) 75%; 2.2 mm Hg (21°C) 85%; 0.0285 mm Hg (20°C) (100%).
**Boiling Point:** 135°C (75%); 158°C (85%); 260°C (100%); at 213°C decomposes.

**Floatability (Water):** Sinks and mixes.
**Odour:** Odourless.
**Flash Point:** Not flammable.
**Vapour Density:** 3.4
**Specific Gravity:** 1.58 (75%); 1.88 (80%); 1.69 (85%); 1.83 (100%) (20°C).

**Colour:** Colourless liquid or transparent solid.
**Explosive Limits:** Not flammable.
**Melting Point:** 21.1°C (85%); 42°C (100%); −17.5°C (75%); 4.6°C (80%).

## HAZARD DATA

### Human Health

**Symptoms:** Inhalation: irritation of mucous membranes, watering of eyes, difficulty breathing, salivation, nausea. Ingestion: pain in swallowing, intense thirst, abdominal pain, nausea; Contact: skin – burning, inflammation, blisters; eyes – burning, watering.
**Toxicology:** Moderately toxic by ingestion and contact.
TLV®- (inhalation) 1 mg/m³.
Short-term Inhalation Limits – 3 mg/m³ (15 min).
$LC_{50}$ - No information.
$TC_{Lo}$ - Inhalation: human = 10 mg/m³.
Delayed Toxicity - No information.
$LD_{50}$ - Oral: rat = 1.53 g/kg

### Fire

**Fire Extinguishing Agents:** Not combustible. Use water in fires involving phosphoric acid. Use water to cool fire-exposed containers and knock down vapours.
**Behaviour in Fire:** Not combustible. When heated to decomposition, emits toxic POx fumes.
**Ignition Temperature:** Not combustible.
**Burning Rate:** Not combustible.

### Reactivity

**With Water:** Soluble with mild exothermic reaction.
**With Common Materials:** Reacts with some metals to liberate flammable hydrogen gas.
**Stability:** Stable.

### Environment

**Water:** Prevent entry into water intakes and waterways. Fish toxicity: 138 ppm/24 h/mosquito fish/TLm/freshwater; Aquatic toxicity rating = 100 to 1 000 ppm/96 h/TLm/freshwater; BOD: None.
**Land-Air:** No information.
**Food Chain Concentration Potential:** None.

## EMERGENCY MEASURES

**Special Hazards**
CORROSIVE.

**Immediate Responses**
Keep non-involved people away from spill site. Issue warning: "CORROSIVE". Avoid contact. Stop or reduce discharge, if this can be done without risk. Contain spill by diking to prevent runoff. Notify supplier and environmental authorities.

**Protective Clothing and Equipment**
In fires, Respiratory protection - self-contained breathing apparatus protective clothing. Otherwise, Goggles - (mono), tight fitting. Gloves - rubber or plastic coated. Boots - high, rubber (pants worn outside boots).

**Fire and Explosion**
Not combustible. Use water in fires involving phosphoric acid. Use water to cool fire-exposed containers and knock down vapours. When heated to decomposition, toxic POx fumes are emitted.

**First Aid**
Move victim out of spill area to fresh air. Call for medical assistance, but start first aid at once. Inhalation: if breathing has stopped give artificial respiration (not mouth-to-mouth method); if laboured, give oxygen. Contact: skin and eyes - remove contaminated clothing and flush skin and eyes with plenty of warm water. Ingestion: give conscious victim water or milk to drink. Keep warm and quiet. If medical assistance is not immediately available, transport victim to hospital doctor or clinic.

## ENVIRONMENTAL PROTECTION MEASURES

**Response**

**Water**
1. Stop or reduce discharge if safe to do so.
2. Contact manufacturer or supplier for advice.
3. If possible, contain discharge by damming or water diversion.
4. Notify environmental authorities to discuss disposal and cleanup of contaminated materials.

**Land-Air**
1. Stop or reduce discharge if safe to do so.
2. Contact manufacturer or supplier for advice.
3. Dike to prevent runoff from rainwater or water application.
4. Remove material with pumps or vacuum equipment and place in appropriate containers.
5. Recover undamaged containers.
6. Neutralize soil with lime.
7. Notify environmental authorities to discuss disposal and cleanup of contaminated materials.

**Disposal**
1. Contact manufacturer or supplier for advice on disposal.
2. Neutralization of $H_3PO_4$ with a suitable alkali (lime) will provide a product which could be buried in a landfill site or placed in other appropriate areas.
3. Contact environmental authorities for advice on disposal.

PHOSPHORIC ACID $H_3PO_4$

# PHOSPHORUS, RED  $P_4$

## IDENTIFICATION

| Common Synonyms | Observable Characteristics | UN No. 1338 Manufacturers |
|---|---|---|
| AMORPHOUS PHOSPHORUS | Reddish-brown powder. Odourless. | Erco Industries Limited, Long Harbour, Nfld., Varennes, Que. |

### Transportation and Storage Information

**Shipping State:** Solid.
**Classification:** Flammable solid.
**Inert Atmosphere:** No requirement.
**Venting:** Open.

**Label(s):** White and red striped label - FLAMMABLE SOLID; Class 4.1.
**Storage Temperature:** Ambient.

**Grades or Purity:** Technical (99.9%).
**Containers and Materials:** Cans, drums; steel.

### Physical and Chemical Characteristics

**Physical State (20°C, 1 atm):** Solid.
**Solubility (Water):** Insoluble in warm water. Extremely low solubility in cold water.
**Molecular Weight:** 123.9 ($P_4$).
**Vapour Pressure:** 1 mm Hg (237°C).
**Boiling Point:** 280°C (ignites at 202 to 280°C).

**Floatability (Water):** Sinks.
**Odour:** Odourless.
**Flash Point:** No information.
**Vapour Density:** 4.8
**Specific Gravity:** 2.2 to 2.3 (20°C).

**Colour:** Reddish-brown.
**Explosive Limits:** No information.
**Melting Point:** Sublimes at 398 to 416°C; melts at 595°C; ignites at 202 to 280°C.

## HAZARD DATA

### Human Health

**Symptoms:** Inhalation: (dust) irritation of respiratory tract. Contact: eyes - irritation.
**Toxicology:** Low toxicity by all routes.
**TLV®** - No information.
**Short-term Inhalation Limits** - No information.

**$LC_{50}$** - No information.
**Delayed Toxicity** - No information.

**$LD_{50}$** - No information.
**$LD_{Lo}$** - Oral: man = 4.41 g/kg

### Fire

**Fire Extinguishing Agents:** Use water to extinguish.
**Behaviour in Fire:** Heat may cause reversion to white (yellow) phosphorus, which is toxic and spontaneously flammable upon contact with air. Burning yields irritating oxides of phosphorus ($PO_x$).
**Ignition Temperature:** 202 to 280°C.
**Burning Rate:** No information.

### Reactivity

**With Water:** No reaction.
**With Common Materials:** Can react with reducing materials.
**Stability:** Stable.

### Environment

**Water:** Prevent entry into water intakes and waterways. Fish toxicity = 0.105 mg/L/48 h/bluegill/TLm/freshwater;
**BOD:** Not available.
**Land-Air:** No information.
**Food Chain Concentration Potential:** None.

## EMERGENCY MEASURES

### Special Hazards
FLAMMABLE.

### Immediate Response
Keep non-involved people away from spill site. Issue warning: "FLAMMABLE". CALL FIRE DEPARTMENT. Stop or reduce discharge, if this can be done without risk. Dike to prevent runoff from rainwater or water application. Notify manufacturer and environmental authorities.

### Protective Clothing and Equipment
In fire, Respiratory protection - self-contained breathing apparatus protective clothing. Otherwise, Goggles - (mono), tight fitting. Gloves - rubber.

### Fire and Explosion
Use water to extinguish. Heat may cause reversion to white (yellow) phosphorus which is toxic and spontaneously flammable upon contact with air. Burning yields irritating oxides of phosphorus ($PO_x$).

### First Aid
Move victim out of spill area to fresh air. Call for medical assistance, but start first aid at once. Inhalation: if breathing has stopped, give artificial respiration; if laboured, give oxygen. Contact: skin and eyes - remove contaminated clothing and wash eyes and skin thoroughly with plenty of water. Ingestion: give milk or water to conscious victim to drink. Keep warm and quiet. If medical assistance is not immediately available, transport victim to hospital, doctor or clinic.

## ENVIRONMENTAL PROTECTION MEASURES

### Response

**Water**
1. Stop or reduce discharge if safe to do so.
2. Contact manufacturer or supplier for advice.
3. If possible, contain discharge by damming or water diversion.
4. Dredge or vacuum pump to remove contaminated bottom sediments.
5. Notify environmental authorities to discuss disposal and cleanup of contaminated materials.

**Land-Air**
1. Stop or reduce discharge if safe to do so.
2. Contact manufacturer or supplier for advice.
3. Dike to prevent runoff from rainwater or water application.
4. Remove material by manual or mechanical means.
5. Recover undamaged containers.
6. Notify environmental authorities to discuss disposal and cleanup of contaminated materials.

### Disposal
1. Contact manufacturer or supplier for advice on disposal.
2. Contact environmental authorities for advice on disposal.

PHOSPHORUS, RED  $P_4$

# PHOSPHORUS, WHITE OR YELLOW   $P_4$

## IDENTIFICATION

UN No. 1381

| Common Synonyms | Observable Characteristics | Manufacturers |
|---|---|---|
| WHITE PHOSPHORUS<br>YELLOW PHOSPHORUS<br>WHITE DRY PHOSPHORUS | White to yellow, waxy solid. Odourless, but $PO_x$ fumes from combustion have a garlic-like odour. | Erco Industries Limited,<br>Varennes, Que.,<br>Long Harbour, Nfld. |

### Transportation and Storage Information

**Shipping State:** Solid.
**Classification:** Spontaneously combustible.
**Inert Atmosphere:** Water cover.
**Venting:** Pressure-vacuum.

**Label(s):** White and red label - SPONTANEOUSLY COMBUSTIBLE; Class 4.2
**Storage Temperature:** Ambient.

**Grades or Purity:** Technical, 99.9%.
**Containers and Materials:** Cans, drums, tank trucks, tank cars (ship under water).

### Physical and Chemical Characteristics

**Physical State** (20°C, 1 atm): Solid.
**Solubility** (Water): 0.0003 g/100 mL (15°C).
**Molecular Weight:** 123.9 ($P_4$).
**Vapour Pressure:** 0.028 mm Hg (21°C).
1 mm Hg (76.6°C).
**Boiling Point:** 280°C.

**Floatability** (Water): Sinks.
**Odour:** Odourless.
**Flash Point:** Ignites spontaneously in air.
**Vapour Density:** 4.4
**Specific Gravity:** 1.82 (20°C).

**Colour:** White to yellow.
**Explosive Limits:** No information.
**Melting Point:** 44°C.

## HAZARD DATA

### Human Health

**Symptoms:** Inhalation: irritation of respiratory tract, cough, weakness, nausea and vomiting, laboured breathing, anemia. Contact: eyes and skin - may cause severe burns. Ingestion: causes nausea, vomiting, jaundice, delirium, coma and death. Symptoms after ingestion may be delayed from a few hours to 3 days.
**Toxicology:** Highly toxic by all routes.
TLV®- (inhalation) 0.1 mg/m³.      $LC_{50}$ - No information.          $LD_{50}$ - Intraperitoneal: rat = 0.1 g/kg
Short-term Inhalation Limits - 0.3 mg/m³    Delayed Toxicity - Severe attack of liver and bones.    $LD_{Lo}$ - Oral: human = 0.0014 g/kg
(15 min).                                   Symptoms may persist from 5 to 15 days.

### Fire

**Fire Extinguishing Agents:** Flood with water. After fire is extinguished cover with wet sand or dirt. Use extreme caution during cleanup, since reignition may occur.
**Behaviour in Fire:** Intense white smoke is formed and highly irritating and toxic $PO_x$ fumes are released.
**Ignition Temperature:** 30°C.     **Burning Rate:** No information.

### Reactivity

**With Water:** No reaction.
**With Common Materials:** Ignites when exposed to air. Reacts vigorously with oxidizing materials. Reacts violently with alkaline hydroxides, halogens, chlorates, iodates, chlorosulfonic acid, nitrates, ammonium nitrate, bromates, bromine, calcium chlorate, cesium, chlorine, chromic acid, chromic anhydride, copper, fluorine, iodine, iron, lead dioxide, lithium, manganese, mercuric oxide, nickel, nitric acid, oxygen, performic acid, platinum, potassium, potassium bromate, potassium chlorate, potassium hydroxide, potassium iodate, potassium permanganate, silver nitrate, sodium, sodium bromate, sodium chlorate, sodium chlorite, sodium hydroxide, sodium iodate, sodium peroxide, sulfur, sulfuric acid, zinc bromate and zinc chlorate.
**Stability:** Ignites spontaneously on contact with air and is very reactive.

### Environment

**Water:** Prevent entry into water intakes and waterways. Aquatic toxicity rating: <1 ppm/96 h/TLm/freshwater; 0.045 ppm/96 h/bluegill/$LC_{50}$/freshwater; 0.025 ppm/96 h/Cod/$LC_{50}$/saltwater.
**Land-Air:** Oral: duck = 3 mg/kg (6 to 33 hours) $LD_{100}$.
**Food Chain Concentration Potential:** None.

## EMERGENCY MEASURES

### Special Hazards

SPONTANEOUSLY COMBUSTIBLE. REACTIVE. Ignites spontaneously when exposed to air.

### Immediate Responses

Keep non-involved people away from spill site. Issue warnings: "FLAMMABLE; POISON". Call Fire Department. Call manufacturer. Stop or reduce discharge, if this can be done without risk. Dike area of spill and cover with water. Notify environmental authorities.

### Protective Clothing and Equipment

In fires or confined spaces, Respiratory protection - self-contained breathing apparatus and totally encapsulated suit. Otherwise, Goggles - (mono), tight fitting. If face shield is used, it must not replace goggles. Gloves - rubber. Boots - high, rubber (pants worn outside boots). Fire-resistant clothing (coveralls, suit, apron, etc.).

### Fire and Explosion

Flood discharge area with water (after diking). When fire is extinguished, cover with wet sand or earth. Cover fires with water spray, dry chemical, sand or earth. Use water to cool fire-exposed containers. Highly irritating and toxic $PO_x$ fumes are released.

### First Aid

Move victim out of spill area to fresh air. Call for medical assistance, but start first aid immediately. Inhalation: if breathing has stopped give artificial respiration; if laboured, give oxygen. Contact: immediately flush skin and eyes with plenty of water while removing contaminated clothing. Keep skin area wet until medical attention is obtained. Ingestion: if victim is conscious, give plenty of milk or water to drink and induce vomiting. If medical assistance is not immediately available, transport victim to hospital, doctor or clinic.

## ENVIRONMENTAL PROTECTION MEASURES

### Response

**Water**
1. Stop or reduce discharge if safe to do so.
2. Contact manufacturer or supplier for advice.
3. If possible, contain discharge by damming or water diversion.
4. Dredge or vacuum pump to remove contaminants, liquids and contaminated bottom sediments
5. Notify environmental authorities to discuss disposal and cleanup of contaminated materials.

**Land-Air**
1. Stop or reduce discharge if safe to do so.
2. Contact manufacturer or supplier for advice.
3. Dike to prevent runoff from rainwater or water application.
4. Remove material by manual or mechanical means. Note: material requires special handling because of its spontaneous ignition in air. Spilled material should be covered at all times with wet sand or dirt.
5. Recover undamaged containers.
6. Notify environmental authorities to discuss disposal and cleanup of contaminated materials.

### Disposal

1. Contact manufacturer or supplier for advice on disposal.
2. Contact environmental authorities for advice on disposal.

PHOSPHORUS, WHITE   $P_4$

# PHOSPHORUS PENTASULFIDE ($P_2S_5)_2$

## IDENTIFICATION

**UN No. 1340**

### Common Synonyms

TETRAPHOSPHORUS DECASULPHIDE
PHOSPHORIC SULPHIDE
PHOSPHORUS PERSULFIDE
THIOPHOSPHORIC ANHYDRIDE
PHOSPHORUS SULPHIDE

### Observable Characteristics

Yellow to green solid. Hydrogen sulfide-like (rotten eggs) odour.

### Manufacturers

No Canadian manufacturer.
Selected U.S. manufacturer:
Hooker Industrial Chemicals Division, Niagara Falls, N.Y., USA.

### Transportation and Storage Information

**Shipping State:** Solid.
**Classification:** Flammable.
**Inert Atmosphere:** No requirement.
**Venting:** Closed.

**Label(s):** Red and white striped label – FLAMMABLE SOLID; Class 4.1, Group II.
**Storage Temperature:** Ambient.

**Grades or Purity:** Regular and reactive, 99+%.
**Containers and Materials:** Cans, fibre or steel drums and tote bins.

### Physical and Chemical Characteristics

**Physical State** (20°C, 1 atm): Solid.
**Solubility (Water):** Reacts evolving $H_2S$ and forming soluble phosphorus compounds.
**Molecular Weight:** ($P_4S_{10}$) 444.5
**Vapour Pressure:** 1 mm Hg (300°C).
**Boiling Point:** 513 to 515°C.

**Floatability (Water):** Sinks and reacts, toxic $H_2S$ gas evolved.
**Odour:** Rotten-eggs (0.00021 to 0.00047 ppm as $H_2S$, odour threshold).
**Flash Point:** Data not available but autoignition temperatures as dust; 260 to 290°C, and liquid; 275°C (approx.).
**Vapour Density:** No information.
**Specific Gravity:** 2.03 (17°C).

**Colour:** Yellow to green.
**Explosive Limits:** Dust can be explosive in certain concentrations. Can be ignited by friction.
**Melting Point:** 275 to 290°C.

## HAZARD DATA

### Human Health

**Symptoms:** Contact of phosphorus pentasulfide with water or moisture produces $H_2S$ which is highly toxic (see hydrogen sulfide). **Inhalation:** dizziness, headache, fatigue, difficulty breathing, irritation of respiratory tract. **Contact:** skin - irritation and burning; eyes - irritation and burning. **Ingestion:** produces symptoms similar to inhalation; however ingestion is extremely serious, as $H_2S$ is formed.
**Toxicology:** Highly toxic by all routes.
TLV - (inhalation) 1 mg/m³.
Short-term Inhalation Limits - 3 mg/m³ (15 min).
$LC_{50}$ - Inhalation: rat = 713 ppm/1 h (as $H_2S$).  $LD_{50}$ - Oral: rat = 0.39 g/kg
Delayed Toxicity - No information.

### Fire

**Fire Extinguishing Agents:** Carbon dioxide, dry chemical, sand.
**Behaviour in Fire:** When heated to decomposition, emits highly toxic $SO_x$ and $PO_x$ fumes. Reacts with water or moisture to produce toxic $H_2S$ and $P_2O_5$ vapours. In contact with moisture, may ignite organic materials. Can ignite spontaneously from heat of reaction with limited amount of moisture.
**Ignition Temperature:** Dust, 260 to 290°C;   **Burning Rate:** No information.
liquid, 275°C.

### Reactivity

**With Water:** Reacts with water exothermically producing $H_2S$ gas and soluble phosphorus compounds (primarily $P_2O_5$). Can ignite spontaneously from heat of reaction with limited amount of moisture.
**With Common Materials:** Reacts with acids in a similar manner to water. Can react vigorously with oxidizing materials.
**Stability:** Stable (within the limits of the foregoing).

### Environment

**Water:** Prevent entry into water intakes and waterways. Fish toxicity: 1.38 ppm/48 h/fathead minnow/TLm/freshwater (as $H_2S$).
**Land-Air:** No information.
**Food Chain Concentration Potential:** No information.

## EMERGENCY MEASURES

**Special Hazards**

FLAMMABLE. CONTACT WITH WATER RELEASES TOXIC AND POISONOUS $H_2S$ AND $P_2O_5$ GASES. MAY IGNITE IN PRESENCE OF MOISTURE.

**Immediate Response**

Keep non-involved people away from spill site. Issue warning: "FLAMMABLE. POISONOUS AND REACTIVE WITH WATER". Call fire department (describe unusual properties - NO WATER). Evacuate area. Contact manufacturer for guidance and assistance. Notify environmental authorities.

**Protective Clothing and Equipment**

Respiratory protection - self-contained breathing apparatus and totally encapsulated suit. Gloves - rubber, neoprene or vinyl. Boots - high, rubber (pants worn outside boots).

**Fire and Explosion**

Use carbon dioxide, dry chemical or sand to extinguish. When heated to decomposition, emits highly toxic fumes of $SO_x$ and $PO_x$. Reacts with water or moisture to produce toxic $H_2S$ and $P_2O_5$ vapours. In contact with moisture, may ignite organic materials.

**First Aid**

Move victim out of spill area to fresh air. Call for medical assistance, but start first aid at once. Contact: eyes and skin - flush affected areas with plenty of water. Inhalation: if breathing has stopped, give artificial respiration (not mouth-to-mouth method); if laboured, give oxygen. Ingestion: give conscious victim water to drink and induce vomiting. If medical assistance is not immediately available, transport victim to hospital, doctor or clinic.

## ENVIRONMENTAL PROTECTION MEASURES

**Response**

**Water**
1. Stop or reduce discharge if safe to do so.
2. Contact manufacturer or supplier for advice.
3. If possible, contain discharge by damming or water diversion.
4. Notify environmental authorities to discuss disposal and cleanup of contaminated materials.

**Land-Air**
1. Stop or reduce discharge if safe to do so.
2. Contact manufacturer or supplier for advice.
3. Dike to prevent runoff from rainwater or water application.
4. Remove materials by manual or mechanical means. Caution must be observed when handling this material.
5. Recover undamaged containers.
6. Notify environmental authorities to discuss disposal and cleanup of contaminated materials.

**Disposal**
1. Contact manufacturer or supplier for advice on disposal.
2. Contact environmental authorities for advice on disposal.

PHOSPHORUS PENTASULFIDE ($P_2S_5$)$_2$

# PHOSPHORUS TRICHLORIDE   PCl₃

## IDENTIFICATION

| Common Synonyms | Observable Characteristics | Manufacturers |
|---|---|---|
| PHOSPHORUS CHLORIDE<br>CHLORIDE OF PHOSPHORUS | Colourless to slightly yellow liquid. Fumes on contact with air. Pungent, irritating, HCl-like odour. | No Canadian manufacturer.<br>Selected U.S. manufacturer: Hooker Industrial Chemicals Division, Niagara Falls, NY, USA |

UN No. 1809

### Transportation and Storage Information

**Shipping State:** Liquid.
**Classification:** Corrosive.
**Inert Atmosphere:** No requirement.
**Venting:** Pressure-vacuum.
**Pump Type:** Gear, centrifugal.
**Note:** PCl₃ is corrosive to steel.

**Label(s):** White and black label - CORROSIVE; Class 8, Group II.
**Storage Temperature:** Ambient.
**Hose Type:** Nickel-lined stainless steel.

**Grades or Purity:** Technical, 98.5% PCl₃ min.
**Containers and Materials:** Drums, tank cars, tank trucks; nickel-lined stainless steel.

### Physical and Chemical Characteristics

**Physical State** (20°C, 1 atm): Liquid.
**Solubility (Water):** Reacts to produce HCl and PH₃ acids.
**Molecular Weight:** 137.3
**Vapour Pressure:** 100 mm Hg (21°C).
**Boiling Point:** 74.2 to 76.1°C.

**Floatability (Water):** Sinks and reacts violently with water to produce HCl and PH₃ acids.
**Odour:** Pungent (like HCl).
**Flash Point:** Not flammable.
**Vapour Density:** 4.8
**Specific Gravity:** 1.57 (20°C).

**Colour:** Colourless to slightly yellow.
**Explosive Limits:** Not flammable.
**Melting Point:** -112°C.

## HAZARD DATA

### Human Health

**Symptoms:** Inhalation: sore throat, coughing, laboured breathing, dizziness, headache. Contact: skin - burning and inflammation; eyes - burning. Ingestion: irritation, nausea, vomiting, burning.
**Toxicology:** Highly toxic by all routes.
TLV - (inhalation) 0.2 ppm; 1.5 mg/m³.
Short-term Inhalation Limits - 0.5 ppm; 3 mg/m³ (15 min).

$LC_{50}$ - Inhalation: rat = 104 ppm/4 h
$LC_{50}$ - Inhalation: guinea pig = 50 ppm/4 h
Delayed Toxicity - No information.

$LD_{50}$ - Oral: rat = 0.55 g/kg

### Fire

**Fire Extinguishing Agents:** Use carbon dioxide or dry chemical.
**Behaviour in Fire:** When heated to decomposition, emits highly toxic chloride and $PO_x$ fumes. Reacts with water or moisture to produce HCl and PH₃.
**Ignition Temperature:** No information.    **Burning Rate:** No information.

### Reactivity

**With Water:** Reacts violently and exothermically with water or moisture producing HCl and PH₃.
**With Common Materials:** Reacts with oxidizing materials. Reacts violently with acetic acid, aluminum, dimethyl sulfoxide, fluorine, hydroxylamine, lead dioxide, nitric acid, nitrous acid, organic matter, potassium and sodium.
**Stability:** Stable (within the limits of the foregoing).

### Environment

**Water:** Prevent entry into water intakes and waterways. Aquatic toxicity rating = 10 to 100 ppm/96 h/TLm/freshwater; BOD: Not available.
**Land-Air:** No information.
**Food Chain Concentration Potential:** None.

## EMERGENCY MEASURES

### Special Hazards

CORROSIVE. REACTIVE WITH WATER.

### Immediate Response

Keep non-involved people away from spill site. Issue warning: "CORROSIVE". Call Fire Department (describe unusual properties - NO WATER). Contact manufacturer for guidance, and assistance if possible. Avoid contact and inhalation. Contain spill area by diking. Stop or reduce discharge if this can be done without risk. Notify environmental authorities.

### Protective Clothing and Equipment

Respiratory protection - self-contained breathing apparatus and totally encapsulated suit. Gloves - rubber. Boots - rubber, high (pants worn outside boots).

### Fire and Explosion

Use carbon dioxide or dry chemical to extinguish. When heated to decomposition, emits highly toxic chloride and $PO_x$ fumes. Reacts with water or moisture to produce HCl and $PH_3$.

### First Aid

Move victim out of spill area to fresh air. Call for medical assistance, but start first aid at once. Contact: skin and eyes - immediately flush with large amounts of water and remove contaminated clothing. Inhalation: if breathing has stopped, give artificial respiration (not mouth-to-mouth method); if laboured, give oxygen. Ingestion: dilute concentration of chemical by giving water or milk to conscious victim to drink. Do not induce vomiting. If medical assistance is not immediately available, transport victim to hospital, doctor or clinic.

## ENVIRONMENTAL PROTECTION MEASURES

### Response

**Water**
1. Stop or reduce discharge if safe to do so.
2. Contact manufacturer or supplier for advice.
3. If possible, contain discharge by damming or water diversion.
4. Dredge or vacuum pump to remove contaminants, liquids and contaminated bottom sediments.
5. Notify environmental authorities to discuss disposal and cleanup of contaminated materials.

**Land-Air**
1. Stop or reduce discharge if safe to do so.
2. Contact manufacturer or supplier for advice.
3. Dike to prevent runoff from rainwater or water application.
4. Remove material by manual or mechanical means. Caution is required when handling materials.
5. Recover undamaged containers.
6. Absorb residual liquid on natural or synthetic sorbents.
7. Notify environmental authorities to discuss disposal and cleanup of contaminated materials.

### Disposal

1. Contact manufacturer or supplier for advice on disposal.
2. Contact environmental authorities for advice on disposal.

PHOSPHORUS TRICHLORIDE  $PCl_3$

# PHTHALIC ANHYDRIDE $C_6H_4(CO)_2O$

## IDENTIFICATION

| Common Synonyms | Observable Characteristics | Manufacturers |
|---|---|---|
| PHTHALANDIONE<br>PHTHALIC ACID ANHYDRIDE<br>PAN<br>1,3 - ISOBENZOFURANDIONE<br>1,2 - BENZENEDICARBOXYLIC ACID ANHYDRIDE | White powder or crystals. Choking odour. | BASF Canada, Cornwall, Ontario. |

**Transportation and Storage Information**

Shipping State: Solid.
Classification: Corrosive.
Inert Atmosphere: No requirement.
Venting: Open (flame arrester).

Label(s): Black and white label - CORROSIVE; Class 8, Group III.
Storage Temperature: Ambient.

Grades or Purity: Commercial, 99.8%.
Containers and Materials: Multiwall paper bags, drums.

**Physical and Chemical Characteristics**

Physical State (20°C, 1 atm): Solid.
Solubility (Water): 1g/100mL (20°C);
Molecular Weight: 148.1
Vapour Pressure: 0.0002 mm Hg (20°C); 0.001 mm Hg (30°C).
Boiling Point: Sublimes 284.5°C.

Floatability (Water): Sinks.
Odour: Choking (0.32 to 0.72 mg/m$^3$, odour threshold).
Flash Point: 152°C (c.c.).
Vapour Density: 5.1
Specific Gravity: 1.53 (20°C).

Colour: White.
Explosive Limits: 1.7 to 10.5%.
Melting Point: 130 to 132°C.

## HAZARD DATA

**Human Health**

Symptoms: Inhalation: sore throat, irritation of nose, eyes and respiratory tract, coughing, difficulty breathing, cyanosis. Contact: skin - itching, irritation, inflammation, brown staining; eyes - stinging, watering, inflammation. Ingestion: burning sensation, stomach cramps, nausea and vomiting, general weakness, dizziness, diarrhea, shock, convulsions, coma.
Toxicology: Moderately toxic by ingestion.
TLV® 1 ppm, 6 mg/m$^3$.
Short-term Inhalation Limits - 4 ppm; 24 mg/m$^3$ (15 min).

LC$_{50}$ - No information.
Delayed Toxicity - No information.

LD$_{50}$ - Oral: rat = 4.02 g/kg

**Fire**

Fire Extinguishing Agents: Use carbon dioxide and/or dry chemical. Water or foam may cause frothing.
Behaviour in Fire: No information. Burning Rate: No information.
Ignition Temperature: 570°C.

**Reactivity**

With Water: Slowly reacts to form benzoic acid.
Stability: Reacts violently with nitric acid.
With Common Materials: Stable.

**Environment**

Water: Prevent entry into potable water intakes. Harmful to aquatic life. Aquatic toxicity rating = 10 to 100 ppm/96 h/TLm/freshwater; Fish toxicity: 756 ppm/96 h/fathead minnow/TLm/freshwater; BOD: 72 to 102%, 5 days.
Land-Air: No information.
Food Chain Concentration Potential: None.

UN No. 2214

## EMERGENCY MEASURES

**Special Hazards**

CORROSIVE.

**Immediate Responses**

Keep non-involved people away from spill site. Issue warning: "CORROSIVE". Call Fire Department. Avoid contact and inhalation. Notify manufacturer or supplier. Dike to contain runoff. Notify environmental authorities.

**Protective Clothing and Equipment**

In fires or confined spaces - Respiratory Protection - self-contained breathing apparatus and totally encapsulated suit. Otherwise, protective outer clothing as required.

**Fire and Explosion**

Use carbon dioxide or dry chemical to extinguish. Water or foam may cause frothing.

**First Aid**

Move victim out of spill site to fresh air. Call for medical assistance, but start first aid at once. <u>Inhalation</u>: if breathing has stopped give artificial respiration; if laboured, give oxygen. <u>Contact</u>: skin - remove contaminated clothing and flush affected areas with plenty of water; eyes - irrigate with plenty of water. <u>Ingestion</u>: give conscious victim plenty of water to drink and induce vomiting. If medical assistance is not immediately available, transport victim to hospital, doctor, or clinic.

## ENVIRONMENTAL PROTECTION MEASURES

**Response**

**Water**
1. Stop or reduce discharge if safe to do so.
2. Contact manufacturer or supplier for advice.
3. If possible, contain discharge by damming or water diversion.
4. Dredge or vacuum pump to remove contaminants, liquids and contaminated bottom sediments.
5. Notify environmental authorities to discuss disposal and cleanup of contaminated materials.

**Land-Air**
1. Stop or reduce discharge if safe to do so.
2. Contact manufacturer or supplier for advice.
3. Dike to prevent runoff from rainwater or water application.
4. Remove material by manual or mechanical means.
5. Recover undamaged containers.
6. Notify environmental authorities to discuss disposal and cleanup of contaminated materials.

**Disposal**
1. Contact manufacturer or supplier for advice on disposal.
2. Contact environmental authorities for advice on disposal.
3. Incinerate (approval of environmental authorities required).

PHTHALIC ANHYDRIDE   $C_6H_4(CO)_2O$

# POLYCHLORINATED BIPHENYLS  $C_{12}H_{10-x}Cl_x$

## IDENTIFICATION

UN No. 2315

### Common Trade Names or Synonyms

PCBs
AROCHLOR
ASKAREL
PYRANOL
INERTEEN

### Observable Characteristics

Clear to pale yellow mobile liquids and some solids. Weak, bitter odour.

### Manufacturers

Previous Canadian supplier: Monsanto Canada Limited, Mississauga, Ont. (PCBs no longer being manufactured)

### Transportation and Storage Information

**Shipping State:** Liquid or solid.
**Classification:** Environmentally dangerous substance.
**Inert Atmosphere:** No requirement.
**Venting:** Open.
**Pump Type:** Most centrifugal types except those with rubber, neoprene, butyl rubber or PVC.

**Label(s):** Black and white label - POISON; Class 9.2.
**Storage Temperature:** Ambient.
**Hose Type:** Most types, except rubber, neoprene, butyl rubber or PVC. These should not be used after coming in contact with PCBs.

**Grades or Purity:** 11 grades (some liquid, some solid) which differ primarily in their chlorine content (20-68% by weight).
**Containers and Materials:** Drums. Electrical equipment.

### Physical and Chemical Characteristics

**Physical State (20°C, 1 atm):** Liquid or solid.
**Solubility (Water):** Low; 0.001 mg/100 mL (freshwater); 0.0004 mg/100 mL (saltwater).
**Molecular Weight:** Variable; 327 (average).
**Vapour Pressure:** 0.006 mm Hg (25°C) (Arochlor 1016); 0.001 mm Hg (0°C); 29 mm Hg (200°C).

**Floatability (Water):** Sinks.
**Odour:** Weak and bitter (odour threshold <0.5 mg/m$^3$).
**Flash Point:** Greater than 140°C.
**Vapour Density:** No information.
**Specific Gravity:** 1.3 to 1.5 (20°C).
**Boiling Point:** 360 to 390°C.

**Colour:** Colourless, to pale yellow.
**Explosive Limits:** Not flammable.
**Melting Point:** No information.

## HAZARD DATA

### Human Health

**Symptoms: Skin Contact and Ingestion:** small pimples, dark pigmentation of exposed areas, followed by blackheads and postules. **Inhalation and Ingestion:** nausea, vomiting, loss of weight, jaundice, edema, abdominal pain, possible coma and death.
**Toxicology:** Severely toxic and irritating. The higher the chlorine content of the compound, the more toxic it is.
TLV® - (skin) 0.5 mg/m$^3$ (54% chlorine)
42% chlorine = 1.0 mg/m$^3$.
Short-term Inhalation Limits - (skin)
2.0 mg/m$^3$ (15 min) (42% Cl); 1.0 mg/m$^3$ (15 min) (54% Cl).

LC$_{50}$ - No information.
TCLo - Inhalation: human = 12 mg/m$^3$
Arochlor 1242 (42% chlorine)
Delayed Toxicity - Possible carcinogen; adverse reproductive effects and atrophy of the liver.

LD$_{50}$ - Oral: rat = 4.47 g/kg Arochlor 1232 (32% chlorine).
LD$_{50}$ - Oral: rat = 4.25 g/kg Arochlor 1242 (42% chlorine).
LD$_{50}$ - Oral: rat = 1.32 g/kg Arochlor 1260 (60% chlorine).

### Fire

**Fire Extinguishing Agents:** Water, foam, dry chemical or carbon dioxide. Use water only if safe to do so.
**Behaviour in Fire:** Highly toxic, irritating gases (chlorides and chlorine) can be evolved.
**Ignition Temperature:** ~1000°C.
**Burning Rate:** No information.

### Reactivity

**With Water:** No reaction.
**With Common Materials:** Reacts with liquid chlorine.
**Stability:** Stable.

### Environment

**Water:** Prevent entry into water intakes and waterways. Harmful to aquatic life in very low concentrations. Fish toxicity: 0.278 mg/L/96 h/bluegill/ TLm/freshwater; 0.005 ppm/336-1 080 h/pinfish/TLm/saltwater; EPA Criterion: 0.0015 µg/L/24 h av./conc. should not exceed 6.2 µg/L (freshwater); 0.024 µg/L/24 h/conc. should not exceed 0.2 µg/L (saltwater); BOD: Very low.
**Land-Air:** LD$_{50}$: 2 000 ppm (mallard duck).
**Food Chain Concentration Potential:** Known to accumulate in fatty tissue.

## EMERGENCY MEASURES

### Special Hazards
POISON.

### Immediate Responses
Keep non-involved people away from spill site. Issue warning: "POISON". Call Fire Department. Avoid contact and inhalation. Contact supplier. Stop or reduce discharge if this can be done without risk. Contain any spillage. Notify environmental authorities.

### Protective Clothing and Equipment
In fires, Respiratory protection - self-contained breathing apparatus and totally encapsulated suit. Goggles, Boots - polyethylene, neoprene, Viton. Note: contaminated clothing should be disposed of in same manner as PCBs.

### Fire and Explosion
Use water, foam, dry chemical or carbon dioxide to extinguish fires. Use water only if safe to do so. In fire, toxic gases (chlorides and chlorine) can be evolved.

### First Aid
Move victim out of spill area to fresh air. Call for doctor, but start first aid at once. Contact - using appropriate gloves, remove contaminated clothing. Clothing should be disposed of in the same manner as PCBs. Wash eyes and affected skin thoroughly with plenty of water. Ingestion: have conscious victim rinse mouth with water. Keep victim warm and quiet. If medical assistance is not immediately available, transport victim to hospital, doctor or clinic.

## ENVIRONMENTAL PROTECTION MEASURES

### Response

**Water**
1. Notify environmental authorities immediately.
2. Stop or reduce discharge if safe to do so.
3. Contact manufacturer or supplier for advice.
4. If possible, contain discharge by damming or water diversion.
5. Dredge or vacuum pump to remove contaminants, liquids and contaminated bottom sediments.
6. Notify environmental authorities to discuss disposal and cleanup of contaminated materials.

**Land-Air**
1. Notify environmental authorities immediately.
2. Stop or reduce discharge if safe to do so.
3. Contact manufacturer or supplier for advice.
4. Contain spill by diking with earth or other barrier.
5. Remove material with pumps or vacuum equipment and place in appropriate containers.
6. Remove material by manual or mechanical means.
7. Absorb residual liquid on natural or synthetic sorbents.
8. Remove contaminated soil for disposal.
9. Notify environmental authorities to discuss disposal and cleanup of contaminated materials.

### Disposal
1. Contact manufacturer or supplier for advice on disposal.
2. Contact environmental authorities for advice on disposal.

POLYCHLORINATED BIPHENYLS $C_{12}H_{10-x}Cl_x$

# POTASH (POTASSIUM CHLORIDE)  KCl

## IDENTIFICATION

### Common Synonyms
POTASSIUM MURIATE
MURIATE OF POTASH

### Observable Characteristics
Colourless to whitish-pink crystals. Odourless.

### Manufacturers
International Minerals and Chemicals, Esterhazy, Saskatchewan.
Potash Corporation of Saskatchewan, Rocanville, Saskatchewan.
Kalium Chemicals, Belle Plaine, Saskatchewan.
Allan Potash Mines, Saskatoon, Saskatchewan.

### Transportation and Storage Information
**Shipping State:** Solid
**Classification:** None.
**Inert Atmosphere:** No requirement.
**Venting:** Open.
**Label(s):** Not regulated.
**Storage Temperature:** Ambient.
**Grades or Purity:** Industrial.
**Containers and Materials:** Multiwall paper bags. Bulk by truck or train; steel.

### Physical and Chemical Characteristics
**Physical State (20°C, 1 atm):** Solid.
**Solubility (Water):** 23.8 g/100 mL (20°C); 56.7 g/100 mL (100°C).
**Molecular Weight:** 74.6
**Vapour Pressure:** No information.
**Boiling Point:** 1420 to 1500°C (sublimes).
**Floatability (Water):** Sinks and mixes.
**Odour:** Odourless.
**Flash Point:** Not flammable.
**Vapour Density:** No information.
**Specific Gravity:** 1.99 (20°C).
**Colour:** Colourless to whitish-pink.
**Explosive Limits:** Not flammable.
**Melting Point:** 770°C.

## HAZARD DATA

### Human Health
**Symptoms:** Inhalation: slight irritation of nose, sneezing. Ingestion: nausea and vomiting. Contact: skin - irritation and inflammation; eyes - irritation, inflammation and watering.
**Toxicology:** Low toxicity by all routes.
**TLV®** No information.
**Short-term Inhalation Limits** - No information.
**LC$_{50}$** - No information.
**Delayed Toxicity** - No information.
**LD$_{50}$ - Oral:** mouse = 0.383 g/kg
**LD$_{50}$ - Oral:** guinea pig = 2.5 g/kg
**LD$_{50}$ - Oral:** Infant = 0.938 g/kg

### Fire
**Fire Extinguishing Agents:** Not combustible. Most fire extinguishing agents may be used on fires involving potash.
**Behaviour in Fire:** Not combustible.
**Ignition Temperature:** Not combustible.
**Burning Rate:** Not combustible.

### Reactivity
**With Water:** No reaction; soluble.
**Stability:** Reacts violently with bromine trifluoride and (potassium permanganate and sulfuric acid).
**With Common Materials:** Stable.

### Environment
**Water:** Prevent entry into water intakes and waterways.
**Land-Air:** No information.
**Food Chain Concentration Potential:** No information.

## EMERGENCY MEASURES

**Special Hazards**

**Immediate Responses**

Keep non-involved people away from spill site. Notify manufacturer or supplier. Dike to prevent runoff from rainwater or water application. Notify environmental authorities.

**Protective Clothing and Equipment**

Outer protective clothing as required.

**Fire and Explosion**

Not combustible. Most fire extinguishing agents may be used on fires involving potash.

**First Aid**

Move victim out of spill area to fresh air. Call for medical assistance, but start first aid at once. Inhalation: if breathing has stopped, give artificial respiration; if laboured, give oxygen. Contact: skin - remove contaminated clothing and flush affected areas with plenty of water; eyes - irrigate with plenty of water. Ingestion: give conscious victim plenty of water to drink and induce vomiting. If medical assistance is not immediately available, transport victim to hospital, doctor or clinic.

## ENVIRONMENTAL PROTECTION MEASURES

**Response**

**Water**
1. Stop or reduce discharge if safe to do so.
2. Contact manufacturer or supplier for advice.
3. If possible, contain discharge by damming or water diversion.
4. Dredge or vacuum pump to remove contaminants, liquids and contaminated bottom sediments.
5. Notify environmental authorities to discuss disposal and cleanup of contaminated materials.

**Land-Air**
1. Stop or reduce discharge if safe to do so.
2. Contact manufacturer or supplier for advice.
3. Dike to prevent runoff from rainwater or water application.
4. Remove material by manual or mechanical means.
5. Recover undamaged containers.
6. Notify environmental authorities to discuss disposal and cleanup of contaminated materials.

**Disposal**
1. Contact manufacturer or supplier for advice on disposal.
2. Contact environmental authorities for advice on disposal.

POTASH (POTASSIUM CHLORIDE)   KCl

# POTASSIUM CARBONATE $K_2CO_3$

## IDENTIFICATION

| Common Synonyms | Observable Characteristics | Manufacturers |
|---|---|---|
| SALT OF TARTAR<br>PEARL ASH<br>POTASH (laboratory usage) | White powder. Odourless. | |

### Transportation and Storage Information

**Shipping State:** Solid.
**Classification:** None.
**Inert Atmosphere:** No requirement.
**Venting:** Open.

**Label(s):** Not regulated.
**Storage Temperature:** Ambient.

**Grades or Purity:** Calcined 80 to 85%, 85 to 90%, 90 to 95%, 96 to 99%; hydrated 80 to 85%.
**Containers and Materials:** Multiwall paper bags, kegs, drums; bulk by train.

### Physical and Chemical Characteristics

**Physical State (20°C, 1 atm):** Solid.
**Solubility (Water):** 112 g/100 mL (20°C); 156 g/100 mL (100°C).
**Molecular Weight:** 138.2
**Vapour Pressure:** No information.
**Boiling Point:** Decomposes, 891°C.

**Floatability (Water):** Sinks and mixes.
**Odour:** Odourless.
**Flash Point:** Not flammable.
**Vapour Density:** No information.
**Specific Gravity:** 2.43 (20°C).

**Colour:** White.
**Explosive Limits:** Not flammable.
**Melting Point:** 891°C.

## HAZARD DATA

### Human Health

**Symptoms:** Inhalation: irritation of nose, eyes and throat, sneezing, difficulty breathing, coughing; Contact: skin - itching, burning and inflammation; eyes - pain, watering. Ingestion: burning sensation, pain, stomach cramps, nausea and vomiting.
**Toxicology:** Moderately toxic by ingestion. Low toxicity by inhalation and contact.    $LD_{50}$ - Oral: rat = 1.87 g/kg
**TLV®** No information.                                    $LC_{50}$ - No information.
**Short-term Inhalation Limits** - No information.          **Delayed Toxicity** - No information.

### Fire

**Fire Extinguishing Agents:** Not combustible. Most fire extinguishing agents may be used on fires involving potassium carbonate.
**Behaviour in Fire:** Not combustible.
**Ignition Temperature:** Not combustible.    **Burning Rate:** Not combustible.

### Reactivity

**With Water:** No reaction; soluble.
**With Common Materials:** Reacts violently with chlorine trifluoride.
**Stability:** Stable.

### Environment

**Water:** Prevent entry into water intakes and waterways.
**Land-Air:** No information.
**Food Chain Concentration Potential:** No information.

## EMERGENCY MEASURES

**Special Hazards**
CORROSIVE.

**Immediate Responses**
Keep non-involved people away from spill site. Call Fire Department. Avoid contact and inhalation. Notify manufacturer or supplier. Dike to prevent runoff. Notify environmental authorities.

**Protective Clothing and Equipment**
Protective outer clothing as required. <u>Boots</u> - high rubber.

**Fire and Explosion**
Not combustible. Most fire extinguishing agents may be used on fires involving potassium carbonate.

**First Aid**
Move victim out of spill site to fresh air. Call for medical assistance, but start first aid at once. <u>Inhalation</u>: if breathing has stopped give artificial respiration; if laboured, give oxygen. <u>Contact</u>: skin - remove contaminated clothing and flush affected areas with plenty of water; eyes - irrigate with plenty of water. <u>Ingestion</u>: give water to conscious victim to drink, do not induce vomiting. If medical assistance is not immediately available, transport victim to hospital, doctor or clinic.

## ENVIRONMENTAL PROTECTION MEASURES

**Response**

**Water**
1. Stop or reduce discharge if safe to do so.
2. Contact manufacturer for advice.
3. If possible, contain discharge by damming or water diversion.
4. Dredge or vacuum pump to remove contaminants, liquids and contaminated bottom sediments.
5. Notify environmental authorities to discuss disposal and cleanup of contaminated materials.

**Land-Air**
1. Stop or reduce discharge if safe to do so.
2. Contact manufacturer or supplier for advice.
3. Dike to prevent runoff from rainwater or water application.
4. Remove material by manual or mechanical means.
5. Recover undamaged containers.
6. Notify environmental authorities to discuss disposal and cleanup of contaminated materials.

**Disposal**
1. Contact manufacturer or supplier for advice on disposal.
2. Contact environmental authorities for advice on disposal.

POTASSIUM CARBONATE $K_2CO_3$

# POTASSIUM HYDROXIDE   KOH

## IDENTIFICATION

UN No. 1813 solid
       1814 liquid

| Common Synonyms | Observable Characteristics | Manufacturers |
|---|---|---|
| CAUSTIC POTASH<br>POTASSIUM HYDRATE<br>LYE<br>POTASSA | 45% KOH solution: Clear, colourless liquid. Odourless. Solid KOH: White solid. Odourless. | Canadian manufacturer:<br>C-I-L Cornwall, Ontario.<br>Canadian supplier:<br>C-I-L Toronto, Ontario.<br>Canada Colours and Chemicals Ltd., Toronto, Ontario.<br>Hooker Chemical, Calgary, Alta.<br>Canadian Occidental Petroleum, Calgary, Alberta.<br><br>Originating from:<br>International Minerals and Chemical, USA<br>Hooker Chemical, USA |

### Transportation and Storage Information

**Shipping State:** Solid, liquid (aqueous solution). **Classification:** Corrosive. **Inert Atmosphere:** No requirement. **Venting:** Open. **Pump Type:** Gear, centrifugal; steel, stainless steel (solutions).

**Label(s):** White and Black label - CORROSIVE, Class 8, Group II. **Storage Temperature:** Ambient. **Hose Type:** Natural rubber, polyethylene, polypropylene, Chemiflex 951 (solutions).

**Grades or Purity:** Solution, 45% KOH, solid, 88 to 92% KOH (ground, flakes, pellets, sticks, lumps). **Containers and Materials:** Solutions - drums, tank cars, tank trucks; steel. Solid - cans and drums (steel).

### Physical and Chemical Characteristics

**Physical State** (20°C, 1 atm): Solid. **Solubility** (Water): 107 g/100 mL (15°C); 178 g/100 mL (100°C). **Molecular Weight:** 56.1 **Vapour Pressure:** 1 mm Hg (719°C). **Boiling Point:** 133°C, (45% KOH); 1320°C (Solid KOH).

**Floatability** (Water): Mixes readily and sinks. **Odour:** Odourless. **Flash Point:** Not flammable. **Vapour Density:** No information. **Specific Gravity:** 1.46, 45% liquid KOH (15.5°C); 2.04, solid (20°C).

**Colour:** Liquid - colourless, clear. Solid - white. **Explosive Limits:** Not flammable. **Melting Point:** 45% solution KOH, -30°C; Solid, 360-380°C (depending on water content).

## HAZARD DATA

### Human Health

**Symptoms:** Contact with solutions can result in severe burns. KOH is a strong irritant. Contact: skin - burns; eyes - burns, severe damage. Inhalation: (of dust or mist) may damage upper respiratory tract and even lung tissue. Ingestion: can result in severe damage to mucous membranes, nausea, vomiting and diarrhea. **Toxicology:** Highly toxic by ingestion and inhalation.
TLV® (inhalation) 2 mg/m³.   $LC_{50}$ - No information.   $LD_{50}$ - Oral: rat = 0.365 g/kg
Short-term Inhalation Limits - No information.   Delayed Toxicity - No information.

### Fire

**Fire Extinguishing Agents:** Not combustible. Most fire extinguishing agents may be used on fires involving KOH. Use water only in flooding amounts. **Behaviour in Fire:** Not combustible. Solid KOH in contact with water may generate sufficient heat to ignite combustible material. May cause ignition on contact with organic chemicals.
**Ignition Temperature:** Not combustible.   **Burning Rate:** Not combustible.

### Reactivity

**With Water:** Dissolves exothermically.
**With Common Materials:** When wet may attack aluminum, tin, lead and zinc to generate $H_2$ gas. Reacts violently with acetic acid, acrolein, acrylonitrile, 1,2-dichloroethylene, maleic anhydride, nitroethane, nitromethane, nitropropane, nitrophenol, N-nitrosomethylurea, phosphorus, tetrachloroethane, tetrahydrofuran and trichloroethylene.
**Stability:** Stable.

### Environment

**Water:** Prevent entry into water intakes and waterways. Fish toxicity: 80 ppm/24 h/mosquito fish/TLm/freshwater; Aquatic toxicity rating = 10 to 100 ppm/96 h/TLm/freshwater; BOD: None.   **Food Chain Concentration Potential:** None.
**Land-Air:** No information.

## EMERGENCY MEASURES

**Special Hazards**

CORROSIVE.

**Immediate Responses**

Keep non-involved people away from spill site. Issue warning; "CORROSIVE". Call Fire Department. Contain spill by diking with earth or other available material. Contact manufacturer for guidance. Avoid contact and inhalation. Stop or reduce discharge, if this can be done without risk. Notify environmental authorities.

**Protective Clothing and Equipment**

Respiratory protection - filter or dust type respirators are normally sufficient protection against dusts and mists. Goggles - (mono), tight fitting. If face shield used, it must not replace goggles. Gloves - rubber. Boots - high, rubber (pants worn outside boots). Outerwear - as required, coveralls, aprons.

**Fire and Explosion**

Most fire extinguishing agents may be used. Water should be used only in flooding amounts.

**First Aid**

Move victim out of spill area to fresh air. Call for medical assistance, but start first aid at once. Inhalation: if breathing has stopped, give artificial respiration; if laboured, give oxygen. Contact: eyes - irrigate with plenty of water; flush with plenty of water while removing contaminated clothing. Ingestion: give large quantities of milk or water to conscious victim to drink. Do not induce vomiting. If medical assistance is not immediately available, transport victim to hospital, doctor or clinic.

## ENVIRONMENTAL PROTECTION MEASURES

**Response**

**Water**
1. Stop or reduce discharge if safe to do so.
2. Contact manufacturer or supplier for advice.
3. If possible, contain discharge by damming or water diversion.
4. If possible, dredge or vacuum pump to remove contaminants, liquid and contaminated bottom sediments.
5. Notify environmental authorities to discuss disposal and cleanup of contaminated materials.

**Land-Air**
1. Stop or reduce discharge if safe to do so.
2. Contact manufacturer or supplier for advice.
3. Dike to prevent runoff from rainwater or water application.
4. Remove material by manual or mechanical means.
5. Recover undamaged containers.
6. Notify environmental authorities to discuss disposal and cleanup of contaminated materials.

**Disposal**
1. Contact manufacturer or supplier for advice on disposal.
2. Contact environmental authorities for advice on disposal.

POTASSIUM HYDROXIDE   KOH

# POTASSIUM SULFATE  $K_2SO_4$

## IDENTIFICATION

### Common Synonyms

### Observable Characteristics
White powder or crystals; odourless.

### Manufacturers
Potash Corp. of Saskatchewan, Cory, Sask. Potcal, Parry Sound, Ontario. Shamrock Chemicals, Port Stanley, Ontario.

### Transportation and Storage Information
**Shipping State:** Solid.
**Classification:** None.
**Inert Atmosphere:** No requirements.
**Venting:** Open.

**Label(s):** Not regulated.
**Storage Temperature:** Ambient.

**Grades or Purity:** Commercial; technical.
**Containers and Materials:** Bags, drums. Bulk by truck or rail.

### Physical and Chemical Characteristics
**Physical State (20°C, 1 atm):** Solid.
**Solubility (Water):** 8.5 g/100 mL $H_2O$ (10°C) 10.0 g/100 mL (20°C); 14.2 g/100 mL (50°C).
**Molecular Weight:** 174.3
**Vapour Pressure:** No information.
**Boiling Point:** 1 689°C.

**Floatability (Water):** Sinks and slowly mixes.
**Odour:** Odourless.
**Flash Point:** Not flammable.
**Vapour Density:** No information.
**Specific Gravity:** 2.7

**Colour:** White.
**Explosive Limits:** Not flammable.
**Melting Point:** 1 069 to 1 072°C, undergoes transition at 588°C.

## HAZARD DATA

### Human Health
**Symptoms:** Inhalation: irritation of nose and eyes. Ingestion: stomach pains; Contact: skin – irritation, eyes – irritation.
**Toxicology:** Low toxicity by all routes.
**TLV®** No information.
**Short-term Inhalation Limits** – No information.

$LC_{50}$ – No information.
Delayed Toxicity – No information.

$LD_{Lo}$ – Oral: human = 0.80 g/kg
$LD_{Lo}$ – Scutaneous: guinea pig = 3.0 g/kg

### Fire
**Fire Extinguishing Agents:** Not combustible; most fire extinguishing agents may be used in fires involving potassium sulfate.
**Behaviour in Fire:** Not combustible.
**Ignition Temperature:** Not combustible.

**Burning Rate:** Not combustible.

### Reactivity
**With Water:** No reaction; slightly soluble.
**With Common Materials:** May react with finely-divided aluminum under certain circumstances.
**Stability:** Stable.

### Environment
**Water:** Prevent entry into water intakes or waterways.
**Land-Air:** No information.
**Food Chain Concentration Potential:** None.

## EMERGENCY MEASURES

### Special Hazards

### Immediate Responses
Keep non-involved people away from spill site. Stop or reduce discharge if this can be done without risk. Notify supplier and environmental authorities.

### Protective Clothing and Equipment
Protective outerwear as required.

### Fire and Explosion
Not combustible. Most fire extinguishing agents may be used in fires involving potassium sulfate.

### First Aid
Move victim out of spill area to fresh air. Call for medical assistance, but start first aid at once. <u>Contact</u>: skin and eyes - remove contaminated clothing and flush affected areas with plenty of water. <u>Ingestion</u>: give conscious victim plenty of water to <u>drink</u>. If medical assistance is not immediately available, transport victim to hospital, doctor or clinic.

## ENVIRONMENTAL PROTECTION MEASURES

### Response

**Water**
1. Stop or reduce discharge if safe to do so.
2. Contact manufacturer or supplier for advice.
3. If possible, contain discharge by damming or water diversion.
4. Dredge or vacuum pump to remove contaminants, contaminated liquids and contaminated bottom sediments.
5. Notify environmental authorities to discuss disposal and cleanup of contaminated materials.

**Land-Air**
1. Stop or reduce discharge if safe to do so.
2. Contact manufacturer or supplier for advice.
3. Contain spill by diking with earth or other material.
4. Remove material by manual or mechanical means.
5. Recover undamaged containers.
6. Notify environmental authorities to discuss disposal and cleanup of contaminated materials.

### Disposal
1. Contact manufacturer or supplier for advice on disposal.
2. Contact environmental authorities for advice on disposal.
3. Material may be buried in a sanitary landfill (environmental authorities' approval required).

POTASSIUM SULFATE $K_2SO_4$

# PROPANE  CH₃CH₂CH₃

## IDENTIFICATION

| Common Synonyms | Observable Characteristics | UN No. 1978 Manufacturers |
|---|---|---|
| LPG (see also Butane) DIMETHYLMETHANE | Colourless gas. Odourless when pure. | Superior Propane Ltd., Toronto, Ont. Consumers Co-op Refineries Ltd., Regina, Sask. Dome Petroleum Ltd., Calgary, Alta. Goliad Oil Canada Ltd., Calgary, Alta. Mobil Oil Canada Ltd., Calgary, Alta. Pacific Petroleums Ltd., Calgary, Alta. Home Oil Ltd., Calgary, Alta. |

### Transportation and Storage Information

**Shipping State:** Liquid (compressed gas).
**Classification:** Flammable gas.
**Inert Atmosphere:** No requirement.
**Venting:** Safety relief.
**Pump Type:** Rotary LPG.

**Label(s):** Red label - FLAMMABLE GAS.
**Storage Temperature:** Ambient.
**Hose Type:** LPG type; reinforced high pressure.

**Grades or Purity:** Commercial, technical 97.5%.
**Containers and Materials:** Cylinders tank cars, tank trucks; steel.

### Physical and Chemical Characteristics

**Physical State** (20°C, 1 atm): Gas.
**Solubility** (Water): Slight, 0.012 g/100 mL (17.8°C).
**Molecular Weight:** 44.1
**Vapour Pressure:** 400 mm Hg (-56°C); 6 400 mm Hg (21°C).
**Boiling Point:** -42.1°C.

**Floatability** (Water): Liquefied propane floats and boils.
**Odour:** Odourless when pure, added mercaptans give 500 to 20 000 ppm, odour threshold, natural gas odour.
**Flash Point:** -104°C (c.c.).
**Vapour Density:** 1.5 (20°C).
**Specific Gravity:** (liquid) 0.58 (-44°C).

**Colour:** Colourless.
**Explosive Limits** 2.1 to 9.5%.
**Melting Point:** -187.7° to 189.9°C.

## HAZARD DATA

### Human Health

**Symptoms:** Inhalation: asphyxiant; rapid irregular breathing, headache, fatigue, nausea and vomiting, convulsions, loss of consciousness; Contact: skin and eyes - with propane liquified causes frostbite, burning sensation.
**Toxicology:** An asphyxiant. Low toxicity.
TLV® (inhalation) Asphyxiant.
Short-term Inhalation Limits - No information.

$LC_{50}$ - Human: no effect 10 000 ppm brief exposures; slight dizziness in a few minutes at 100 000 ppm.
Delayed Toxicity - No information.

$LD_{50}$ - No information.

### Fire

**Fire Extinguishing Agents:** Stop or reduce discharge if safe to do so. Do not attempt to extinguish fire until leak has been shut off. Use water spray to cool tanks exposed to fire.
**Behaviour in Fire:** When exposed to heat and flame, containers may rupture. Flash back may occur along vapour trail.
**Ignition Temperature:** 432°C.
**Burning Rate:** 8.2 mm/min.

### Reactivity

**With Water:** No reaction.
**With Common Materials:** Reacts vigorously with strong oxidizing agents. Reacts violently with chlorine dioxide.
**Stability:** Stable.

### Environment

**Water:** Prevent entry into water intakes and waterways. Aquatic toxicity rating > 1 000 ppm/96 h/TLm/freshwater.
**Land-Air:** No information.
**Food Chain Concentration Potential:** None.

## EMERGENCY MEASURES

**Special Hazards**
FLAMMABLE.

**Immediate Responses**
Keep non-involved people away from spill site. Issue warning: "FLAMMABLE". CALL FIRE DEPARTMENT. Eliminate all ignition sources. Contact supplier or manufacturer. Flash back may occur along vapour trail. Do not attempt to extinguish fire until leak has been shut off. Contact environmental authorities.

**Protective Clothing and Equipment**
In fire or confined spaces - Respiratory protection - self-contained breathing apparatus. Goggles - (mono), tight fitting should be worn, to protect from liquid propane, which could cause eye injury from frostbite burn.

**Fire and Explosion**
Stop or reduce discharge if safe to do so. Do not attempt to extinguish fire until leak has been controlled. Let fire burn. Use water spray to cool fire-exposed containers.

**First Aid**
Move victim out of spill area to fresh air. Call for medical assistance, but start first aid at once. Inhalation: if breathing has stopped give artificial respiration; if laboured, give oxygen. Contact: eyes and skin-if exposed to liquid propane, immediately remove contaminated clothing, irrigate eyes and flush skin with water. Treat as for frostbite. Do not rub affected areas. If medical assistance is not immediately available, transport victim to hospital, doctor or clinic.

## ENVIRONMENTAL PROTECTION MEASURES

**Response**

**Water**
1. Stop or reduce discharge if safe to do so.
2. Contact manufacturer or supplier for advice.
3. Notify environmental authorities to discuss disposal and cleanup of contaminated materials.

**Land-Air**
1. Stop or reduce discharge if safe to do so.
2. Contact manufacturer or supplier for advice.
3. Recover undamaged containers.
4. Notify environmental authorities to discuss disposal and cleanup of contaminated materials.

**Disposal**
1. Contact manufacturer or supplier for advice on disposal.
2. Contact environmental authorities for advice on disposal.
3. Burn or flare at spill site (under knowledgeable supervision).

PROPANE  $CH_3CH_2CH_3$

# PROPYLENE  CH$_3$CH:CH$_2$

## IDENTIFICATION

| Common Synonyms | Observable Characteristics | UN No. 1077 Manufacturers |
|---|---|---|
| PROPENE<br>METHYLETHYLENE<br>METHYLETHENE | Colourless gas. Mild, aromatic odour. | Petrosar Limited, Corunna, Ont.<br>Esso Chemical Canada, Sarnia, Ont.<br>Petromont, Varennes, Que. |

### Transportation and Storage Information

**Shipping State:** Liquid (compressed gas).
**Classification:** Flammable gas.
**Inert Atmosphere:** No requirement.
**Venting:** Safety relief.
**Pump Type:** Rotary LPG.
**Label(s):** Red label – FLAMMABLE GAS.
**Storage Temperature:** Ambient.
**Hose Type:** LPG types reinforced high pressure.
**Grades or Purity:** Technical min. 95%; Pure min. 99%.
**Containers and Materials:** Cylinders, tank cars, tank trucks; steel.

### Physical and Chemical Characteristics

**Physical State (20°C, 1 atm):** Gas.
**Solubility (Water):** 0.02 g/100 mL (20°C).
**Molecular Weight:** 42.1
**Vapour Pressure:** 7600 mm Hg (20°C); 400 mm Hg (-61°C); 3800 mm Hg (-5°C).
**Boiling Point:** -47.7°C.
**Floatability (Water):** Liquid floats and boils.
**Odour:** Mild, aromatic (22.5 to 67.6 ppm, odour threshold).
**Flash Point:** -108°C (c.c.).
**Vapour Density:** 1.5 (20°C).
**Specific Gravity:** (liquid) 0.51 (-48°C).
**Colour:** Colourless.
**Explosive Limits:** 2.0 to 11.1%.
**Melting Point:** -185.2°C.

## HAZARD DATA

### Human Health

**Symptoms:** Inhalation: asphyxiant, headache, fatigue, mental confusion, nausea and vomiting, convulsions, loss of consciousness. <u>Contact</u>: skin and eyes – with liquified, may cause frostbite.
**Toxicology:** Low toxicity. Asphyxiant.
**TLV®** Asphyxiant.
**Short-term Inhalation Limits** – No information.
**LC$_{50}$** – No information.
**Delayed Toxicity** – No information.
**LD$_{50}$** – No information.

### Fire

**Fire Extinguishing Agents:** Stop or reduce discharge if safe to do so. Do not attempt to extinguish fire until leak has been shut off. Use water to cool fire-exposed containers.
**Behaviour in Fire:** Exposed containers may rupture. Flash back may occur along vapour trail.
**Ignition Temperature:** 455°C.
**Burning Rate:** 8 mm/min (liquid).

### Reactivity

**With Water:** No reaction.
**With Common Materials:** Can react vigorously with oxidizing materials. Reacts violently with nitrogen dioxide, nitrogen tetroxide, nitrous oxide.
**Stability:** Stable.

### Environment

**Water:** Prevent entry into water intakes and waterways. Aquatic toxicity rating >1 000 ppm/96h/TLm/freshwater;
**Land-Air:** Sweet pea: declination in seedlings, 1000 ppm, 3 days; tomato: 50 ppm, 2 days growth deformation of petiole.
**Food Chain Concentration Potential:** None.

## EMERGENCY MEASURES

**Special Hazards**

FLAMMABLE.

**Immediate Responses**

Keep non-involved people away from spill site. Issue warning: "FLAMMABLE". CALL FIRE DEPARTMENT. Eliminate all sources of ignition. Contact supplier or manufacturer for assistance. Flash back may occur along vapour trail. Control of discharge should be attempted only by experienced persons. Do not attempt to extinguish fire until leak has been shut off. Notify environmental authorities.

**Protective Clothing and Equipment**

In fires or confined areas; Respiratory protection - self-contained breathing apparatus. Otherwise, Goggles - (mono), tight fitting, to protect against liquid (and frost burn). Gloves - rubber. Outerwear - as required.

**Fire and Explosion**

Stop or reduce discharge if safe to do so. Use water spray to cool fire-exposed containers. Stop flow of gas before attempting to extinguish fire.

**First Aid**

Move victim out of spill area to fresh air. Call for medical assistance, but start first aid at once. Inhalation: if breathing has stopped, give artificial respiration; if laboured, give oxygen. Contact: skin and eyes (liquid) remove contaminated clothing, irrigate eyes and flush skin with water; treat as for frostbite. Do not rub affected areas. If medical assistance is not immediately available, transport victim to hospital, doctor or clinic.

## ENVIRONMENTAL PROTECTION MEASURES

**Response**

**Water**
1. Stop or reduce discharge if safe to do so.
2. Contact manufacturer or supplier for advice.
3. Notify environmental authorities to discuss disposal and cleanup of contaminated materials.

**Land-Air**
1. Stop or reduce discharge if safe to do so.
2. Contact manufacturer or supplier for advice.
3. Recover undammaged containers.
4. Notify environmental authorities to discuss disposal and cleanup of contaminated materials.

**Disposal**
1. Contact manufacturer or supplier for advice on disposal.
2. Contact environmental authorities for advice on disposal.
3. Burn or flare at spill site (under knowledgeable supervision).

PROPYLENE   $CH_3CH{:}CH_2$

# PROPYLENE GLYCOL  $CH_3CH(OH)CH_2OH$

## IDENTIFICATION

### Common Synonyms
1,2 PROPYLENE GLYCOL
- METHYLENE GLYCOL
- MONOPROPYLENE GLYCOL
- α-PROPYLENE GLYCOL
- METHYL GLYCOL
- TRIMETHYL GLYCOL

1,3 PROPYLENE GLYCOL
- 2-DIOXYGLYCOL
- TRIMETHYLENE GLYCOL
- β-PROPYLENE GLYCOL
- 1,3-HYDROXYPROPANE

### Observable Characteristics
Colourless liquids.
Practically Odourless.

### Manufacturers
Dow Chemical Canada Inc., Sarnia, Ontario.
Suppliers:
Hall Chemical, Montreal, Quebec.

Originating from: Oxirane, USA

### Transportation and Storage Information
**Shipping State:** Liquid
**Classification:** None.
**Inert Atmosphere:** No requirement.
**Venting:** Open (flame arrester).
**Pump Type:** Centrifugal or positive displacement; steel or stainless steel.

**Label(s):** Not regulated.
**Storage Temperature:** Ambient.
**Hose Type:** Seamless stainless steel.

**Grades or Purity:** Technical, 95+%; pure, 99%.
**Containers and Materials:** Drums, tank cars, tank trucks; steel.

### Physical and Chemical Characteristics
**Physical State (20°C, 1 atm):** Liquid.
**Solubility (Water):** Soluble in all proportions.
**Molecular Weight:** 76.1
**Vapour Pressure:** (1,2) 0.2 mm Hg (20°C).
**Boiling Point:** (1,2) 188.2°C; (1,3) 210 to 211°C.

**Floatability (Water):** Sinks in freshwater; may float in saltwater.
**Odour:** Odourless (practically).
**Flash Point:** (1,2) 99°C (c.c.).
**Vapour Density:** (1,3) (1,2) 2.5
**Specific Gravity:** 1.04 (1,2); 1.05 (1,3) (20°C).

**Colour:** Colourless.
**Explosive Limits:** (1,2) 2.6 to 12.5%.
**Melting Point:** (1,2) -85°C.

## HAZARD DATA

### Human Health
**Symptoms:** Inhalation: intoxication, nose and eye irritation, staggering, headache, mental confusion, nausea and vomiting, drowsiness, stupor. Ingestion: gastrointestinal irritation and symptoms similar to inhalation. Contact: skin - may be absorbed; eyes - redness, pain.

**Toxicology:** Low toxicity by all routes.
**TLV®:** No information.
**Short-term Inhalation Limits** - No information.

$LC_{50}$ - No information.
Delayed Toxicity - No information.

$LD_{50}$ - Oral: rat = 20 g/kg (1,2)
$LD_{50}$ - Oral: mouse = 4.77 g/kg (1,3)
$TD_{Lo}$ - Oral: child = 70 g/kg (1,2)

### Fire
**Fire Extinguishing Agents:** Use alcohol foam, dry chemical or $CO_2$. Water may be ineffective, but may be used to cool fire-exposed containers.
**Behaviour in Fire:** No information.
**Ignition Temperature:** (1,2) 371°C.

**Burning Rate:** 1.5 mm/min

### Reactivity
**With Water:** No reaction; soluble.
**Stability:** Can react with oxidizing materials.
**With Common Materials:** Stable.

### Environment
**Water:** Prevent entry into water intakes and waterways. Aquatic toxicity rating >1 000 ppm/96 h/TLm/freshwater; BOD: 96%, 5 days.
**Land-Air:** No information.
**Food Chain Concentration Potential:** None.

## EMERGENCY MEASURES

**Special Hazards**

COMBUSTIBLE.

**Immediate Responses**

Keep non-involved people away from spill site. Call Fire Department. Call manufacturer for advice. Dike to prevent runoff. Notify environmental authorities.

**Protective Clothing and Equipment**

Outer protective clothing as required.

**Fire and Explosion**

Use alcohol foam. Water may be ineffective, but should be used to cool fire-exposed containers.

**First Aid**

Move victim out of spill area to fresh air. Call for medical assistance, but start first aid at once. <u>Inhalation</u>: if breathing has stopped, give artificial respiration; if laboured, give oxygen. <u>Contact</u>: skin - remove contaminated clothing and flush affected areas with plenty of water; eyes - irrigate with plenty of water. <u>Ingestion</u>: give plenty of water to conscious victim to drink and induce vomiting. If medical assistance is not immediately available, transport victim to doctor, hospital or clinic.

## ENVIRONMENTAL PROTECTION MEASURES

**Response**

**Water**
1. Stop or reduce discharge if safe to do so.
2. Contact manufacturer or supplier for advice.
3. If possible, contain discharge by damming or water diversion.
4. Dredge or vacuum pump to remove contaminants, liquids and contaminated bottom sediments.
5. Notify environmental authorities to discuss disposal and cleanup of contaminated materials.

**Land-Air**
1. Stop or reduce discharge if safe to do so.
2. Contact manufacturer or supplier for advice.
3. Contain spill by diking with earth or other barrier.
4. Remove material with pumps or vacuum equipment and place in appropriate containers.
5. Recover undamaged containers.
6. Absorb residual liquid on natural or synthetic sorbents.
7. Notify environmental authorities to discuss disposal and cleanup of contaminated materials.

**Disposal**
1. Contact manufacturer or supplier for advice on disposal.
2. Contact environmental authorities for advice on disposal.
3. Incinerate (approval of environmental authorities required).

PROPYLENE GLYCOL    $CH_3CH(OH)CH_2OH$

# PROPYLENE OXIDE   $CH_3CHCH_2O$

## IDENTIFICATION

UN No. 1280

**Common Synonyms**
METHYLOXIRANE
1,2-PROPYLENE OXIDE
1,2-EPOXYPROPANE
PROPENEOXIDE
METHYL ETHYLYLENE OXIDE

**Observable Characteristics**
Colourless liquids. Ethereal odour.

**Manufacturers**
Dow Chemical Canada Inc., Sarnia, Ontario.

### Transportation and Storage Information

**Shipping State:** Liquid.
**Classification:** Flammable liquid.
**Inert Atmosphere:** Inerted.
**Venting:** Safety-relief.
**Pump Type:** Rotary or centrifugal; steel, stainless steel.

**Label(s):** Red and white label - FLAMMABLE LIQUID; Class 3.1, Group I.
**Storage Temperature:** Ambient.
**Hose Type:** Seamless stainless steel.

**Grades or Purity:** Technical.
**Containers and Materials:** Drums, tank cars; steel.

### Physical and Chemical Characteristics

**Physical State (20°C, 1 atm):** Liquid.
**Solubility (Water):** 65 g/100 mL (30°C); 40.5 g/100 mL (20°C).
**Molecular Weight:** 58.1
**Vapour Pressure:** 400 mm Hg (18°C).
**Boiling Point:** 34°C.

**Floatability (Water):** Floats and mixes.
**Odour:** Ethereal (9.9 to 35.0 ppm, odour threshold)
**Flash Point:** -37°C (c.c.).
**Vapour Density:** 2.0
**Specific Gravity:** 0.83 (20°C).

**Colour:** Colourless.
**Explosive Limits:** 2.8 to 37.0%.
**Melting Point:** -104 to -112°C.

## HAZARD DATA

### Human Health

**Symptoms:** <u>Inhalation:</u> headache, coughing, nausea, vomiting, unconsciousness, mild depression of central nervous system and lung irritation. <u>Contact:</u> skin - inflammation, burns; eyes - very irritating, watering, corneal burns, inflammation. <u>Ingestion:</u> abdominal pain, nausea and vomiting, diarrhea, convulsions.
**Toxicology:** Moderately toxic by all routes.   $LC_{50}$ - Inhalation: mouse = 1 740 ppm/4h.   $LD_{50}$ - Oral: rat = 0.93 g/kg
TLV®- 20 ppm, 50 mg/m³.   Delayed Toxicity - Suspected carcinogen.
Short-term Inhalation Limits - No information.

### Fire

**Fire Extinguishing Agents:** Use dry chemical, alcohol foam or carbon dioxide. Water may be ineffective but should be used to cool fire-exposed containers and disperse vapours. Do not extinguish fire until leak is shut off. Contact with certain materials may cause polymerization which may result in container rupture.
**Behaviour in Fire:** Flashback may occur along vapour trail.   **Burning Rate:** 3.3 mm/min.
**Ignition Temperature:** 449°C.

### Reactivity

**With Water:** No reaction; soluble.
**Stability:** Can react vigorously with oxidizing materials. May polymerize violently with anhydrous chlorides of iron, tin, aluminum; peroxides of iron or aluminum and alkaline metal hydroxides. Reacts violently with ammonium hydroxide, chlorosulfonic acid, hydrochloric acid, hydrofluoric acid, nitric acid, oleum and sulfuric acid.
**With Common Materials:** Stable, within the limits of the foregoing.

### Environment

**Water:** Prevent entry into potable water intakes and waterways. Harmful to aquatic life. Aquatic toxicity rating >1 000 ppm/96 h/TLm/freshwater; BOD: 75%, 5 days.
**Land-Air:** No information.
**Food Chain Concentration Potential:** None.

## EMERGENCY MEASURES

**Special Hazards**
FLAMMABLE.

**Immediate Responses**
Keep non-involved people away from spill site. Issue warning, "FLAMMABLE". Call Fire Department. Eliminate all ignition sources. Dike to prevent runoff. Notify manufacturer. Notify environmental authorities.

**Protective Clothing and Equipment**
Respiratory protection - self-contained breathing apparatus and totally encapsulated suit. Clothing and boots: rubber.

**Fire and Explosion**
Use dry chemical, alcohol foam or carbon dioxide to extinguish. Water may be ineffective, but should be used to cool fire-exposed containers and disperse vapours. Flashback may occur along vapour trail. Contact with certain materials may cause polymerization which may result in container rupture.

**First Aid**
Move victim out of spill site to fresh air. Call for medical assistance, but start first aid at once. Inhalation: if breathing has stopped, give artificial respiration; if laboured, give oxygen. Contact: skin - remove contaminated clothing, and flush affected areas with plenty of water; eyes - irrigate with plenty of water for at least 15 minutes. Ingestion: give plenty of water to conscious victim to drink. Induce vomiting. If medical assistance is not immediately available, transport victim to hospital, doctor or clinic.

## ENVIRONMENTAL PROTECTION MEASURES

**Response**

**Water**
1. Stop or reduce discharge if safe to do so.
2. Contact manufacturer or supplier for advice.
3. If possible, contain discharge by damming or water diversion.
4. Dredge or vacuum pump to remove contaminants, liquids and contaminated bottom sediments.

**Land-Air**
1. Stop or reduce discharge if safe to do so.
2. Contact manufacturer or supplier for advice.
3. Contain spill by diking with earth or other barrier.
4. Remove material with pumps or vacuum equipment and place in appropriate containers.
5. Recover undamaged containers.
6. Adsorb residual liquid on natural or synthetic sorbents.
7. Notify environmental authorities to discuss disposal and cleanup of contaminated materials.

**Disposal**
1. Contact manufacturer or supplier for advice on disposal.
2. Contact environmental authorities for advice on disposal.
3. Incinerate (approval of environmental authorities required).

PROPYLENE OXIDE    $CH_3CHCH_2O$

## SODIUM  Na

## IDENTIFICATION

UN No. 1428 sodium metal
1429 sodium metal dispersions in organic liquids

| Common Synonyms | Observable Characteristics | Manufacturers |
|---|---|---|
| SODIUM METAL<br>METALLIC SODIUM<br>NATRIUM | Silvery-white. Solid, turning greyish-white upon exposure to air. Odourless. | Canadian suppliers:<br>Ethyl Corporation of Canada Ltd., Corunna, Ont., Toronto, Ont. |

### Transportation and Storage Information

**Shipping State:** Solid.
**Classification:** Flammable solid (dangerous when wet).
**Inert Atmosphere:** Under naphthas or parrafins for shipping and storage.
**Venting:** Pressure-vacuum.

**Label(s):** Blue label - FLAMMABLE; Class 4.3, Group II.
**Storage Temperature:** Ambient.

**Grades or Purity:** Commercial, technical; 99.9%.
**Containers and Materials:** Drums and smaller containers (solid form); under naphthas or parrafins. May also be shipped as a dispersion in organic liquids.

### Physical and Chemical Characteristics

**Physical State** (20°C, 1 atm): Solid.
**Solubility (water):** Reacts violently forming $H_2$ gas and NaOH.
**Molecular Weight:** 23
**Vapour Pressure:** 0.014 mm Hg (300°C); 1.6 mm Hg (400°C).
**Boiling Point:** 881 to 893°C.

**Floatability (Water):** Reacts violently with water, producing $H_2$ gas and NaOH.
**Odour:** Odourless.
**Flash Point:** Not flammable.
**Vapour Density:** No information.
**Specific Gravity:** 0.97 (20°C); liquid, 0.93.

**Colour:** Silvery-white; greyish-white upon exposure to air.
**Explosive Limits:** Not flammable (Note melting point). Hydrogen: 4.1 to 74.2% (for hydrogen).
**Melting Point:** 97.8°C.

## HAZARD DATA

### Human Health

**Symptoms: Inhalation:** (NaOH fumes) irritation of nose, eyes and throat, difficulty breathing, coughing. **Ingestion:** (improbable) immediate burning sensation, vomiting, stomach cramps, rapid breathing, diarrhea, loss of consciousness. **Contact:** skin - painful ulcerations irreparable tissue damage, shock; eyes - very painful, intense watering, burns and ulcerations.
**Toxicology:** Highly toxic upon contact.
**TLV®** Inhalation: 2 mg/m$^3$ (as NaOH).   LC$_{50}$ - No information.   LD$_{50}$ - No information.
**Short-term Inhalation Limits** - No information.   **Delayed Toxicity** - No information.

### Fire

**Fire Extinguishing Agents:** Soda Ash, dry sodium chloride or dry graphite may be used (written in order of preference). DO NOT USE WATER.
**Behaviour in Fire:** Burns violently accompanied by explosions which cause splattering of the material. Vapour ignites spontaneously at room temperature.
**Ignition Temperature:** >115°C (as hydrogen evolved   **Burning Rate:** No information.
from contact with water or moisture).

### Reactivity

**With Water:** VIOLENT - generates hydrogen and is very exothermic. Reaction with water also produces sodium hydroxide (NaOH).
**With Common Materials:** Reacts violently with aluminum bromide, aluminum chloride, aluminum fluoride, ammonium nitrate, antimony halides, arsenic halides, bismuth halides, carbon dioxide, carbon tetrachloride, chlorinated hydrocarbons, chlorine, chloroform, chromium halides and oxides, copper halides and oxides, 1,2-dichloroethylene, dichloromethane, hydrofluoric acid, hydrogen peroxide, hydrogen sulfide, iodine, iron, iron halides, lead oxide, maleic anhydride, manganese halides, mercury halides, methyl chloride, monoammonium phosphate, nitric acid, nitrous oxide, phosgene, phosphorus, phosphorus halides and oxides, potassium oxides, silver halides, sulfur, sulfur halides and oxides, sulfuric acid, tin halides, tetrachloroethane, trichloroethylene, vanadium halides and zinc halides.
**Stability:** Reactive with moisture, air and many common compounds.

### Environment

**Water:** Sodium with water produces NaOH which raises the pH of a water body; this can be harmful to aquatic life; BOD: None.
**Land-Air:** No information.
**Food Chain Concentration Potential:** None.

## EMERGENCY MEASURES

**Special Hazards**

FLAMMABLE, REACTS VIOLENTLY WITH WATER, LIBERATING HYDROGEN. VERY REACTIVE.

**Immediate Responses**

Keep non-involved people away from spill site. Issue warning: "FLAMMABLE AND REACTIVE WITH WATER". CALL FIRE DEPARTMENT (warn them of the unusual properties - no water). Contact manufacturer for assistance. Contain spill area by diking with earth or other available material. Avoid contact and inhalation of fumes. Notify environmental authorities.

**Protective Clothing and Equipment**

In fires or confined spaces - Respiratory protection - self-contained breathing apparatus. Otherwise, Goggles - (mono), tight fitting. If face shield is used it must not replace goggles. Gloves - rubber. Boots - safety, rubber (pants worn outside boots). Outerwear - flame-proofed overalls or coveralls.

**Fire and Explosion**

Use soda ash, dry sodium chloride or dry graphite to extinguish. DO NOT USE WATER. Burns violently, accompanied by explosions which cause splattering of the material. Vapour ignites spontaneously at 121°C. Vapour droplets ignite spontaneously at room temperature. Pools of molten sodium ignite at 204 to 427°C.

**First Aid**

Move victim out of spill area to fresh air. Call for medical assistance, but start first aid at once. Contact: eyes - irrigate with plenty of water; skin: remove particles of sodium quickly but carefully and flush affected area with a copious amount of water for at least 30 minutes. Remove contaminated clothing at once. Inhalation: in case of exposure to fumes from fire, give oxygen if breathing is difficult, give artificial respiration if breathing has stopped. Ingestion: (improbable) give water to conscious victim to rinse mouth. Do not induce vomiting. If medical assistance is not immediately available, transport victim to hospital, doctor or clinic.

## ENVIRONMENTAL PROTECTION MEASURES

**Response**

**Water**
1. Stop or reduce discharge if safe to do so.
2. Contact manufacturer or supplier for advice.
3. If possible, contain discharge by damming or water diversion.
4. Notify environmental authorities to discuss disposal and cleanup of contaminated materials.

**Land-Air**
1. Stop or reduce discharge if safe to do so.
2. Contact manufacturer or supplier for advice.
3. Contain spill by diking with earth or other barrier.
4. Remove material by manual or mechanical means. Avoid contact with water or moisture.
5. Recover undamaged containers.
6. Notify environmental authorities to discuss disposal and cleanup of contaminated materials.

**Disposal**
1. Contact manufacturer or supplier for advice on disposal.
2. Contact environmental authorities for advice on disposal.

SODIUM   Na

# SODIUM ALUMINATE  $Na_2Al_2O_3$ or $NaAlO_2$

## IDENTIFICATION

UN No. 2812 anhydrous
1819 solution

**Common Synonyms**
META-SODIUM ALUMINATE
ALUMINUM SODIUM OXIDE

**Observable Characteristics**
White powder. Odourless.

**Manufacturers**
Handy Chemicals,
Laprairie, Quebec.

### Transportation and Storage Information

**Shipping State:** Solid or liquid (aqueous solution).
**Classification:** Solution - Corrosive.
**Inert Atmosphere:** No requirement.
**Venting:** Open.
**Pump Type:** No information.

**Label(s):** Solution - Black and white label - CORROSIVE.
**Storage Temperature:** Ambient.
**Hose Type:** No information.

**Grades or Purity:** Technical or 27°Bé solution.
**Containers and Materials:** Solid - multi-wall paper sacks, drums. Solution - drums and bulk; steel or stainless steel.

### Physical and Chemical Characteristics

**Physical State:** Solid.
**Solubility (Water):** Very soluble.
**Molecular Weight:** 82 (Na Al $O_2$).
**Vapour Pressure:** No information.
**Boiling Point:** >1 800°C.

**Floatability (Water):** Sinks and mixes.
**Odour:** Odourless.
**Flash Point:** Not flammable.
**Vapour Density:** No information.
**Specific Gravity:** >1.5 (20°C); 27°Bé solution 1.23

**Colour:** White.
**Explosive Limits:** Not flammable.
**Melting Point:** 1650°C

## HAZARD DATA

### Human Health

**Symptoms:** Inhalation: (dust) irritation of respiratory tract. Ingestion: sore throat, abdominal pain, diarrhea. Contact: skin - redness and pain; eyes - watering, redness and pain.
**Toxicology:** Moderately toxic by all routes.
**TLV®** - No information.
**Short-term Inhalation Limits** - No information.

$LC_{50}$ - No information.
**Delayed Toxicity** - No information.

$LD_{50}$ - No information.

### Fire

**Fire Extinguishing Agents:** Not combustible. Most fire extinguishing agents may be used in fires involving sodium aluminate.
**Behaviour in Fire:** Not combustible.
**Ignition Temperature:** Not combustible.

**Burning Rate:** Not combustible.

### Reactivity

**With Water:** No reaction; soluble.
**With Common Materials:** No information.
**Stability:** Stable.

### Environment

**Water:** Prevent entry into water intakes and waterways.
**Land-Air:** No information.
**Food Chain Concentration Potential:** No information.

# EMERGENCY MEASURES

**Special Hazards**

CORROSIVE.

**Immediate Responses**

Keep non-involved people away from spill site. Issue warning: "CORROSIVE". Call manufacturer or supplier for advice. Dike to prevent runoff from rainwater or water application. Notify environmental authorities.

**Protective Clothing and Equipment**

Protective outer clothing as required.

**Fire and Explosion**

Not combustible. Most fire extinguishing agents may be used in fires involving sodium aluminate.

**First Aid**

Move victim out of spill site to fresh air. Call for medical assistance, but start first aid at once. Inhalation: if breathing has stopped, give artificial respiration; if laboured, give oxygen. Contact: skin - remove contaminated clothing and flush affected areas with plenty of water; eyes - irrigate with plenty of water. Ingestion: give conscious victim plenty of water to drink. If medical assistance is not immediately available, transport victim to doctor, hospital or clinic.

# ENVIRONMENTAL PROTECTION MEASURES

**Response**

**Water**
1. Stop or reduce discharge if safe to do so.
2. Contact manufacturer or supplier for advice.
3. If possible, contain discharge by damming or water diversion.
4. Dredge or vacuum pump to remove contaminants, liquids and contaminated bottom sediments.
5. Notify environmental authorities to discuss disposal and cleanup of contaminated materials.

**Land-Air**
1. Stop or reduce discharge if safe to do so.
2. Contact manufacturer or supplier for advice.
3. Dike to prevent runoff from rainwater or water application.
4. Remove material by manual or mechanical means.
5. Recover undamaged containers.
6. Notify environmental authorities to discuss disposal and cleanup of contaminated materials.

**Disposal**
1. Contact manufacturer or supplier for advice on disposal.
2. Contact environmental authorities for advice on disposal.

SODIUM ALUMINATE   $Na_2Al_2O_3$ or $NaAlO_2$

# SODIUM ARSENITE  $NaAsO_2$

## IDENTIFICATION

UN No. 1686 Aqueous solutions
2027 Anhydrous

**Manufacturers**
No Canadian Manufacturers.

### Common Synonyms
SODIUM META ARSENITE
ARSENEOUS ACID MONOSODIUM SALT

### Observable Characteristics
White to grey-white powder. Odourless.

### Transportation and Storage Information
**Shipping State:** Solid, aqueous solutions.
**Classification:** Poison.
**Inert Atmosphere:** No requirement.
**Venting:** Pressure-vacuum.
**Pump Type:** No information.

**Label(s):** Black and white label – POISON; Group 6.
**Storage Temperature:** Ambient.
**Hose Type:** No information:

**Grades or Purity:** Pure; Technical (55 to 98%); Solution 75% arseneous oxide.
**Containers and Materials:**

### Physical and Chemical Characteristics
**Physical State (20°C, 1 atm):** Solid.
**Solubility (Water):** Soluble.
**Molecular Weight:** 129.9
**Vapour Pressure:** No information.
**Boiling Point:** Decomposes.

**Floatability (Water):** Sinks and mixes.
**Odour:** Odourless.
**Flash Point:** Not flammable.
**Vapour Density:** No information.
**Specific Gravity:** 1.87 (20°C).

**Colour:** White to greyish-white.
**Explosive Limits:** Not flammable.
**Melting Point:** 615°C.

## HAZARD DATA

### Human Health
**Symptoms:** Inhalation and Ingestion: irritation of stomach and intestines, abdominal pain, nausea and vomiting, mental confusion, weakness, diarrhea, coma. Contact: skin – irritation and burning; eyes – irritation and burning.
**Toxicology:** Highly toxic by ingestion and contact.
TLV® (as As) 0.2 mg/m$^3$       $LC_{50}$ – No information.          $LD_{50}$ – Oral: rat = 0.041 g/kg
Short-term Inhalation Limits – No information.    Delayed Toxicity – No information.    $LD_{50}$ – Skin: rat = 0.15 g/kg

### Fire
**Fire Extinguishing Agents:** Not combustible. Most fire extinguishing agents may be used. Water should be used sparingly to avoid contamination of area with material.
**Behaviour in Fire:** Not combustible.          **Burning Rate:** Not combustible.
**Ignition Temperature:** Not combustible.

### Reactivity
**With Water:** No reaction; soluble.
**With Common Materials:** No information.
**Stability:** Stable.

### Environment
**Water:** Prevent entry into water intakes and waterways. Fish toxicities: 14 to 39 mg/L/96 h/rainbow trout/$LC_{50}$/freshwater; 21 to 42 mg/L/96 h/$LC_{50}$/freshwater.
**Land-Air:** Waterfowl = 32 mg/kg, $LD_{50}$.
**Food Chain Concentration Potential:** No information.

# EMERGENCY MEASURES

**Special Hazards**

POISON.

**Immediate Responses**

Keep non-involved people away from spill site. Issue warning "POISON". Avoid contact and inhalation. Notify manufacturer or supplier. Dike to prevent runoff. Notify environmental authorities.

**Protective Clothing and Equipment**

Respiratory protection - self-contained breathing apparatus and totally-encapsulated suit.

**Fire and Explosion**

Not combustible. Most fire extinguishing agents may be used on fires involving sodium arsenite. Water should be used sparingly to avoid contamination of area with material.

**First Aid**

Move victim out of spill area to fresh air. Call for medical assistance, but start first aid at once. <u>Inhalation:</u> if breathing has stopped, give artificial respiration; if laboured, give oxygen. <u>Contact:</u> skin - remove contaminated clothing and flush affected areas with plenty of water; eyes - irrigate with plenty of water. <u>Ingestion:</u> give conscious victim plenty of water to drink and induce vomiting. If medical assistance is not immediately available, transport victim to hospital, doctor or clinic.

# ENVIRONMENTAL PROTECTION MEASURES

**Response**

**Water**
1. Stop and reduce discharge if safe to do so.
2. Contact manufacturer or supplier for advice.
3. If possible, contain discharge by damming or water diversion.
4. Dredge or vacuum pump to remove contaminants, liquids and contaminated bottom sediments.
5. Notify environmental authorities to discuss disposal and cleanup of contaminated materials.

**Land-Air**
1. Stop or reduce discharge if safe to do so.
2. Contact manufacturer or supplier for advice.
3. Dike to prevent runoff from rainwater or water application.
4. Removal of material by manual or mechanical means.
5. Recover undamaged containers.
6. Remove contaminated soil for disposal.
7. Notify environmental authorities to discuss disposal and cleanup of contaminated materials.

**Disposal**
1. Contact manufacturer or supplier for advice on disposal.
2. Contact environmental authorities for advice on disposal.

SODIUM ARSENITE   $NaAsO_2$

# SODIUM BOROHYDRIDE  NaBH$_4$

## IDENTIFICATION

| Common Synonyms | Observable Characteristics | UN No. 1426 |
|---|---|---|
| BOROHYDRIDE<br>SODIUM TETRAHYDROBORATE | White to grey-white powder or lumps.<br>Odourless. | **Manufacturers**<br>No Canadian Manufacturers.<br>Mallinckrodt Ind. Chem.,<br>St. Louis, MO. |

### Transportation and Storage Information

**Shipping State:** Solid.
**Classification:** Flammable solid.
**Inert Atmosphere:** No requirement.
**Venting:** Closed.

**Label(s):** Blue and white label - FLAMMABLE SOLID; Class 4.3, Group I.
**Storage Temperature:** Ambient.

**Grades or Purity:** 95 to 98% minimum purity; dry powder pellets; 12% solution in 43% aqueous sodium hydroxide.
**Containers and Materials:** Bottles, polyethylene - lined metal drums.

### Physical and Chemical Characteristics

**Physical State** (20°C, 1 atm): Solid.
**Solubility** (Water): 55 g/100 mL (25°C); reacts with hot water producing $H_2$ and $Na_2B_4O_7 \cdot 10H_2O$.
**Molecular Weight:** 37.8
**Vapour Pressure:** No information.
**Boiling Point:** Decomposes (400°C).

**Floatability** (Water): Sinks and mixes or reacts.
**Odour:** Odourless.
**Flash Point:** Not flammable, but the hydrogen released from decomposition is.
**Vapour Density:** No information.
**Specific Gravity:** 1.07 (20°C).

**Colour:** White to grey-white.
**Explosive Limits:** Not flammable, but the hydrogen released from decomposition is.
**Melting Point:** Decomposes (400°C).

## HAZARD DATA

### Human Health

**Symptoms:** <u>Inhalation</u>: sore throat, shortness of breath, coughing, headache; <u>Contact</u>: eyes - redness, pain; skin-redness; <u>Ingestion</u>: sore throat, abdominal pain, diarrhea, headache, dizziness.
**Toxicology:** Highly toxic by ingestion.
**TLV®** - (as sodium borate anhydrous) 1 mg/m$^3$; as sodium borate decahydrate 5 mg/m$^3$.
Short-term Inhalation Limits - No information.

$LC_{50}$ - No information.
Delayed Toxicity - No information.

$LD_{50}$ - Intraperitoneal:  rat = 0.018 g/kg

### Fire

**Fire Extinguishing Agents:** Graphite, limestone, soda ash, sodium chloride powders. DO NOT USE water, carbon dioxide, or halogenated extinguishing agents.
**Behaviour in Fire:** Decomposes and produces highly flammable hydrogen gas.
**Ignition Temperature:** Not combustible by itself -  **Burning Rate:** Not combustible.
hydrogen is ignited at 500°C.

### Reactivity

**With Water:** Soluble; reacts with hot water to produce hydrogen gas and sodium borate.
**With Common Materials:** Can react with oxidizing materials.
**Stability:** Stable.

### Environment

**Water:** prevent entry into water intakes and waterways. As sodium borate: Fish toxicity: 3 000 to 3 300 ppm/tns/minnow/minimum lethal dose/freshwater; 8 200 ppm/48 h/mosquito fish/TLm/freshwater.
**Land-Air:** No information.
**Food Chain Concentration Potential:** None.

# EMERGENCY MEASURES

## Special Hazards

FLAMMABLE - flammable gas released ($H_2$) upon contact with water.

## Immediate Responses

Keep non-involved people away from spill site. Issue warning "FLAMMABLE". CALL FIRE DEPARTMENT. Notify manufacturer or supplier. Dike to prevent runoff. Notify environmental authorities.

## Protective Clothing and Equipment

In fires Respiratory protection - self-contained breathing apparatus. Otherwise, Goggles, Gloves rubber; boots - rubber (pants worn outside boots) and protective outer clothing as required.

## Fire and Explosion

Use graphite, limestone, soda ash and sodium chloride powders to extinguish. Do not use water, carbon dioxide, or halogenated extinguishing agents. Decomposes and produces highly flammable hydrogen gas.

## First Aid

Move victim out of spill site to fresh air. Call for medical assistance, but start first aid at once. Inhalation: if breathing has stopped, give artificial respiration; if laboured give oxygen. Contact: skin - remove contaminated clothing, and flush affected areas with plenty of water; eyes - irrigate with plenty of water. Ingestion: give plenty of water to conscious victim to drink and induce vomiting. If medical assistance is not immediately available, transport victim to hospital, doctor or, clinic.

# ENVIRONMENTAL PROTECTION MEASURES

## Response

**Water**
1. Stop and reduce discharge if safe to do so.
2. Contact manufacturer or supplier for advice.
3. If possible, contain discharge by damming or water diversion.
4. Dredge or vacuum pump to remove contaminants, liquids and contaminated bottom sediments.
5. Notify environmental authorities to discuss disposal and cleanup of contaminated materials.

**Land-Air**
1. Stop or reduce discharge if safe to do so.
2. Contact manufacturer or supplier for advice.
3. Dike to prevent runoff from rainwater or water application.
4. Remove material by manual or mechanical means.
5. Recover undamaged containers.
6. Notify environmental authorities to discuss disposal and cleanup of contaminated materials.

## Disposal

1. Contact manufacturer or supplier for advice on disposal.
2. Contact environmental authorities for advice on disposal.

SODIUM BOROHYDRIDE   $NaBH_4$

# SODIUM CARBONATE   Na$_2$CO$_3$ (anhydrous)

## IDENTIFICATION

| Common Synonyms | Observable Characteristics | Manufacturers |
|---|---|---|
| CALCINED SODA<br>SODA ASH<br>SODA MONOHYDRATE<br>CRYSTAL CARBONATE<br>CARBONIC ACID, DISODIUM SALT | White to grey crystalline solid or powder.<br>Odourless. | Allied Chemical Limited,<br>Amherstburg, Ontario. |

### Transportation and Storage Information

**Shipping State:** Solid.
**Classification:** None.
**Inert Atmosphere:** No requirement.
**Venting:** Open.

**Label(s):** Not regulated.
**Storage Temperature:** Ambient.

**Grades or Purity:** Dense (58%), light (58%), extra light, natural and refined.
**Containers and Materials:** Bags, barrels, drums and bulk by truck or train; steel.

### Physical and Chemical Characteristics

**Physical State** (20°C, 1 atm): Solid.
**Solubility** (Water): 7.1 g/100 mL (0°C); 22 g/100 mL (20°C); 45.5 g/100 mL (100°C).
**Molecular Weight:** 106.0
**Vapour Pressure:** No information.
**Boiling Point:** Slowly begins to decompose at 400°C.

**Floatability** (Water): Sinks and mixes.
**Odour:** Odourless.
**Flash Point:** Not flammable.
**Vapour Density:** No information.
**Specific Gravity:** 2.53 (20°C).

**Colour:** White to grey.
**Explosive Limits:** Not flammable.
**Melting Point:** 851°C.

## HAZARD DATA

### Human Health

**Symptoms:** Inhalation: irritation of respiratory tract, coughing, sneezing, difficulty breathing. Contact: with dust causes eye and skin irritation. Excessive contact can cause "soda ulcers" and perforation of nasal septum. Ingestion: of large amounts is corrosive to the gastrointestinal tract, causing cramps, vomiting, diarrhea and possible circulatory collapse.
**Toxicology:** Moderately toxic by ingestion. Low toxicity by inhalation and contact.
**TLV®** - No information.    **LC$_{50}$** - No information.    **LD$_{50}$** - Intraperitoneal: mouse = 0.117 g/kg
**Short-term Inhalation Limits** - No information.    **Delayed Toxicity** - No information.    **LD$_{Lo}$** - Oral: rat = 4 g/kg

### Fire

**Fire Extinguishing Agents:** Not combustible. Most fire extinguishing agents may be used on fires involving sodium carbonate.
**Behaviour in Fire:** Not combustible. Begins to decompose at 400°C producing CO$_2$ gas.
**Ignition Temperature:** Not combustible.    **Burning Rate:** Not combustible.

### Reactivity

**With Water:** No reaction; moderately soluble.
**With Common Materials:** Can react violently with aluminum phosphorus pentoxide and fluoride sulfuric acid.
**Stability:** Stable.

### Environment

**Water:** Prevent entry into water intakes and waterways. Fish toxicity: 265 mg/L/48 h/Daphnia magna/TLm/freshwater; BOD: No information.
**Land-Air:** No information.
**Food Chain Concentration Potential:** None.

## EMERGENCY MEASURES

**Special Hazards**

**Immediate Responses**

Keep non-involved people away from spill site. Avoid contact and inhalation of dust. Stop or reduce discharge, if this can be done without risk. Notify supplier. Notify environmental authorities.

**Protective Clothing and Equipment**

If dust is present wear dust respirator, industrial (tight fitting) goggles, gloves and coveralls.

**Fire and Explosion**

Not combustible. Most fire extinguishing agents may be used on fires involving sodium carbonate.

**First Aid**

Move victim out of spill area to fresh air. Call for medical assistance, but start first aid at once. Inhalation: if breathing has stopped give artificial respiration; if laboured, give oxygen. Contact: skin - remove contaminated clothing and flush affected areas with plenty of water; eyes - irrigate with plenty of water. Ingestion: give warm water to conscious victim to drink. Do not induce vomiting. If medical attention is not immediately available, transport victim to hospital, clinic or doctor.

## ENVIRONMENTAL PROTECTION MEASURES

**Response**

**Water**
1. Stop or reduce discharge if safe to do so.
2. Contact manufacturer or supplier for advice.
3. If possible, contain discharge by damming or water diversion.
4. Dredge or vacuum pump to remove contaminants, liquids and contaminated bottom sediments.
5. Notify environmental authorities to discuss disposal and cleanup of contaminated materials.

**Land-Air**
1. Stop or reduce discharge if safe to do so.
2. Contact manufacturer or supplier for advice.
3. Dike to prevent runoff from rainwater or water application.
4. Remove material by manual or mechanical means.
5. Recover undamaged containers.
6. Notify environmental authorities to discuss cleanup and disposal of contaminated materials.

**Disposal**

1. Contact manufacturer or supplier for advice on disposal.
2. Contact environmental authorities for advice on disposal.
3. Contaminated materials may be buried in a secured landfill site (approval of environmental authorities required).

SODIUM CARBONATE   $Na_2CO_3$ (anhydrous)

# SODIUM CHLORATE  NaClO₃

## IDENTIFICATION

**Common Synonyms**
CHLORATE OF SODA
CHLORIC SALT OF SODIUM

**Observable Characteristics**
Colourless to white crystalline solid, colourless in solution. Commercial material may be pale yellow.

**UN No.** 1495 solid
2428 solution

**Manufacturers**
Erco Industries Ltd.,
Buckingham, Que., Vancouver, B.C., Thunder Bay, Ontario.
PPG Industries (CPI), Beauharnois, Que.
Canadian Occidental Petroleum, Squamish, B.C., Nanaimo, B.C.
B.C. Chemicals, Prince George, B.C.

### Transportation and Storage Information

**Shipping State:** Solid, liquid (aqueous solution).
**Classification:** Oxidizer.
**Inert Atmosphere:** No requirement.
**Venting:** Open.
**Pump Type:** For solutions, most types.

**Label(s):** Yellow label - OXIDIZER; Class 5.1, Group II.
**Storage Temperature:** Ambient.
**Hose Type:** Polyethylene, polypropylene, natural rubber.

**Grades or Purity:** Commercial, 99% NaClO₃; technical, 99.5% minimum.
**Containers and Materials:** Anhydrous: drums, steel, plastic-lined steel. Solutions: drums, tank cars; stainless steel, plastic-lined steel.

### Physical and Chemical Characteristics

**Physical State** (20°C, 1 atm): Solid or liquid (solution)
**Solubility** (Water): 72 g/100 mL (-15°C); 79 g/100 mL (0°C); 101 g/100 mL (20°C); 126 g/100 mL (40°C).
**Molecular Weight:** 106.4
**Vapour Pressure:** Not pertinent.
**Boiling Point:** Decomposes at 300°C releasing oxygen.

**Floatability** (Water): Mixes and sinks.
**Odour:** Odourless.
**Flash Point:** Not flammable.
**Vapour Density:** No information.
**Specific Gravity:** Solid: 2.49 (15°C); Saturated Solution: 1.43 (20°C); 1.38 (-15°C).

**Colour:** Colourless to white (or pale yellow).
**Explosive Limits** Not flammable.
**Melting Point:** 248 to 261°C.

## HAZARD DATA

### Human Health

**Symptoms:** Inhalation: sneezing, small nasal ulcerations, irritation of respiratory tract. **Contact:** Prolonged exposure to dust may result in irritation to the skin, mucous membranes and eyes. **Ingestion:** vomiting, abdominal pain, cyanosis, diarrhea, unconsciousness and death.
**Toxicology:** Moderately toxic by ingestion. Low toxicity by inhalation and contact. LD₅₀ - Oral: rat = 1.2 g/kg
TLV®- No information.  LC₅₀ - No information.
Short-term Inhalation Limits - No information.  Delayed Toxicity - No information.

### Fire

**Fire Extinguishing Agents:** Not combustible by itself but causes combustion. In fires involving sodium chlorate, use water in flooding amounts.
**Behaviour in Fire:** Upon decomposition, releases toxic chloride fumes. Upon heating, can release oxygen, making extinguishment difficult.
**Ignition Temperature:** Not combustible, but causes combustion of other materials.

### Reactivity

**With Water:** No reaction; soluble.
**With Common Materials:** In contact with strong acids, can release toxic chlorine and chlorine dioxide gases. Reacts with aluminum, ammonium thiosulfate, arsenic, arsenic trioxide, carbon, charcoal, copper, manganese dioxide, metal sulfides, organic acids, organic matter, phosphorus, potassium cyanide, sulfur, sulfuric acid, thiocyanates and zinc.
**Stability:** Stable within the limits of the foregoing.

### Environment

**Water:** Prevent entry into water intakes and waterways. Harmful to aquatic life. Aquatic toxicity rating = >1 000 ppm/96 h/TLm/freshwater; Fish toxicities: 3.8 ppm/*Scenedesmus*/threshold toxicity/freshwater; 4 200 ppm/24 h/rainbow trout/LC₅₀/freshwater; 3 157 ppm/24 h/channel catfish/LC₅₀/freshwater.
**Land-Air:** Detrimental to vegetation - is used as a weed killer.
**Food Chain Concentration Potential:** None.

## EMERGENCY MEASURES

### Special Hazards
OXIDIZER.

### Immediate Responses
Keep non-involved people away from spill site. Issue warning "OXIDIZER". Call Fire Department. Notify manufacturer. Avoid contact and inhalation. Contain spill by diking to prevent runoff. Stop or reduce discharge, if this can be done without risk. Notify environmental authorities.

### Protective Clothing and Equipment
In fires - Respiratory protection - self-contained breathing apparatus. Otherwise, if dust is present, respiratory should be used. <u>Goggles</u> - (mono), tight fitting, or face shield. <u>Gloves</u> - rubber. <u>Outer clothing</u> - nonflammable coveralls, etc. <u>Boots</u> - rubber; avoid contact of contaminated clothing with ignition sources.

### Fire and Explosion
Not combustible by itself, but can cause most organic material to ignite. Use water in flooding amounts. Clothing impregnated with sodium chlorate is highly flammable when dry.

### First Aid
Move victim out of spill site to fresh air. Call for medical assistance, but start first aid at once. <u>Inhalation</u>: if breathing has stopped, give artificial respiration; if laboured, give oxygen. <u>Contact</u>: eyes - irrigate immediately with plenty of water; skin - remove contaminated clothing and flush affected areas with plenty of water. <u>Ingestion</u>: give milk or water to conscious victim to drink and induce vomiting. Repeat several times. Keep warm and quiet. If medical assistance is not immediately available, transport victim to hospital, doctor or clinic.

## ENVIRONMENTAL PROTECTION MEASURES

### Response

**Water**
1. Stop or reduce discharge if safe to do so.
2. Contact manufacturer or supplier for advice.
3. If possible, contain discharge by damming or water diversion.
4. Notify environmental authorities to discuss disposal and cleanup of contaminated materials.

**Land-Air**
1. Stop or reduce discharge if safe to do so.
2. Contact manufacturer or supplier for advice.
3. Dike to prevent runoff from rainwater or water application.
4. Remove material by manual or mechanical means (for solid spills).
5. Remove undamaged containers.
6. Notify environmental authorities to discuss disposal and cleanup of contaminated materials.

### Disposal
1. Contact manufacturer or supplier for advice on disposal.
2. Contact environmental authorities for advice on disposal.

SODIUM CHLORATE   $NaClO_3$

# SODIUM CHLORIDE  NaCl

## IDENTIFICATION

### Common Synonyms
COMMON SALT
HALITE
ROCK SALT
SEA SALT

### Observable Characteristics
Colourless to white powder, lumps.
Odourless.

### Manufacturers
Domtar, Goderich, Ontario.
Canadian Salt, Ojibway, Ontario.
Dow Chemical Canada
Fort Saskatchewan, Alberta.

### Transportation and Storage Information
**Shipping State:** Solid.
**Classification:** None.
**Inert Atmosphere:** No requirement.
**Venting:** Open.
**Label(s):** Not regulated.
**Storage Temperature:** Ambient.
**Grades or Purity:** Variable.
**Containers and Materials:** Multiwall paper bags, barrels, drums, bulk by train or truck; steel.

### Physical and Chemical Characteristics
**Physical State** (20°C, 1 atm): Solid.
**Solubility** (Water): 35.8 g/100 mL (20°C) 39.4 g/100 mL (100°C).
**Molecular Weight:** 58.4
**Vapour Pressure:** 1 mm Hg (865°C).
**Boiling Point:** 1413°C.
**Floatability** (Water): Sink and mixes.
**Odour:** Odourless.
**Flash Point:** Not flammable.
**Vapour Density:** No information.
**Specific Gravity:** 2.17 (25°C).
**Colour:** Colourless to white.
**Explosive Limits:** Not flammable.
**Melting Point:** 801°C.

## HAZARD DATA

### Human Health
**Symptoms:** Inhalation: slight irritation of nose, sneezing; Ingestion: disagreeable taste, nausea and vomiting; Contact: skin - irritation and inflammation; eyes - irritation, watering, inflammation.
**Toxicology:** Low toxicity by all routes.
**TLV**®  No information.
**Short-term Inhalation Limits** - No information.
$LD_{50}$ - No information.
**Delayed Toxicity** - No information.
$LD_{50}$ - Oral: rat = 3.0 g/kg
$TL_{LO}$ - Oral: human = 12.36 g/kg

### Fire
**Fire Extinguishing Agents:** Not combustible. Most fire extinguishing agents may be used in fires involving sodium chloride.
**Behaviour in Fire:** Not combustible.
**Ignition Temperature:** Not combustible.
**Burning Rate:** Not combustible.

### Reactivity
**With Water:** No reaction; soluble.
**With Common Materials:** Can react violently with bromine trifluoride and lithium.
**Stability:** Stable.

### Environment
**Water:** Prevent entry into water intakes and waterways. Fish toxicity: 17.5 g/L/96 h/mosquito fish/TLm/freshwater; 2.9 to 11.7/192 to 216 h/sticklebacks/$LC_{50}$/freshwater; 3.7 g/L/64 h/Daphnia magna/$LC_{50}$/freshwater. BOD: None. Concentrations of 2.6 to 5.0 g/L restricts yields of many crops. Greater than 5.1 g/L affects most crops.
**Land-Air:** Salt concentrations of 1.3 to 2.6 g/L may affect sensitive crops.
**Food Chain Concentration Potential:** None.

## EMERGENCY MEASURES

**Special Hazards**

**Immediate Responses**

Keep non-involved people away from spill site. Notify manufacturer or supplier. Dike to prevent runoff from rainwater or water application. Notify environmental authorities.

**Protective Clothing and Equipment**

Outer protective clothing as required.

**Fire and Explosion**

Not combustible. Most fire extinguishing agents may be used on fires involving sodium chloride.

**First Aid**

Move victim out of spill area to fresh air. Call for medical assistance, but start first aid at once. <u>Inhalation</u>: move victim to fresh air. <u>Contact</u>: skin - flush affected areas with water; eyes - irrigate with water. <u>Ingestion</u>: give conscious victim plenty of water to drink and induce vomiting. Transport victim to hospital, doctor, or clinic if necessary.

## ENVIRONMENTAL PROTECTION MEASURES

**Response**

**Water**
1. Stop or reduce discharge if safe to do so.
2. Contact manufacturer or supplier for advice.
3. Notify environmental authorities to discuss disposal and cleanup of contaminated materials.

**Land-Air**
1. Stop or reduce discharge if safe to do so.
2. Contact manufacturer or supplier for advice.
3. Dike to prevent runoff from rainwater or water application.
4. Remove material by manual or mechanical means.
5. Recover undamaged containers.
6. Notify environmental authorities to discuss cleanup and disposal of contaminated materials.

**Disposal**
1. Contact manufacturer or supplier for advice on disposal.
2. Contact environmental authorities for advice on disposal.
3. Contaminated materials may be buried in a secured landfill site (approval of environmental authorities required).

SODIUM CHLORIDE   NaCl

# SODIUM CYANIDE  NaCN

## IDENTIFICATION     UN No. 1689

| Common Synonyms and Trade Names | Observable Characteristics | Manufacturers | |
|---|---|---|---|
| CYANIDE OF SODIUM<br>CYANOGRAN (Du Pont)<br>CYANOIDS (Kraft) | White, crystalline solid, powder or granules. Odourless when dry; when wet gives typical cyanide almond-like odour. | Canadian supplier:<br>Canadian Industries Ltd.,<br>Montreal, Quebec. | Originating from:<br>ICI, United Kingdom. |

### Transportation and Storage Information

Shipping State: Solid.
Classification: Poison.
Inert Atmosphere: No requirement.
Venting: Closed.

Label(s): White label – POISON;
Class 6.1, Group I.
Storage Temperature: Ambient.

Grades or Purity: 97 to 99%.
Containers and Materials: Drums; steel.

### Physical and Chemical Characteristics

Physical State (20°C, 1 atm): Solid.
Solubility (Water): 48 g/100 mL (10°C); 82 g/100 mL (35°C).
Molecular Weight: 49.0
Vapour Pressure: 1 mm Hg (817°C).
Boiling Point: 1496°C.

Floatability: (Water): Sinks and mixes.
Odour: Odourless when dry; when wet, gives typical cyanide almond-like odour.
Flash Point: Not flammable.
Vapour Density: 0.93 as HCN (25°C).
Specific Gravity: 1.6 (25°C).

Colour: White.
Explosive Limits: Not flammable.
Melting Point: 560 to 564°C.

## HAZARD DATA

### Human Health

**Symptoms:** Inhalation: headache, dizziness, nausea, rapid breathing, anguish, convulsions, foaming at mouth, prolonged coma, death. Ingestion: symptoms similar to inhalation. Contact: skin – absorbed with symptoms similar to inhalation; eyes – irritation, watering and symptoms similar to inhalation.
**Toxicology:** Highly toxic by all routes.
TLV® - (skin) 5 mg/m$^3$ (as CN).
Short-term Inhalation Limits – No information.

$LC_{50}$ - Inhalation: rat = 484 ppm/1 h (as HCN).
$LC_{Lo}$ - Inhalation: human = 120 mg/m$^3$/1 h (as HCN); = 200 mg/m$^3$/10 min (as HCN).
Delayed Toxicity - No information.

$LD_{50}$ - Oral: rat = 0.0064 g/kg
$LD_{Lo}$ - Oral: human = 0.0029 g/kg

### Fire

Fire Extinguishing Agents: Not combustible; however, water may be used on fires involving sodium cyanide.
Behaviour in Fire: Not combustible. In fires may evolve toxic fumes.
Ignition Temperature: Not combustible.     Burning Rate: Not combustible.

### Reactivity

With Water: Contact with water or moist air may produce HCN; soluble.
With Common Materials: Contact with acids or weak alkalis produces poisonous and flammable HCN gas. Reacts violently with nitrates, nitrites and other oxidizing agents, and chlorates.
Stability: Stable when dry.

### Environment

Water: Prevent entry into water intakes and waterways. Fish toxicity: 0.15 ppm/96 h/bluegill/TLm/freshwater; 0.25 ppm/48 h/prawn/$LC_{50}$/saltwater; BOD: 6%, 7 days (theoretical).
Land-Air: No information.
Food Chain Concentration Potential: No information.

## EMERGENCY MEASURES

### Special Hazards
POISON. Contact with acids or weak alkalis liberates HCN.

### Immediate Responses
Keep non-involved people away from spill site. Issue warning: "POISON". Avoid contact and inhalation. Contact supplier or manufacturer for guidance. Stop or reduce discharge, if this can be done without risk. Notify environmental authorities.

### Protective Clothing and Equipment
Respiratory protection - self-contained breathing apparatus and totally encapsulated suit. Gloves - rubber.

### Fire and Explosion
Not combustible. In fires involving sodium cyanide water may be used to extinguish; however, area should be diked to prevent runoff.

### First Aid
Move victim out of spill area to fresh air. Call for medical assistance, but start first aid at once. Inhalation: if breathing has stopped, give artificial respiration (not mouth-to-mouth method); if laboured, give oxygen. Contact: skin - remove contaminated clothing, and flush affected areas with plenty of water; eyes - irrigate with plenty of water. Ingestion: induce vomiting in conscious victim, and repeat until vomitus is clear. If medical assistance is not immediately available, transport victim to hospital, doctor or clinic.

## ENVIRONMENTAL PROTECTION MEASURES

### Response

**Water**
1. Stop or reduce discharge if safe to do so.
2. Contact manufacturer or supplier for advice.
3. If possible, contain discharge by damming or water diversion.
4. If possible, dredge or vacuum pump to remove contaminants, liquids and contaminated bottom sediments.
5. Notify environmental authorities to discuss disposal and cleanup of contaminated materials.

**Land-Air**
1. Stop or reduce discharge if safe to do so.
2. Contact manufacturer or supplier for advice.
3. Dike to prevent runoff from rainwater or water application.
4. Remove material by manual or mechanical means. Avoid contact with material.
5. Recover undamaged containers.
6. Remove contaminated soil for disposal.
7. Notify environmental authorities to discuss disposal and cleanup of contaminated materials.

### Disposal
1. Contact manufacturer or supplier for advice on disposal.
2. Contact environmental authorities for advice on disposal.

SODIUM CYANIDE   NaCN

# SODIUM DICHLOROISOCYANURATE    NaNC(O)NCIC(O)NCICO

## IDENTIFICATION

### Common Synonyms
NaDCC
POOL CHLORINATOR
1-SODIUM-3,5-DICHLORO-S-TRIAZINE-2,4,6-TRIONE
ISOCYANURIC ACID, DICHLORO SODIUM SALT
SODIUM DICHLOROISOCYANURATE (dihydrate).

### Observable Characteristics
White powder. Bleach-like odour.

### Manufacturers
No Canadian Manufacturers.
Monsanto, St. Louis, MO.
FMC, Philadelphia, PA.

### Transportation and Storage Information
**Shipping State:** Solid.
**Classification:** None.
**Inert Atmosphere:** No requirement.
**Venting:** Pressure-vacuum.

**Label(s):** Not regulated.
**Storage Temperature:** Cool, ambient.

**Grades or Purity:** Technical, 39 to 60% available chlorine.
**Containers and Materials:** Plastic-lined fibre drums, drums; steel, stainless steel.

### Physical and Chemical Characteristics
**Physical State (20°C, 1 atm):** Solid.
**Solubility (Water):** Soluble in all proportions, reacts to release chlorine.
**Molecular Weight:** 220.0 (anhydrous).
**Vapour Pressure:** No information.
**Boiling Point:** Decomposes 230°C.

**Floatability (Water):** Floats, mixes and reacts.
**Odour:** Bleach-like.
**Flash Point:** Not flammable.
**Vapour Density:** (as $Cl_2$) 2.5 (0°C).
**Specific Gravity:** 0.96 (20°C).

**Colour:** White.
**Explosive Limits:** Not flammable.
**Melting Point:** Decomposes 230°C.

## HAZARD DATA

### Human Health
**Symptoms:** Inhalation: (dust) irritation of respiratory tract and mucous membranes; inhalation (of chlorine)-burning, irritation, difficulty breathing. Ingestion: irritation of gastrointestinal tract, difficulty breathing.
**Toxicology:** Moderately toxic by all routes.
TLV® (as chlorine) 1 ppm; 3 $mg/m^3$.
Short-term Inhalation Limits - (as chlorine) 3 ppm; 9 $mg/m^3$ (15 min).

$LC_{50}$ - No information.
Delayed Toxicity - Symptoms may appear some time after exposure.

$LD_{50}$ - Oral: rat = 1.40 g/kg

### Fire
**Fire Extinguishing Agents:** Not combustible. Use carbon dioxide or dry chemical to extinguish fires involving sodium dichloroisocyanurate. Water should only be used in flooding amounts.
**Behaviour in Fire:** When heated to decomposition, emits carbon monoxide and chloride fumes. Contact with water produces chlorine gas. Contact with organic material may cause spontaneous combustion.
**Ignition Temperature:** Not combustible.
**Burning Rate:** Not combustible.

### Reactivity
**With Water:** Reacts to release chlorine gas.
**With Common Materials:** Reacts with many organic materials, to initiate spontaneous combustion. Can react violently with ammonia, ammonium salts, urea and hydrated salts.

### Environment
**Water:** Prevent entry into water intakes and waterways. Fish toxicity = 0.07 to 0.15 ppm/96 h/fathead minnow/$LC_{50}$/freshwater (as Cl); 0.14 to 0.29 ppm/96 h/rainbow trout/$TL_{50}$/freshwater.
**Land-Air:** No information.
**Food Chain Concentration Potential:** None.

## EMERGENCY MEASURES

### Special Hazards
OXIDIZER. CORROSIVE. Reacts with water to release chlorine.

### Immediate Responses
Keep non-involved people away from spill site. Call Fire Department. Avoid contact and inhalation. Notify manufacturer or supplier. Notify environmental authorities.

### Protective Clothing and Equipment
In fires or confined spaces Respiratory protection - self-contained breathing apparatus and totally encapsulated suit. Otherwise, outer protective clothing as required and self-contained breathing apparatus.

### Fire and Explosion
Not combustible. Use dry chemical or carbon dioxide to extinguish fires involving sodium dichloroisocyanurate. Water should only be used in flooding amounts. When heated to decomposition emits carbon monoxide and chloride fumes.

### First Aid
Move victim out of spill area to fresh air. Call for medical assistance, but start first aid at once. **Inhalation:** if breathing has stopped, give artificial respiration; if laboured, give oxygen. **Contact:** skin - remove contaminated clothing and flush affected areas with water; eyes - irrigate with water. **Ingestion:** give conscious victim water to drink and induce vomiting. If medical assistance is not immediately available, transport victim to hospital, doctor or clinic.

## ENVIRONMENTAL PROTECTION MEASURES

### Response

**Water**
1. Stop or reduce discharge if safe to do so.
2. Contact manufacturer or supplier for advice.
3. If possible, contain spill by damming or water diversion.
4. Dredge or vacuum pump to remove contaminants, liquids and contaminated bottom sediments.
5. Notify environmental authorities to discuss disposal and cleanup of contaminated materials.

**Land-Air**
1. Stop or reduce discharge if safe to do so.
2. Contact manufacturer or supplier for advice.
3. Dike to prevent runoff from rainwater or water application.
4. Remove material by manual or mechanical means.
5. Recover undamaged containers.
6. Notify environmental authorities to discuss disposal and cleanup of contaminated materials.

### Disposal
1. Contact manufacturer or supplier for advice on disposal.
2. Contact environmental authorities for advice on disposal.

SODIUM DICHLOROISOCYANURATE    NaNC(O)NCIC(O)NCICO

# SODIUM DICHROMATE  $Na_2Cr_2O_7 \cdot 2H_2O$

UN No. 1479

## IDENTIFICATION

**Common Synonyms**
SODIUM ACID CHROMATE
SODIUM BICHROMATE
SODIUM BICHROMATE DIHYDRATE

**Observable Characteristics**
Red-orange crystalline solid.
Odourless.

**Manufacturers**
No Canadian manufacturers.
Selected U.S. manufacturer:
Essex Chemical Company,
Clifton, NJ, USA

Canadian supplier:
Canadian Industries Limited,
Toronto, Ontario.
Allied Chemical Ltd.,
Toronto, Ont., Montreal, Quebec.

### Transportation and Storage Information

**Shipping State:** Solid, liquid (aqueous solution).
**Classification:** Oxidizer.
**Inert Atmosphere:** No requirement.
**Venting:** Open.
**Pump Type:** Gear, centrifugal; steel, stainless steel (for solutions only).

**Label(s):** Yellow label - OXIDIZER; Class 5.1, Group II.
**Storage Temperature:** Ambient.
**Hose Type:** Natural rubber, polyethylene, polypropylene.

**Grades or Purity:** Technical, crystalline (98.8-99%). Technical, solution (69-70%).
**Containers and Materials:** Dry - multiwall paper bags; fibre drums. Solution - tank cars, tank trucks; steel.

### Physical and Chemical Characteristics

**Physical State** (20°C, 1 atm): Solid.
**Solubility** (Water): 238 g/100 mL (0°C); 508 g/100 mL (80°C).
**Molecular Weight:** 298.1
**Vapour Pressure:** No information.
**Boiling Point:** Decomposes (400°C).

**Floatability** (Water): Sinks and mixes.
**Odour:** Odourless.
**Flash Point:** Not flammable.
**Vapour Density:** No information.
**Specific Gravity:** 2.52 (13°C); 2.34 (25°C).

**Colour:** Red-orange.
**Explosive Limits:** Not flammable.
**Melting Point:** 357°C (anhydrous). 69 to 70% solution, -48.2°C.

## HAZARD DATA

### Human Health

**Symptoms:** Inhalation: irritation of nose and eyes, tingling and burning sensation in respiratory tract, sneezing. Contact: extremely corrosive to skin and mucous membranes. May cause rash or external ulcers on skin. Ingestion: of large quantities may be fatal, burning sensation in mouth and throat, nausea and vomiting.
**Toxicology:** Highly toxic by contact and ingestion.
TLV® - (inhalation) 0.05 mg/m³ (as Cr).
Short-term Inhalation Limits - No information.

LC$_{50}$ - No information.
Delayed Toxicity - Suspected carcinogen.

LD$_{50}$ - Intraperitoneal: rodent = 0.16 g/kg
LD$_{Lo}$ - Intraperitoneal: guinea pig = 0.335 g/kg

### Fire

**Fire Extinguishing Agents:** Not combustible. In fires involving sodium dichromate, flood with water.
**Behaviour in Fire:** Not combustible but may ignite combustible materials upon contact. Decomposes to produce oxygen when heated.
**Ignition Temperature:** Not combustible.
**Burning Rate:** Not combustible.

### Reactivity

**With Water:** No reaction; soluble. Sodium dichromate is a strong oxidizer and may ignite organic materials.
**With Common Materials:** Reacts violently with hydrazine.
**Stability:** Stable.

### Environment

**Water:** Prevent entry into water intakes and waterways. Harmful to aquatic life. Fish toxicity: 145 ppm/24 h/bluegill/TLm/freshwater; 10 ppm/48 h/Daphnia magna/TLm/freshwater; Aquatic Toxicity rating = 100 to 1 000 ppm/96 h/TLm; BOD: None.
**Land-Air:** No information.
**Food Chain Concentration Potential:** None.

## EMERGENCY MEASURES

**Special Hazards**

OXIDIZER.

**Immediate Responses**

Keep non-involved people away from spill site. Issue warning: "OXIDIZER". Avoid contact and inhalation. Stop or reduce discharge if this can be done without risk. Dike to prevent runoff. Notify manufacturer or supplier and environmental authorities.

**Protective Clothing and Equipment**

Goggles - (mono), tight fitting. Gloves - rubber. Dust respirator. Protective work clothing - rubber apron, coveralls. <u>Boots</u> - high, rubber (pants outside boots).

**Fire and Explosion**

Not combustible. In fires involving this material, flood discharge area with water.

**First Aid**

Move victim out of spill site to fresh air. Call for medical assistance, but start first aid at once. <u>Inhalation</u>: if breathing has stopped, give artificial respiration; if laboured give oxygen. <u>Contact</u>: eyes - irrigate with plenty of water; skin - flush affected areas with plenty of water and remove contaminated clothing. <u>Ingestion</u>: do not induce vomiting; however, vomiting may occur spontaneously. If medical assistance is not immediately availble, transport victim to hospital, doctor or clinic.

## ENVIRONMENTAL PROTECTION MEASURES

**Response**

**Water**
1. Stop or reduce discharge if safe to do so.
2. Contact manufacturer or supplier for advice.
3. If possible, contain discharge by damming or water diversion.
4. Dredge or vacuum pump to remove contaminants, liquids and contaminated bottom sediments.
5. Notify environmental authorities to discuss disposal and cleanup of contaminated materials.

**Land-Air**
1. Stop or reduce discharge if safe to do so.
2. Contact manufacturer or supplier for advice.
3. Dike to prevent runoff from rainwater or water application.
4. Remove material by manual or mechanical means.
5. Recover undamaged containers.
6. Remove contaminated soil for disposal.
7. Notify environmental authorities to discuss disposal and cleanup of contaminated materials.

**Disposal**
1. Contact manufacturer or supplier for advice on disposal.
2. Contact environmental authorities for advice on disposal.

SODIUM DICHROMATE     $Na_2Cr_2O_7 \cdot 2H_2O$

SODIUM DITHIONITE  $Na_2S_2O_4 \cdot xH_2O$ (x = 0 to 2)

## IDENTIFICATION

UN No. 1384

### Common Synonyms

SODIUM BISULFITE
SODIUM HYDROSULFITE
SODIUM SULFOXYLATE

### Observable Characteristics

White to yellow powder or crystals. Sulfurous odour.

### Manufacturers

No Canadian manufacturer.
Canadian supplier:
Vir Chem of Canada, Montreal, Que.

Originating from:
Virginia Chem, USA
Royce Chemical, USA (direct)

### Transportation and Storage Information

**Shipping State:** Solid.
**Classification:** Flammable solid; spontaneous combustible.
**Inert Atmosphere:** No requirement.
**Venting:** Closed.

**Label(s):** Red and white label - FLAMMABLE; Class 4.2, Group II. Red, black and white label - SPONTANEOUS COMBUSTIBLE.
**Storage Temperature:** Ambient.

**Grades or Purity:** Technical.
**Containers and Materials:** Barrels, kegs, boxes; glass or metal lined. Drums; steel.

### Physical and Chemical Characteristics

**Physical State** (20°C, 1 atm): Solid.
**Solubility** (Water): 25 g/100 mL (20°C).
**Molecular Weight:** 174.1 (solute).
**Vapour Pressure:** No information.
**Boiling Point:** Decomposes, 52 to 55°C.

**Floatability** (Water): Sinks and mixes.
**Odour:** Faint sulfurous odour.
**Flash Point:** Not flammable by itself.
**Vapour Density:** No information.
**Specific Gravity:** 1.2 (20°C).

**Colour:** White to yellow.
**Explosive Limits:** Not flammable but can react with water.
**Melting Point:** Decomposes, 52 to 55°C.

## HAZARD DATA

### Human Health

**Symptoms:** Inhalation: sore throat, coughing, shortness of breath. Contact: eyes and skin - redness and pain. Ingestion: causes abdominal pain and nausea.
**Toxicology:** Moderately toxic by skin contact, ingestion and inhalation.
TLV® - No information.        LC$_{50}$ - No information.        LD$_{50}$ - No information.
Short-term Inhalation Limits - No information.        Delayed Toxicity - No information.

### Fire

**Fire Extinguishing Agents:** Use dry sand, carbon dioxide, or dry chemical; water should only be used in flooding amounts. The addition of small amounts of water or contact with a highly humid atmosphere can cause fire.
**Behaviour in Fire:** In fires, sulfur dioxide is released.        **Burning Rate:** No information.
**Ignition Temperature:** Decomposes, 52 to 55°C.

### Reactivity

**With Water:** Small amounts of water or contact with a highly humid atmosphere can cause ignition. In large amounts of water; soluble.
**With Common Materials:** Reacts with acids to form sulfur oxides.
**Stability:** Stable, within the limits of the foregoing.

### Environment

**Water:** Prevent entry into water intakes and waterways. Harmful to aquatic life. BOD: No information.
**Land-Air:** No information.
**Food Chain Concentration Potential:** No information.

## EMERGENCY MEASURES

### Special Hazards
FLAMMABLE (spontaneously combustible).

### Immediate Responses
Keep non-involved people away from spill area. Issue warning: "FLAMMABLE". CALL FIRE DEPARTMENT. Warn about water and moisture hazard. Contact manufacturer or supplier for guidance. Contain spill area by diking with earth or other available material. Dike to contain runoff. Notify environmental authorities.

### Protective Clothing and Equipment
Respiratory protection - in fire, use self-contained breathing apparatus. Goggles - (mono), tight fitting. Gloves - rubber. Boots - high, rubber (pants worn outside boots). Protective clothing - as required - coveralls, acid suit, etc.

### Fire and Explosion
Extinguishing agents: dry sand, dry chemical, carbon dioxide. Use water only in flooding amounts. In fires, $SO_2$ is released.

### First Aid
Move victim out of spill site to fresh air. Call for medical assistance, but start first aid at once. Inhalation: if breathing has stopped, give artificial respiration; if laboured, give oxygen. Contact: eyes - irrigate with plenty of water; skin - flush with plenty of water. Ingestion: give large amounts of water or milk to conscious victim. Keep warm and quiet. If medical assistance is not immediately available, transport victim to hospital, doctor or clinic.

## ENVIRONMENTAL PROTECTION MEASURES

### Response

**Water**
1. Stop or reduce discharge if safe to do so.
2. Contact manufacturer or supplier for advice.
3. If possible, contain discharge by damming or water diversion.
4. Dredge or vacuum pump to remove contaminants, liquids and contaminated bottom sediments.
5. Notify environmental authorities to discuss disposal and cleanup of contaminated materials.

**Land-Air**
1. Stop or reduce discharge if safe to do so.
2. Contact manufacturer or supplier for advice.
3. Dike to prevent runoff from rainwater or water application.
4. Remove material by manual or mechanical means.
5. Recover undamaged containers.
6. Notify environmental authorities to discuss disposal and cleanup of contaminated materials.

### Disposal
1. Contact manufacturer or supplier for advice on disposal.
2. Contact environmental authorities for advice on disposal.

SODIUM DITHIONITE   $Na_2S_2O_4 \cdot xH_2O$ (x = 0 to 2)

# SODIUM HYDROSULFIDE   $NaHS \cdot xH_2O$ (solution) (x = 0 to 3)

## IDENTIFICATION

**UN No.** 2923 solid >25% water of crystallization
2318 solid <25% water of crystallization
NA2922 solution

### Common Synonyms
SODIUM SULPHYDRATE
SODIUM BISULFIDE
SODIUM HYDROGEN SULFIDE
SODIUM MERCAPTAN

### Observable Characteristics
Colourless to yellow crystals. Clear to light straw-coloured (yellow to black at higher temperatures). Offensive, rotten-eggs odour ($H_2S$).

### Manufacturers
Cornwall Chemical Ltd., Montreal, Quebec. Canadian Industries Ltd., (CIL), Montreal, Quebec.

### Transportation and Storage Information
**Shipping State:** Solid or liquid (aqueous solution).
**Classification:** Spontaneous combustible; corrosive.
**Inert Atmosphere:** No requirement.
**Venting:** Pressure-vacuum.
**Pump Type:** Gear, centrifugal; steel, stainless steel (solutions only).

**Label(s):** Black and white label - CORROSIVE (>25% water of crystallization). Red, black and white label - SPONTANEOUS COMBUSTIBLE (<25% water of crystallization).
**Storage Temperature:** Ambient.
**Hose Type:** Natural rubber, Hypalon, polyethylene, polypropylene.

**Grades or Purity:** Anhydrous 70 to 72% NaHS. Solutions 40 to 44% NaHS.
**Containers and Materials:** Drums, tank cars, tank trucks; steel, stainless steel (aluminum, zinc, copper, brass, bronze are not resistant).

### Physical and Chemical Characteristics
**Physical State** (20°C, 1 atm): Liquid (typically shipped as solution).
**Solubility** (Water): Soluble in all proportions.
**Molecular Weight:** 56.1 (solute).
**Vapour Pressure:** No information.
**Boiling Point:** 100 to 122°C (solution).

**Floatability** (Water): Sinks and mixes.
**Odour:** Rotten-egg odour (0.047 ppm).
**Flash Point:** Not flammable.
**Vapour Density:** No information.
**Specific Gravity:** 1.31 (15.5°C).

**Colour:** Clear, light straw colour (yellow to black at higher temperatures).
**Explosive Limits:** Not flammable (4.3 to 46% ($H_2S$)).
**Melting Point:** 17°C; (solution) loses water; 55°C (anhydrous).

## HAZARD DATA

### Human Health
**Symptoms:** Corrosive alkaline liquid. **Contact:** eyes and skin - causes severe burns. **Ingestion:** causes severe burning and corrosion of the gastrointestinal tract, pain in throat and abdomen, nausea and vomiting, diarrhea. In severe cases, collapse, unconsciousness and paralysis of respiration may occur. **Inhalation:** of NaHS mist will cause irritation of the respiratory tract and possibly systemic poisoning. Vapours are extremely dangerous as the sense of smell is rapidly lost on exposure to high concentrations or prolonged exposure to dilute concentrations of $H_2S$ in air.
**Toxicology:** Highly toxic by all routes.
TLV® (inhalation) 10 ppm; 14 mg/m$^3$ (as $H_2S$).  LC$_{50}$ - Inhalation: mouse = 673 ppm/1 h ($H_2S$).  LD$_{50}$ - No information.
Short-term Inhalation Limits - 15 ppm; 21 mg/m$^3$ for 15 min (as $H_2S$).   Delayed Toxicity - No information.

### Fire
**Fire Extinguishing Agents:** Not combustible. Most fire extinguishing agents may be used. Water should be used in flooding amounts. However, flammable and toxic $H_2S$ is liberated when heated.
**Behaviour in Fire:** Not combustible, but releases flammable $H_2S$ when heated.
**Ignition Temperature:** Not combustible.  **Burning Rate:** Not combustible.

### Reactivity
**With Water:** No reaction; soluble.
**With Common Materials:** NaHS is stable while alkaline. Reacts with acids to release hydrogen sulfide gas. Reacts violently with diazonium salts.
**Stability:** Stable.

### Environment
**Water:** Prevent entry into water intakes and waterways. Fish toxicity: 206 mg/L/96 h/mosquito fish/TLm/freshwater.
**Land-Air:** No information.
**Food Chain Concentration Potential:** No information.

## EMERGENCY MEASURES

### Special Hazards

CORROSIVE. Upon heating releases $H_2S$ (poisonous and flammable). <25% water of crystallization, spontaneously combustible.

### Immediate Responses

Keep non-involved people away from spill site. Issue warnings: "CORROSIVE, SPONTANEOUSLY COMBUSTIBLE". Call Fire Department. Avoid contact or inhalation. Contact manufacturer for advice. If $H_2S$ fumes are present, evacuate people downwind. Contain spill by diking with earth or other available material. Attempt to control discharge, if this can be done without risk. Notify environmental authorities.

### Protective Clothing and Equipment

Respiratory protection - if $H_2S$ present, use self-contained breathing apparatus. Goggles - (mono), tight fitting. If face shield used, it must not replace goggles. Gloves - rubber. Boots - high, rubber (pants worn outside boots). Outerwear - as required, acid suit, aprons, etc., rubber or vinyl.

### Fire and Explosion

Not combustible by itself. $H_2S$ is released upon heating; when containing <25% water of crystallization contact with water or moisture may cause spontaneous combustion. Most fire extinguishing agents may be used. Water should be used in flooding amounts.

### First Aid

Move victim out of spill area to fresh air. Call for medical assistance, but start first aid at once. Inhalation: if breathing has stopped, give artificial respiration; if laboured, give oxygen. Ingestion: if victim is conscious give large amounts of milk or water. Induce vomiting. Contact: eyes - immediately irrigate with plenty of water; skin - immediately flush affected areas with water and remove contaminated clothing. If medical assistance is not immediately available, transport victim to hospital, doctor or clinic.

## ENVIRONMENTAL PROTECTION MEASURES

### Response

**Water**
1. Stop or reduce discharge if safe to do so.
2. Contact manufacturer or supplier for advice.
3. If possible, contain discharge by damming or water diversion.
4. Dredge or vacuum pump to remove contaminants, liquids and contaminated bottom sediments.
5. Notify environmental authorities to discuss disposal and cleanup of contaminated materials.

**Land-Air**
1. Stop or reduce discharge if safe to do.
2. Contact manufacturer or supplier for advice.
3. Contain spill by diking with earth or other barrier.
4. If liquid, remove material with pumps or vacuum equipment and place in appropriate containers.
4a. If solid, remove material by manual or mechanical means.
5. Recover undamaged containers.
6. Absorb residual liquid on natural or synthetic sorbents.
7. Notify environmental authorities to discuss disposal and cleanup of contaminated materials.

### Disposal

1. Contact manufacturer or supplier for advice on disposal.
2. Contact environmental authorities for advice or disposal.

SODIUM HYDROSULFIDE   $NaHS \cdot xH_2O$ (solution) (x = 0 to 3)

# SODIUM HYDROXIDE  NaOH

## IDENTIFICATION

UN No. 1823 solid
1824 solution

### Common Synonyms
CAUSTIC SODA
CAUSTIC
LYE
SODA LYE
SODIUM HYDRATE
WHITE CAUSTIC

### Observable Characteristics
Clear to slightly turbid liquid, with a clear to slightly coloured (white to yellow) appearance. Odourless. Anhydrous solid: White to slightly coloured pellets or flakes.

### Manufacturers
Canadian Industries Limited, Becancour, Que.
Dow Chemical Canada Inc., Sarnia, Ont., Fort Saskatchewan, Alta.
Canadian Occidental Petroleum Ltd., Vancouver, B.C.
FMC, Squamish, B.C.

### Transportation and Storage Information

**Shipping State:** Solid and liquid (aqueous solution).
**Classification:** Corrosive.
**Inert Atmosphere:** No requirement.
**Venting:** Open.
**Pump Type:** Gear, centrifugal; steel, stainless steel (for solutions).

**Grades or Purity:** Solution, 50%, 73% NaOH. Dry - (anhydrous, 99+% NaOH), flake, powder solid.
**Containers and Materials:** Liquid - tank cars, tank trucks. Dry - cans, drums, hopper cars.
**Label(s):** White and black label - CORROSIVE; Class 8, Group II.
**Storage Temperature:** Ambient.
**Hose Type:** Natural rubber, Hypalon, polyethylene, polypropylene, flexible stainless steel (solutions).

### Physical and Chemical Characteristics

**Physical State** (20°C, 1 atm): Solid.
**Solubility (Water):** Soluble, 42 g/100 mL (0°C); 111 g/100 mL (25°C); 347 g/100 mL (100°C).
**Molecular Weight:** 40.0 (solute).
**Vapour Pressure:** 1 mm Hg (739°C); 42 mm Hg (1000°C).
**Boiling Point:** 50% solution: 142 to 148°C; 73% solution: 188 to 195°C; 1 390°C (anhydrous).
**Floatability (Water):** Sinks and mixes.
**Odour:** Odourless.
**Flash Point:** Not flammable.
**Vapour Density:** No information.
**Specific Gravity:** 213 (anhydrous); 50% sol'n, 1.53; 73% sol'n, 2.00 (15.5°C).
**Colour:** Liquid - clear to slightly coloured. Solid (dry) - white.
**Explosive Limits:** Not flammable.
**Melting Point:** Anhydrous 318°C; 50% solution, 12 to 15°C; 73% solution, 63°C.

## HAZARD DATA

### Human Health

**Symptoms: Contact:** skin - severe burns, often resulting in deep ulceration and ultimate scarring, can result from contact with solid or liquid forms. Eyes - solid or liquid; very rapidly causes severe damage. **Inhalation:** of dust or mist may cause damage to upper respiratory tract and even to lung tissue proper. **Ingestion:** (liquid or solid) severe damage to mucous membranes or deeper tissues. If perforation occurs, subsequent severe scar formation may occur. Penetration into vital areas may be fatal.
**Toxicology:** Highly toxic upon contact and ingestion.
TLV®- (inhalation) 2 mg/m$^3$ (particulate).      LC$_{50}$ - No information.      LD$_{Lo}$ - Oral: rabbit = 0.5 g/kg (10% solution).
Short-term Inhalation Limits - No information.   Delayed Toxicity - None known.

### Fire

**Fire Extinguishing Agents:** Not combustible. Most fire extinguishing agents may be used in fires involving sodium hydroxide. Use water in flooding amounts. Anhydrous form in contact with water may generate sufficient heat to ignite combustible materials. May cause ignition on contact with organic chemicals.
**Behaviour in Fire:** Not combustible.    **Burning Rate:** Not combustible.
**Ignition Temperature:** Not combustible.

### Reactivity

**With Water:** Anhydrous form dissolves with great heat evolution. Boils and splatters.
**With Common Materials:** Reacts violently with acetaldehyde, acetic acid, acetic anhydride, acrolein, acrylonitrile, allyl alcohol, allyl chloride, aluminum, chlorohydrin, chloronitrotoluenes, chlorosulfonic acid, 1,2-dichloroethylene, ethylene cyanohydrin, glyoxyl, hydrochloric acid, hydrofluoric acid, hydroquinone, maleic anhydride, nitric acid, nitroparaffins, oleum, phosphorus, phosphorus pentoxide, sulfuric acid, tetrahydrofuran and trichloroethylene.
**Stability:** Stable.

### Environment

**Water:** Prevent entry into water intakes and waterways. Harmful to aquatic life in high concentrations. Fish toxicity: 25 ppm/24 h/brook trout/LC100/freshwater; 99 ppm/48 h/bluegill/TLm/freshwater; 10 to 33 ppm/48 h/shrimp/LC50/saltwater; BOD: None.
**Land-Air:** No information.
**Food Chain Concentration Potential:** None.

## EMERGENCY MEASURES

**Special Hazards**

CORROSIVE.

**Immediate Responses**

Keep non-involved people away from spill site. Issue warning: "CORROSIVE". Call Fire Department. Contain spill by diking with earth or other available material. Contact manufacturer for advice. Avoid contact and inhalation. Stop or reduce discharge, if this can be done without risk. Notify environmental authorities.

**Protective Clothing and Equipment**

Respiratory protection - suitable respirator. Goggles - (mono), tight fitting. If face shield used, it must not replace goggles. Gloves - rubber. Boots - high, rubber (pants should be worn outside boots). Outerwear - as required; coveralls, aprons, suits - rubber, vinyl.

**Fire and Explosion**

Not combustible. Most fire extinguishing agents may be used in fires involving sodium hydroxide. Use water in flooding amounts. Anhydrous form in contact with water may generate sufficient heat to ignite combustible materials. May cause ignition on contact with organic materials.

**First Aid**

Move victim out of spill site to fresh air. Call for medical assistance, but start first aid at once. Contact: eyes - irrigate with plenty of water; skin - flush with plenty of water, while removing contaminated clothing. Continue washing (eyes and skin) for an additional 30 minutes if considered necessary. Ingestion: give large quantities of milk or water to conscious victim only, to dilute chemical. Vomiting may occur spontaneously but should not be induced. If medical assistance is not immediately available, transport victim to hospital, doctor or clinic.

## ENVIRONMENTAL PROTECTION MEASURES

**Response**

**Water**
1. Stop or reduce discharge if safe to do so.
2. Contact manufacturer or supplier for advice.
3. If possible, contain discharge by damming or water diversion.
4. Dredge or vacuum pump to remove contaminants, liquids and contaminated bottom sediments.
5. Notify environmental authorities to discuss disposal and cleanup of contaminated materials.

**Land-Air**
1. Stop or reduce discharge if safe to do so.
2. Contact manufacturer or supplier for advice.
3. Dike to prevent runoff from rainwater or water application.
4. Remove material by manual or mechanical means.
5. Recover undamaged containers.
6. Notify environmental authorities to discuss disposal and cleanup of contaminated materials.

**Disposal**
1. Contact manufacturer or supplier for advice on disposal.
2. Contact environmental authorities for advice on disposal.

SODIUM HYDROXIDE   NaOH

# SODIUM HYPOCHLORITE  NaOCl

## IDENTIFICATION

UN No. 1791

### Common Synonyms/Trade Names
CHLOROX
LIQUID BLEACH

### Observable Characteristics
Green to yellow watery liquid. Chlorine-type odour (like bleach). Pure sodium hypochlorite is a powder which is unstable in air, and rarely shipped.

### Manufacturers
Bristol-Myers Canada, Toronto, Ont.
Canadian Industries Ltd., Becancour, Que.
MacMillan Bloedel, Nanaimo, B.C.
Prince Albert Pulp, Prince Albert, Sask.

**Grades or Purity:** Technical >7% available chlorine; solutions are typically 7 to 15% chlorine.
**Containers and Materials:** Carboys, drums and tanks cars.

### Transportation and Storage Information
**Shipping State:** Liquid (aqueous solution).
**Classification:** Corrosive liquid.
**Inert Atmosphere:** No requirement.
**Venting:** Pressure-vacuum.
**Pump Type:** Steel, stainless steel.

**Label(s):** Black and white label - CORROSIVE LIQUID; Class 8, Group II.
**Storage Temperature:** Ambient.
**Hose Type:** Natural rubber, Hypalon, polyethylene, polypropylene, flexible stainless steel.

### Physical and Chemical Characteristics
**Physical State (20°C, 1 atm):** Solid.
**Solubility (Water):** Solutions are soluble in all proportions (26.0 g/100 mL (0°C) (anhydrous).
**Molecular Weight:** 74.4 (solute).
**Vapour Pressure:** No information.
**Boiling Point:** Decomposes at temperatures >100°C.

**Floatability (Water):** Sinks and mixes.
**Odour:** Like chlorine bleach.
**Flash Point:** Not flammable (solutions).
**Vapour Density:** No information.
**Specific Gravity:** 1.00 to 1.025 at 20°C (solution; depending on concentration).

**Colour:** Green to yellow.
**Explosive Limits:** Anhydrous explosive in air; solutions not flammable.
**Melting Point:** 5% solution, -6°C.

## HAZARD DATA

### Human Health
**Symptoms:** Contact: eyes and skin - irritation, burning and inflammation. Inhalation: of fumes causes severe pulmonary irritation with coughing and choking followed by pulmonary edema. Ingestion: irritation and corrosion of mucous membranes with pain and vomiting, stomach cramps, and diarrhea.
**Toxicology:** Moderately toxic by inhalation and ingestion.
TLV - 1 ppm; 3 mg/mL (as chlorine).       LC$_{50}$ - No information.        LD$_{50}$ - No information.
Short-term Inhalation Limits - 3 ppm; 9 mg/mL    Delayed Toxicity - None known.
(15 min) (as chlorine).

### Fire
**Fire Extinguishing Agents:** Solutions are not combustible; however, in fires involving sodium hypochlorite solution, most fire extinguishing materials may be used. In fires involving anhydrous form, water should only be used in flooding amounts.
**Behaviour in Fire:** Decomposes in fire, generating toxic and corrosive chlorine gas. NaOCl·5H$_2$O highly unstable, anhydrous form explosive in air.
**Ignition Temperature:** Solutions not combustible.    **Burning Rate:** Solutions not combustible.

### Reactivity
**With Water:** No reaction. Soluble. Anhydrous form reacts violently with water.
**With Common Materials:** Concentrated forms may cause combustible materials to ignite. Reacts violently with amines, ammonium acetate, ammonium carbonate, ammonium nitrate, ammonium phosphate, cellulose and ethyleneimine.
**Stability:** Solutions are stable. (Anhydrous and pentahydrate are not stable.)

### Environment
**Water:** Prevent entry into water intakes and waterways. Harmful to aquatic life. Fish toxicity: 5.9 ppm/96 h/fathead minnow/LC$_{50}$/freshwater (4 to 6% aqueous solution); 52 ppm/96 h/grass shrimps/LC$_{50}$/freshwater (4 to 6% aqueous solution); BOD: No information.
**Land-Air:** No information.
**Food Chain Concentration Potential:** None.

## EMERGENCY MEASURES

**Special Hazards**

CORROSIVE. Concentrated and other forms can be strong oxidizers.

**Immediate Responses**

Keep non-involved people away from spill site. Issue warning: "CORROSIVE". Call Fire Department. Stop discharge if safe to do so. Contain spill by diking. Contact manufacturer. Contact environmental authorities.

**Protective Clothing and Equipment**

In fires, Respiratory protection - self-containing breathing apparatus; otherwise rubber gloves and goggles, protective clothing - with pants worn outside boots.

**Fire and Explosion**

Solutions not combustible. Most fire extinguishing agents may be used in fire involving NaOCl. Cool fire-exposed containers with water. Pentahydrate and anhydrous forms are not stable. Decomposes in fires generating toxic and corrosive chlorine gas.

**First Aid**

Move victim out of spill site to fresh air. Call for medical assistance, but start first aid at once. Contact: skin - remove contaminated clothing and flush affected areas with plenty of water; eyes - flush with water. Ingestion: if victim is conscious, give milk or water to drink. Keep victim warm. Inhalation: if breathing has stopped, give artificial respiration (not mouth-to-mouth method); if laboured, give oxygen. If medical assistance is not immediately available, transport victim to hospital, doctor or clinic.

## ENVIRONMENTAL PROTECTION MEASURES

**Response**

**Water**
1. Stop or reduce discharge if safe to do so.
2. Contact manufacturer or supplier for advice.
3. If possible, contain discharge by damming or water diversion.
4. Notify environmental authorities to discuss disposal and cleanup of contaminated materials.

**Land-Air**
1. Stop or reduce discharge if safe to do so.
2. Contact manufacturer or supplier for advice.
3. Contain spill by diking with earth or other barrier.
4. Remove material with pumps or vacuum equipment and place in appropriate containers.
5. Recover undamaged containers.
6. Absorb residual liquid on natural or synthetic sorbents.
7. Notify environmental authorities to discuss disposal and cleanup of contaminated materials.

**Disposal**
1. Contact manufacturer or supplier for advice on disposal.
2. Contact environmental authorities for advice on disposal.

SODIUM HYPOCHLORITE   NaOCl

# SODIUM NITRATE  NaNO$_3$

## IDENTIFICATION

UN No. 1498

### Common Synonyms
CHILE SALTPETER
SODA NITER
NITRATINE

### Observable Characteristics
Colourless to white crystals or powder. Odourless.

### Manufacturers
No Canadian manufacturer.
Selected U.S. manufacturer:
Olin Corporation,
Lake Charles, Louisiana, USA
Canadian supplier:
Canadian Industries Limited,
Toronto, Ont. (and branches across Canada).

### Transportation and Storage Information

**Shipping State:** Solid.
**Classification:** Oxidizing material.
**Inert Atmosphere:** No requirement.
**Venting:** Open.

**Label(s):** Yellow label - OXIDIZER; Class 5.1, Group III.
**Storage Temperature:** Ambient.

**Grades or Purity:** Technical, synthetic - 99.4% NaNO$_3$ min.
**Containers and Materials:** Polyethylene-lined paper bags. Carloads or bulk.

### Physical and Chemical Characteristics

**Physical State** (20°C, 1 atm): Solid.
**Solubility** (Water): 73 g/100 mL (0°C); 92.1 g/100 mL (25°C); 180 g/100 mL (100°C).
**Molecular Weight:** 85.01
**Vapour Pressure:** No information.
**Boiling Point:** Decomposes at 380°C.

**Floatability** (Water): Sinks and mixes.
**Odour:** Odourless.
**Flash Point:** Not flammable.
**Vapour Density:** No information.
**Specific Gravity:** 2.26 (25°C).

**Colour:** Colourless to white.
**Explosive Limits:** Explodes at 537°C.
**Melting Point:** 307°C.

## HAZARD DATA

### Human Health

**Symptoms: Contact:** reddening of skin and eyes. **Inhalation:** of toxic fumes induces coughing and shortness of breath. **Ingestion:** dizziness, abdominal cramps, vomiting, convulsions, cyanosis, loss of consciousness.
**Toxicology:** Highly toxic by ingestion.
TLV®- No information.
Short-term Inhalation Limits - No information.

LC$_{50}$ - No information.
Delayed Toxicity - No information.

LD$_{50}$ - No information.
LD$_{Lo}$ - Oral: rat = 0.20 g/kg

### Fire

**Fire Extinguishing Agents:** Not directly combustible; explodes. In fires involving sodium nitrate, use flooding amounts of water.
**Behaviour in Fire:** May explode at 537°C. Decomposes with heat to produce toxic NO$_x$ fumes.
**Ignition Temperature:** 537°C.
**Burning Rate:** No information.

### Reactivity

**With Water:** No reaction; soluble.
**With Common Materials:** Strong oxidizer, can ignite many organic materials. Reacts violently with antimony, cyanides, sodium hypophosphite, sulfur and charcoal.
**Stability:** May explode when detonated or heated to 537°C.

### Environment

**Water:** Prevent entry into water intakes or waterways. Fish toxicity: 665 mg/L/96 h/Daphnia magna/TLm/freshwater; BOD: Not available.
**Land-Air:** No information.
**Food Chain Concentration Potential:** No information.

## EMERGENCY MEASURES

### Special Hazards
OXIDIZER.

### Immediate Responses
Keep non-involved people away from spill site. Issue warning: "OXIDIZER". Call Fire Department. Stop discharge if safe to do so. Remove unbroken bags or drums from spill area. Dike to contain spill. Contact manufacturer. Notify environmental authorities.

### Protective Clothing and Equipment
Respiratory protection - in fires, self-contained breathing apparatus. Otherwise; Gloves - rubber. Safety glasses and work clothing, coveralls, apron, etc.

### Fire and Explosion
Not directly combustible; explodes. Emits toxic $NO_x$ fumes upon heating. Use flooding amounts of water to extinguish.

### First Aid
Move victim out of spill site to fresh air. Call for medical assistance, but start first aid at once. Contact: skin - remove clothing and flush with large amounts of water. Inhalation: if in fire, fumes of nitrous oxide may be present. If breathing has stopped, give artificial respiration. If breathing is difficult, give oxygen. Ingestion: give milk or water to conscious victim. Induce vomiting. If medical assistance is not quickly available, transport victim to hospital, doctor or clinic.

## ENVIRONMENTAL PROTECTION MEASURES

### Response

**Water**
1. Stop or reduce discharge if safe to do so.
2. Contact manufacturer or supplier for advice.
3. If possible, contain discharge by damming or water diversion.
4. Dredge or vacuum pump to remove contaminants, liquids and contaminated bottom sediments.
5. Notify environmental authorities to discuss cleanup and disposal of contaminated materials.

**Land-Air**
1. Stop or reduce discharge if safe to do so.
2. Contact manufacturer or supplier for advice.
3. Dike to prevent runoff from rainwater or water application.
4. Remove material by manual or mechanical means.
5. Recover undamaged containers.
6. Notify environmental authorities to discuss cleanup and disposal of contaminated materials.

### Disposal
1. Contact manufacturer or supplier for advice on disposal.
2. Contact environmental authorities for advice on disposal.

SODIUM NITRATE    $NaNO_3$

# SODIUM PENTACHLOROPHENATE  $NaC_6Cl_5O$

## IDENTIFICATION

UN No. 2567

### Common Synonyms

SODIUM PENTACHLOROPHENOLATE
DOWICIDE-G
SODIUM PENTACHLOROPHENOL
SODIUM PENTACHLOROPHENOXIDE
PENTACHLOROPHENOL, SODIUM SALT

### Observable Characteristics

White to light-brown powder. Slight, characteristic odour (phenolic or carbolic odour).

### Manufacturers

No Canadian manufacturers.
Canadian suppliers:
P. Leiner & Sons, Toronto, Ont.
Monsanto Canada, Mississauga, Ont.
Originating from:
W.R. Grace & Co. USA, Monsanto, USA

### Transportation and Storage Information

**Shipping State:** Solid.
**Classification:** Poison.
**Inert Atmosphere:** No requirement.
**Venting:** Open.

**Label(s):** White label - POISON; Class 6.1, Group II.
**Storage Temperature:** Ambient.

**Grades or Purity:** Technical.
**Containers and Materials:** Bags, drums, carlots.

### Physical and Chemical Characteristics

**Physical State** (20°C, 1 atm): Solid.
**Solubility** (Water): 33 g/100 mL (25°C.)
**Molecular Weight:** 288.4
**Vapour Pressure:** 0.0002 mm Hg (25°C).
**Boiling Point:** No information.

**Floatability** (Water): Sinks and mixes.
**Odour:** Slight, characteristic (phenolic or carbolic).
**Flash Point:** Not flammable.
**Vapour Density:** 9.2
**Specific Gravity:** 2.0 (25°C).

**Colour:** White to light brown.
**Explosive Limits:** No information.
**Melting Point:** No information.

## HAZARD DATA

### Human Health

**Symptoms:** Inhalation: nose irritation, nausea and vomiting, coughing, high fever, profuse sweating, rapid difficult breathing; cyanosis, convulsions, loss of consciousness, death due to heart failure, risk of severe pulmonary edema if victim survives. Ingestion: same as inhalation. Contact: skin and eyes - irritation, inflammation and blisters.

**Toxicology:** Highly toxic by all routes.
$TLV^{®}$ (skin) 0.5 mg/m³ (as pentachlorophenol).
Short-term Inhalation Limits - (skin) - 1.5 mg/m³ (15 min).

$LC_{50}$ - No information.
Delayed Toxicity - No information.

$LD_{50}$ - **Oral:** rat = 0.21 g/kg
$LD_{50}$ - **Subcutaneous:** rat = 0.072 g/kg

### Fire

**Fire Extinguishing Agents:** Not combustible. Most fire extinguishing agents may be used in fires involving sodium pentachlorophenate.
**Behaviour in Fire:** When heated to decomposition, toxic fumes of chlorides are evolved.
**Ignition Temperature:** No information. **Burning Rate:** No information.

### Reactivity

**With Water:** No reaction; soluble.
**With Common Materials:** No information.
**Stability:** Stable.

### Environment

**Water:** Prevent entry into water intakes or waterways. Highly toxic to aquatic life: 20 ppm/algae/growth stopped immediately/freshwater; 0.2-0.6 ppm/19 species fish/lethal range/freshwater; Aquatic toxicity rating: <1 ppm/96 h/TLm/freshwater; BOD: No information.
**Land-Air:** Toxic to fungi and plants.
**Food Chain Concentration Potential:** Suspected.

## EMERGENCY MEASURES

**Special Hazards**

POISON.

**Immediate Responses**

Keep non-involved people away from spill site. Issue warning: "POISON". Avoid contact and inhalation of dust. Stay upwind. If water is used, dike area to prevent runoff. Notify supplier. Notify environmental authorities.

**Protective Clothing and Equipment**

In fires, Respiratory protection - self-contained breathing apparatus. Otherwise use dust respirator. Acid suit (jacket and pants) or coveralls. Goggles -(mono), tight fitting. Gloves - neoprene. Boots - high, rubber (pants worn over boots).

**Fire and Explosion**

Not flammable; in fires involving sodium pentachlorophenate; use water, dry chemical, foam or carbon dioxide to extinguish. Cool fire-exposed containers with water. In fire, wear self-contained breathing apparatus to avoid toxic decomposition products (chlorides).

**First Aid**

Move victim out of spill site to fresh air. Call for medical assistance, but start first aid immediately. **Inhalation:** give artificial respiration if breathing has stopped (not mouth-to-mouth method); give oxygen if breathing is laboured. **Contact:** skin and eyes - remove dust-contaminated clothing and wash eyes and affected skin with plenty of warm water for at least 15 min. **Ingestion:** give warm water to conscious victim and induce vomiting. Repeat several times. Keep warm and quiet. If medical assistance is not immediately available, transport victim to hospital, doctor or clinic.

## ENVIRONMENTAL PROTECTION MEASURES

**Response**

**Water**
1. Stop or reduce discharge if safe to do so.
2. Contact manufacturer or supplier for advice.
3. If possible, contain discharge by damming or water diversion.
4. Dredge or vacuum pump to remove contaminants, liquids and contaminated bottom sediments.
5. Notify environmental authorities to discuss disposal and cleanup of contaminated materials.

**Land-Air**
1. Stop or reduce discharge if safe to do so.
2. Contact manufacturer or supplier for advice.
3. Dike to prevent runoff from rainwater or water application.
4. Remove material by manual or mechanical means.
5. Recover undamaged containers.
6. Absorb residual liquid on natural or synthetic sorbents.
7. Remove contaminated soil for disposal.
8. Notify environmental authorities to discuss disposal and cleanup of contaminated materials.

**Disposal**
1. Contact manufacturer or supplier for advice on disposal.
2. Contact environmental authorities for advice on disposal.

SODIUM PENTACHLOROPHENATE   $NaC_6Cl_5O$

# SODIUM PHOSPHATE (dibasic)   $Na_2HPO_4$

## IDENTIFICATION

UN No. NA9147

### Common Synonyms

DSP
PHOSPHATE OF SODA
DISODIUM HYDROGEN PHOSPHATE
DISODIUM PHOSPHATE
DISODIUM MONOHYDROGEN PHOSPHATE
SODIUM ORTHOPHOSPHATE, SECONDARY
SECONDARY SODIUM PHOSPHATE

### Observable Characteristics

Commonly shipped in various forms:
anydrous $Na_2HPO_4$
dihydrate $Na_2HPO_4 \cdot 2H_2O$
heptahydrate $Na_2HPO_4 \cdot 7H_2O$
dodecahydrate $Na_2HPO_4 \cdot 12H_2O$

Colourless, translucent powders and crystals. Odourless.

### Manufacturers

Erco Industries,
Port Maitland, Ontario,
Buckingham, Que.

### Transportation and Storage Information

**Shipping State:** Solid.
**Classification:** Class 9.2
**Inert Atmosphere:** No requirement.
**Venting:** Open.

**Label(s):** Not regulated.
**Storage Temperature:** Ambient.

**Grades or Purity:** Commercial.
**Containers and Materials:** Paper bags, drums, barrels.

### Physical and Chemical Characteristics

**Physical State** (20°C, 1 atm): Solid.
**Solubility** (Water): dihydrate 100 g/100 mL (50°C), 117 g/100 mL (80°C); anhydrous 1.63 g/100 mL (0°C); heptahydrate 104 g/100 mL (40°C); dodecahydrate 4.2 g/100 mL (0°C), 87 g/100 mL (40°C).
**Molecular Weight:** anhyd. 142; dihydrate 178; hepta. 268; dodeca. 358.
**Vapour Pressure:** No information.
**Boiling Point:** Anhydrous at 240°C is converted to sodium pyrophosphate; dihydrate at 92.5°C loses $2H_2O$; hepta hydrate at 48°C loses $5H_2O$; dodecahydrate at 100°C loses $12H_2O$.

**Floatability** (Water): Sinks and mixes.
**Odour:** Odourless.
**Flash Point:** Not flammable.
**Vapour Density:** No information.
**Specific Gravity:** dihydrate 2.07 (15°C); hepta. 1.68 (25°C); dodeca. 1.52 (25°C).

**Colour:** Colourless to translucent.
**Explosive Limits:** Not flammable.
**Melting Point:** Anhydrous 240°C converted to sodium pyrophosphate; dihydrate 95°C loses $2H_2O$; heptahydrate 48°C loses $5H_2O$; dodecahydrate 35.1°C loses $5H_2O$.

## HAZARD DATA

### Human Health

**Symptoms:** Inhalation: of dust may irritate nose and throat. Ingestion: injures mouth, throat and gastrointestinal tract resulting in vomiting, cramps and diarrhea; pain and burning in mouth may occur. Contact: eyes – local irritation to chronic damage; skin – dermatitis.
**Toxicology:** Low toxicity by ingestion.
**TLV®** - No information.
**Short-term Inhalation Limits** - No information.

$LC_{50}$ - No information.
**Delayed Toxicity** - No information.

$LD_{50}$ - Oral: rat = 12.9 g/kg (hepta).
$LD_{Lo}$ - Intraperitoneal: rat = 1.0 g/kg (anhydrous).

### Fire

**Fire Extinguishing Agents:** Not combustible. Most fire extinguishing agents may be used in fires involving sodium phosphate (dibasic).
**Behaviour in Fire:** Anhydrous forms toxic $PO_x$ upon decomposition in fires.
**Ignition Temperature:** Not combustible.
**Burning Rate:** Not combustible.

### Reactivity

**With Water:** No reaction; soluble.
**With Common Materials:** Reacts violently with magnesium.
**Stability:** Stable.

### Environment

**Water:** Prevent entry into water intakes and waterways. Harmful to aquatic life. Fish toxicity: 126 ppm/92 h/Daphnia magna/TLm/freshwater.
**Land-Air:** No information.
**Food Chain Concentration Potential:** None.

## EMERGENCY MEASURES

**Special Hazards**

**Immediate Responses**

Keep non-involved people away from site. Avoid contact and inhalation of dust. Dike to prevent runoff. Notify manufacturer or supplier for advice. Notify environmental authorities.

**Protective Clothing and Equipment**

In fires, Respiratory protection - self-contained breathing apparatus; otherwise, safety goggles or face shield, dust mask, rubber gloves, rubber boots, and protective clothing as required.

**Fire and Explosion**

Not combustible. Most fire extinguishing agents may be used in fires involving sodium phosphate (dibasic). Toxic $PO_x$ fumes are formed upon decomposition in fires.

**First Aid**

Move victim out of spill site to fresh air. Call for medical assistance, but start first aid at once. Dust Contact: eyes - flush with water. Inhalation: if breathing has stopped give artificial respiration; if laboured, give oxygen. Solid Contact: skin - remove contaminated clothing and flush affected areas with water. Ingestion: give conscious victim plenty of water to drink and induce vomiting. If medical assistance is not immediately available, transport victim to hospital, doctor or clinic.

## ENVIRONMENTAL PROTECTION MEASURES

**Response**

**Water**
1. Stop or reduce discharge if safe to do so.
2. Contact manufacturer or supplier for advice.
3. If possible, contain discharge by damming or water diversion.
4. Dredge or vacuum pump to remove contaminants, liquids and contaminated bottom sediments.
5. Notify environmental authorities to discuss disposal and cleanup of contaminated materials.

**Land-Air**
1. Stop or reduce discharge if safe to do so.
2. Contact manufacturer or supplier for advice.
3. Dike to prevent runoff from rainwater or water application.
4. Remove material by manual or mechanical means.
5. Recover undamaged containers.
6. Notify environmental authorities to discuss disposal and cleanup of contaminated materials.

**Disposal**
1. Contact manufacturer or supplier for advice on disposal.
2. Contact environmental authorities for advice on disposal.

SODIUM PHOSPHATE (dibasic)   $Na_2HPO_4$

# SODIUM PHOSPHATE (monobasic)   $NaH_2PO_4$

## IDENTIFICATION

### Common Synonyms
MSP
MONOSODIUM DIHYDROGEN PHOSPHATE
MONOSODIUM PHOSPHATE
SODIUM BIPHOSPHATE
SODIUM ACID PHOSPHATE
SODIUM ORTHOPHOSPHATE, primary
(also shipped as a hydrate $NaH_2PO_4 \cdot H_2O$)

### Observable Characteristics
White, crystalline powder. Odourless.
Monohydrate - large translucent crystals.

### Manufacturers
Erco Industries,
Port Maitland, Ont.
Buckingham, Que.

### Transportation and Storage Information
**Shipping State:** Solid.
**Classification:** Not regulated.
**Inert Atmosphere:** No requirement.
**Venting:** Open.
**Label(s):** Not regulated.
**Storage Temperature:** Ambient.
**Grades or Purity:** Technical; commercial.
**Containers and Materials:** Paper bags, drums and barrels.

### Physical and Chemical Characteristics
**Physical State** (20°C, 1 atm): Solid.
**Solubility** (Water): 59.9 g/100 mL (0°C); 427 g/100 mL (100°C) (monohydrate); anhydrous and dihydrate very soluble.
**Molecular Weight:** 120 (anhydrous); 138 (monohydrate).
**Vapour Pressure:** No information.
**Boiling Point:** monohydrate - loses water at 100°C; decomposes at 138°C; anhydrous form forms sodium acid pyrophosphate at 225 to 250°C; and sodium metaphosphate at 350 to 400°C.
**Floatability** (Water): Sinks and mixes.
**Odour:** Odourless.
**Flash Point:** Not flammable.
**Vapour Density:** No information.
**Specific Gravity:** 2.04 (monohydrate).
**Colour:** White.
**Explosive Limits:** Not flammable.
**Melting Point:** Decomposes at 204°C; monohydrate loses water at 100°C.

## HAZARD DATA

### Human Health
**Symptoms:** Inhalation: of dust may irritate nose and throat. Ingestion: injures mouth, throat and gastrointestinal tract resulting in vomiting, cramps and diarrhea; pain and burning in mouth may occur. Contact: eyes - local irritation to chronic damage, skin - dermatitis.
**Toxicology:** Low by all routes.
**TLV®** No information.   **LC$_{50}$** - No information.   **LD$_{50}$** - Intramuscular:  rat = 0.25 g/kg
**Short-term Inhalation Limits** - No information.   **Delayed Toxicity** - No information.

### Fire
**Fire Extinguishing Agents:** Not combustible. Most fire extinguishing agents may be used in fires involving sodium phosphate (monobasic).
**Behaviour in Fire:** At high temperatures, forms highly toxic $PO_x$ fumes.
**Ignition Temperature:** Not combustible.   **Burning Rate:** Not combustible.

### Reactivity
**With Water:** None; soluble (forms a weak acidic solution with water).
**With Common Materials:** Reacts violently with magnesium.
**Stability:** Stable.

### Environment
**Water:** Prevent entry into water intakes and waterways. Harmful to aquatic life. Fish toxicity:  126 ppm/72 h/Daphnia magna/TLm/freshwater.
**Land-Air:** No information.
**Food Chain Concentration Potential:** None.

## EMERGENCY MEASURES

### Special Hazards

### Immediate Responses
Keep non-involved people away from site. Avoid contact and inhalation of dust. Dike to prevent runoff. Notify manufacturer. Notify environmental authorities.

### Protective Clothing and Equipment
In fires, Respiratory protection - self-contained breathing apparatus; otherwise, safety goggles or face shield, dust mask, rubber gloves, rubber boots, and protective clothing as required.

### Fire and Explosion
Not combustible. Most fire extinguishing agents may be used in fires involving sodium phosphate (monobasic). At high temperatures $PO_x$ fumes are emitted.

### First Aid
Move victim out of spill site to fresh air. Call for medical assistance, but start first aid at once. Dust: Contact: eyes - flush with water. Inhalation: if breathing has stopped, give artificial respiration; if laboured, give oxygen. Solid: Contact: skin - remove contaminated clothing and flush affected areas with water. Ingestion: induce vomiting in conscious victim; give water to drink and keep warm. If medical assistance is not immediately available, transport victim to hospital, doctor or clinic.

## ENVIRONMENTAL PROTECTION MEASURES

### Response

**Water**
1. Stop or reduce discharge if safe to do so.
2. Contact manufacturer or supplier for advice.
3. If possible, contain discharge by damming or water diversion.
4. Dredge or vacuum pump to remove contaminants, liquids and contaminated bottom sediments.
5. Notify environmental authorities to discuss disposal and cleanup of contaminated materials.

**Land-Air**
1. Stop or reduce discharge if safe to do so.
2. Contact manufacturer or supplier for advice.
3. Dike to prevent runoff from rainwater or water application.
4. Remove material by manual or mechanical means.
5. Recover undamaged containers.
6. Notify environmental authorities to discuss disposal and cleanup of contaminated materials.

### Disposal
1. Contact manufacturer or supplier for advice on disposal.
2. Contact environmental authorities for advice on disposal.

SODIUM PHOSPHATE (monobasic)   $NaH_2PO_4$

# SODIUM PHOSPHATE (tribasic)  $Na_3PO_4 \cdot 12H_2O$

## IDENTIFICATION

UN No. NA9148

### Common Synonyms

TSP
SODIUM ORTHOPHOSPHATE
SODIUM PHOSPHATE
TRIBASIC SODIUM PHOSPHATE
TRISODIUM ORTHOPHOSPHATE

**Forms:** dodecahydrate $Na_3PO_4 \cdot 12H_2O$
**Note:** Information on this sheet is for the dodecahydrate, the most commonly shipped form.

### Observable Characteristics

Crystals. Odourless. Clear to white.

### Manufacturers

Erco Industries.
Port Maitland, Ontario,
Buckingham, Quebec.

### Transportation and Storage Information

**Shipping State:** Solid.
**Classification:** Class 9.2.
**Inert Atmosphere:** No requirement.
**Venting:** Open.

**Label(s):** Not required.
**Storage Temperature:** Ambient.

**Grades or Purity:** Commercial, CP.
**Containers and Materials:** Barrels and bags.

### Physical and Chemical Characteristics

**Physical State** (20°C, 1 atm): Solid.
**Solubility** (Water): 1.5 g/100 mL (0°C); 28.3 g/100 mL (15°C); 157 g/100 mL (70°C).
**Molecular Weight:** 380.12
**Vapour Pressure:** No information.
**Boiling Point:** 73 to 77°C decomposes; loses $12H_2O$ at 100°C.

**Floatability** (Water): Sinks and mixes.
**Odour:** Odourless.
**Flash Point:** Not flammable.
**Vapour Density:** No information.
**Specific Gravity:** 1.62 (20°C).

**Colour:** Clear to white.
**Explosive Limits:** Not flammable.
**Melting Point:** Decomposes at 73 to 77°C.

## HAZARD DATA

### Human Health

**Symptoms:** Inhalation: nose, eyes and throat, irritation; sneezing, difficulty in breathing, coughing. Ingestion: burning sensation in mouth, pain in swallowing, stomach cramps. Contact: skin - itching, burning sensation, inflammation; eyes - irritation and burning.
**Toxicology:** Moderately toxic by ingestion and inhalation.
**TLV®** - No information.   $LC_{50}$ - No information.   $LD_{50}$ - Oral: rat = 7.4 g/kg (hepta.)
**Short-term Inhalation Limits** - No information.   **Delayed Toxicity** - No information.

### Fire

**Fire Extinguishing Agents:** Not combustible. Most fire extinguishing agents may be used in fires involving sodium phosphate (tribasic).
**Behaviour in Fire:** Not combustible, but at high temperatures degrades and gives off toxic $PO_x$ fumes.
**Ignition Temperature:** Not combustible.   **Burning Rate:** Not combustible.

### Reactivity

**With Water:** No reaction; soluble. Solutions are caustic.
**With Common Materials:** Reacts violently with magnesium.
**Stability:** Stable.

### Environment

**Water:** Prevent entry into water intakes and waterways. Harmful to aquatic life; Fish toxicity: 126 ppm/72 h/Daphnia magna/TLm/freshwater.
**Land-Air:** No information.
**Food Chain Concentration Potential:** None.

## EMERGENCY MEASURES

**Special Hazards**

**Immediate Responses**

Keep non-involved people away from site. Avoid contact and inhalation of dust. Dike to prevent runoff. Notify manufacturer or supplier. Notify environmental authorities.

**Protective Clothing and Equipment**

In fires, Respiratory protection - self-contained breathing apparatus; otherwise, safety goggles or face shield, dust mask, rubber gloves, rubber boots, and protective clothing as required.

**Fire and Explosion**

Not combustible. Most fire extinguishing agents may be used in fires involving sodium phosphate (tribasic). Emits toxic $PO_x$ fumes upon heating.

**First Aid**

Move victim out of spill site to fresh air. Call for medical assistance, but start first aid at once. Dust Contact: eyes - flush with water. Inhalation: if breathing has stopped, give artificial respiration; if laboured, give oxygen. Solid Contact: skin - remove contaminated clothing and flush affected areas with water. Ingestion: give plenty of water to conscious victim and induce vomiting. If medical assistance is not immediately available, transport victim to hospital, doctor or clinic.

## ENVIRONMENTAL PROTECTION MEASURES

**Response**

**Water**
1. Stop or reduce discharge if safe to do so.
2. Contact manufacturer or supplier for advice.
3. If possible, contain discharge by damming or water diversion.
4. Dredge or vacuum pump to remove contaminants, liquids and contaminated bottom sediments.
5. Notify environmental authorities to discuss disposal and cleanup of contaminated materials.

**Land-Air**
1. Stop or reduce discharge if safe to do so.
2. Contact manufacturer or supplier for advice.
3. Dike to prevent runoff from rainwater or water application.
4. Remove material by manual or mechanical means.
5. Recover undamaged containers.
6. Notify environmental authorities to discuss disposal and cleanup of contaminated materials.

**Disposal**
1. Contact manufacturer or supplier for advice on disposal.
2. Contact environmental authorities for advice on disposal.

SODIUM PHOSPHATE (tribasic)   $Na_3PO_4 \cdot 12H_2O$

# SODIUM SILICATE  $Na_2O \cdot nSiO_2 \cdot xH_2O$ (n = 1 to 5)

## IDENTIFICATION

### Common Synonyms
SILICATE OF SODA
WATER GLASS
SODIUM METASILICATE (n=1).

### Observable Characteristics
Colourless to greenish-white powder. Odourless. Solutions are also colourless to green-white and viscous in appearance.

### Manufacturers
National Silicates Ltd., Toronto, Ont.
Valleyfield, Que.

### Transportation and Storage Information
**Shipping State:** Solid or liquid (aqueous solution).
**Classification:** Not regulated.
**Inert Atmosphere:** No requirement.
**Venting:** Open.
**Pump Type:** Gear, centrifugal steel, stainless steel (for solutions).

**Label(s):** Not regulated.
**Storage Temperature:** Ambient.
**Hose Type:** Hypalon, polyethylene, polypropylene, natural rubber.

**Grades or Purity:** 40, 47, 52° Bé.
**Containers and Materials:** Drums, tank cars, tank trucks; steel.

### Physical and Chemical Characteristics
**Physical State** (20°C, 1 atm): Solid.
**Solubility (Water):** Soluble to very soluble depending on form.
**Molecular Weight:** Varies (122 to 288).
**Vapour Pressure:** No information.
**Boiling Point:** Solutions decompose at about 100°C.

**Floatability (Water):** Sinks and mixes.
**Odour:** Odourless.
**Flash Point:** Not flammable.
**Vapour Density:** No information.
**Specific Gravity:** (solution) 1.1 to 1.8 (20°C); (solid) 2.4 (metasilicate).

**Colour:** Colourless to greenish-white.
**Explosive Limits:** Not flammable.
**Melting Point:** Variable, solutions decompose at about 100°C. Solids: disilicate 874°C; metasilicate 1 088°C; orthosilicate 1 018°C.

## HAZARD DATA

### Human Health
**Symptoms:** Contact: skin - burning and inflammation; eyes - pain, watering. Ingestion: burning, nausea, vomiting and stomach cramps. Inhalation: sneezing, difficulty in breathing, coughing fits and chemical bronchitis.
**Toxicology:** Low toxicity by all routes.
**TLV®:** No information.
**Short-term Inhalation Limits** - No information.

$LC_{50}$ - No information.
Delayed Toxicity - No information.

$LD_{50}$ - Oral: mouse = 1.10 g/kg
$LD_{50}$ - Oral: rat = 1.30 g/kg

### Fire
**Fire Extinguishing Agents:** Not combustible. Most fire extinguishing agents may be used on fires involving sodium silicate.
**Behaviour in Fire:** Not combustible.
**Ignition Temperature:** Not combustible.

**Burning Rate:** Not combustible.

### Reactivity
**With Water:** No reaction, soluble.
**With Common Materials:** Reacts violently with fluorine.
**Stability:** Stable.

### Environment
**Water:** Prevent entry into water intakes and waterways. Harmful to aquatic life in high concentrations. Fish toxicity: 2 320 ppm/96 h/mosquito fish/TLm/ freshwater; 216 ppm/96 h/Daphnia magna/TLm/lake water; BOD: None.
**Land-Air:** No information.
**Food Chain Concentration Potential:** None.

## EMERGENCY MEASURES

**Special Hazards**

**Immediate Responses**

Keep non-involved people away from spill site. Contain spill by diking if there is water runoff from fire or rain. Avoid contact and inhalation. Notify manufacturer. Notify environmental authorities.

**Protective Clothing and Equipment**

Safety goggles - tight fitting. Rubber gloves. Protective work clothing, as required.

**Fire and Explosion**

Not combustible. Most fire extinguishing agents may be used on fires involving sodium silicate.

**First Aid**

Move victim out of spill site to fresh air. Call for medical assistance, but start first aid at once. Contact: remove contaminated clothing and wash eyes and skin thoroughly with plenty of water. Ingestion: give water or milk to conscious victim. If medical assistance is not immediately available, transport victim to hospital, doctor or clinic.

## ENVIRONMENTAL PROTECTION MEASURES

**Response**

**Water**
1. Stop or reduce discharge if safe to do so.
2. Contact manufacturer or supplier for advice.
3. If possible, contain discharge by damming or water diversion.
4. Dredge or vacuum pump to remove contaminants, liquids and contaminated bottom sediments.
5. Notify environmental authorities to discuss disposal and cleanup of contaminated materials.

**Land-Air**
1. Stop or reduce discharge if safe to do so.
2. Contact manufacturer or supplier for advice.
3. Dike to prevent runoff from rainwater or water application.
4. Remove material by manual or mechanical means.
5. Recover undamaged containers.
6. Notify environmental authorities to discuss disposal and cleanup of contaminated materials.

**Disposal**
1. Contact manufacturer or supplier for advice on disposal.
2. Contact environmental authorities for advice on disposal.
3. Contaminated materials may be buried in a secured landfill site (with environmental authorities' approval).

SODIUM SILICATE   $Na_2O \cdot nSiO_2 \cdot xH_2O$ (n = 1 to 5)

# SODIUM SULFATE  $Na_2SO_4 \cdot xH_2O$ (x = 0, 7, 10)

## IDENTIFICATION

**Common Synonyms**
DISODIUM SULFATE
GLAUBER'S SALT ($Na_2SO_4 \cdot 10H_2O$)(decahydrate)
SALT CAKE
Also occurs in two basic mineral forms: thenardite and metathenardite; heptahydrate and decahydrate forms.

**Observable Characteristics**
White crystals or powder. Odourless.

**Manufacturers**
Saskatchewan Minerals, Chaplin, Sask.
Ormiston Mining and Smelting, Ormiston, Fox Valley, Sask.,
Alberta Sulfate, Metiskow, Alta.

### Transportation and Storage Information
**Shipping State:** Solid.
**Classification:** None.
**Inert Atmosphere:** No requirement.
**Venting:** Open.

**Label(s):** Not regulated.
**Storage Temperature:** Ambient.

**Grades or Purity:** Technical and CP.
**Containers and Materials:** Bags, drums or bulk.

### Physical and Chemical Characteristics
**Physical State** (20°C, 1 atm): Solid.
**Solubility** (Water): thenardite 4.7 g/100 mL (0°C); 42.7 g/100 mL (100°C); metathenardite 48.8 g/100 mL (40°C); heptahydrate 19.5 g/100 mL (0°C); 44 g/100°mL (20°C); decahydrate 11 g/100 mL (0°C); 36 g/100 mL (15°C); 92.7 g/100 mL (30°C).
**Molecular Weight:** 142 anhydrous; 268 hepta.; 322 decahydrate.
**Vapour Pressure:** No information.
**Boiling Point:** Loses $10H_2O$ at 100°C (decahydrate).

**Floatability** (Water): Sinks and mixes.
**Odour:** Odourless.
**Flash Point:** Not flammable.
**Vapour Density:** No information.
**Specific Gravity:** thenardite 2.66, metathenardenite 2.70, decahydrate 1.46 (25°C).

**Colour:** White.
**Explosive Limits:** Not flammable.
**Melting Point:** Hepta turns to anhydrous at 24.4°C; anhydrous 884°C; decahydrate 32.4°C.

## HAZARD DATA

### Human Health
**Symptoms: Ingestion:** low toxicity. **Contact:** skin - not irritating, but prolonged contact with concentrated solutions or powders should be avoided; eyes - mechanical irritation and watering.
**Toxicology:** Low toxicity by all routes.
**TLV®:** No information.
**Short-term Inhalation Limits** - No information.

$LC_{50}$ - No information.
**Delayed Toxicity** - No information.

$LD_{50}$ - Oral: mouse = 5.99 g/kg (anhydrous)

### Fire
**Fire Extinguishing Agents:** Not combustible. Most fire extinguishing agents may be used in fires involving sodium sulfate.
**Behaviour in Fire:** Not combustible.
**Ignition Temperature:** Not combustible.

**Burning Rate:** Not combustible.

### Reactivity
**With Water:** No reaction; soluble.
**With Common Materials:** Reacts violently with aluminum under certain circumstances.
**Stability:** Stable.

### Environment
**Water:** Prevent entry into water intakes and waterways. Fish toxicities: 13,500 ppm/96 h/bluegill/$LC_{50}$/freshwater; 5 200 ppm/48 h/Daphnia magna/$LC_{100}$/freshwater; 16,500 ppm/96 h/mosquito fish/TLm/freshwater.
**Land-Air:** 7 500 mg/L/15 days/poultry/$LC_{33}$.
**Food Chain Concentration Potential:** No information.

## EMERGENCY MEASURES

**Special Hazards**

**Immediate Responses**

Keep non-involved people away from spill site. Avoid contact and inhalation. Dike to prevent runoff from water application or rainwater. Notify manufacturer. Notify environmental authorities.

**Protective Clothing and Equipment**

Safety goggles - (tight fitting). Rubber gloves. Protective clothing, as required.

**Fire and Explosion**

Not combustible. Most fire extinguishing agents may be used in fires involving sodium silicate.

**First Aid**

Move victim out of spill site to fresh air. Call for medical assistance, but start first aid at once. Contact: remove contaminated clothing and wash eyes and skin thoroughly with plenty of water. Ingestion: give water or milk to conscious victim. If medical assistance is not immediately available, transport victim to hospital, doctor or clinic.

## ENVIRONMENTAL PROTECTION MEASURES

**Response**

**Water**
1. Stop or reduce discharge if safe to do so.
2. Contact manufacturer or supplier for advice.
3. If possible, contain discharge by damming or water diversion.
4. Dredge or vacuum pump to remove contaminants, liquids and contaminated bottom sediments.
5. Notify environmental authorities to discuss disposal and cleanup of contaminated materials.

**Land-Air**
1. Stop or reduce discharge if safe to do so.
2. Contact manufacturer or supplier for advice.
3. Dike to prevent runoff from rainwater or water application.
4. Remove material by manual or mechanical means.
5. Recover undamaged containers.
6. Notify environmental authorities to discuss disposal and cleanup of contaminated materials.

**Disposal**
1. Contact manufacturer or supplier for advice on disposal.
2. Contact environmental authorities for advice on disposal.
3. Contaminated materials may be buried in a secured landfill site (with environmental authorities' approval).

SODIUM SULFATE   $Na_2SO_4 \cdot xH_2O$ (x = 0, 7, 10)

# SODIUM SULFITE   $Na_2SO_3$

## IDENTIFICATION

**Common Synonyms**
Found in two forms: anhydrous and heptahydrate. ($Na_2SO_3 \cdot 7H_2O$)

**Observable Characteristics**
White crystals or powder. Odourless.

**Manufacturers**
International Paper Consolidated. Matane, Que.
Abitibi-Price, Kenogami, Que.
Bathurst, Bathurst, NB.
Lake Utopia Paper, St-George, N.B.

### Transportation and Storage Information

**Shipping State:** Solid.
**Classification:** None.
**Inert Atmosphere:** No requirement.
**Venting:** Open.

**Label(s):** Not regulated.
**Storage Temperature:** Ambient.

**Grades or Purity:** Technical anhydrous; 90 ±1%.
**Containers and Materials:** Bags and drums.

### Physical and Chemical Characteristics

**Physical State (20°C, 1 atm):** Solid.
**Solubility (Water):** Anhydrous 12.5 g/100 mL (0°C); 28.3 g/100 mL (80°C); heptahydrate: 32.8 g/100 mL (0°C); 196 g/100 mL (40°C).
**Molecular Weight:** 126 (anhyd.), 252 (heptahydrate).
**Vapour Pressure:** No information.
**Boiling Point:** Decomposes (hydrate loses $7H_2O$ at 150°C).

**Floatability (Water):** Sinks and mixes.
**Odour:** Odourless.
**Flash Point:** Not flammable.
**Vapour Density:** No information.
**Specific Gravity:** 2.63 at 15°C (anhydrous); 1.54 at 15°C (heptahydrate).

**Colour:** White.
**Explosive Limits:** Not flammable.
**Melting Point:** Decomposes.

## HAZARD DATA

### Human Health

**Symptoms: Inhalation:** (dust) sore throat, coughing, shortness of breath. **Contact:** skin and eyes - causes redness, pain, inflammation. **Ingestion:** violent and chronic diarrhea, circulatory disturbances, central nervous system depression and death.
**Toxicology:** Moderately toxic by ingestion.
**TLV -** No information.
**Short-term Inhalation Limits -** No information.

**LC50 -** No information.
**Delayed Toxicity -** No information.

**LD50 - Intravenous:** rat = 0.115 g/kg (anhydrous).
**LDLo - Oral:** rabbit = 2.83 g/kg (anhydrous).
**LD50 - Intraperitoneal:** mouse = 0.277 g/kg (hepta).

### Fire

**Fire Extinguishing Agents:** Not combustible. In fires involving sodium sulfite, most fire extinguishing agents may be used.
**Behaviour in Fire:** Not combustible.
**Ignition Temperature:** Not combustible.
**Burning Rate:** Not combustible.

### Reactivity

**With Water:** No reaction; soluble.
**With Common Materials:** No information.
**Stability:** Stable.

### Environment

**Water:** Prevent entry into water intakes and waterways. Hazardous to aquatic life in high concentrations. Fish toxicity: 2 600 ppm/24, 48 and 96 h/mosquito fish/TLm/freshwater; BOD: 12%.
**Land-Air:** No information.
**Food Chain Concentration Potential:** None.

## EMERGENCY MEASURES

**Special Hazards**

**Immediate Responses**

Keep non-involved people away from spill site. Avoid contact. Dike area to prevent runoff from rainwater. Contact manufacturer. Contact environmental authorities.

**Protective Clothing and Equipment**

Safety goggles - tight fitting. Rubber gloves. Protective clothing as required.

**Fire and Explosion**

Not combustible. Most fire extinguishing agents may be used in fires involving sodium sulfite.

**First Aid**

Move victim out of spill site to fresh air. Call for medical assistance, but start first aid at once. Contact: remove contaminated clothing and wash with plenty of water. Ingestion: give water or milk to conscious victim. If medical help is not immediately available, transport victim to clinic, doctor or hospital.

## ENVIRONMENTAL PROTECTION MEASURES

**Response**

**Water**
1. Stop or reduce discharge if safe to do so.
2. Contact manufacturer or supplier for advice.
3. If possible, contain discharge by damming or or water diversion.
4. Dredge or vacuum pump to remove contaminants, liquids and contaminated bottom sediments.
5. Notify environmental authorities to discuss disposal and cleanup of contaminated materials.

**Land-Air**
1. Stop or reduce discharge if safe to do so.
2. Contact manufacturer or supplier for advice.
3. Dike to prevent runoff from rainwater or water application.
4. Remove material by manual or mechanical means.
5. Recover undamaged containers.
7. Notify environmental authorities to discuss disposal and cleanup of contaminated materials.

**Disposal**
1. Contact manufacturer or supplier for advice on disposal.
2. Contact environmental authorities for advice on disposal.
3. Contaminated materials may be buried in a secured landfill site (with approval of environmental authorities).

SODIUM SULFITE   $Na_2SO_3$

# STEARIC ACID $CH_3(CH_2)_{16}CO_2H$

## IDENTIFICATION

### Common Synonyms
HYDROFOL ACID
OCTADECANOIC ACID
n-OCTADECYLIC ACID
STEAROPHANIC ACID
1-HEPTADECACARBOXYLIC ACID

### Observable Characteristics
White to pale yellow solid with a slight fatty odour.

### Manufacturers
Proctor and Gamble Canada, Hamilton, Ont.
Emery Industries, Toronto, Ontario.
Canada Packers, Toronto, Ontario.
Armak Chemicals, Saskatoon, Saskatchewan.

### Transportation and Storage Information
**Shipping State:** Solid.
**Classification:** Not regulated.
**Inert Atmosphere:** No requirement.
**Venting:** Open.
**Label(s):** Not regulated.
**Storage Temperature:** Ambient.

**Grades or Purity:** USP; Commercial; triple pressed; double pressed. Most commercial "stearic acid" is 45% palmitic acid, 50% stearic acid, and 5% oleic acid.
**Containers and Materials:** Cans, barrels, bags, bulk by truck and train.

### Physical and Chemical Characteristics
**Physical State** (20°C, 1 atm): Solid.
**Solubility** (Water): 0.034 g/100 mL (25°C); 0.1 g/100 mL (87°C).
**Molecular Weight:** 284.5
**Vapour Pressure:** 1.0 mm Hg (173.7°C).
**Boiling Point:** 361 to 383°C (may also decompose at these temperatures).
**Floatability** (Water): Floats.
**Odour:** Tallow-like; 20 ppm odour threshold.
**Flash Point:** 196°C (c.c.).
**Vapour Density:** 9.8
**Specific Gravity:** 0.94 at 20°C.
**Colour:** White to pale yellow.
**Explosive Limits:** No information.
**Melting Point:** 69 to 72°C.

## HAZARD DATA

### Human Health
**Symptoms:** Generally considered nontoxic. Inhalation of dust irritates nose and throat. Dust causes mild eye irritation.
**Toxicology:** Low toxicity by all routes.
**TLV®** No information.
**Short-term Inhalation Limits** - No information.
**LC$_{50}$** - No information.
**Delayed Toxicity** - No information.
**LD$_{50}$** - Intravenous: rat = 0.022 g/kg
**LD$_{50}$** - Intravenous: mouse = 0.056 g/kg

### Fire
**Fire Extinguishing Agents:** Dry chemical or carbon dioxide. Water or foam may cause frothing. Use water spray to keep fire-exposed containers cool.
**Behaviour in Fire:** No severe toxic products are produced by combustion.
**Ignition Temperature:** 395°C.
**Burning Rate:** No information.

### Reactivity
**With Water:** No reaction.
**With Common Materials:** No information.
**Stability:** Stable.

### Environment
**Water:** Prevent entry into water intakes and waterways. Harmful to aquatic life. BOD: 144%, 5 days.
**Land-Air:** No information.
**Food Chain Concentration Potential:** None.

## EMERGENCY MEASURES

**Special Hazards**

COMBUSTIBLE.

**Immediate Responses**

Keep non-involved people away from spill site. Call Fire Department. Avoid contact with solid and dust. Stop discharge if safe to do so. Notify manufacturer. Notify environmental authorities.

**Protective Clothing and Equipment**

Safety goggles - tight fitting. Rubber gloves. Protective clothing as required.

**Fire and Explosion**

Dry chemical or carbon dioxide. Water or foam may cause frothing.

**First Aid**

Move victim out of spill site to fresh air. Call for medical assistance, but start first aid at once. Dust Contact: eyes - flush with water. Inhalation: if breathing has stopped, give artificial respiration; if laboured, give oxygen. Solid Contact: remove contaminated clothing and flush affected areas with water; eyes - flush with water. Ingestion: if victim is conscious, give milk or water to drink and induce vomiting. If medical assistance is not immediately available, transport victim to hospital, doctor or clinic.

## ENVIRONMENTAL PROTECTION MEASURES

**Response**

**Water**
1. Stop or reduce discharge if safe to do so.
2. Contact manufacturer or supplier for advice.
3. If possible, contain discharge by booming.
4. If floating, skim and remove.
5. Notify environmental authorities to discuss disposal and cleanup of contaminated materials.

**Land-Air**
1. Stop or reduce discharge if safe to do so.
2. Contact manufacturer or supplier for advice.
3. Contain spill by diking with earth or other barrier.
4. Remove material by manual or mechanical means.
5. Recover undamaged containers.
6. Notify environmental authorities to discuss disposal and cleanup of contaminated materials.

**Disposal**
1. Contact manufacturer or supplier for advice on disposal.
2. Contact environmental authorities for advice on disposal.
3. Contaminated materials may be buried in a secured landfill site (with environmental authorities' approval).

STEARIC ACID   $CH_3(CH_2)_{16}CO_2H$

# STYRENE MONOMER (inhibited)   $C_6H_5CH{:}CH_2$

## IDENTIFICATION

UN No. 2055

### Common Synonyms
VINYLBENZENE
PHENYLETHYLENE
CINNAMENE
ETHENYLBENZENE
STYROL

### Observable Characteristics
Colourless to yellowish liquid. Sweet odour.

### Manufacturers
Polysar, Sarnia, Ontario.
Dow Chemical Canada Inc., Sarnia, Ontario.

### Transportation and Storage Information
**Shipping State:** Liquid.
**Classification:** Flammable liquid.
**Inert Atmosphere:** No requirement.
**Venting:** Open (flame arrester).
**Pump Type:** Centrifugal, gear; steel; most materials not containing copper.

**Label(s):** Red label - FLAMMABLE; Class 3.2, Group II.
**Storage Temperature:** Ambient.
**Hose Type:** Fluorelastomer, stainless steel, Teflon.

**Grades or Purity:** Commercial, 99.5% min; technical, 99.2%; polymer, 99.6%.
**Containers and Materials:** Tank cars, tank trucks, drums, cans; steel, galvanized iron, black iron.

### Physical and Chemical Characteristics
**Physical State** (20°C, 1 atm): Liquid.
**Solubility** (Water): 0.028 g/100 mL (15°C); 0.03 g/100 mL (20°C); 0.04 g/100 mL (40°C).
**Molecular Weight:** 104.2
**Vapour Pressure:** 4.3 mm Hg (15°C); 9.5 mm Hg (30°C).
**Boiling Point:** 145.2°C.

**Floatability** (Water): Floats.
**Odour:** Sweet odour (0.15 to 25 ppm, odour threshold).
**Flash Point:** 37°C (o.c.); 32°C (c.c.).
**Vapour Density:** 3.6
**Specific Gravity:** 0.91 (25°C).

**Colour:** Colourless to yellow.
**Explosive Limits:** 1.1 to 6.1%.
**Melting Point:** -31°C.

## HAZARD DATA

### Human Health
**Symptoms: Contact:** eyes - irritation, possibly slight transient corneal injury; skin - moderate irritation upon prolonged or repeated contact. Not likely to be absorbed through skin in toxic amounts. **Ingestion:** Irritation of lips, mouth and throat; pain in swallowing, abdominal pain, nausea and vomiting, diarrhea, shock, convulsions may occur. **Inhalation:** irritation of mucous membranes, difficulty in breathing, coughing, bluish face and lips, dizziness, fatigue, pneumonia.
**Toxicology:** Moderate toxicity by inhalation or ingestion.
TLV®- (inhalation) 50 ppm; 215 mg/m³.
Short-term Inhalation Limits - 100 ppm; 425 mg/m³ (15 min).
$LC_{50}$ - No information.
$TC_{Lo}$ - Inhalation: human = 376 to 600 ppm
Delayed Toxicity - Possible kidney damage from ingestion. Suspected carcinogen.
$LD_{50}$ - Oral: rat = 5.0 g/kg

### Fire
**Fire Extinguishing Agents:** Use dry chemical, foam or carbon dioxide; water may be ineffective but may be used to keep fire-exposed containers cool and knock down vapours.
**Behaviour in Fire:** At elevated temperatures, polymerization may occur. Fire-exposed containers may rupture. Flashback may occur along vapour trail.
**Ignition Temperature:** 490°C.
**Burning Rate:** 5.2 mm/min.

### Reactivity
**With Water:** No reaction.
**With Common Materials:** Reacts vigorously with oxidizers and metallic halides. Reacts violently with chlorosulfonic acid, oleum and sulfuric acid.
**Stability:** Stable.

### Environment
**Water:** Prevent entry into water intakes and waterways. Fish toxicity: 22 ppm/96 h/bluegill/TLm/freshwater; 26 mg/L/24 h/goldfish/$LD_{50}$/freshwater; 68 mg/L/24 h/brine shrimp/TLm/saltwater; Aquatic toxicity rating = 10 to 100 ppm/96 h/TLm/freshwater; BOD: 55 to 195%, 5 days.
**Land-Air:** No information.
**Food Chain Concentration Potential:** None.

## EMERGENCY MEASURES

### Special Hazards
FLAMMABLE. May polymerize violently upon heating.

### Immediate Responses
Keep non-involved people away from spill area. Issue warning: "FLAMMABLE". Call Fire Department. Eliminate all ignition sources. Avoid contact and inhalation. Call manufacturer or supplier for guidance. Contain spill by diking. Stop or reduce discharge, if this can be done without risk. Notify environmental authorities.

### Protective Clothing and Equipment
In fires, Respiratory protection - self-contained breathing apparatus; otherwise, Goggles - (mono), tight fitting; Gloves - rubber. Boots - high, rubber (pants worn outside boots). Protective clothing - coveralls. In major spill or fire - totally encapsulated suit.

### Fire and Explosion
Dry chemical, foam or CO$_2$ to extinguish. At elevated temperatures, such as in fire, polymerization may occur. Fire-exposed container, may rupture. Fire may flash back along vapour trail.

### First Aid
Move victim out of spill site to fresh air. Call for medical assistance, but start first aid at once. Contact: eyes - irrigate with plenty of water immediately for at least 15 minutes; skin - flush with plenty of water for at least 15 minutes; at same time, remove contaminated clothing. Inhalation: if breathing stops, give artificial respiration; if laboured, give oxygen. Do not induce vomiting. If medical assistance is not immediately available, transport victim to hospital, doctor or clinic.

## ENVIRONMENTAL PROTECTION MEASURES

### Response

**Water**
1. Stop or reduce discharge if safe to do so.
2. Contact manufacturer or supplier for advice.
3. If possible, contain discharge by booming.
4. If floating, skim and remove.
5. Notify environmental authorities to discuss disposal and cleanup of contaminated materials.

**Land-Air**
1. Stop or reduce discharge if safe to do so.
2. Contact manufacturer or supplier for advice.
3. Contain spill by diking with earth or other barrier. Pump water into area to prevent soil penetration.
4. Remove material with pumps or vacuum equipment and place in appropriate containers.
5. Remove material by manual or mechanical means.
6. Recover undamaged containers.
7. Notify environmental authorities to discuss disposal and cleanup of contaminated materials.

### Disposal
1. Contact manufacturer or supplier for advice on disposal.
2. Contact environmental authorities for advice on disposal.

STYRENE MONOMER (inhibited)   $C_6H_5CH{:}CH_2$

# SULFUR  $S_8$

## IDENTIFICATION

UN No. 1350
2448 molten

### Common Synonyms
SULPHUR
BRIMSTONE
FLOWERS OF SULFUR
SULFUR FLOUR

### Observable Characteristics
Yellowish powder or lumps. Pure is odourless; with impurities, faint rotten-eggs type odour.

### Manufacturers
Aquitaine of Canada, Ram River, Alta.
Chevron Standard, Kayob, Alta.
Shell Canada, Waterton, Alta.
Amoco/Texasgulf, Windfall, Alta.

### Transportation and Storage Information
**Shipping State:** Solid, liquid (molten).
**Classification:** Not regulated.
**Inert Atmosphere:** No requirement.
**Venting:** No requirement.
**Pump Type:** Centrifugal; steel, stainless steel. Relevant for molten material only.

**Label(s):** Voluntary red and white striped label - FLAMMABLE; Class 4.1, Group III.
**Storage Temperature:** Ambient (molten, 142°C).
**Hose Type:** Flexible steel or stainless steel. Molten material only.

**Grades or Purity:** Technical, crude, refined, high purity.
**Containers and Materials:** Bags, barrels, bulk, carlots or truck loads.

### Physical and Chemical Characteristics
**Physical State** (20°C, 1 atm): Solid.
**Solubility** (Water): Insoluble.
**Molecular Weight:** 256.5 (atomic weight: 32.1).
**Vapour Pressure:** <0.0001 mm Hg (20°C); 0.11 mm Hg (140°C); 1 mm Hg (184°C); 10 mm Hg (244°C).
**Boiling Point:** 444.6°C.

**Floatability** (Water): Sinks (finely divided powder may trap air and remain floating).
**Odour:** Faint rotten-egg odour (impure); pure form odourless.
**Flash Point:** Pure sulfur, 188 to 207°C; impure sulfur as low as 168°C.
**Vapour Density:** 8.9
**Specific Gravity:** 1.92 to 2.07 (20°C); 1.80 (molten) (120°C).

**Colour:** Yellow.
**Explosive Limits:** 35 to 1 400 g/m³ dust.
**Melting Point:** 113 to 122°C.

## HAZARD DATA

### Human Health
**Symptoms:** Inhalation: sore throat, coughing. Contact: eyes and skin - (dust) irritation (burns with molten). Ingestion: irritation of mouth, sore throat, diarrhea.
**Toxicology:** Low toxicity by all routes. $LC_{50}$ - No information. $LD_{50}$ - No information.
TLV® - No information. Delayed Toxicity - No information.
Short-term Inhalation Limits - No information.

### Fire
**Fire Extinguishing Agents:** Use dry chemical, carbon dioxide or water spray. Avoid straight streams of water which will scatter molten sulfur and dust.
**Behaviour in Fire:** Produces toxic $SO_x$ fumes. Dust may explode if there is an ignition source.
**Ignition Temperature:** 232 to 266°C (190°C; dust). **Burning Rate:** No information.

### Reactivity
**With Water:** No reaction.
**With Common Materials:** May react with oxidizing materials under some conditions. Reacts violently with halogens, carbides, zinc, tin, sodium, nickel, phosphorus, potassium, calcium, aluminum, ammonia, ammonium nitrate, potassium permanganate, ammonium perchlorate, barium halites, bromates, calcium chlorate, calcium hypochlorite, calcium iodate, charcoal, chlorates, chlorine oxides, chromic anhydride, fluorine, hydrocarbons, iodates, lead halides, lead dioxide, lithium, magnesium halides, mercury oxides, perchlorates, potassium halides, potassium perchlorate, silver halides, sodium halides, sodium nitrate and zinc halides.
**Stability:** Stable.

### Environment
**Water:** Prevent entry into water intakes and waterways. Harmful to aquatic life in high concentrations. Fish toxicity: 10 000 ppm/96 h/mosquito fish/TLm/freshwater; 16 000 ppm/5 h/goldfish/$LC_{100}$/freshwater (as colloidal suspension); 2 100 ppm/1 h/goldfish/$LC_{100}$ freshwater; BOD: No information.
**Land-Air:** No information.
**Food Chain Concentration Potential:** None.

## EMERGENCY MEASURES

**Special Hazards**
COMBUSTIBLE.

**Immediate Responses**
Keep non-involved persons away from spill site. Issue warning; "COMBUSTIBLE". Call Fire Department. Call supplier for advice. Avoid contact and inhalation. Stop discharge if this can be done without risk. Notify environmental authorities.

**Protective Clothing and Equipment**
In fires, Respiratory protection - self-contained breathing apparatus; otherwise; Goggles - (mono), tight fitting. Gloves - boots - coveralls.

**Fire and Explosion**
Use dry chemical, carbon dioxide or water spray. Avoid straight streams of water. Toxic $SO_x$ fumes produced in fires.

**First Aid**
Move victim out of spill site to fresh air. Call for medical assistance, but start first aid at once. Contact: eyes - irrigate with plenty of warm water for at least 15 minutes. Inhalation: remove to fresh air. If breathing has stopped, give artificial respiration; if laboured, give oxygen. Ingestion: give water to conscious victim to drink. I-duce vomiting. If medical assistance is not immediately available, transport victim to hospital, doctor or clinic.

## ENVIRONMENTAL PROTECTION MEASURES

**Response**

**Water**
1. Stop or reduce discharge if safe to do so.
2. Contact manufacturer or supplier for advice.
3. If possible, contain discharge by damming or water diversion.
4. Dredge or vacuum pump to remove contaminants, liquids and contaminated bottom sediments.
5. Notify environmental authorities to discuss disposal and cleanup of contaminated materials.

**Land-Air**
1. Stop or reduce discharge if safe to do so.
2. Contact manufacturer or supplier for advice.
3. Contain spill by diking with earth or other barrier.
4. Remove material by manual or mechanical means.
5. Recover undamaged containers.
6. Notify environmental authorities to discuss disposal and cleanup of contaminated materials.

**Disposal**
1. Contact manufacturer or supplier for advice on disposal.
2. Contact environmental authorities for advice on disposal.

SULFUR $S_8$

# SULFUR DIOXIDE $SO_2$

## IDENTIFICATION

**UN No. 1079**

### Common Synonyms
SULFUROUS ACID, ANHYDRIDE
SULFUROUS OXIDE

### Observable Characteristics
Colourless liquid or gas. Strong, pungent odour.

### Manufacturers
Canadian Industries Ltd.,
Copper Cliff, Ont.
Cominco Ltd.,
Trail, B.C.

### Transportation and Storage Information
**Shipping State:** Gas or liquid (compressed gas).
**Classification:** Poison.
**Inert Atmosphere:** No requirement.
**Venting:** Safety-relief valve.
**Pump Type:** No information.

**Label(s):** White label - POISON; Class 2.3.
**Storage Temperature:** Ambient.
**Hose Type:** Flexible steel or stainless steel.

**Grades or Purity:** Commercial, 99.9% $SO_2$; refrigeration, 99.98% $SO_2$.
**Containers and Materials:** Cylinders, ton containers, tank cars; steel.

### Physical and Chemical Characteristics
**Physical State** (20°C, 1 atm)**:** Gas.
**Solubility** (Water)**:** 23 g/100 mL (0°C); 0.58 g/100 mL (90°C).
**Molecular Weight:** 64.1
**Vapour Pressure:** 2 538 mm Hg (21°C).
**Boiling Point:** -10°C.

**Floatability** (Water)**:** Mixes (liquid $SO_2$ sinks).
**Odour:** Strong, pungent (3 ppm, odour threshold).
**Flash Point:** Not flammable.
**Vapour Density:** 2.3 (0°C).
**Specific Gravity:** (liquid) 1.45 (-10°C).

**Colour:** Colourless.
**Explosive Limits:** Not flammable.
**Melting Point:** -75.8°C.

## HAZARD DATA

### Human Health
**Symptoms: Inhalation:** sore throat, coughing, shortness of breath, laboured breathing, bronchitis, respiratory paralysis. **Contact: eyes** - blurred vision, redness, pain; **skin** - redness, pain, burns. Contact with liquid produces frostbite.
**Toxicology:** Highly toxic by all routes.
TLV® (inhalation) 2 ppm; 5 mg/m³.
Short-term Inhalation Limits - 5 ppm; 10 mg/m³ (15 min).

$LC_{50}$ - No information.
$LC_{Lo}$ - Human: Inhalation = 400 ppm (1 min).
$TC_{Lo}$ - Human: Inhalation = 4 ppm (1 min).
Delayed Toxicity - No information.

$LD_{50}$ - No information.

### Fire
**Fire Extinguishing Agents:** Not combustible. Stop flow of gas before attempting to extinguish fire. Most fire extinguishing agents may be used on fires involving sulfur dioxide. In fires involving $SO_2$, cool fire-exposed containers with water.
**Behaviour in Fire:** Not combustible.
**Ignition Temperature:** Not combustible.
**Burning Rate:** Not combustible.

### Reactivity
**With Water:** Soluble; reacts to form sulfurous acid.
**With Common Materials:** Reacts violently with acrolein, aluminum, chlorates, chromium, ferrous oxide, fluorine, manganese, potassium chlorate, sodium carbide and stannous oxide.
**Stability:** Stable.

### Environment
**Water:** Prevent entry into water intakes and waterways. Harmful to aquatic life. Fish toxicity: 5 ppm/1 h/trout/$LC_{100}$/freshwater; 16 ppm/1 h/sunfish/$LC_{100}$/freshwater; BOD: No information.
**Land-Air:** Concentrations >1 ppm injurious to plant foliage.
**Food Chain Concentration Potential:** None.

## EMERGENCY MEASURES

**Special Hazards**

POISON.

**Immediate Responses**

Keep non-involved people away from spill site. Issue warning: "POISON". Call Fire Department. Contact manufacturer for advice and assistance. Shut off leak; if safe to do so. In cold weather, if liquid $SO_2$ is discharging, it may be possible to safely dike the spill, wearing full protective equipment. Notify environmental authorities.

**Protective Clothing and Equipment**

Respiratory protection - use self-contained breathing apparatus. Gloves - rubber. Outerwear - coveralls or gas tight suit in high gas concentrations.

**Fire and Explosion**

Not combustible. Stop flow of gas before attempting to extinguish fire. Most fire extinguishing agents may be used on fires involving sulfur dioxide. In fires involving sulfur dioxide, cool fire-exposed containers with water.

**First Aid**

Move victim out of spill site to fresh air. Call for medical assistance, but start first aid at once. Contact: eyes - irrigate with plenty of water for at least 15 minutes. If pain persists, irrigate for another 15 minutes; skin - flush skin immediately with plenty of soapy water for at least 15 minutes. At same time remove contaminated clothing. Treat as for frostbite. Inhalation: if breathing has stopped, start artificial respiration at once. If breathing is laboured, administer oxygen. If medical assistance is not immediately available, transport victim to hospital, doctor or clinic.

## ENVIRONMENTAL PROTECTION MEASURES

**Response**

**Water**
1. Stop or reduce discharge if safe to do so.
2. Contact manufacturer or supplier for advice.
3. If possible, contain discharge by damming or water diversion.
4. Notify environmental authorities to discuss disposal and cleanup of contaminated materials.

**Land-Air**
1. Stop or reduce discharge if safe to do so.
2. Contact manufacturer or supplier for advice.
3. Contain spill by diking with earth or other barrier, if possible.
4. Recover undamaged containers.
5. Notify environmental authorities to discuss disposal and cleanup of contaminated materials.

**Disposal**
1. Contact manufacturer or supplier for advice on disposal.
2. Contact environmental authorities for advice on disposal.

SULFUR DIOXIDE  $SO_2$

# SULFURIC ACID  $H_2SO_4$

## IDENTIFICATION

UN No. 1832

### Common Synonyms
HYDROGEN SULPHATE
FERTILIZER ACID
BATTERY ACID
DIPPING ACID
(See Oleum for concentrated solutions).

### Observable Characteristics
Colourless to brown liquid. Sharp, penetrating odour.

### Manufacturers
(CIL) Canadian Industries Ltd., Copper Cliff, Ont.
ESSO Chemicals Canada, Redwater, Alta.
Texasgulf Canada, Timmins, Ontario.
Western Co-operative Fertilizers, Calgary, Alta.

### Transportation and Storage Information
**Shipping State:** Liquid.
**Classification:** Corrosive liquid.
**Inert Atmosphere:** No requirement.
**Venting:** Open.
**Pump Type:** Centrifugal; alloy 20 (for 70% and up).

**Label(s):** White and black label - CORROSIVE; Class 8, Group II.
**Storage Temperature:** Ambient.
**Hose Type:** Chemiflex 951 (polypropylene), flexible stainless steel - rigid pipe and swivel joints.

**Grades or Purity:** Commercial, 52° Bé 65.1% $H_2SO_4$; 58° Bé 74.4% $H_2SO_4$; 60° Bé 77.7% $H_2SO_4$; 66° Bé 93.2% $H_2SO_4$.
**Containers and Materials:** Bottles, carboys, (lined) drums, tank trucks, tank cars; stainless steel.

### Physical and Chemical Characteristics
**Physical State (20°C, 1 atm):** Liquid.
**Solubility (Water):** Soluble in all proportions.
**Molecular Weight:** 98.1 (pure).
**Vapour Pressure:** ~1 mm Hg at 38°C for 66° Bé.
**Boiling Point:** (66° Bé), ~281°C.

**Floatability (Water):** Sinks and mixes. May react vigorously.
**Odour:** Sharp, penetrating (1 mg/m³ odour threshold).
**Flash Point:** Not flammable.
**Vapour Density:** 2.8 (20°C) ($SO_3$).
**Specific Gravity:** 52° Bé 1.56; 58° Bé 1.67; 60° Bé 1.71; 66° Bé 1.84.

**Colour:** Clear to dark brown.
**Explosive Limits:** Not flammable.
**Melting Point:** (66° Bé), -32°C; 100% 10.4°C; (52° Bé), -40°C; 58° Bé, -44°C; (60° Bé), -8°C.

## HAZARD DATA

### Human Health
**Symptoms:** Highly concentrated sulfuric acid is rapidly destructive to body tissues on contact. Contact: skin - dermatitis and burns; eyes - rapidly causes severe damage, and possible loss of sight. Inhalation: of concentrated vapour or mist will cause damage to the upper respiratory tract and lung tissue, sore throat, coughing, laboured breathing. Ingestion: sore throat, abdominal pain, nausea and vomiting.
**Toxicology:**
TLV®- (inhalation) 1 mg/m³.  $LC_{50}$ - Inhalation: guinea pig = 18 mg/m³.  $LD_{50}$ - Oral: rat = 2.14 g/kg
Short-term Inhalation Limits - No information.  Delayed Toxicity - None known.

### Fire
**Fire Extinguishing Agents:** Not combustible. Use dry chemical to fight adjacent fires.
**Behaviour in Fire:** Not combustible. In fires, toxic $SO_x$ fumes may be released. May react with metals producing flammable $H_2$ gas.
**Ignition Temperature:** Not combustible.  **Burning Rate:** Not combustible.

### Reactivity
**With Water:** Soluble. Concentrated solutions may react violently producing toxic $SO_x$ fumes.
**With Common Materials:** Powerful oxidizer; concentrated solutions may ignite organic materials. Can react violently with acetic anhydride, acetonitrile, acrolein, acrylonitrile, allyl alcohol, allyl chloride, ammonium hydroxide, aniline, n-butyraldehyde, carbides, chlorates, chlorosulfonic acid, epichlorohydrin, ethylenediamine, ethylene glycol, hydrochloric acid, hydrofluoric acid, iron, isoprene, metals (powdered), perchlorates, phosphorus, potassium-t-butoxide, potassium permanganate, propylene oxide, pyridine, sodium, sodium carbonate, sodium chlorate, sodium hydroxide, steel, styrene monomer, vinyl acetate and zinc chlorate.
**Stability:** Stable (within the limits of the foregoing).

### Environment
**Water:** Prevent entry into water intakes and waterways. Harmful to aquatic life. Fish toxicity: 10 to 24.5 mg/L/24 h/bluegill/lethal/freshwater; 45.2 ppm/48 h/prawn/$LC_{50}$/saltwater; 138 ppm/4 h/goldfish/lethal/freshwater; 80 to 90 ppm/48 h/shrimp/$LC_{50}$/saltwater; BOD: None.
**Land-Air:** No information.
**Food Chain Concentration Potential:** None.

## EMERGENCY MEASURES

**Special Hazards**

CORROSIVE. Reactive with water and other common materials.

**Immediate Responses**

Keep non-involved people away from spill site. Issue warning: "CORROSIVE". Call Fire Department. Eliminate all ignition sources. Contact manufacturer for advice. Contain spill by diking with earth or other material. Avoid inhalation and contact. Stop or reduce discharge if this can be done without risk. Notify environmental authorities.

**Protective Clothing and Equipment**

Respiratory protection - self-contained breathing apparatus and totally encapsulated suit. Gloves - gauntlet type, rubber, vinyl, Boots - high, rubber or neoprene (pants worn outside boots).

**Fire and Explosion**

Not combustible. Avoid use of water. Toxic $SO_x$ fumes are released in fires.

**First Aid**

Move victim out of spill area to fresh air. Call for medical assistance, but start first aid at once. Inhalation: if breathing has stopped, give artificial respiration (not mouth-to-mouth method); if laboured, give oxygen. Contact: eyes - irrigate immediately with plenty of water; skin - flush with plenty of water, and remove contaminated clothing. Ingestion: give plenty of water to conscious victim to drink to reduce acid concentration. Do not induce vomiting. If medical assistance is not immediately available, transport victim to hospital, doctor or clinic.

## ENVIRONMENTAL PROTECTION MEASURES

**Response**

**Water**
1. Stop or reduce discharge if safe to do so.
2. Contact manufacturer or supplier for advice.
3. If possible, contain discharge by damming or water diversion.
4. Notify environmental authorities to discuss disposal and cleanup of contaminated materials.

**Land-Air**
1. Stop or reduce discharge if safe to do so.
2. Contact manufacturer or supplier for advice.
3. Contain spill by diking with earth or other barrier.
4. Remove material with pumps or vacuum equipment and place in appropriate containers.
5. Remove material by manual or mechanical means.
6. Remove contaminated soil for disposal or neutralize with lime.
7. Notify environmental authorities to discuss disposal and cleanup of contaminated materials.

**Disposal**
1. Contact manufacturer or supplier for advice on disposal.
2. Contact environmental authorities for advice on disposal.

SULFURIC ACID  $H_2SO_4$

# SULFURYL CHLORIDE $SO_2Cl_2$

## IDENTIFICATION

UN No. 1834

### Common Synonyms
SULFONYL CHLORIDE
SULFURIC OXYCHLORIDE
CHLOROSULFURIC ACID
SULFURIC CHLORIDE

### Observable Characteristics
Colourless to yellow.
Acrid, choking odour.

### Manufacturers

### Transportation and Storage Information
**Shipping State:** Liquid.
**Classification:** Corrosive.
**Inert Atmosphere:** No requirement.
**Venting:** Pressure-vacuum.

**Label(s):** Black and white label - CORROSIVE.
**Storage Temperature:** Ambient.

**Grades or Purity:** Technical 99%.
**Containers and Materials:** Drums, carbuoys; plastic-lined or steel.

### Physical and Chemical Characteristics
**Physical State:** Liquid.
**Solubility (Water):** Reacts to form hydrochloric and sulfuric acids.
**Molecular Weight:** 135.0
**Vapour Pressure:** 99.8 mm Hg (18°C).
**Boiling Point:** 69.1°C.

**Floatability (Water):** Sinks and reacts to form hydrochloric and sulfuric acids.
**Odour:** Acrid, choking.
**Flash Point:** Not flammable.
**Vapour Density:** 4.6
**Specific Gravity:** 1.67 (20°C).

**Colour:** Colourless to yellow.
**Explosive Limits:** Not flammable.
**Melting Point:** -54.1°C.

## HAZARD DATA

### Human Health
**Symptoms:** <u>Inhalation</u>: sore throat, shortness of breath, laboured breathing. <u>Ingestion</u>: severe burns, abdominal pain. <u>Contact</u>: skin - burns, inflammation; eyes-burns, blurred vision, inflammation.
**Toxicology:** Highly toxic by all routes.
TLV® (as $SO_2$) 2 ppm, 5 mg/m$^3$;
(as Cl) 1 ppm, 3 mg/m$^3$.
Short-term Inhalation Limits - No information.

$LC_{50}$ - No information.
Delayed Toxicity - No information.

$LD_{50}$ - No information.

### Fire
**Fire Extinguishing Agents:** Not combustible. Most fire extinguishing agents may be used, except water. Contact with water may produce a violent reaction with production of hydrochloric and sulfuric acids. Water may be used to cool fire-exposed containers.
**Behaviour in Fire:** When heated to decomposition emits highly toxic $SO_x$ and chloride fumes.
**Ignition Temperature:** Not combustible.
**Burning Rate:** Not combustible.

### Reactivity
**With Water:** Reacts to produce hydrochloric and sulfuric acids. Reaction may be violent.
**With Common Materials:** Reacts violently with lead dioxide.
**Stability:** Stable.

### Environment
**Water:** Prevent entry into water intakes and waterways. Aquatic toxicity rating = 10 to 100 ppm/96 h/TLm/freshwater.
**Land-Air:** No information.
**Food Chain Concentration Potential:** None.

## EMERGENCY MEASURES

**Special Hazards**

CORROSIVE.

**Immediate Responses**

Keep non-involved people away from spill site. Issue warning "CORROSIVE". Avoid contact and inhalation. Call manufacturer or supplier. Dike to contain material. Notify environmental authorities.

**Protective Clothing and Equipment**

Respiratory protection - self-contained breathing apparatus and totally encapsulated suit.

**Fire and Explosion**

Not combustible. Most fire extinguishing agents may be used on fires involving sulfuryl chloride, except water. Contact with water may produce a violent reaction with production of hydrochloric and sulfuric acids. Water may be used to cool fire-exposed containers. When heated to decomposition, emits highly toxic $SO_x$ and chloride fumes.

**First Aid**

Move victim out of spill area to fresh air. Call for medical assistance, but start first aid at once. Inhalation: if breathing has stopped, give artificial respiration (not mouth-to-mouth method); if laboured, give oxygen. Contact: skin - remove contaminated clothing and flush affected areas with plenty of water; eyes -irrigate with water. Ingestion: give water to conscious victim to drink. If medical assistance is not immediately available, transport victim to hospital, doctor or clinic.

## ENVIRONMENTAL PROTECTION MEASURES

**Response**

**Water**
1. Stop or reduce discharge if safe to do so.
2. Contact manufacturer or supplier for advice.
3. If possible, contain discharge by damming or water diversion.
4. Dredge or vacuum pump to remove contaminants, liquids and contaminated bottom sediments.
5. Notify environmental authorities to discuss disposal and cleanup of contaminated materials.

**Land-Air**
1. Stop or reduce discharge if safe to do so.
2. Contact manufacturer or supplier for advice.
3. Contain spill by diking with earth or other barrier.
4. Remove material with pumps or vacuum equipment and place in appropriate containers.
5. Recover undamaged containers.
6. Absorb residual liquid on natural or synthetic sorbents.
7. Notify environmental authorities to discuss disposal and cleanup of contaminated materials.

**Disposal**
1. Contact manufacturer or supplier for advice on disposal.
2. Contact environmental authorities for advice on disposal.

SULFURYL CHLORIDE   $SO_2Cl_2$

# TALL OIL

## IDENTIFICATION

### Common Synonyms
TALLOL
LIQUID ROSIN

(A mixture of rosin and fatty acids produced from wood pulp.)

### Observable Characteristics
Yellow to brown liquids. Acrid odour.

### Manufacturers
Hercules Canada, Burlington, Ontario.
B.C. Chemical, Prince George, B.C.
Great Lakes Forest Products, Thunder Bay, Ontario.
Boise Cascade, Fort Frances, Ontario.
Prince Albert Pulp, Prince Albert, Saskatchewan.

### Transportation and Storage Information
**Shipping State:** Liquid.
**Classification:** None.
**Inert Atmosphere:** No requirement.
**Venting:** Open (flame arrester).
**Pump Type:** Most types.

**Label(s):** Not regulated.
**Storage Temperature:** Ambient.
**Hose Type:** Most types.

**Grades or Purity:** Crude, refined.
**Containers and Materials:** Drums, tank cars, tank trucks; steel.

### Physical and Chemical Characteristics
**Physical State (20°C, 1 atm):** Liquid.
**Solubility (Water):** Insoluble.
**Molecular Weight:** Variable.
**Vapour Pressure:** No information.
**Boiling Point:** High.

**Floatability (Water):** Floats.
**Odour:** Acrid.
**Flash Point:** 182°C (c.c.)
**Vapour Density:** No information.
**Specific Gravity:** 0.95 to 0.99 (20°C).

**Colour:** Yellow to brown.
**Explosive Limits:** No information.
**Melting Point:** Variable.

## HAZARD DATA

### Human Health
**Symptoms:** Inhalation: dizziness, rapid shallow breathing, respiratory tract irritation. Ingestion: nausea, vomiting, diarrhea, shallow respiration, unconsciousness. Contact: skin - irritation; eyes - irritation.
**Toxicology:** Low toxicity by all routes.
**TLV®** No information.
**Short-term Inhalation Limits -** No information.

$LC_{50}$ - No information.
**Delayed Toxicity -** No information.

$LD_{50}$ - No information.

### Fire
**Fire Extinguishing Agents:** Foam, dry chemical or carbon dioxide, water may be ineffective but may be used to cool fire-exposed containers.
**Behaviour in Fire:** No information.
**Ignition Temperature:** No information.

**Burning Rate:** No information.

### Reactivity
**With Water:** No reaction.
**With Common Materials:** Can react with oxidizing agents.
**Stability:** Stable.

### Environment
**Water:** Prevent entry into water intakes and waterways.
**Land-Air:** No information.
**Food Chain Concentration Potential:** None.

# EMERGENCY MEASURES

**Special Hazards**
COMBUSTIBLE.

**Immediate Responses**
Keep non-involved people away from spill site. Notify manufacturer or supplier. Dike to contain material. Stop or reduce discharge if safe to do so. Notify environmental authorities.

**Protective Clothing and Equipment**
Outer protective clothing as required.

**Fire and Explosion**
Use foam, dry chemical or carbon dioxide to extinguish. Water may be ineffective, but may be used to cool fire-exposed containers.

**First Aid**
Move victim out of spill site to fresh air. Call for medical assistance, but start first aid at once. **Ingestion:** if breathing has stopped give artificial respiration; if laboured, give oxygen. **Contact:** skin - remove contaminated clothing and flush affected areas with plenty of water; eyes - irrigate with plenty of water. **Ingestion:** give plenty of water to conscious victim to drink and induce vomiting. If medical assistance is not immediately available, transport victim to hospital, doctor or clinic.

# ENVIRONMENTAL PROTECTION MEASURES

**Response**

**Water**
1. Stop or reduce discharge if safe to do so.
2. Contact manufacturer or supplier for advice.
3. If possible, contain discharge by booming.
4. If floating, skim and remove.
5. Notify environmental authorities to discuss disposal and cleanup of contaminated materials.

**Land-Air**
1. Stop or reduce discharge if safe to do so.
2. Contact manufacturer or supplier for advice.
3. Contain spill by diking with earth or other barrier.
4. Remove material with pumps or vacuum equipment and place in appropriate containers.
5. Recover undamaged containers.
6. Absorb residual liquid on natural or synthetic sorbents.
7. Notify environmental authorities to discuss disposal and cleanup of contaminated materials.

**Disposal**
1. Contact manufacturer or supplier for advice on disposal.
2. Contact environmental authorities for advice on disposal.
3. Incinerate (approval of enviromental authorities required).

TALL OIL

# TEREPHTHALIC ACID  $C_6H_4(COOH)_2$

## IDENTIFICATION

| Common Synonyms | Observable Characteristics | Manufacturers |
|---|---|---|
| 1,4-BENZENEDICARBOXYL ACID<br>p-BENZENEDICARBOXYLIC ACID<br>TPA<br>p-PHTHALIC ACID | White crystals or powder. Slight acidic odour. | No Canadian manufacturers<br>Canadian supplier:<br>Millhaven Fibres,<br>Millhaven, Ontario.<br>Originating from:<br>Amoco, USA |

| Transportation and Storage Information | | |
|---|---|---|
| **Shipping State:** Solid.<br>**Classification:** None.<br>**Inert Atmosphere:** No requirement.<br>**Venting:** Open. | **Label(s):** Not regulated.<br>**Storage Temperature:** Ambient. | **Grades or Purity:** Commercial, fibre.<br>**Containers and Materials:** Bags, barrels, bulk, truck, carlots. |

| Physical and Chemical Characteristics | | |
|---|---|---|
| **Physical State** (20°C, 1 atm): Solid.<br>**Solubility** (Water): 0.002 g/100 mL (25°C).<br>**Molecular Weight:** 166.1<br>**Vapour Pressure:** 0.5 mm Hg (120°C).<br>**Boiling Point:** Sublimes >300°C. | **Floatability** (Water): Sinks.<br>**Odour:** Slight acidic.<br>**Flash Point:** 260°C (o.c.).<br>**Vapour Density:** 5.74 (21°C).<br>**Specific Gravity:** 1.5 (20°C). | **Colour:** White.<br>**Explosive Limits:** (dust) 0.05 g/L (LEL).<br>**Melting Point:** Sublimes >300°C. |

## HAZARD DATA

### Human Health
**Symptoms: Contact:** skin - slight irritation, redness, pain. **Inhalation:** of dust or vapours irritating, sore throat, coughing.
**Toxicology:** Low toxicity by all routes.
**TLV®** No information.   **LC$_{50}$** - No information.   **LD$_{50}$** - Oral: rat = 18.8 g/kg
**Short-term Inhalation Limits** - No information.   **Delayed Toxicity** - No information.   **LD$_{50}$** - Intraperitoneal: mouse = 1.43 g/kg

### Fire
**Fire Extinguishing Agents:** Use dry chemical or carbon dioxide. Water or foam may cause frothing.
**Behaviour in Fire:** No information.   **Burning Rate:** No information.
**Ignition Temperature:** 496°C.

### Reactivity
**With Water:** No reaction.
**With Common Materials:** May react with oxidizing agents.
**Stability:** Stable.

### Environment
**Water:** Prevent entry into water intakes and waterways.
**Land-Air:** Decomposition by soil microorganisms in 2 days.
**Food Chain Concentration Potential:** No information.

## EMERGENCY MEASURES

**Special Hazards**
COMBUSTIBLE.

**Immediate Responses**
Keep non-involved people away from spill site. If burning, call Fire Department. Avoid contact and inhalation. Contact manufacturer. Contact environmental authorities.

**Protective Clothing and Equipment**
<u>Safety goggles</u> - tight fitting. Rubber gloves. Protective clothing as required.

**Fire and Explosion**
Extinguish fire with dry chemical or carbon dioxide. Water or foam may cause frothing.

**First Aid**
Move victim out of spill site to fresh air. Call for medical assistance, but start first aid at once. <u>Contact</u>: eyes - flush with water; skin - remove contaminated clothing and flush affected areas with water. <u>Inhalation</u>: move to fresh air. <u>Ingestion</u>: give water to conscious victim to drink. Transport victim to hospital, doctor or clinic.

## ENVIRONMENTAL PROTECTION MEASURES

**Response**

**Water**
1. Stop or reduce discharge if safe to do so.
2. Contact manufacturer or supplier for advice.
3. If possible, contain discharge by damming or water diversion.
4. Dredge or vacuum pump to remove contaminants, liquids and contaminated bottom sediments.
5. Notify environmental authorities to discuss disposal and cleanup of contaminated materials.

**Land-Air**
1. Stop or reduce discharge if safe to do so.
2. Contact manufacturer or supplier for advice.
3. Contain spill by diking with earth or other barrier.
4. Remove material by manual or mechanical means.
5. Recover undamaged containers.
6. Notify environmental authorities to discuss disposal and cleanup of contaminated materials.

**Disposal**
1. Contact manufacturer or supplier for advice on disposal.
2. Contact environmental authorities for advice on disposal.

TEREPHTHALIC ACID   $C_6H_4(COOH)_2$

# TERPHENYLS  ($(C_6H_5)_2C_6H_4$ (o,m,p))

## IDENTIFICATION

### Common Synonyms
DIPHENYLBENZENE
BENZENE-DIPHENYL

### Observable Characteristics
Colourless or light yellow solids.
Odourless.

### Manufacturers
No Canadian manufacturers.

### Transportation and Storage Information
**Shipping State:** Solid.
**Classification:** None.
**Inert Atmosphere:** No requirement.
**Venting:** Open.

**Label(s):** Not regulated.
**Storage Temperature:** Ambient.

**Grades or Purity:** Technical.
**Containers and Materials:** Bags, drums.

### Physical and Chemical Characteristics
**Physical State** (20°C, 1 atm): Solid.
**Solubility** (Water): Insoluble.
**Molecular Weight:** 230.3
**Vapour Pressure:** No information.
**Boiling Point:** o, 332°C; m, 365°C; p, 405°C.

**Floatability** (Water): Sinks.
**Odour:** Odourless.
**Flash Point:** o, 163°C; m, 191°C; p, 207°C (o.c.)
**Vapour Density:** 7.95
**Specific Gravity:** o, 1.14; m, 1.16; p, 1.24 (20°C).

**Colour:** Colourless to light yellow.
**Explosive Limits:** No information.
**Melting Point:** o, 58°C; m, 86 to 89°C; p, 213°C.

## HAZARD DATA

### Human Health
**Symptoms: Contact:** eyes and skin - irritating. **Ingestion:** burning sensation in throat. **Inhalation:** burning sensation, lung damage.
**Toxicology:** Moderately toxic by ingestion or inhalation.
**TLV** - 0.05 ppm; 5 mg/m$^3$.
**Short-term Inhalation Limits** - No information.
**LC$_{50}$** - No information.
**Delayed Toxicity** - Possible injury to liver and kidneys.
**LD$_{50}$** - No information.
**LD$_{Lo}$** - Oral: rat = 0.5 g/kg

### Fire
**Fire Extinguishing Agents:** Use dry chemical or carbon dioxide; water or foam may cause frothing.
**Behaviour in Fire:** No information.
**Ignition Temperature:** Not available.
**Burning Rate:** No information.

### Reactivity
**With Water:** No reaction.
**With Common Materials:** Reacts with strong oxidizers.
**Stability:** Stable.

### Environment
**Water:** Prevent entry into water intakes and waterways.
**Land-Air:** No information.
**Food Chain Concentration Potential:** No information.

## EMERGENCY MEASURES

**Special Hazards**

POISONOUS, COMBUSTIBLE.

**Immediate Responses**

Keep non-involved people away from spill site. If burning, call Fire Department. Avoid contact and inhalation. Contact manufacturer. Contact environmental authorities.

**Protective Clothing and Equipment**

In fires or enclosed spaces; Respiratory protection - self-contained breathing apparatus and totally encapsulated suit. Safety goggles - tight fitting. Rubber gloves. Protective clothing as required.

**Fire and Explosion**

Use dry chemical or carbon dioxide to extinguish; water or foam may cause frothing.

**First Aid**

Move victim out of spill site to fresh air. Call for medical assistance, but start first aid at once. Contact: eyes - flush with water immediately; skin - flush with water, remove contaminated clothing. Inhalation: give artificial respiration if necessary (not mouth-to-mouth method). Ingestion: do not induce vomiting. If medical assistance is not immediately available, transport victim to doctor, clinic or hospital.

## ENVIRONMENTAL PROTECTION MEASURES

**Response**

Water
1. Stop or reduce dischagrge if safe to do so.
2. Contact manufacturer or supplier for advice.
3. If possible, contain discharge by damming or or water diversion.
4. Dredge or vacuum pump to remove contaminants, liquids and contaminated bottom sediments.
5. Notify environmental authorities to discuss disposal and cleanup of contaminated materials.

Land-Air
1. Stop or reduce dischagrge if safe to do so.
2. Contact manufacturer or supplier for advice.
3. Contain spill by diking with earth or or other barrier.
4. Remove material by manual or mechanical means.
5. Recover undamaged containers.
6. Remove contaminated soil for disposal.
7. Notify environmental authorities to discuss disposal and cleanup of contaminated materials.

**Disposal**

1. Contact manufacturer or supplier for advice on disposal.
2. Contact environmental authorities for advice on disposal.

TERPHENYLS   $(C_6H_5)_2C_6H_4(o,m,p)$

# TETRAETHYL LEAD  ($C_2H_5$)$_4$Pb

## IDENTIFICATION

UN No. 1649

### Common Synonyms
MOTOR FUEL ANTIKNOCK COMPOUND
TEL
TETRAETHYL PLUMBANE
TML, TETRAMETHYL LEAD

### Observable Characteristics
Liquid, usually containing a coloured dye (red, blue) (colourless in pure state). Sweet, pleasant odour.

### Manufacturers
Dupont of Canada Ltd., Maitland, Ont.
Ethyl Corporation of Canada Ltd., Corunna, Ont.

### Transportation and Storage Information
**Shipping State:** Liquid.
**Classification:** Poison.
**Inert Atmosphere:** No requirement.
**Venting:** Pressure-vacuum.
**Pump Type:** Most types.

**Label(s):** White label - POISON; Class 6.1, Group I.
**Storage Temperature:** Ambient.
**Hose Type:** Most types.
**Containers and Materials:** Drums, trailers, tank cars, tank trucks; steel.

**Grades or Purity:** Technical (98%). "Ethyl" TML motor mix; tetramethyl lead 50.8%, ethylene dibromide 17.9%, ethylene dichloride 18.8%, toluene 12.2%. "Ethyl" TEL motor mix; tetraethyl lead 61.5%, ethylene dibromide 17.9%, ethylene dichloride 18.8%, kerosene 1.2%. "Ethyl" TEL aviation mix; tetraethyl lead 61.4%, ethylene dibromide 35.7%, kerosene 2.5%.

### Physical and Chemical Characteristics
**Physical State** (20°C, 1 atm): Liquid.
**Solubility** (Water): Insoluble.
**Molecular Weight:** 323.4
**Vapour Pressure:** 0.15 mm Hg (20°C); 1.0 mm Hg (38.4°C) (technical); 7.5 mm Hg (20°C) TEL aviation mix; 35.9 mm Hg (20°C) TEL motor mix; 39.6 mm Hg (20°C) TML motor mix.
**Boiling Point:** Technical decomposes 198 to 202°C; TEL aviation mix 138°C; TEL motor mix 105°C; TML motor mix 97°C.

**Floatability** (Water): Sinks.
**Odour:** Sweet, pleasant.
**Flash Point:** 85°C (o.c.); 93°C (c.c.) (technical); 32°C (o.c.) (TML); 118°C (o.c.) (TEL motor mix); 121°C (o.c.) (TEL aviation).
**Vapour Density:** 7 to 11.2 (technical); 4.9 (TML); 3.7 (TEL motor mix); 6.6 (TEL aviation).
**Specific Gravity:** Technical 1.7; TEL (motor), TML, 1.59; TEL aviation 1.74 (all at 20°C).

**Colour:** Red, blue or orange depending on dye added.
**Explosive Limits:** No information.
**Melting Point:** -137°C (technical).

## HAZARD DATA

### Human Health
**Symptoms:** Inhalation and Ingestion: restlessness, headache, trembling hands, pale complexion, nausea and vomiting, insomnia and nightmares. Contact: skin - (absorbed) itching, inflammation, blisters; eyes - irritation, watering inflammation. Extreme exposure could be fatal.
**Toxicology:** Highly toxic by all routes.
TLV® (skin) 0.1 mg/m³ tetraethyl (as Pb); 0.15 mg/m³ (tetramethyl).
Short-term Inhalation Limits - 0.3 mg/m³ tetraethyl 15 min (skin)(as Pb); 0.5 mg/m³ tetramethyl (as Pb).

LC$_{50}$ - Inhalation: rat = 850 mg/m³ (60 min)
Delayed Toxicity - Lead poisoning.

LD$_{50}$ - Via skin: rat = 0.015 g/kg
LD$_{Lo}$ - Oral: rat = 0.017 g/kg

### Fire
**Fire Extinguishing Agents:** Use water mist, dry chemical, foam or carbon dioxide to extinguish. Use water to cool fire-exposed containers and disperse vapours.
**Behaviour in Fire:** Heating to decomposition produces toxic lead vapours. Burning (TEL mixes) in air is only sustained in presence of burning combustibles. TML will undergo self-sustained burning.
**Ignition Temperature:** Closed containers may explode above 80°C.
**Burning Rate:** No information.

### Reactivity
**With Water:** No reaction.
**With Common Materials:** Can react vigorously with oxidizing materials.
**Stability:** Commercial grades are stable. Technical grade is only stable below 80°C.

### Environment
**Water:** Prevent entry into water intakes and waterways. Fish toxicity: 0.2 mg/L/96 h/bluegill/TLm/freshwater; Aquatic toxicity rating = <1 ppm/96 h/TLm/freshwater; 1.4 mg/L/48 h/bluegill/TLm/freshwater; 2 mg/L/24 h/bluegill/TLm/freshwater; BOD: No information.
**Land-Air:** No information.

## EMERGENCY MEASURES

### Special Hazards
POISON.

### Immediate Responses
Keep non-involved people away from spill site. Issue warning: "POISON". Call Fire Department. Evacuate area and keep all persons upwind. Avoid contact and inhalation. Notify manufacturer and request assistance. Dike spill area to prevent liquid from entering water intakes, watercourses or sewers. Stop or reduce discharge, if this can be done without risk. Notify environmental authorities.

### Protective Clothing and Equipment
Respiratory protection - self-contained breathing apparatus and totally encapsulated protective clothing. <u>Gloves</u> - neoprene, rubber. <u>Boots</u> - high, rubber (pants worn outside boots).

### Fire and Explosion
Use water mist, dry chemical, foam or carbon dioxide to extinguish. Use water to cool fire-exposed containers and disperse vapours. Heating to decomposition produces toxic vapours. Burning (TEL mixes) in air is only sustained in presence of burning combustibles. TML will undergo self-sustained burning.

### First Aid
Move victim out of spill site to fresh air. Call for medical assistance, but start first aid at once. <u>Contact</u>: while removing contaminated clothing, immediately flush skin and eyes with plenty of warm water for at least 15 minutes. Speed is of utmost importance. <u>Ingestion</u>: in case of swallowing, vomiting should be induced for conscious victim only, give milk or water to drink. <u>Inhalation</u>: if breathing has stopped, give artificial respiration (not mouth-to-mouth method); if laboured, give oxygen. If medical assistance is not immediately available, transport victim to hospital, doctor or clinic.

## ENVIRONMENTAL PROTECTION MEASURES

### Response

**Water**
1. Stop or reduce discharge if safe to do so.
2. Contact manufacturer or supplier for advice.
3. If possible, contain discharge by damming or water diversion.
4. Dredge or vacuum pump to remove contaminants, liquids and contaminated bottom sediments.
5. Notify environmental authorities to discuss disposal and cleanup of contaminated materials.

**Land-Air**
1. Stop or reduce discharge if safe to do so.
2. Contact manufacturer or supplier for advice.
3. Contain spill by diking with earth or other barrier.
4. Remove material by manual or mechanical means.
5. Recover undamaged containers.
6. Absorb residual liquid on natural or synthetic sorbents.
7. Remove contaminated soil for disposal.
8. Notify environmental authorities to discuss disposal and cleanup of contaminated materials.

### Disposal
1. Contact manufacturer or supplier for advice on disposal.
2. Contact environmental authorities for advice on disposal.

TETRAETHYL LEAD $(C_2H_5)_4Pb$

# TITANIUM DIOXIDE $TiO_2$

## IDENTIFICATION

### Common Synonyms
TITANIC OXIDE
TITANIC ANHYDRIDE
TITANIUM WHITE

### Observable Characteristics
White powder. Odourless.

### Manufacturers
NL Chemical Canada, Varennes, Quebec.
Tioxide of Canada, Tracy, Quebec.

### Transportation and Storage Information
**Shipping State:** Solid.
**Classification:** Not regulated.
**Inert Atmosphere:** No requirement.
**Venting:** No requirement.

**Label(s):** Not regulated.
**Storage Temperature:** Ambient.

**Grades or Purity:** Technical and pure.
**Containers and Materials:** Fibre drums, paper bags; bulk by truck or train.

### Physical and Chemical Characteristics
**Physical State (20°C, 1 atm):** Solid.
**Solubility (Water):** Insoluble.
**Molecular Weight:** 79.9
**Vapour Pressure:** No information.
**Boiling Point:** 2 500 to 3 000°C.

**Floatability (Water):** Sinks (finely divided powder may trap air and remain floating).
**Odour:** Odourless.
**Flash Point:** Not flammable.
**Vapour Density:** No information.
**Specific Gravity:** 4.26 (20°C).

**Colour:** White
**Explosive Limits:** Not flammable.
**Melting Point:** 1 830 to 1 850°C.

## HAZARD DATA

### Human Health
**Symptoms:** Inhalation: coughing (nuisance particle) - eyes, redness. Ingestion: generally nontoxic.
**Toxicology:** Relatively nontoxic.
**TLV®** (dust) 10 mg/m$^3$.     **LC$_{50}$** - No information.     **LD$_{50}$** - No information.
Short-term Inhalation Limits - 20 mg/m$^3$ for     **Delayed Toxicity** - None.     **LDLo** - Intramuscular: rat = 0.3 g/kg
15 min (dust).

### Fire
**Fire Extinguishing Agents:** Not combustible. Most fire extinguishing agents may be used in fires involving titanium dioxide.
**Behaviour in Fire:** Not combustible. Titanium dioxide is very inert even in a fire.
**Ignition Temperature:** Not combustible.     **Burning Rate:** Not combustible.

### Reactivity
**With Water:** No reaction.
**With Common Materials:** Reacts violently with lithium.
**Stability:** Stable.

### Environment
**Water:** Prevent entry into water intakes and waterways. Aquatic toxicity rating >1 000 ppm/96 h/TLm/freshwater.
**Land-Air:** No information.
**Food Chain Concentration Potential:** No information.

## EMERGENCY MEASURES

**Special Hazards**

**Immediate Responses**

Notify manufacturer. Notify environmental authorities.

**Protective Clothing and Equipment**

Respiratory protection - dust respirators should be worn if required; otherwise, goggles and protective clothing as required.

**Fire and Explosion**

Not combustible. Most fire extinguishing agents may be used on fires containing titanium dioxide.

**First Aid**

Move victim out of spill site to fresh air. Call for medical assistance, but start first aid at once. Inhalation: remove victim to fresh air, allow to rest. Contact: eyes or skin - rinse with plenty of water; remove contaminated clothing. Ingestion: give water to conscious victim to drink. If necessary, transport victim to doctor, clinic or hospital.

## ENVIRONMENTAL PROTECTION MEASURES

**Response**

**Water**
1. Stop or reduce discharge if safe to do so.
2. Contact manufacturer or supplier for advice.
3. If possible, contain discharge by damming or water diversion.
4. Dredge or vacuum pump to remove contaminants, liquids and contaminated bottom sediments.
5. Notify environmental authorities to discuss disposal and cleanup of contaminated materials.

**Land-Air**
1. Stop or reduce discharge if safe to do so.
2. Contact manufacturer or supplier for advice.
3. Remove material by manual or mechanical means.
4. Recover undamaged containers.
5. Notify environmental authorities to discuss disposal and cleanup of contaminated materials.

**Disposal**
1. Contact manufacturer or supplier for advice on disposal.
2. Contact environmental authorities for advice on disposal.
3. Can be placed in a landfill site (approval of environmental authorities required).

TITANIUM DIOXIDE   $TiO_2$

# TOLUENE  $C_6H_5CH_3$

## IDENTIFICATION

UN No. 1294

### Common Synonyms

TOLUOL
METHYLBENZENE
METHYLBENZOL
PHENYLMETHANE

### Observable Characteristics

Clear, colourless liquid. Aromatic odour.

### Manufacturers

Finachem Canada, Montreal, Quebec.
Sunchem, Sarnia, Ontario.
Esso Chemical Canada, Sarnia, Ontario.
Shell Canada, Corunna, Ontario.
Dow Chemical Canada Inc., Sarnia, Ont.

### Transportation and Storage Information

**Shipping State:** Liquid.
**Classification:** Flammable.
**Inert Atmosphere:** No requirement.
**Venting:** Open (flame arrester) or pressure-vacuum.
**Pump Type:** Gear, centrifugal; steel, all - iron, stainless steel.

**Label(s):** Red label - FLAMMABLE LIQUID; Class 3.2, Group II.
**Storage Temperature:** Ambient.
**Hose Type:** Polyethylene, Viton, butyl, flexible stainless steel.

**Grades or Purity:** Industrial, 94+%, nitration, 99.8+%.
**Containers and Materials:** Drums, tank cars, tank trucks; steel.

### Physical and Chemical Characteristics

**Physical State** (20°C, 1 atm): Liquid.
**Solubility** (Water): 0.047 g/100 mL (16°C); 0.087 g/100 mL (20°C).
**Molecular Weight:** 92.2
**Vapour Pressure:** 10 mm Hg (6.4°C); 22 mm Hg (20°C); 36.7 mm Hg (30°C); 40 mm Hg (31.8°C).
**Boiling Point:** 110.6°C.

**Floatability** (Water): Floats.
**Odour:** Aromatic (0.17 to 40 ppm, odour threshold).
**Flash Point:** 12.8°C (o.c.); 4°C (c.c.).
**Vapour Density:** 3.1
**Specific Gravity:** 0.87 (20°C).

**Colour:** Colourless.
**Explosive Limits:** 1.2 to 7.1%.
**Melting Point:** -95°C.

## HAZARD DATA

### Human Health

**Symptoms: Inhalation:** Headache, nausea, loss of appetite, rapidly developing pulmonary edema. **Ingestion:** vomiting, nausea, irritability, central nervous system depression, depression of bone marrow, depressed respiration. **Contact:** eyes - irritation; skin - may be absorbed.
**Toxicology:** Moderately toxic by contact. Moderately toxic by ingestion and inhalation.
TLV® - (skin) 100 ppm; 375 mg/m³.
Short-term Inhalation Limits - (skin) 150 ppm; 560 mg/m³.

$LC_{50}$ - Inhalation: mouse = 5 320 ppm/8 h
$LC_{Lo}$ - Inhalation: rat = 4 000 ppm/4 h
Delayed Toxicity - No information.

$LD_{50}$ - Oral: rat = 5 g/kg
$LD_{50}$ - skin: rabbit = 14 g/kg

### Fire

**Fire Extinguishing Agents:** Use carbon dioxide, dry chemical or foam. Water may be ineffective but may be used to cool fire-exposed containers.
**Behaviour in Fire:** Flashback may occur along vapour trail.
**Ignition Temperature:** 480°C.
**Burning Rate:** 5.7 mm/min.

### Reactivity

**With Water:** No reaction.
**With Common Materials:** Can react vigorously with oxidizing materials. Reacts violently with (nitric and sulfuric acids), nitrogen tetroxide and silver perchlorite.
**Stability:** Stable.

### Environment

**Water:** Prevent entry into water intakes and waterways. Aquatic toxicity rating: 10 to 100 ppm/96 h/TLm/freshwater; Fish toxicity: 1 180 mg/L/96 h/sunfish/TLm/freshwater; 24 mg/L/24 h/bluegill/TLm/freshwater; 1 340 mg/L/24 h/mosquito fish/TLm/freshwater; BOD: 86%, 5 days.
**Land-Air:** No information.
**Food Chain Concentration Potential:** No information.

## EMERGENCY MEASURES

**Special Hazards**

FLAMMABLE.

**Immediate Responses**

Keep non-involved people away from spill site. Issue warning: "FLAMMABLE". CALL FIRE DEPARTMENT. Eliminate all ignition sources. Call manufacturer or supplier for advice. Avoid contact and inhalation. Contain spill by diking. Prevent entry into water intakes and sewers. Stop or reduce discharge if this can be done without risk. Notify environmental authorities.

**Protective Clothing and Equipment**

Respiratory protection - In fires or enclosed spaces, self-contained breathing apparatus; otherwise, chemical goggles - (mono), tight fitting. Face shield must not replace goggles. Gloves - rubber. Boots - rubber, high (pants worn outside boots). Outer protective clothing as required. Totally encapsulated protective clothing in fires or enclosed spaces.

**Fire and Explosion**

Use dry chemical, carbon dioxide or alcohol foam. Water may be ineffective, but may be used to cool fire-exposed containers. Flash back may occur along vapour trail.

**First Aid**

Move victim out of spill site to fresh air. Call for medical assistance, but start first aid at once. Contact: skin and eyes - wash affected parts with soap and water. Flush eyes with copious quantities of water for at least 15 minutes. Ingestion: do not induce vomiting. Give water or milk to conscious victim. Inhalation: if breathing has stopped, give artificial respiration. If breathing is laboured, give oxygen. If medical assistance is not immediately available, transport victim to hospital, doctor or clinic.

## ENVIRONMENTAL PROTECTION MEASURES

**Response**

**Water**
1. Stop or reduce discharge if safe to do so.
2. Contact manufacturer or supplier for advice.
3. If possible, contain discharge by booming.
4. If floating, skim and remove.
5. Notify environmental authorities to discuss disposal and cleanup of contaminated materials.

**Land-Air**
1. Stop or reduce discharge if safe to do so.
2. Contact manufacturer or supplier for advice.
3. Contain spill by diking with earth or other barrier.
4. Remove material with pumps or vacuum equipment and place in appropriate containers.
5. Recover undamaged containers.
6. Absorb residual liquid on natural or synthetic sorbents.
7. Notify environmental authorities to discuss disposal and cleanup of contaminated materials.

**Disposal**
1. Contact manufacturer or supplier for advice on disposal.
2. Contact environmental authorities for advice on disposal.
3. Can be taken to chemical incinerator (approval of environmental authorities required).

TOLUENE    $C_6H_5CH_3$

# TOLUENE DIISOCYANATE  $1\text{-}CH_3C_6H_3(NCO)_2\text{-}2,4$

UN No. 2078

## IDENTIFICATION

### Common Synonyms
2,4-TOLULENE DIISOCYANATE
DI-ISO-CYANATOTOLUENE
2,4-TOLYLENE DIISOCYANATE
META-TOLYLENE DIISOCYANATE
TOLUENE 2,4-DIISOCYANATE
ISOCYANIC ACID, METHYLPHENYLINE ESTER
2,4 - TDI
VORANATE T-80™

### Observable Characteristics
Water white to pale yellow liquid with sharp, pungent odour.

### Manufacturers
No Canadian Manufacturers.
Canadian Suppliers:
BASF Canada, Montreal, Que.
Dow Chemical Canada Inc.
Mobay Canada, Toronto
Olin Chemical Canada

Originating from:
BASF Wyandotte, USA,
Dow Chemical, USA, Freeport, TX.
Mobay USA, Pittsburgh.
Olin, USA, Lake Charles LA.

### Transportation and Storage Information
**Shipping State:** Liquid.
**Classification:** Poisonous liquid.
**Inert Atmosphere:** Inerted.
**Venting:** Pressure-vacuum.
**Pump Type:** Centrifugal or positive displacement; steel or stainless steel.

**Label(s):** White label - POISONOUS LIQUID; Class 6.1; Groups I or II.
**Storage Temperature:** Ambient.
**Hose Type:** Seamless stainless steel, Teflon, Viton.

**Grades or Purity:** Commercial, 2,4-TDI 99%; 2,4-TDI 80% and 20% 2,6-TDI; 2,4-TDI 65% and 35% 2,6-TDI.
**Containers and Materials:** Drums, tank cars, tank trucks; steel.

### Physical and Chemical Characteristics
**Physical State:** Liquid.
**Solubility (Water):** Reacts producing $CO_2$ and other products.
**Molecular Weight:** 174.2
**Vapour Pressure:** 0.01 mm Hg (20°C); 1 mm Hg (80°C).
**Boiling Point:** 251°C.

**Floatability (Water):** Sinks and reacts.
**Odour:** Sharp pungent (0.4 to 2.4 ppm, odour threshold).
**Flash Point:** 132°C (o.c).
**Vapour Density:** 6.0
**Specific Gravity:** 1.22 at 25°C.

**Colour:** Water white to pale yellow.
**Explosive Limits:** 0.9 to 9.5%.
**Melting Point:** 20 to 22°C. 80/20 14°C (freezing point).

## HAZARD DATA

### Human Health
**Symptoms: Inhalation:** Potent sensitizer and lung irritant if inhaled. May produce bronchospasm (asthma), pneumonitis, bronchitis and pulmonary edema. Nocturnal cough and shortness of breath. **Contact:** skin - irritation; eyes - severe irritation. **Ingestion:** corrosive to digestive tract causing severe irritation, nausea, vomiting and diarrhea.
**Toxicology:** Highly toxic by inhalation, moderately toxic by ingestion.
TLV®- 0.005 ppm, 0.04 mg/m$^3$   LC$_{50}$ - Inhalation: mouse = 10 ppm/4 h   LD$_{50}$ - Oral: rat = 5.8 g/kg
Short-term Inhalation Limits - 0.02 ppm,   Delayed Toxicity - No information.
0.15 mg/m$^3$ for 15 min.

### Fire
**Fire Extinguishing Agents:** Use water spray, dry chemical or carbon dioxide. Water or foam may cause frothing. Cool exposed tanks with water.
**Behaviour in Fire:** When heated to decomposition, emits highly toxic fumes.
**Ignition Temperature:** No information.   **Burning Rate:** No information.

### Reactivity
**With Water:** Reacts violently to form carbon dioxide gas and other products.
**With Common Materials:** Reacts with acids, amines, bases and alcohols. May react with metals and compounds of copper, tin and mercury.
**Stability:** Stable.

### Environment
**Water:** Prevent entry into water intakes and waterways. Aquatic toxicity rating = 1 to 10 ppm/96 h/TLm/freshwater.
**Land-Air:** No information.
**Food Chain Concentration Potential:** None.

## EMERGENCY MEASURES

**Special Hazards**

POISON.

**Immediate Responses**

Keep non-involved people away from spill site. Issue warning: "POISON". Call Fire Department. Keep upwind. Avoid contact and inhalation. Evacuate from downwind. Dike to prevent runoff. Call manufacturer or supplier. Call environmental authorities.

**Protective Clothing and Equipment**

Respiratory protection - self-contained breathing apparatus and totally encapsulated suit.

**Fire and Explosion**

Use water spray, dry chemical or carbon dioxide to extinguish. Water or foam may cause frothing. When heated to decomposition, emits highly toxic fumes.

**First Aid**

Move victim out of spill site to fresh air. Call for medical assistance, but start first aid at once. Inhalation: if breathing has stopped, give artificial respiration (not mouth-to-mouth method); if laboured, give oxygen. Contact: skin - remove contaminated clothing and flush affected areas with plenty of water for at least 15 minutes; eyes - irrigate with water; Ingestion: give water to conscious victim to drink. If medical assistance is not immediately available, transport victim to hospital, doctor or clinic.

## ENVIRONMENTAL PROTECTION MEASURES

**Response**

**Water**

1. Stop or reduce discharge if safe to do so.
2. Contact manufacturer or supplier for advice.
3. Notify environmental authorities to discuss disposal and cleanup of contaminated materials.

**Land-Air**

1. Stop or reduce discharge if safe to do so.
2. Contact manufacturer or supplier for advice.
3. Contain spill by diking with earth or other barrier.
4. Remove material with pumps or vacuum equipment and place in appropriate containers (caution: avoid contact and inhalation).
5. Recover undamaged containers.
6. Notify environmental authorities to discuss disposal and cleanup of contaminated materials.

**Disposal**

1. Contact manufacturer or supplier for advice on disposal.
2. Contact environmental authorities for advice on disposal.

TOLUENE DIISOCYANATE   $1\text{-}CH_3C_6H_3(NCO)_2\text{-}2,4$

# TRIALLATE $C_{10}H_{16}Cl_3NOS$

## IDENTIFICATION

**Common Synonyms**
S-2,3,3-TRICHLOROALLYL DIISOPROPYLTHIOCARBAMATE

**Common Trade Names**
AVADEX

(A herbicide used for pre- or post-emergent control of wild oats).

**Observable Characteristics**
Light brown liquid or grey-brown solid.

**UN No. 2761**
Danger Group According to Percentage of Active Substance: Group III, Liquid 30 to 100%

**Manufacturers**
Monsanto Company, Winnipeg, Manitoba

### Transportation and Storage Information

**Shipping State:** Solid or liquid (formulation).
**Classification:** None.
**Inert Atmosphere:** No requirement.
**Venting:** Open.
**Pump Type:** No information.

**Label(s):** Not regulated.
**Storage Temperature:** Ambient.
**Hose Type:** No information.

**Grades or Purity:** Various, as shown below.
**Containers and Materials:** Glass bottles; cans, drums; steel.

### Physical and Chemical Characteristics

**Physical State** (20°C, 1 atm): Solid (technical).
**Solubility** (Water): 0.0004 g/100 mL (25°C); EC is dispersible in water.
**Molecular Weight:** 304.7
**Vapour Pressure:** 0.00012 mm Hg (25°C) (technical).
**Boiling Point:** 136°C (at 1 mm Hg) (technical) (decomposes >200°C).

**Floatability** (Water): Sinks; EC mixes.
**Odour:** No information.
**Flash Point:** 95°C (o.c.).
**Vapour Density:** No information.
**Specific Gravity:** 1.27 (25°C) technical.

**Colour:** Liquid-light brown; Solid-grey brown.
**Explosive Limits:** Not flammable.
**Melting Point:** 29 to 30°C (technical).

## HAZARD DATA

### Human Health

**Symptoms:** Ingestion: CNS depression, headache and rash.
**Toxicology:** Low toxicity by all routes.
**TLV®:** No information. **LC$_{50}$** – No information.
**Short-term Inhalation Limits** – No information. **Delayed Toxicity** – No information.

**LD$_{50}$ - Oral:** rat = 1.47 g/kg
**LD$_{50}$ - Skin:** rabbit = 2.23 g/kg

### Fire

**Fire Extinguishing Agents:** Foam, carbon dioxide or dry chemical.
**Behaviour in Fire:** Releases toxic fumes.
**Ignition Temperature:** No information. **Burning Rate:** No information.

### Reactivity

**With Water:** No reaction.
**With Common Materials:** No information.
**Stability:** Stable.

### Environment

**Water:** Prevent entry into water intakes and waterways. Fish toxicities: 9.6 ppm/96 h/rainbow trout/TLm/freshwater; 4.9 ppm/96 h/bluegill/TLm/freshwater.
**Land-Air:** LD$_{50}$ - Oral: Duck = >5000 ppm. LC$_{50}$ - Oral: Quail =>5000 ppm.
**Food Chain Concentration Potential:** No information.

## EMERGENCY MEASURES

**Special Hazards**
POISON.

**Immediate Responses**
Keep non-involved people away from spill site. Stop or reduce discharge if safe to do so. Notify manufacturer or supplier. Dike to contain material or water runoff. Notify environmental authorities.

**Protective Clothing and Equipment**
In fires or confined spaces - Respiratory protection - self-contained breathing apparatus and totally encapsulated suit. Otherwise, approved pesticide respirator and impervious outer clothing.

**Fire and Explosion**
Use carbon dioxide, foam or dry chemical to extinguish. Releases toxic fumes in fires.

**First Aid**
Move victim out of spill area to fresh air. Call for medical assistance, but start first aid at once. Inhalation: if breathing has stopped, give artificial respiration (not mouth-to-mouth method); if laboured, give oxygen. Contact: skin - remove contaminated clothing and flush affected areas with plenty of water; eyes - irrigate with plenty of water. Ingestion: give water to conscious victim to drink and induce vomiting; in the case of petroleum distillates, do not induce vomiting for fear of aspiration and chemical pneumonia. If medical assistance is not immediately available, transport victim to hospital, doctor, or clinic.

## ENVIRONMENTAL PROTECTION MEASURES

**Response**

**Water**
1. Stop or reduce discharge if safe to do so.
2. Contact manufacturer or supplier for advice.

**Floats**
3. If possible contain discharge by booming.
4. If floating, skim and remove.

**Sinks or mixes**
3. If possible contain discharge by damming or water diversions.
4. Dredge or vacuum pump to remove contaminants, liquids and contaminated bottom sediments.

5. Notify environmental authorities to discuss disposal and cleanup of contaminated materials.

**Land-Air**
1. Stop or reduce discharge if safe to do so.
2. Contact manufacturer or supplier for advice.
3. Contain spill by diking with earth or other barrier.
4. If liquid, remove material with pumps or vacuum equipment and place in appropriate containers.
5. If solid, remove material by manual or mechanical means.
6. Recover undamaged containers.
7. Absorb residual liquid on natural or synthetic sorbents.
8. Remove contaminated soil for disposal.
9. Notify environmental authorities to discuss cleanup and disposal of contaminated materials.

**Disposal**
1. Contact manufacturer or supplier for advice on disposal.
2. Contact environmental authorities for advice on disposal.

**Available Formulations**

**Technical Grade:** Purity: 90+%.
Properties: Combustible, insoluble in water.

**Formulations:**

**Type:**
EC - emulsifiable concentrate
GR - granular

**Purity:**
- typically 35%
- typically 10% in inerts (clay)

**Properties:**
- dispersible in water

**TRIALLATE** $C_{10}H_{16}Cl_3NOS$

# TRICHLORFON  $(CH_3O)_2P(O)CH(OH)CCl_3$

## IDENTIFICATION

### Common Synonyms
2,2,2-TRICHLORO-1-HYDROXYETHYL-PHOSPHONATE
TRICHLORPHON

### Common Trade Names
DYLOX, DUTOX
(An insecticide used for the control of a variety of pests.)

### Observable Characteristics
White or brownish powder or liquid.

**UN No. 2783**
Danger Group According to Percentage of Active Substance: Group III, Solid 2 to 20%
Liquid 0.5 to 20%

### Manufacturers
Interprovincial Co-ops Ltd., Saskatoon, Sask.
Chemagro Ltd., Mississauga, Ontario.
Chipman Inc, Stoney Creek, Ontario.
Les produits Marquette, Longueuil, Québec.

### Transportation and Storage Information

**Shipping State:** Solid or liquid (formulation).
**Classification:** None.
**Inert Atmosphere:** No requirement.
**Venting:** Open.
**Pump Type:** No information.

**Label(s):** Not regulated.
**Storage Temperature:** Ambient.
**Hose Type:** No information.

**Grades or Purity:** Various, as described below.
**Containers and Materials:** Glass bottles; cans, drums; steel.

### Physical and Chemical Characteristics

**Physical State** (20°C, 1 atm): Solid (technical).
**Solubility** (Water): 15.4 g/100 mL (25°C) technical; EC is dispersible in water, SN and SP are soluble.
**Molecular Weight:** 257.4
**Vapour Pressure:** 0.0000078 mm Hg (20°C) technical.
**Boiling Point:** 100°C (0.1 mm Hg) technical.

**Floatability** (Water): Sink and mix.
**Odour:** No information.
**Flash Point:** Not flammable.
**Vapour Density:** No information.
**Specific Gravity:** 1.73 (20°C) technical.

**Colour:** White or brownish.
**Explosive Limits:** Not flammable.
**Melting Point:** 83 to 84°C (technical).

## HAZARD DATA

### Human Health
**Symptoms:** Inhalation, Skin Absorption or Ingestion: nausea, salivation, tearing, abdominal cramps, vomiting, sweating, muscular tremors, cyanosis.
**Toxicology:** Highly toxic by ingestion. Moderately toxic by inhalation and skin contact.
**TLV®** No information.  **LC$_{50}$ - Inhalation:** rat = 1.3 mg/m³  **LD$_{50}$ - Oral:** rat = 0.45 g/kg
**Short-term Inhalation Limits** - No information.  **Delayed Toxicity** - Suspected mutagen.  **LD$_{50}$ - Slutancoes** rat = 0.40 g/kg

### Fire
**Fire Extinguishing Agents:** Use foam, carbon dioxide or dry chemical.
**Behaviour in Fire:** Releases toxic fumes.  **Burning Rate:** No information.
**Ignition Temperature:** No information.

### Reactivity
**With Water:** May hydrolyse to dichlorovos in alkaline conditions.
**With Common Materials:** Decomposes to dichlorvos in alkaline solutions.
**Stability:** Stable.

### Environment
**Water:** Prevent entry into water intakes and waterways. Fish toxicities: 0.0081 mg/L/48 h/Daphnia magna/LC$_{50}$/freshwater; 1.2 to 2.4 mg/L/96 h/rainbow trout/LC$_{50}$/freshwater; 6.7 to 9.2 mg/L/96 h/fathead minnow/LC$_{50}$/freshwater; 2.6 to 3.7 mg/L/96 h/bluegill/LC$_{50}$/freshwater.
**Land-Air:** LD$_{50}$ - Oral: Chicken = 0.125 g/kg; Oral: Wild bird = 0.04 g/kg
**Food Chain Concentration Potential:** Suspected to convert to more toxic dichlorvos.

## EMERGENCY MEASURES

**Special Hazards**
POISON.

**Immediate Responses**
Keep non-involved people away from spill site. Stop or reduce discharge if safe to do so. Notify manufacturer or supplier. Dike to contain material or water runoff. Notify environmental authorities.

**Protective Clothing and Equipment**
In fires or confined spaces - Respiratory Protection - self-contained breathing apparatus and totally encapsulated suit. Otherwise, approved pesticide respirator and impervious outer clothing.

**Fire and Explosion**
Use carbon dioxide, foam or dry chemical to extinguish. Releases toxic fumes in fires.

**First Aid**
Move victim out of spill area to fresh air. Call for medical assistance, but start first aid at once. Inhalation: if breathing has stopped, give artificial respiration (not mouth-to-mouth method); if laboured, give oxygen. Contact: skin - remove contaminated clothing and flush affected areas with plenty of water; eyes - irrigate with plenty of water. Ingestion: give water to conscious victim and induce vomiting; in the case of petroleum distillates, do not induce vomiting for fear of aspiration and chemical pneumonia. If medical assistance is not immediately available, transport victim to hospital, doctor, or clinic.

## ENVIRONMENTAL PROTECTION MEASURES

**Response**

**Water**
1. Stop or reduce discharge if safe to do so.
2. Contact manufacturer or supplier for advice.

**Floats**
3. If possible contain discharge by booming.
4. If floating, skim and remove.

**Sinks or mixes**
3. If possible contain discharge by damming or water diversions.
4. Dredge or vacuum pump to remove contaminants, liquids and contaminated bottom sediments.

5. Notify environmental authorities to discuss disposal and cleanup of contaminated materials.

**Land-Air**
1. Stop or reduce discharge if safe to do so.
2. Contact manufacturer or supplier for advice.
3. Contain spill by diking with earth or other barrier.
4. If liquid, remove material with pumps or vacuum equipment and place in appropriate containers.
5. If solid, remove material by manual or mechanical means.
6. Recover undamaged containers.
7. Absorb residual liquid on natural or synthetic sorbents.
8. Remove contaminated soil for disposal.
9. Notify environmental authorities to discuss cleanup and disposal of contaminated materials.

**Disposal**
1. Contact manufacturer or supplier for advice on disposal.
2. Contact environmental authorities for advice on disposal.

**Available Formulations**

**Technical Grade:** Purity: 90+%.
Properties: moderately soluble in water, combustible.

**Formulations:**

| Type: | Purity: | Properties: |
|---|---|---|
| EC - emulsifiable concentrate | - typically 80% | - dispersible in water |
| SN - solution | - typically 85% | - miscible in water |
| SP - soluble powder | - typically 80% | - miscible in water |
| GR - granular | - typically 5%, remainder is inert | |

Other Possible Ingredients Found in Formulations: dichlorvos, oxydemeton methyl.

**TRICHLORFON** $(CH_3O)_2P(O)CH(OH)CCl_3$

# 1,1,1-TRICHLOROETHANE   $CH_3CCl_3$

## IDENTIFICATION

UN No. 2831

**Common Synonyms**
TRICHLOROETHANE
METHYLCHLOROFORM
CHLOROTHENE
AEROTHENE

**Observable Characteristics**
Colourless, liquid with a sweetish odour.

**Manufacturers**
Dow Chemical Canada, Inc., Sarnia, Ontario.

**Transportation and Storage Information**
**Shipping State:** Liquid.
**Classification:** Poisonous liquid.
**Inert Atmosphere:** No requirement.
**Venting:** Pressure-vacuum
**Pump Type:** Steel, stainless steel (not aluminum); centrifugal or positive displacement.

**Label(s):** White label - POISONOUS LIQUID; Class 6.1, Group III.
**Storage Temperature:** Ambient.
**Hose Type:** Seamless stainless steel, Teflon (not natural rubber).

**Grades or Purity:** Commercial and technical.
**Containers and Materials:** Drums and tanks cars; steel (not aluminum).

**Physical and Chemical Characteristics**
**Physical State** (20°C, 1 atm): Liquid.
**Solubility** (Water): 0.44 g/100 mL (20°C).
**Molecular Weight:** 133.4
**Vapour Pressure:** 100 mm Hg (20°C); 127 mm Hg (25°C); 155 mm Hg (30°C).
**Boiling Point:** 74.1°C.

**Floatability** (Water): Sinks.
**Odour:** Sweetish; 100 to 700 ppm, odour threshold.
**Flash Point:** Not flammable under ordinary temperatures and pressures.
**Vapour Density:** 4.6
**Specific Gravity:** 1.34 at 20°C.

**Colour:** Colourless.
**Explosive Limits** 8 to 10.5% (at elevated temperatures).
**Melting Point:** -30 to -33°C.

## HAZARD DATA

**Human Health**
**Symptoms: Contact:** skin - irritation and dermatitis; eyes - pain and irritation. **Inhalation or ingestion:** dizziness, nausea, fainting, unconsciousness and respiratory depression.
**Toxicology:** Moderately toxic by ingestion.
TLV• - 350 ppm; 1 900 mg/m³.
Short-term Inhalation Limits - 450 ppm; 2 450 mg/m³ (15 min).

$LC_{50}$ - No information.
$LC_{Lo}$ - Inhalation: man = 4 900 ppm/10 min.
Delayed Toxicity - No information.

$LD_{50}$ - Oral: rat = 10.3 g/kg
$LD_{50}$ - Oral: mouse = 11.2 g/kg

**Fire**
**Fire Extinguishing Agents:** Use dry chemical, foam or carbon dioxide. Water may be used to cool fire-exposed containers.
**Behaviour in Fire:** Toxic and irritating gases are generated in fires ($Cl_2$, HCl, phosgene, etc.).
**Ignition Temperature:** 537°C. **Burning Rate:** 2.9 mm/min.

**Reactivity**
**With Water:** No reaction.
**With Common Materials:** Reacts violently with acetone, oxygen, sodium metal and sodium hydroxide.
**Stability:** Stable.

**Environment**
**Water:** Prevent entry into water intakes and waterways. Hazardous to aquatic life. Fish toxicity: 75 to 150 ppm/tns/pinfish/TLm/saltwater; Aquatic toxicity rating = 10 to 100 ppm/96 h/TLm/freshwater.
**Land-Air:** No information.
**Food Chain Concentration Potential:** None.

## EMERGENCY MEASURES

**Special Hazards**

POISON.

**Immediate Responses**

Keep non-involved people away from spill site. Issue warning: "POISON". Call Fire Department. Avoid contact and inhalation. Dike to prevent runoff of material. Contact manufacturer. Contact environmental authorities.

**Protective Clothing and Equipment**

Respiratory protection - self-contained breathing apparatus and totally encapsulated protective clothing.

**Fire and Explosion**

Use dry chemical, foam or carbon dioxide to extinguish fires. Use water to keep fire-exposed containers cool. Toxic and irritating gases ($Cl_2$, $HCl$, phosgene) are generated in fires.

**First Aid**

Move victim out of spill site to fresh air. Call for medical assistance, but start first aid at once. Contact: skin - remove contaminated clothing and wash affected area with soap and water; eyes - irrigate with water. Inhalation: give artificial respiration if necessary. Ingestion: give water to drink; do not induce vomiting. Transport to hospital; advise doctor not to give adrenalin or other stimulants.

## ENVIRONMENTAL PROTECTION MEASURES

**Response**

Water
1. Stop or reduce discharge if safe to do so.
2. Contact manufacturer or supplier for advice.
3. If possible, contain discharge by damming or water diversion.
4. Dredge or vacuum pump to remove contaminants, liquids and contaminated bottom sediments.
5. Notify environmental authorities to discuss disposal and cleanup of contaminated materials.

Land-Air
1. Stop or reduce discharge if safe to do so.
2. Contact manufacturer or supplier for advice.
3. Contain spill by diking with earth or other barrier.
4. Remove material with pumps or vacuum equipment and place in appropriate containers.
5. Recover undamaged containers.
6. Adsorb residual liquid on natural or synthetic sorbents.
7. Remove contaminated soil for disposal.
8. Notify environmental authorities to discuss disposal and cleanup of contaminated materials.

**Disposal**

1. Contact manufacturer or supplier for advice on disposal.
2. Contact environmental authorities for advice on disposal.

1,1,1-TRICHLOROETHANE   $CH_3CCl_3$

# TRIFLURALIN  $F_3C(NO_2)_2C_6H_2N(C_3H_7)_2$

## IDENTIFICATION

### Common Synonyms
α,α,α-TRIFLUORO-2,6-DINITRO-N,N-DIPROPYL-p-TOLUIDINE

### Common Trade Names
TREFLAN

### Observable Characteristics
Orange or brown solid, brown liquid. Slight odour.

### Manufacturers
Interprovincial Co-ops Ltd., Saskatoon, Sask.
Eli Lilly and Co. (Canada Ltd.), Toronto, Ontario.
Chevron Chemical (Canada Ltd.), Burlington, Ontario.
Saskatchewan Wheat Pool, Regina, Saskatchewan.

(A herbicide used for the pre-emergent control of grasses and broadleaf plants.)

### Transportation and Storage Information
**Shipping State:** Solid or liquid (formulation).
**Classification:** None.
**Inert Atmosphere:** No requirement.
**Venting:** Open.
**Pump Type:** No information.
**Label(s):** Not regulated.
**Storage Temperature:** Ambient.
**Hose Type:** No information.
**Grades or Purity:** Various as described below.
**Containers and Materials:** Glass bottles; cans, drums; steel.

### Physical and Chemical Characteristics
**Physical State** (20°C, 1 atm): Solid (technical).
**Solubility** (Water): 0.0001 g/100 mL (technical); EC is dispersible.
**Molecular Weight:** 335
**Vapour Pressure:** 0.0002 (30°C) technical.
**Boiling Point:** 139-149°C (at mm Hg) technical.
**Floatability** (Water): Sinks, EC is dispersible.
**Odour:** Slight.
**Flash Point:** EC may be flammable.
**Vapour Density:** No information.
**Specific Gravity:** 1.29 (25°C) technical.
**Colour:** Orange to brown.
**Explosive Limits:** EC may be flammable.
**Melting Point:** 49°C (technical).

## HAZARD DATA

### Human Health
**Symptoms:** Inhalation: respiratory tract irritation, convulsions, coma; Contact: skin - irritation. Ingestion: irritation of gastrointestinal tract.
**Toxicology:** Moderate toxicity by all routes. $LC_{50}$ - No information. $LD_{50}$ - Oral: mouse = 5.0 g/kg
**TLV®** No information. Delayed Toxicity - Suspected carcinogen.
**Short-term Inhalation Limits** - No information.

### Fire
**Fire Extinguishing Agents:** Use foam, carbon dioxide or dry chemical.
**Behaviour in Fire:** Releases toxic fumes.
**Ignition Temperature:** No information.
**Burning Rate:** No information.

### Reactivity
**With Water:** No reaction.
**With Common Materials:** No information.
**Stability:** Stable.

### Environment
**Water:** Prevent entry into water intakes and waterways. Fish toxicity: 0.026 to 0.062 mg/L/96 h/rainbow trout/$LC_{50}$/freshwater; 0.083 to 0.134 mg/L/96 h/fathead minnow/$LC_{50}$/freshwater; 0.047 to 0.070 mg/L/96 h/bluegill/$LC_{50}$/freshwater; 0.320 to 1.0 mg/L/96 h/Daphnia magna/$LC_{50}$/freshwater.
**Land-Air:** $LD_{50}$ - Oral: Chicken = 72.0 g/kg; (no effect level).
**Food Chain Concentration Potential:** None.

## EMERGENCY MEASURES

### Special Hazards
POISON.

### Immediate Responses
Keep non-involved people away from spill site. Stop or reduce discharge if safe to do so. Notify manufacturer or supplier. Dike to contain material or water runoff. Notify environmental authorities.

### Protective Clothing and Equipment
In fires or confined spaces - Respiratory Protection - self-contained breathing apparatus and totally encapsulated suit. Otherwise, approved pesticide respirator and impervious outer clothing.

### Fire and Explosion
Use carbon dioxide, foam or dry chemical to extinguish. Releases toxic fumes in fires.

### First Aid
Move victim out of spill area to fresh air. Call for medical assistance, but start first aid at once. Inhalation: if breathing has stopped, give artificial respiration (not mouth-to-mouth method); if laboured, give oxygen. Contact: skin - remove contaminated clothing and flush affected areas with plenty of water; eyes - irrigate with plenty of water. Ingestion: give water to conscious victim to drink and induce vomiting; in the case of petroleum distillates, do not induce vomiting for fear of aspiration and chemical pneumonia. if medical assistance is not immediately available, transport victim to hospital, doctor, or clinic.

## ENVIRONMENTAL PROTECTION MEASURES

### Response

**Water**
1. Stop or reduce discharge if safe to do so.
2. Contact manufacturer or supplier for advice.

**Floats**
3. If possible contain discharge by booming.
4. If floating, skim and remove.

**Sinks or mixes**
3. If possible contain discharge by damming or water diversions.
4. Dredge or vacuum pump to remove contaminants, liquids and contaminated bottom sediments.
5. Notify environmental authorities to discuss disposal and cleanup of contaminated materials.

**Land-Air**
1. Stop or reduce discharge if safe to do so.
2. Contact manufacturer or supplier for advice.
3. Contain spill by diking with earth or other barrier.
4. If liquid, remove material with pumps or vacuum equipment and place in appropriate containers.
5. If solid, remove material by manual or mechanical means.
6. Recover undamaged containers.
7. Absorb residual liquid on natural or synthetic sorbents.
8. Remove contaminated soil for disposal.
9. Notify environmental authorities to discuss cleanup and disposal of contaminated materials.

### Disposal
1. Contact manufacturer or supplier for advice on disposal.
2. Contact environmental authorities for advice on disposal.

### Available Formulations

**Technical Grade:** Purity: 95+%.
Properties: combustible, insoluble in water.

**Formulations:**

| Type: | Purity: | Properties: |
|---|---|---|
| EC - emulsifiable concentrate | - typically 35% remainder aromatic petroleum distillates | - dispersible in water, combustible and possible flammable. |
| GR - granular | - typically 5% | - combustible |

TRIFLURALIN    $F_3C(NO_2)_2C_6H_2N(C_3H_7)_2$

# TRINITROTOLUENE $C_6H_2(CH_3)(NO_2)_3$

## IDENTIFICATION

**Common Synonyms**
METHYLTRINITROBENZENE
TNT

**Observable Characteristics**
Pale yellow crystals or flakes. Wet is a yellow slurry or sludge. Odourless.

UN No. 0209 dry
1356 wetted (containing 10% water)

**Manufacturers**
Canadian Industries Ltd.,
Montreal, Quebec.

### Transportation and Storage Information

**Shipping State:** Solid.
**Classification:** Explosive.
**Inert Atmosphere:** No requirement.
**Venting:** Open.

**Label(s):** Orange and black label – EXPLOSIVE; Class 1.1.D.
**Storage Temperature:** Ambient.

**Grades or Purity:** Technical.
**Containers and Materials:** Wooden cases or kegs.

### Physical and Chemical Characteristics

**Physical State** (20°C, 1 atm): Solid.
**Solubility** (Water): 0.02 g/100 mL (15°C); 0.07 g/100 mL (100°C).
**Molecular Weight:** 227.1
**Vapour Pressure:** 0.043 mm Hg (81°C).
**Boiling Point:** Explodes at 240°C.

**Floatability** (Water): Sinks.
**Odour:** Odourless.
**Flash Point:** Explodes.
**Vapour Density:** No information.
**Specific Gravity:** 1.65 (20°C).

**Colour:** Pale yellow.
**Explosive Limits:** Explodes if vigorously shocked or heated to 240°C.
**Melting Point:** 81°C.

## HAZARD DATA

### Human Health

**Symptoms:** Inhalation, Skin absorption, or Ingestion: jaundice, dermatis, cyanosis pallour, loss of appetite, nausea.
**Toxicology:** Moderately toxic by ingestion and contact.
TLV® (skin) 0.5 mg/m³.
Short-term Inhalation Limits – (skin) 3 mg/m³ (15 min).
$LC_{50}$ – No information.
Delayed Toxicity – Damage to liver, bones and kidneys.
$LD_{50}$ – No information.
$LD_{Lo}$ – Oral: rat = 0.7 g/kg

### Fire

**Fire Extinguishing Agents:** Fight fires from explosive-resistant location. Use water to cool fire-exposed containers.
**Behaviour in Fire:** Will explode. In explosion releases toxic $NO_x$ fumes.
**Ignition Temperature:** Explodes 240°C.
**Detonation velocity:** (6 900 m/s).

### Reactivity

**With Water:** No reaction.
**With Common Materials:** Can react vigorously with reducing materials.
**Stability:** Strong shock or elevated temperature (240°C) will detonate TNT.

### Environment

**Water:** Prevent entry into water intakes and waterways. Toxic to aquatic life.
**Land–Air:** No information.
**Food Chain Concentration Potential:** No information.

## EMERGENCY MEASURES

### Special Hazards
EXPLOSIVE. Can detonate at 240°C or with shock.

### Immediate Responses
Keep non-involved people away from spill site and evacuate people to a safe distance as soon as possible. Issue warning "EXPLOSIVE". CALL FIRE DEPARTMENT. Enforce strict evacuation of area. Contact manufacturer. Contact environmental authorities.

### Protective Clothing and Equipment
Respiratory protection - after explosion has occured, self-contained breathing apparatus; otherwise, face shield rubber gloves, boots and protective clothing as required.

### Fire and Explosion
Explosive, fight fires from explosive-resistant location. Use water spray to cool fire-exposed containers.

### First Aid
Move victim out of spill site to fresh air. Call for medical assistance, but start first aid at once. <u>Contact</u>: eyes - irrigate immediately; skin - wash with soapy water immediately. <u>Ingestion</u>: give water to drink and induce vomiting. If medical assistancce is not immediately available, transport victim to doctor, hospital or clinic.

## ENVIRONMENTAL PROTECTION MEASURES

### Response

**Water**
1. Evacuate area and control spill from an explosion resistant location.
2. Stop or reduce discharge if safe to do so.
3. Contact manufacturer or supplier for advice.
4. Notify environmental authorities to discuss disposal and cleanup of contaminated materials.

**Land-Air**
1. Evacuate area and control spill from an explosion resistant location.
2. Stop or reduce discharge if safe to do so.
3. Contact manufacturer or supplier for advice.
4. Notify environmental authorities to discuss disposal and cleanup of contaminated materials.

### Disposal
1. Contact manufacturer or supplier for advice on disposal.
2. Contact environmental authorities for advice on disposal.

TRINITROTOLUENE   $C_6H_2(CH_3)(NO_2)_3$

# TURPENTINE $C_{10}H_{16}$ (primarily)

## IDENTIFICATION

UN No. 1299

### Common Synonyms
SPIRITS OF TURPENTINE
TURPS
GUM TURPENTINE
WOOD TURPENTINE
GUMTHUS
Mixture of several compounds, principally pinene, diterpene.

### Observable Characteristics
Colourless liquid, with penetrating, unpleasant odour.

### Manufacturers
Record Chemical, Napierville, Quebec.

### Transportation and Storage Information
**Shipping State:** Liquid.
**Classification:** Flammable liquid.
**Inert Atmosphere:** No requirement.
**Venting:** Open (flame arrester).
**Pump Type:** Steel, stainless steel.

**Label(s):** Red label – FLAMMABLE LIQUID; Class 3.2, Group III.
**Storage Temperature:** Ambient.
**Hose Type:** Most types (except rubber).

**Grades or Purity:** Variety.
**Containers and Materials:** Bottles, cans, drums, tank cars and trucks; steel.

### Physical and Chemical Characteristics
**Physical State** (20°C, 1 atm): Liquid.
**Solubility** (Water): Insoluble.
**Molecular Weight:** Variable, 136 (average).
**Vapour Pressure:** 5 mm Hg (25°C).
**Boiling Point:** 154 to 170°C.

**Floatability** (Water): Floats.
**Odour:** Aromatic, penetrating, unpleasant.
**Flash Point:** 35°C (c.c.).
**Vapour Density:** 4.8
**Specific Gravity:** 0.85 to 0.88 (15°C).

**Colour:** Colourless.
**Explosive Limits:** 0.8% (LEL).
**Melting Point:** -55°C.

## HAZARD DATA

### Human Health
**Symptoms:** Inhalation: vapour causes headache, confusion, respiratory distress. If liquid is taken into lungs, causes severe pneumonitis. **Contact:** skin - irritation. Ingestion: irritates entire digestive system and may injure kidneys.
**Toxicology:** Highly toxic by inhalation.
TLV® - (skin) 100 ppm; 560 mg/m³.
Short-term Inhalation Limits - 150 ppm; 840 mg/m³ (15 min).

$LC_{50}$ - No information.
$TC_{Lo}$ - Inhalation: human = 175 ppm

$LD_{50}$ - Oral: rat = 5.8 g/kg
$LD_{50}$ - Inhalation: mouse = 0.06 g/kg

### Fire
**Fire Extinguishing Agents:** Use foam, dry chemical or carbon dioxide. Water may be ineffective, but should be used to keep fire-exposed containers cool.
**Behaviour in Fire:** Releases acrid fumes in fire.
**Ignition Temperature:** 253°C.
**Burning Rate:** 2.4 mm/min.

### Reactivity
**With Water:** No reaction.
**With Common Materials:** Can react with oxidizers. Reacts violently with calcium hypochlorite, chlorine, chromic anhydride, hexachloromelamine, stannic chloride and trichloromelamine.
**Stability:** Stable.

### Environment
**Water:** Prevent entry into water intakes and waterways. Harmful to aquatic life. **Fish toxicity:** 100 ppm/tns/fish/toxic/freshwater; Aquatic toxicity rating = 10 to 100 ppm/96 h/TLm/freshwater.
**Land-Air:** No information.
**Food Chain Concentration Potential:** None.

## EMERGENCY MEASURES

**Special Hazards**

FLAMMABLE.

**Immediate Responses**

Keep non-involved people away from spill site. Issue warning: "FLAMMABLE". CALL FIRE DEPARTMENT. Avoid contact and inhalation. Dike to prevent runoff. Notify manufacturer. Notify environmental authorities.

**Protective Clothing and Equipment**

In fires and confined spaces - Respiratory protection - self-contained breathing apparatus; otherwise, face shield, boots and protective clothing as required.

**Fire and Explosion**

To extinguish use foam, dry chemical or carbon dioxide. Water may be ineffective, but may be used to keep fire-exposed containers cool.

**First Aid**

Move victim out of spill site to fresh air. Call for medical assistance, but start first aid at once. Contact: eyes - irrigate; skin - wash with soapy water and remove contaminated clothing. Ingestion: do not induce vomiting. Inhalation: if breathing has stopped, give artificial respiration; if laboured, give oxygen. If medical assistance is not immediately available, transport victim to doctor, clinic or hospital.

## ENVIRONMENTAL PROTECTION MEASURES

**Response**

**Water**
1. Stop or reduce discharge if safe to do so.
2. Contact manufacturer or supplier for advice.
3. If possible, contain discharge by booming.
4. If floating, skim and remove.
5. Notify environmental authorities to discuss disposal and cleanup of contaminated materials.

**Land-Air**
1. Stop or reduce discharge if safe to do so.
2. Contact manufacturer or supplier for advice.
3. Contain spill by diking with earth or other barrier.
4. Remove material with pumps or vacuum equipment and place in appropriate containers.
5. Recover undamaged containers.
6. Absorb residual liquid on natural or synthetic sorbents.
7. Notify environmental authorities to discuss cleanup and disposal of contaminated materials.

**Disposal**
1. Contact manufacturer or supplier for advice on disposal.
2. Contact environmental authorities for advice on disposal.
3. Can be burnt in an incinerator (approval of environmental authorities required).

TURPENTINE  $C_{10}H_{16}$ (primarily)

# UREA  NH$_2$CONH$_2$

## IDENTIFICATION

| Common Synonyms | Observable Characteristics | Manufacturers |
|---|---|---|
| CARBAMIDE<br>CARBONYLDIAMIDE | White crystals or powder. Odourless, or slight ammonia odour. | Canadian Fertilizers, Medicine Hat, Alta.<br>Cominco, Carseland, Alta.<br>Canadian Industries Ltd., Courtright, Ontario. |

### Transportation and Storage Information

**Shipping State:** Solid.
**Classification:** None.
**Inert Atmosphere:** No requirement.
**Venting:** Open.
**Label(s):** Not regulated.
**Storage Temperature:** Ambient.
**Grades or Purity:** Technical, CP, fertilizer (45 to 46% N), feed grade (42% N).
**Containers and Materials:** Bags, drums, bulk, trucks, train. Also shipped as solution in tank cars or tank trucks.

### Physical and Chemical Characteristics

**Physical State** (20°C, 1 atm): Solid.
**Solubility** (Water): 78 g/100 mL (5°C); 119.3 g/100 mL (25°C).
**Molecular Weight:** 60.1
**Vapour Pressure:** No information.
**Boiling Point:** Decomposes above 133°C.
**Floatability** (Water): Sinks and mixes.
**Odour:** Odourless, or slight ammonia odour.
**Flash Point:** Not combustible or flammable.
**Vapour Density:** No information.
**Specific Gravity:** 1.34 at 20°C (solid).
**Colour:** White.
**Explosive Limits:** Not flammable or combustible.
**Melting Point:** 133 to 135°C.

## HAZARD DATA

### Human Health

**Symptoms: Inhalation:** (dust) sore throat, coughing, shortness of breath. **Contact:** skin or eyes - causes redness and irritation. **Ingestion:** sore throat, abdominal pain.
**Toxicology:** Moderately toxic by inhalation and contact.
**TLV®** - No information.
**Short-term Inhalation Limits** - No information.
**LC$_{50}$** - No information.
**Delayed Toxicity** - None.
**LD$_{50}$** - No information.
**LD$_{Lo}$** - Cutaneous: rabbit = 3.0 g/kg

### Fire

**Fire Extinguishing Agents:** Not combustible. Most fire extinguishing agents may be used in fires involving urea.
**Behaviour in Fire:** Not combustible, but urea decomposes at >133°C generating toxic NO$_x$ fumes and ammonia gas.
**Ignition Temperature:** Not combustible.
**Burning Rate:** Not combustible.

### Reactivity

**With Water:** No reaction; soluble.
**With Common Materials:** Reacts violently with hypochlorites.
**Stability:** Stable.

### Environment

**Water:** Prevent entry into water intakes and waterways. Fish toxicity: 16 000 to 30 000 mg/L/24 h/creek chub/critical range/freshwater; Aquatic toxicity rating >1 000 ppm/96 h/TLm/freshwater; BOD: 9 to 100%, 5 days (temperature dependent).
**Land-Air:** No information.
**Food Chain Concentration Potential:** None.

## EMERGENCY MEASURES

**Special Hazards**

**Immediate Responses**

Keep non-involved people away from spill site. Dike to prevent runoff from rainwater or water application. Contact manufacturer. Contact environmental authorities.

**Protective Clothing and Equipment**

In fires, Respiratory protection - self-contained breathing apparatus; otherwise, face shield and protective clothing as required.

**Fire and Explosion**

Not combustible; most fire extinguishing agents may be used in fires involving urea. Urea decomposes above 133°C producing toxic $NO_x$ fumes and ammonia gas.

**First Aid**

Move victim out of spill site to fresh air. Call for medical assistance, but start first aid at once. <u>Contact</u>: skin - remove contaminated clothing and wash affected area with soap and water; eyes - irrigate with water. <u>Ingestion</u>: give water to drink, do not induce vomiting. If necessary, transport victim to hospital, doctor or clinic.

## ENVIRONMENTAL PROTECTION MEASURES

**Response**

**Water**
1. Stop or reduce discharge if safe to do so.
2. Contact manufacturer or supplier for advice.
3. If possible, contain discharge by damming or water diversion.
4. Notify environmental authorities to discuss disposal and cleanup of contaminated materials.

**Land-Air**
1. Stop or reduce discharge if safe to do so.
2. Contact manufacturer or supplier for advice.
3. Dike to prevent runoff from rainwater or water application.
4. Remove material by manual or mechanical means.
5. Recover undamaged containers.
6. Notify environmental authorities to discuss disposal and cleanup of contaminated materials.

**Disposal**
1. Contact manufacturer or supplier for advice on disposal.
2. Contact environmental authorities for advice on disposal.
3. May be buried in landfill site (environmental authorities approval required).

UREA  $NH_2CONH_2$

# VANADIUM PENTOXIDE   $V_2O_5$

## IDENTIFICATION

UN No. 2862

### Common Synonyms
VANADIC ACID ANHYDRIDE
VANADIUM OXIDE

### Observable Characteristics
Yellow to red crystalline powder.
Odourless.

### Manufacturers
No Canadian manufacturer.
U.S. supplier
Monsanto Co.,
St-Louis, MO, USA

### Transportation and Storage Information
**Shipping State:** Solid.
**Classification:** Poison.
**Inert Atmosphere:** No requirement.
**Venting:** No requirement.

**Label(s):** White label - POISON; Class 6.1, Group II.
**Storage Temperature:** Ambient.

**Grades or Purity:** Commercial, technical, CP.
**Containers and Materials:** Drums, multiwall paper bags.

### Physical and Chemical Characteristics
**Physical State** (20°C, 1 atm): Solid.
**Solubility** (Water): 0.8 g/100 mL (20°C).
**Molecular Weight:** 181.9
**Vapour Pressure:** No information.
**Boiling Point:** Decomposes 1 750°C.

**Floatability** (Water): Sinks.
**Odour:** Odourless.
**Flash Point:** Not flammable.
**Vapour Density:** No information.
**Specific Gravity:** 3.36 (18°C).

**Colour:** Yellow to red.
**Explosive Limits:** Not flammable.
**Melting Point:** 690°C.

## HAZARD DATA

### Human Health
**Symptoms: Contact:** skin - redness and itch; eyes - redness, pain, watering. **Ingestion:** metal taste, vomiting, pallour, tremors, gastrointestinal disturbances. **Inhalation:** coughing, chest pain, respiratory system irritation and greenish-black discolouration of tongue.
**Toxicology:** Highly toxic by all routes.
TLV® (dust or fume) 0.05 mg/m³.
$LC_{50}$ - No information.
$LC_{Lo}$ - Inhalation: rat = 70 mg/m³ (2 h)
Delayed Toxocity - No information.
$LD_{50}$ - Oral: rat = 0.01 g/kg
Short-term Inhalation Limits - No information.

### Fire
**Fire Extinguishing Agents:** Not combustible. Most fire extinguishing agents can be used in fires involving vanadium pentoxide.
**Behaviour in Fire:** Not combustible.
**Ignition Temperature:** Not combustible.
**Burning Rate:** Not combustible.

### Reactivity
**With Water:** No reaction.
**With Common Materials:** Can react violently with (Ca + S + $H_2O$), chlorine trifluoride and lithium.
**Stability:** Stable.

### Environment
**Water:** Prevent entry into water intakes and waterways. **Fish toxicity:** 13 ppm/96 h/fathead minnow/TLm/softwater; BOD: None.
**Land-Air:** No information.
**Food Chain Concentration Potential:** No information.

## EMERGENCY MEASURES

**Special Hazards**

POISON.

**Immediate Responses**

Keep non-involved people away from spill site. Issue warning: "POISON". Avoid contact or inhalation of fumes or dust. Stop or reduce discharge, if possible. Notify supplier. Notify environmental authorities.

**Protective Clothing and Equipment**

Respiratory protection - self-contained breathing apparatus. <u>Gloves</u> - rubber or PVC. Outer protective clothing as required.

**Fire and Explosion**

Not combustible. Most fire extinguishing agents may be used in fires involving vanadium pentoxide.

**First Aid**

Move victim out of spill site to fresh air. Call for medical assistance, but start first aid at once. <u>Contact:</u> wash eyes and affected skin thoroughly with plenty of water. <u>Ingestion:</u> give conscious victim plenty of water do not induce vomiting. <u>Inhalation:</u> give artificial respiration if required (not mouth-to-mouth method). If medical assistance is not immediately available, transport victim to hospital, doctor or clinic.

## ENVIRONMENTAL PROTECTION MEASURES

**Response**

**Water**
1. Stop or reduce discharge if safe to do so.
2. Contact manufacturer or supplier for advice.
3. If possible, contain discharge by damming or water diversion.
4. Dredge or vacuum pump to remove contaminants, liquids and contaminated bottom sediments.
5. Notify environmental authorities to discuss disposal and cleanup of contaminated materials.

**Land-Air**
1. Stop or reduce discharge if safe to do so.
2. Contact manufacturer or supplier for advice.
3. Contain spill by diking with earth or other barrier.
4. Remove material by manual or mechanical means.
5. Recover undamaged containers.
6. Remove contaminated soil for disposal.
7. Notify environmental authorities to discuss disposal and cleanup of contaminated materials.

**Disposal**
1. Contact manufacturer or supplier for advice on disposal.
2. Contact environmental authorities for advice on disposal.

VANADIUM PENTOXIDE $V_2O_5$

# VINYL ACETATE   CH$_3$COOCH:CH$_2$

## IDENTIFICATION

UN No. 1301

### Common Synonyms
ACETIC ACID VINYL ESTER
ACETIC ACID ETHENYL
VAC
VAM (Vinyl Acetate Monomer)

### Observable Characteristics
Colourless liquid. Unpleasant in high concentrations; sweet pleasant in small quantities.

### Manufacturers
Celanese Canada Ltd., Edmonton, Alta.

### Transportation and Storage Information
**Shipping State:** Liquid.
**Classification:** Flammable liquid.
**Inert Atmosphere:** No requirement.
**Venting:** Pressure-vacuum.
**Pump Type:** Carbon or stainless steel (explosion-proof).

**Label(s):** Red label - FLAMMABLE LIQUID; Class 3.2, Group II.
**Storage Temperature:** Ambient.
**Hose Type:** Polyethylene, butyl, neoprene, flexible steel or stainless steel.

**Grades or Purity:** Technical Grade A, 99.8% (diphenylamine inhibited); grade H, 99.8% (hydroquinone inhibited).
**Containers and Materials:** Cans, drums, tank cars, tank trucks; steel.

### Physical and Chemical Characteristics
**Physical State** (20°C, 1 atm): Liquid.
**Solubility** (Water): 2.5 g/100 mL (20°C).
**Molecular Weight:** 86.1
**Vapour Pressure:** 83 mm Hg (20°C); 100 mm Hg (23.3°C); 115 mm Hg (25°C).
**Boiling Point:** 72-73°C.

**Floatability** (Water): Floats.
**Odour:** Sweet, unpleasant in high concentration (0.12 to 0.55 ppm, odour threshold).
**Flash Point:** -5°C (o.c.), -8°C (c.c.).
**Vapour Density:** 3.0
**Specific Gravity:** 0.93 (20°C).

**Colour:** Colourless.
**Explosive Limits:** 2.6 to 13.4%.
**Melting Point:** -93 to -100°C.

## HAZARD DATA

### Human Health
**Symptoms: Contact:** skin and eyes - liquid and vapours irritating. **Inhalation:** high concentrations produce anaesthetic effect. **Ingestion:** nausea and vomiting.
**Toxicology:** Moderately toxic by all routes.
TLV®- (inhalation) 10 ppm; 30 mg/m$^3$.
Short-term Inhalation Limits - 20 ppm; 60 mg/m$^3$ (15 min).

LC$_{50}$ - No information.
LC$_{Lo}$ - Inhalation: rat = 4 000 ppm (4 h)
Delayed Toxicity - No information.

LD$_{50}$ - Oral: rat = 2.92 g/kg

### Fire
**Fire Extinguishing Agents:** Use dry chemical or alcohol foam or carbon dioxide. Water may be ineffective, but may be used to cool fire-exposed containers.
**Behaviour in Fire:** Heat initiates polymerization which could proceed violently. Fire may cause violent rupture of tank. Flashback may occur along vapour trail.
**Ignition Temperature:** 402°C.
**Burning Rate:** 3.8 mm/min.

### Reactivity
**With Water:** No reaction; slightly soluble.
**With Common Materials:** Reacts violently with nitric acid, sulfuric acid, oleum, chlorosulfonic acid, ethylene diamine, hydrochloric acid and peroxides, and hydrofluoric acid.
**Stability:** Stable if at room temperature and inhibited.

### Environment
**Water:** Prevent entry into water intakes and waterways. Fouling to shoreline. Harmful to aquatic life in very low concentrations. Fish toxicity: 18 ppm/96 h/bluegill/TLm/freshwater; Aquatic toxicity rating = 10 to 100 ppm/96 h/TLm/freshwater; 19 to 39 mg/L/96 h/TLm/fathead minnow/freshwater; BOD: 62%, 5 days (theoretical).
**Land-Air:** No information.
**Food Chain Concentration Potential:** None.

## EMERGENCY MEASURES

### Special Hazards
FLAMMABLE. Heat initiates polymerization which could proceed violently.

### Immediate Responses
Keep non-involved people away from spill site. Issue warning: "FLAMMABLE". CALL FIRE DEPARTMENT. Eliminate all sources of ignition. Avoid contact and inhalation. Stay upwind and use water spray to control vapour. Stop or reduce discharge, if this can be done without risk. Contact supplier for guidance. Dike runoff from rainwater or water application. Notify environmental authorities.

### Protective Clothing and Equipment
Respiratory protection - self-contained breathing apparatus. Goggles - (mono), tight fitting. If face shield is used, it should not replace goggles. Suit - (jacket and pants) or coveralls, rubber or plastic. Boots - high, rubber (pants worn outside boots). Gloves - rubber or plastic.

### Fire and Explosion
Use dry chemical, carbon dioxide or alcohol foam. Use water spray to knock down vapours and cool fire-exposed containers. Heat initiates polymerization which could proceed violently. Fire may cause violent rupture of tank. Flash back may occur along vapour trail.

### First Aid
Move victim out of spill area to fresh air. Call for medical assistance, but start first aid at once. Inhalation: give artificial respiration if breathing has stopped. Give oxygen if breathing is laboured. Contact: remove contaminated clothing and wash eyes and skin with plenty of warm water for at least 15 minutes. Ingestion: give conscious victim water to drink. If medical assistance is not immediately available, transport victim to hospital, clinic or doctor.

## ENVIRONMENTAL PROTECTION MEASURES

### Response

**Water**
1. Stop or reduce discharge if safe to do so.
2. Contact manufacturer or supplier for advice.
3. If possible, contain discharge by booming.
4. If floating, skim and remove.
5. Notify environmental authorities to discuss disposal and cleanup of contaminated materials.

**Land-Air**
1. Stop or reduce discharge if safe to do so.
2. Contact manufacturer or supplier for advice.
3. Contain spill by diking with earth or other material.
4. Remove material with pumps or vacuum equipment and place in appropriate containers.
5. Recover undamaged containers.
6. Absorb residual liquid on natural or synthetic sorbents.
7. Notify environmental authorities to discuss disposal and cleanup of contaminated materials.

### Disposal
1. Contact manufacturer or supplier for advice on disposal.
2. Contact environmental authorities for advice on disposal.
3. Spray into a chemical waste incinerator (approval of environmental authorities required).

VINYL ACETATE  $CH_3COOCH:CH_2$

# VINYL CHLORIDE   $H_2C{:}CHCl$

## IDENTIFICATION

UN No. 1086

### Common Synonyms
VINYL CHLORIDE MONOMER
VCM
CHLOROETHYLENE
CHLOROETHENE

### Observable Characteristics
Colourless, gas or liquid. Sweet odour.

### Manufacturers
Dow Chemical Canada Inc., Sarnia, Ont., Fort Saskatchewan, Alta.

### Transportation and Storage Information
**Shipping State:** Liquid (compressed gas).
**Classification:** Flammable.
**Inert Atmosphere:** No requirement.
**Venting:** Safety-relief (under pressure); pressure-vacuum (atmospheric pressure).
**Pump Type:** Steel or stainless steel; positive displacement or gas, explosion-proof.

**Label(s):** Red label - FLAMMABLE GAS; Class 2.1.
**Storage Temperature:** Ambient.
**Hose Type:** Teflon, Viton A; flexible stainless steel.

**Grades or Purity:** Commercial or technical, 99+% (inhibitor, phenol 40 to 100 ppm may be added).
**Containers and Materials:** Cylinders, tank cars (steel or stainless steel).

### Physical and Chemical Characteristics
**Physical State** (20°C, 1 atm): Gas.
**Solubility** (Water): 0.006 g/100 mL (10°C); 0.003 g/100 mL (25°C).
**Molecular Weight:** 62.5
**Vapour Pressure:** 240 mm Hg (-40°C); 580 mm Hg (-20°C); 2 660 mm Hg (25°C).
**Boiling Point:** -13.4 to -13.9°C.

**Floatability** (Water): Liquid, floats.
**Odour:** Sweet odour (260 to 4 000 ppm, odour threshold).
**Flash Point:** -78°C (o.c.).
**Vapour Density:** 2.2
**Specific Gravity:** 0.97 (-20°C); 0.91 (15°C).

**Colour:** Colourless.
**Explosive Limits:** 3.6 to 33%.
**Melting Point:** -153 to -160°C.

## HAZARD DATA

### Human Health
**Symptoms: Inhalation:** irritation to respiratory tract, dizziness, anaesthesia, difficulty breathing, lung irritation, headache, paralysis. **Contact:** skin - drying, freezing, inflammation; eyes - irritation, watering, inflammation. **Ingestion:** nausea, vomiting, drowsiness, loss of consciousness, narcosis, shock.
**Toxicology:** High toxicity upon inhalation and contact.
TLV®- (inhalation) 5 ppm; 10 mg/m³.
Short-term Inhalation Limits - No information.

$LC_{50}$ - No information.
$LC_{Lo}$ - Inhalation: rat = 6 000 ppm/4 h
$TD_{Lo}$ - Inhalation: rat = 250 ppm/39 weeks
Delayed Toxicity - Recognized carcinogen.

$LD_{50}$ - Oral: rat = 0.5 g/kg
$TD_{Lo}$ - Oral: rat = 34 g/kg

### Fire
**Fire Extinguishing Agents:** Do not put out fire until leak has been shutoff. Water may be used to cool fire-exposed containers.
**Behaviour in Fire:** Fire may cause violent rupture of tank. Burning releases hydrogen chloride gas. Flashback may occur along vapour trail. Under high temperatures or in contact with certain catalytic impurities, may violently polymerize.
**Ignition Temperature:** 472°C.
**Burning Rate:** 4.3 mm/min.

### Reactivity
**With Water:** No reaction.
**With Common Materials:** Can react vigorously with oxidizing materials.
**Stability:** Stable.

### Environment
**Water:** Prevent entry into water intakes and waterways. Aquatic toxicity rating: >1 000 ppm/96 h/TLm/freshwater; BOD: No information.
**Land-Air:** No information.
**Food Chain Concentration Potential:** None.

## EMERGENCY MEASURES

**Special Hazards**

FLAMMABLE.

**Immediate Responses**

Keep non-involved people away from spill site. Issue warning: "FLAMMABLE". CALL FIRE DEPARTMENT. Eliminate all ignition sources. Contact manufacturer for guidance. Stay upwind and use water spray to control vapour. Dike runoff. Stop or reduce discharge if this can be done without risk. Notify environmental authorities.

**Protective Clothing and Equipment**

Respiratory protection - self-contained breathing apparatus. Boots - high, rubber (pants worn outside boots). Acid suit - (jacket and pants), "slicker suit" - neoprene, or coveralls. Gloves - rubber.

**Fire and Explosion**

Do not put out fire until leak has been shut off. Use water to cool fire-exposed containers. Fire may cause violent rupture of tank. Flash back may occur along vapour trail. Under high temperatures or in contact with catalytic impurities, may violently polymerize.

**First Aid**

Move victim from spill site to fresh air. Call for medical assistance, but start first aid at once. **Inhalation:** give artificial respiration if breathing has stopped. Give oxygen if breathing is laboured. **Contact:** remove contaminated clothing and wash eyes and skin with plenty of warm water for at least 15 minutes. **Ingestion:** unlikely with vinyl chloride but should this happen, give warm water. Keep victim warm and quiet. If medical assistance is not immediately available, transport victim to hospital, doctor or clinic.

## ENVIRONMENTAL PROTECTION MEASURES

**Response**

**Water**
1. Stop or reduce discharge if safe to do so.
2. Contact manufacturer or supplier for advice.
3. If possible, contain discharge by damming or water diversion.
4. Notify environmental authorities to discuss disposal and cleanup of contaminated materials.

**Land-Air**
1. Stop or reduce discharge if safe to do so.
2. Contact manufacturer or supplier for advice.
3. Dike to prevent runoff from rainwater or water application.
4. Remove material with pumps or vacuum equipment and place in appropriate containers.
5. Recover undamaged containers.
6. Notify environmental authorities to discuss disposal and cleanup of contaminated materials.

**Disposal**
1. Contact manufacturer or supplier for advice on disposal.
2. Contact environmental authorities for advice on disposal.

VINYL CHLORIDE  $H_2C{:}CHCl$

# XYLENES  (o-, m-, p-) $C_6H_4(CH_3)_2$

## IDENTIFICATION

UN No. 1307

### Common Synonyms
1,2; 1,3; 1,4 – DIMETHYLBENZENE
XYLOLS
o-, m-, p-dimethylbenzene

Xylenes are typically a mixture of the three isomers, ortho, meta and para, with the last two being the predominant portion.

### Observable Characteristics
Colourless liquid. Sweet aromatic odour.

### Manufacturers
Finachem, Montreal, Quebec.
Sunchem, Sarnia, Ontario.
Shell Canada, Corunna, Ontario.

### Transportation and Storage Information
**Shipping State:** Liquid.
**Classification:** Flammable.
**Inert Atmosphere:** No requirement.
**Venting:** Open or pressure-vacuum with flame arrester.
**Pump Type:** Gear or centrifugal. Steel or stainless steel.

**Label(s):** Red label - FLAMMABLE LIQUID; Class 3.2, Group II.
**Storage Temperature:** Ambient.
**Hose Type:** Material should be resistant to aromatic chemical attack.

**Grades or Purity:** Technical, 99.2%; pure, 99.9%; nitration grade - xylene thinner.
**Containers and Materials:** Bottles, cans, drums, tank cars, tank trucks; steel.

### Physical and Chemical Characteristics
**Physical State** (20°C, 1 atm): Liquid.
**Solubility** (Water): (o-) 0.018 g/100 mL (20°C); (p-) 0.02 g/100 mL (25°C).
**Molecular Weight:** 106.2
**Vapour Pressure:** (o-) 5 mm Hg (20°C); 9 mm Hg (30°C); (m-) 6 mm Hg (20°C); 11 mm Hg (30°C); (p-) 6.5 mm Hg (20°C); 12 mm Hg (30°C).
**Boiling Point:** 144.4°C (o-); 139°C (m-); 138.4 (p-).

**Floatability** (Water): Floats.
**Odour:** Sweet aromatic odour (0.05 to 0.27 ppm, odour threshold).
**Flash Point:** (m-) 27°C; (o-) 32°C; (p-) 27°C.
**Vapour Density:** 3.7 (o-); 3.66 (m-); 3.65 (p-).
**Specific Gravity:** 0.88 (o-); 0.86 (m-); 0.86 (p-); (20°C).

**Colour:** Colourless.
**Explosive Limits:** (m-, p-) 1.1 to 7%; (o-) 1.0 to 6.0%.
**Melting Point:** -25°C (o-); -48 (m-); 13.0°C (p-).

## HAZARD DATA

### Human Health
**Symptoms:** Inhalation: dizziness, coughing, nausea and vomiting, fatigue, drowsiness, narcosis; Contact: eyes - redness, watering, inflammation; skin - dryness, cracking, may be absorbed. Ingestion: gastrointestinal irritation, dizziness, fatigue, loss of consciousness.
**Toxicology:** Moderately toxic by inhalation.
TLV® (skin) 100 ppm; 435 mg/m³.
Short-term Inhalation Limits - (skin) 150 ppm; 655 mg/m³ (15 min).
Delayed Toxicity - No information.

$LC_{50}$ - No information.
$LC_{Lo}$ - Inhalation: rat = 8 000 ppm/4 h (m-).
$LC_{Lo}$ - Inhalation: rat = 4 912 ppm/24 h (p-).

$LD_{50}$ - Oral: rat = 5.0 g/kg (p-)

### Fire
**Fire Extinguishing Agents:** Use dry chemical, carbon dioxide, foam. Water spray may be used to cool fire-exposed containers.
**Behavior in Fire:** Containers may rupture in fires. Flash back may occur along vapour trail.
**Ignition Temperature:** 463°C (o-); 527°C (m-); 528°C (p-).
**Burning Rate:** 5.8 mm/min.

### Reactivity
**With Water:** No reaction.
**With Common Materials:** Can react with oxidizing materials.
**Stability:** Stable.

### Environment
**Water:** Prevent entry into water intakes and waterways. Aquatic toxicity rating = 10 to 100 mg/L/96 h/TLm/freshwater; ortho 13 mg/L/24 h/goldfish/$LD_{50}$; meta 16 mg/L/24 h/goldfish/$LD_{50}$; para 18 mg/L/24 h/bluegill/$LC_{50}$/freshwater; 13.5 ppm/96 h/goldfish/$LC_{50}$/freshwater; 8.2 ppm/96 h/rainbow trout/$LC_{50}$/freshwater; BOD: 64 to 235%, 5 days.
**Land-Air:** 800 to 2 400 ppm threshold for common crops.
**Food Chain Concentration Potential:** No information.

## EMERGENCY MEASURES

**Special Hazards**

FLAMMABLE.

**Immediate Responses**

Keep non-involved people away from spill site. Issue warning: "FLAMMABLE". CALL FIRE DEPARTMENT. Eliminate all sources of ignition. Avoid skin contact and inhalation. Stop or reduce discharge if this can be done without risk. Dike to contain spill and prevent runoff. Notify supplier and environmental authorities.

**Protective Clothing and Equipment**

Respiratory protection - in fires or confined areas, use self-contained breathing apparatus and totally encapsulated suit. Otherwise; Gloves - rubber. Eye protection - optional - goggles or face shield. Boots - rubber (pants worn outside boots).

**Fire and Explosion**

Use dry chemical, carbon dioxide or foam to extinguish. Water spray may be used to cool fire-exposed containers. Flash back may occur along vapour trail.

**First Aid**

Move victim out of spill site to fresh air. Call for medical assistance, but start first aid at once. Contact: remove contaminated clothing, wash affected parts with plenty of water (and soap if available). Flush eyes with plenty of water. Inhalation: give artificial respiration if breathing has stopped, give oxygen if breathing is laboured. Ingestion: do not induce vomiting. If medical assistance is not immediately available, transport victim to hospital, doctor or clinic.

## ENVIRONMENTAL PROTECTION MEASURES

**Response**

Water
1. Stop or reduce discharge if safe to do so.
2. Contact manufacturer or supplier for advice.
3. If possible, contain discharge by booming.
4. If floating, skim and remove.
5. Notify environmental authorities to discuss disposal and cleanup of contaminated materials.

Land-Air
1. Stop or reduce discharge if safe to do so.
2. Contact manufacturer or supplier for advice.
3. Contain spill by diking with earth or other material.
4. Remove material with pumps or vacuum equipment and place in appropriate containers.
5. Recover undamaged containers.
6. Absorb residual liquid on natural or synthetic sorbents.
7. Notify environmental authorities to discuss disposal and cleanup of contaminated materials.

**Disposal**

1. Contact manufacturer or supplier for advice on disposal.
2. Contact environmental authorities for advice on disposal.
3. Incinerate (approval of environmental authorities required).

XYLENES   (o-, m-, p-) $C_6H_4(CH_3)_2$

# YELLOW CAKE  $UO_2(OH)_{2-x}(O-y^+)_x$

$x = 0, 1, 2$
$y+ = Na^+, Mg^+$ or $NH_4^+$

## IDENTIFICATION

| Common Synonyms | Observable Characteristics | UN No. 2912 |
| --- | --- | --- |
| URANIUM CONCENTRATE<br>SYNTHETIC CARNOLITE | Yellow to yellow-brown powder.<br>Odourless. | **Manufacturers**<br>Eldorado Nuclear, Ottawa, Ont., Eldorado, Sask.<br>Dennison Mines, Rio Algom Ltd., Elliot Lake, Ont.<br>Madawaska Mines, Bancroft, Ont. |

### Transportation and Storage Information

**Shipping State:** Solid.
**Classification:** Radioactive.
**Inert Atmosphere:** No requirement.
**Venting:** Open.

**Label(s):** Yellow, white and black label - RADIOACTIVE SUBSTANCE; LSA 1.
**Storage Temperature:** Ambient.

**Grades or Purity:** Various, by % Uranium.
**Containers and Materials:** Drums; steel.

### Physical and Chemical Characteristics

**Physical State** (20°C, 1 atm): Solid.
**Solubility** (Water): Insoluble.
**Molecular Weight:** Variable.
**Vapour Pressure:** No information.
**Boiling Point:** No information.

**Floatability** (Water): Sinks.
**Odour:** Odourless.
**Flash Point:** Not flammable.
**Vapour Density:** No information.
**Specific Gravity:** Variable >1.

**Colour:** Yellow to yellow-brown.
**Explosive Limits:** Not flammable.
**Melting Point:** No information.

## HAZARD DATA

### Human Health

**Symptoms:** Inhalation: (dust) severe respiratory tract irritation. Ingestion: nausea, vomiting. <u>Contact</u>: skin - irritation, yellow colouration of skin.
**Toxicology:** Moderately toxic by all routes.
TLV® - 0.2 mg/m³ (as uranium)
Short-term Inhalation Limits - 0.6 mg/m³ (15 min). (as uranium)

$LC_{50}$ - No information.
Delayed Toxicity - No information.

$LD_{50}$ - No information.

### Fire

**Fire Extinguishing Agents:** Not combustible. Most fire extinguishing agents may be used in fires involving yellow cake.
**Behaviour in Fire:** Not combustible.
**Ignition Temperature:** Not combustible.
**Burning Rate:** Not combustible.

### Reactivity

**With Water:** No reaction.
**With Common Materials:** No information.
**Stability:** Radioactivity relatively low level of gamma radiation.

### Environment

**Water:** Prevent entry into water intakes and waterways.
**Land-Air:** No information.
**Food Chain Concentration Potential:** No information.

## EMERGENCY MEASURES

**Special Hazards**

RADIOACTIVE

**Immediate Responses**

Keep non-involved people away from spill site. Notify Atomic Energy Control Board. Notify manufacturer or supplier. Notify environmental authorities.

**Protective Clothing and Equipment**

Outer protective clothing as required.

**Fire and Explosion**

Not combustible. Most fire extinguishing agents may be used in fires involving yellow cake.

**First Aid**

Move victim out of spill site to fresh air. Call for medical assistance, but start first aid at once. Inhalation: if breathing has stopped give artificial respiration; if laboured, give oxygen. Contact: skin - remove contaminated clothing and flush affected areas with plenty of water; eyes - irrigate with plenty of water. Ingestion: give conscious victim plenty of water to drink and induce vomiting. If medical assistance is not immediately available, transport victim to hospital, doctor or clinic.

## ENVIRONMENTAL PROTECTION MEASURES

**Response**

Water
1. Notify the Atomic Energy Control Board.
2. Stop or reduce discharge if safe to do so.
3. Contact manufacturer or supplier for advice.
4. If possible, contain discharge by damming or water diversion.
5. Dredge or vacuum pump to remove contaminants, liquids and contaminated bottom sediments.
6. Notify environmental authorities to discuss disposal and cleanup of contaminated materials.

Land-Air
1. Notify the Atomic Energy Control Board.
2. Stop or reduce discharge if safe to do so.
3. Contact manufacturer or supplier for advice.
4. Remove material by manual or mechanical means.
5. Recover undamaged containers.
6. Remove contaminated soil for disposal.
7. Notify environmental authorities to discuss disposal and cleanup of contaminated materials.

**Disposal**

1. Notify the Atomic Energy Control Board.
2. Contact manufacturer or supplier for advice on disposal.
3. Contact environmental authorities for advice on disposal.

YELLOW CAKE  $UO_2(OH)_{2-x}(O-y^+)_x$

$x = 0,1,2$
$y^+ = Na^+, Mg^+ \text{ or } NH_4^+$

# ZINC  Zn

## IDENTIFICATION

### Common Synonyms
GRANULAR POWDER ZINC
ZINC DUST

### Observable Characteristics
Shiny white metal with bluish-grey lustre or blue-grey to black powder or dust. Odourless.

### Manufacturers
Cominco, Trail, B.C.
Canadian Electrolytic Zinc, Valleyfield, Que.
Texasgulf Canada, Hoyle, Ontario.

### Transportation and Storage Information
**Shipping State:** Solid.
**Classification:** None.
**Inert Atmosphere:** No requirement.
**Venting:** No requirement.

**Label(s):** Not regulated.
**Storage Temperature:** Ambient.

**Grades or Purity:** Zinc dust, 99.0%+; zinc powder, 99.0%; intermediate 99.5%; high grade 89.9%.
**Containers and Materials:** Cans, drums, bulk by truck or rail.

### Physical and Chemical Characteristics
**Physical State** (20°C, 1 atm): Solid.
**Solubility** (Water): 0.00007 g/100 mL (25°C).
**Molecular Weight:** 65.4
**Vapour Pressure:** 0.13 mm Hg (487°C).
**Boiling Point:** 907°C.

**Floatability** (Water): Sinks.
**Odour:** Odourless.
**Flash Point:** Not flammable.
**Vapour Density:** No information.
**Specific Gravity:** 7.1 (25°C).

**Colour:** Blue-grey to black.
**Explosive Limits:** Dust forms explosive mixtures with air.
**Melting Point:** 420°C.

## HAZARD DATA

### Human Health
**Symptoms:** Inhalation: (dust) coughing, fever and muscular aches.
**Toxicology:** Low toxicity by inhalation.
$TLV^®$ (inhalation) 5 $mg/m^3$ (as ZnO fume).
Short-term Inhalation Limits - 10 $mg/m^3$ (as ZnO fume).

$LC_{50}$ - No information.
$TC_{Lo}$ - Inhalation: human = 124 $mg/m^3$ (50 min).
Delayed Toxicity - No information.

$LD_{50}$ - No information.

### Fire
**Fire Extinguishing Agents:** In fires use carbon dioxide, dry chemical, or sand.
**Behaviour in Fire:** When heated, releases highly toxic zinc fumes. Zinc dust may form explosive mixtures in air.
**Ignition Temperature:** 460°C (dust layer); **Burning Rate:** No information.
680°C (dust cloud).

### Reactivity
**With Water:** No reaction.
**With Common Materials:** Dust reacts with acids, chlorites, chlorine, fluorine, hydrazine, potassium nitrate, sulfur, cadmium, chlorates, chromic anhydride, nitric acid, performic acid, potassium chromate, potassium peroxide, selenium, sodium chlorate, sodium peroxide, and ammonium nitrate. In contact with acidic or alkali solutions can result in the evolution of hydrogen.
**Stability:** Stable.

### Environment
**Water:** Prevent entry into water intakes and waterways. Toxic to aquatic life in low concentrations (freshwater 0.026 to 0.09 mg/L; saltwater 0.14 to 0.22 mg/L; $Zn^{2+}$/48 h/rainbow trout/TLm/freshwater; BOD: Not available.
**Land-Air:** No information.
**Food Chain Concentration Potential:** No information.

## EMERGENCY MEASURES

**Special Hazards**

Dust concentrations may be explosive.

**Immediate Responses**

Keep non-involved people away from spill site. If zinc dust is involved, caution should be taken in confined area because of explosive potential. Stop discharge if possible. Contact supplier or manufacturer. Notify environmental authorities.

**Protective Clothing and Equipment**

Respiratory protection - in fires or confined areas use self-contained breathing apparatus; otherwise, protective clothing as necessary.

**Fire and Explosion**

Use carbon dioxide and dry chemical.

**First Aid**

Move victim out of spill site to fresh air. Call for medical assistance, but start first aid at once. <u>Inhalation:</u> if breathing has stopped, give artificial respiration; if laboured give oxygen. Victim should be taken to hospital, doctor or clinic.

## ENVIRONMENTAL PROTECTION MEASURES

**Response**

**Water**
1. Stop or reduce discharge if safe to do so.
2. Contact manufacturer or supplier for advice.
3. If possible, contain discharge by booming.
4. Dredge or vacuum pump to remove contaminants, liquids and contaminated bottom sediments.
5. Notify environmental authorities to discuss disposal and cleanup of contaminated materials.

**Land-Air**
1. Stop or reduce discharge if safe to do so.
2. Contact manufacturer or supplier for advice.
3. Contain spill by diking with earth or other barrier.
4. Remove material by manual or mechanical means.
5. Recovery undamaged containers.
6. Remove contaminated soil for disposal.
7. Notify environmental authorities to discuss disposal and cleanup of contaminated materials.

**Disposal**
1. Contact manufacturer or supplier for advice on disposal.
2. Contact environmental authorities for advice on disposal.

ZINC  Zn

# ZINC CHLORIDE  $ZnCl_2$

## IDENTIFICATION

| | | |
|---|---|---|
| **Common Synonyms** | **Observable Characteristics** | **UN No.** 8159 solution / 8160 anhydrous |
| BUTTER OF ZINC (granular) | White, granular, crystal or powder or clear solution. Odourless. | **Manufacturers** |
| ZINC DICHLORIDE SOLUTION | | Canadian Industries Ltd., |
| ZINC MURIATE SOLUTION | | Montreal, Quebec. |

**Transportation and Storage Information**

- **Shipping State:** Solid or liquid (solution).
- **Classification:** Corrosive.
- **Inert Atmosphere:** No requirement.
- **Venting:** Open.
- **Pump Type:** Gear, centrifugal; rubber or plastic lined; stainless steel (solutions only).
- **Label(s):** White and black label - CORROSIVE; Class 8, Group III.
- **Storage Temperature:** Ambient.
- **Hose Type:** Natural rubber, polyethylene, polypropylene, Chemiflex 951 (solutions only).
- **Grades or Purity:** Solid; CP or technical; Solutions, 50% or 62.5%.
- **Containers and Materials:** Drums, tank cars, tank trucks; lined stainless steel or fibreglass.

**Physical and Chemical Characteristics**

- **Physical State** (20°C, 1 atm): Solid.
- **Solubility:** 432 g/100 mL (25°C); 614 g/100 mL (100°C).
- **Molecular Weight:** 136.3
- **Vapour Pressure:** 1 mm Hg (428°C).
- **Boiling Point:** 732°C.
- **Floatability (Water):** Sinks and mixes.
- **Odour:** Odourless.
- **Flash Point:** Not flammable.
- **Vapour Density:** No information.
- **Specific Gravity:** Solution, 1.61 (15.5°C); Solid, 2.9 (25°C).
- **Colour:** White, solid or clear solution.
- **Explosive Limits:** Not flammable.
- **Melting Point:** 283 to 290°C.

## HAZARD DATA

**Human Health**

**Symptoms:** **Contact:** may cause severe skin irritation and serious eye injury. **Ingestion:** swallowing large amount may cause abdominal pains, nausea, diarrhea and shock. **Inhalation:** irritation of nose and throat, headache, cough, chest pain, fever, nausea.
**Toxicology:** Highly toxic by inhalation. Moderately toxic by ingestion and contact.
TLV® (inhalation) 1 mg/m$^3$ (as $ZnCl_2$ fume).  $LC_{50}$ - No information.  $LD_{50}$ - Oral: rat = 0.35 g/kg
Short-term Inhalation Limits - 2 mg/m$^3$ (15 min)  $TC_{Lo}$ - Inhalation: man = 4 800 mg/m$^3$ (30 min).
(as $ZnCl_2$ fume).  Delayed Toxicity - No information.

**Fire**

- **Fire Extinguishing Agents:** Not combustible. Most fire extinguishing agents can be used on fires involving zinc chloride.
- **Behaviour in Fire:** Not combustible, but when heated to decomposition, emits corrosive hydrogen chloride gas.
- **Ignition Temperature:** Not combustible.
- **Burning Rate:** Not combustible.

**Reactivity**

- **With Water:** No reaction; soluble.
- **With Common Materials:** In contact with most acids, corrosive hydrogen chloride is produced. Reacts violently with potassium and is corrosive to most metals.
- **Stability:** Stable.

**Environment**

- **Water:** Prevent entry into water intakes and waterways. Harmful to aquatic life in waterways. Harmful to aquatic life in low concentrations. Fish toxicity: 7.2 ppm/96 h/bluegill/TLm/freshwater; 28 ppm/48 h/zebrafish/TLm/saltwater; 0.10 ppm/48 h/Daphnia magna/$LC_{50}$/continuous flow; BOD: None.
- **Land-Air:** No information.
- **Food Chain Concentration Potential:** None.

## EMERGENCY MEASURES

### Special Hazards
CORROSIVE.

### Immediate Responses
Keep non-involved people away from spill site. Issue warning: "CORROSIVE". Stop or reduce discharge if this can be done without risk. Avoid contact. Contain spill by diking to prevent entry into watercourses and sewers. Notify manufacturer or supplier. Notify environmental authorities.

### Protective Clothing and Equipment
Respiratory protection - in fires or confined spaces use self-contained breathing apparatus; otherwise, Goggles - (mono), tight fitting. If face shield is used, it must not replace goggles. Gloves - rubber. Boots - high, rubber (pants worn outside boots). Outerwear as required, coveralls, apron, etc. For dusts (granular product), dust respirator may be needed.

### Fire and Explosion
Not combustible. In fires involving $ZnCl_2$, most fire extinguishing agents may be used. When heated to decomposition, may emit corrosive hydrogen chloride gas.

### First Aid
Move victim out of spill site to fresh air. Call for medical assistance, but start first aid at once. Inhalation: remove victim to fresh air, get medical attention. Contact: eyes - irrigate with plenty of water for at least 15 minutes; skin - wash exposed skin with plenty of water; at same time, remove contaminated clothing. Ingestion: If victim is conscious, give large amounts of milk or water. If medical assistance is not quickly available, transport victim to hospital, doctor or clinic.

## ENVIRONMENTAL PROTECTION MEASURES

### Response

**Water**
1. Stop or reduce discharge if safe to do so.
2. Contact manufacturer or supplier for advice.
3. If possible, contain discharge by damming or water diversion.
4. Dredge or vacuum pump to remove contaminants, liquids and contaminants bottom sediments.
5. Notify environmental authorities to discuss disposal and cleanup of contaminated materials.

**Land-Air**
1. Stop or reduce discharge if safe to do so.
2. Contact manufacturer or supplier for advice.
3. Dike to prevent runoff from rainwater or water application.
4. Remove material by manual or mechanical means.
5. Recover undamaged containers.
6. Remove contaminated soil for disposal.
7. Notify environmental authorities to discuss disposal and cleanup of contaminated materials.

### Disposal
1. Contact manufacturer or supplier for advice on disposal.
2. Contact environmental authorities for advice on disposal.

ZINC CHLORIDE $ZnCl_2$

# ZINC OXIDE   ZnO

## IDENTIFICATION

### Common Synonyms
ZINC WHITE
CHINESE WHITE
ZINCITE

### Observable Characteristics
White to yellow-white or grey powder. Odourless.

### Manufacturers
Zochem, Bramalea, Ontario.
Pigment and Chemical, Milton, Ontario.
Produits Chemiques, GH, St. Hyacinthe, Quebec.

### Transportation and Storage Information
**Shipping State:** Solid.
**Classification:** None.
**Inert Atmosphere:** No requirement.
**Venting:** Open.
**Label(s):** Not regulated.
**Storage Temperature:** Ambient.
**Grades or Purity:** Lead-free; leaded.
**Containers and Materials:** Boxes, drums, multi-wall paper bags.

### Physical and Chemical Characteristics
**Physical State** (20°C, 1 atm): Solid.
**Solubility** (Water): 0.00016 mg/100 mL (29°C).
**Molecular Weight:** 81.4
**Vapour Pressure:** No information.
**Boiling Point:** Sublimes >1 800°C.
**Floatability** (Water): Sinks.
**Odour:** Odourless.
**Flash Point:** Not flammable.
**Vapour Density:** 2.0
**Specific Gravity:** 5.6 (20°C).
**Colour:** White to yellow-white or grey.
**Explosive Limits:** Not combustible. Not flammable.
**Melting Point:** 1 975°C (sublimes at temperatures >1 800°C).

## HAZARD DATA

### Human Health
**Symptoms:** <u>Inhalation:</u> (fumes) fever, chills, nausea, vomiting, muscular aches, and weakness. <u>Ingestion:</u> metallic taste in mouth, nausea and vomiting.
**Toxicology:** Moderately toxic upon inhalation or ingestion.
TLV® 5 mg/m³ (ZnO fumes).
Short-term Inhalation Limits – 10 mg/m³ (15 min) (ZnO fumes).
$LC_{50}$ – No information.
Delayed Toxicity – No information.
$TC_{Lo}$ – Inhalation = 600 mg/m³ (30 min).
$LD_{50}$ – Oral: rat = 0.63 g/kg

### Fire
**Fire Extinguishing Agents:** Not combustible; most fire extinguishing agents may be used in fires involving ZnO.
**Behaviour in Fire:** Not combustible; toxic ZnO fumes can be formed in fires.
**Ignition Temperature:** Not combustible.
**Burning Rate:** Not combustible.

### Reactivity
**With Water:** No reaction.
**With Common Materials:** Reacts violently with magnesium.
**Stability:** Stable.

### Environment
**Water:** Prevent entry into water intakes and waterways.
**Land-Air:** No information.
**Food Chain Concentration Potential:** No information.

## EMERGENCY MEASURES

### Special Hazards

### Immediate Responses
Keep non-involved people away from spill site. Call manufacturer or supplier for advice. Stop or reduce discharge if safe to do so. Contact environmental authorities.

### Protective Clothing and Equipment
In fires, Respiratory protection - self-contained breathing apparatus; otherwise, suitable dust respirator and protective clothing as required.

### Fire and Explosion
Not combustible; in fires involving zinc oxide; most fire extinguishing agents may be used.

### First Aid
Move victim out of spill site to fresh air. Call for medical assistance, but start first aid at once. Inhalation: give artificial respiration if necessary. Contact: skin - remove contaminated clothing and flush affected areas with water; eyes - irrigate with large amounts of water. Ingestion: give water to drink. If medical assistance is not immediately available, transport victim to doctor, clinic or hospital.

## ENVIRONMENTAL PROTECTION MEASURES

### Response

**Water**
1. Stop or reduce discharge if safe to do so.
2. Contact manufacturer or supplier for advice.
3. If possible, contain discharge by damming or water diversion.
4. Dredge or vacuum pump to remove contaminants, liquids and contaminated bottom sediments.
5. Notify environmental authorities to discuss disposal and cleanup of contaminated materials.

**Land-Air**
1. Stop or reduce discharge if safe to do so.
2. Contact manufacturer or supplier for advice.
3. Contain spill by diking with earth or other barrier.
4. Remove material by manual or mechanical means.
5. Recover undamaged containers.
6. Remove contaminated soil for disposal.
7. Notify environmental authorities to discuss disposal and cleanup of contaminated materials.

### Disposal
1. Contact manufacturer or supplier for advice on disposal.
2. Contact environmental authorities for advice on disposal.

ZINC OXIDE   ZnO

# ZINC SULFATE   $ZnSO_4 \cdot xH_2O$ (x = 0, 1, 7)

## IDENTIFICATION

**UN No. NA9161**

### Common Synonyms

ZINC SULFATE MONOHYDRATE
ZINC SULFATE HEPTAHYDRATE
- White Vitriol
- White Copperas
- Zinc Vitriol
ZINC SULPHATE

### Observable Characteristics

Anhydrous form; colourless, crystals. Hydrous forms; White, crystalline powder. Odourless.

### Manufacturers

Cominco, Trail, B.C.
Canadian Electrolytic Zinc, Valleyfield, Quebec.
Texasgulf Canada, Timmins, Ontario.

### Transportation and Storage Information

**Shipping State:** Solid.
**Classification:** None.
**Inert Atmosphere:** No requirement.
**Venting:** Open.

**Label(s):** Not regulated, Class 9.2, Group I.
**Storage Temperature:** Ambient.

**Grades or Purity:** Commercial, Technical.
**Containers and Materials:** Bags, fibre drums, drums; steel.

### Physical and Chemical Characteristics

**Physical State** (20°C, 1 atm): Solid.
**Solubility** (Water): 96.5 g/100 mL (20°C); (hepta).
**Molecular Weight:** 161.4 (anhydrous); 179.4 (mono); 287.5 (hepta).
**Vapour Pressure:** No information.
**Boiling Point:** Anhydrous 600°C (decomposes); heptahydrate loses $H_2O$ 280°C; monohydrate loses $H_2O$ >238°C.

**Floatability** (Water): Sinks.
**Odour:** Odourless.
**Flash Point:** Not flammable.
**Vapour Density:** No information.
**Specific Gravity:** Heptahydrate 1.96 (25°C); anhydrous 3.5 (25°C).

**Colour:** Colourless to white.
**Explosive Limitss:** Not flammable.
**Melting Point:** (hepta) 100°C; loses water at 280°C; monohydrate loses water at 238°C; all decompose above 500°C.

## HAZARD DATA

### Human Health

**Symptoms:** Inhalation: sore throat, coughing, shortness of breath, laboured breathing. Contact: eyes and skin - redness and pain. Ingestion: abdominal spasms, vomiting and diarrhea.
**Toxicology:** Moderately toxic by inhalation and contact.
**TLV®** No information.
**Short-term Inhalation Limits** - No information.

$LC_{50}$ - No information.
Delayed Toxicity - No information.

$LD_{50}$ - Intraperitoneal: mouse = 0.029 g/kg
$LD_{Lo}$ - Oral: rat = 2.2 g/kg (hepta)

### Fire

**Fire Extinguishing Agents:** Not combustible; most fire extinguishing agents may be used in fires involving zinc sulfate.
**Behaviour in Fire:** Not combustible.
**Ignition Temperature:** Not combustible.
**Burning Rate:** Not combustible.

### Reactivity

**With Water:** No reaction.
**With Common Materials:** No information.
**Stability:** Stable.

### Environment

**Water:** Prevent entry into water intakes and waterways. Fish toxicity: 4.6 ppm/96 h/rainbow trout/$LC_{50}$/freshwater; 0.3 mg/L (Zn)/120 h/stickleback/lethal/freshwater; BOD: Not available.
**Land-Air:** No information.
**Food Chain Concentration Potential:** No information.

## EMERGENCY MEASURES

**Special Hazards**

**Immediate Responses**

Keep non-involved people away from spill site. Dike spill area, particularly if there is any danger of water runoff from fire fighting or rain. Lightly wet down dry spillage, if there is any danger of dust drift from wind. Notify environmental authorities. Notify supplier or manufacturer.

**Protective Clothing and Equipment**

Wear dust mask if necessary. Other protective equipment; gloves, goggles, coveralls, if required.

**Fire and Explosion**

Not combustible; in fires involving zinc sulfate, most fire extinguishing agents may be used.

**First Aid**

Move victim out of spill site to fresh air. Call for medical assistance, but start first aid at once. <u>Inhalation</u>: give artificial respiration, if necessary. <u>Contact</u>: skin - remove contaminated clothing and flush affected areas with water; eyes - irrigate with large amounts of water. <u>Ingestion</u>: give water to drink. If medical assistance is not immediately available, transport victim to doctor, clinic or hospital.

## ENVIRONMENTAL PROTECTION MEASURES

**Response**

**Water**
1. Stop or reduce discharge if safe to do so.
2. Contact manufacturer or supplier for advice.
3. If possible, contain discharge by damming or water diversion.
4. Dredge or vacuum pump to remove contaminants, liquids and contaminated bottom sediments.
5. Notify environmental authorities to discuss disposal and cleanup of contaminated materials.

**Land-Air**
1. Stop or reduce discharge if safe to do so.
2. Contact manufacturer or supplier for advice.
3. Contain spill by diking with earth or other barrier.
4. Remove material by manual or mechanical means.
5. Recover undamaged containers.
6. Remove contaminated soil for disposal.
7. Notify environmental authorities to discuss disposal and cleanup of contaminated materials.

**Disposal**
1. Contact manufacturer or supplier for advice on disposal.
2. Contact environmental authorities for advice on disposal.

ZINC SULFATE  $ZnSO_4 \cdot xH_2O$ (x = 0,1,7)